Omics Technologies
and
Crop Improvement

Omics Technologies
and
Crop Improvement

EDITED BY **Noureddine Benkeblia**

CRC Press
Taylor & Francis Group
Boca Raton London New York

CRC Press is an imprint of the
Taylor & Francis Group, an **informa** business

CRC Press
Taylor & Francis Group
6000 Broken Sound Parkway NW, Suite 300
Boca Raton, FL 33487-2742

First issued in paperback 2019

© 2015 by Taylor & Francis Group, LLC
CRC Press is an imprint of Taylor & Francis Group, an Informa business

No claim to original U.S. Government works

ISBN-13: 978-1-4665-8668-0 (hbk)
ISBN-13: 978-0-367-37829-5 (pbk)

Library of Congress Cataloging-in-Publication Data

Omics technologies and crop improvement / edited by Noureddine Benkeblia.
       pages cm
    Includes bibliographical references and index.
    ISBN 978-1-4665-8668-0
    1. Agricultural biotechnology. 2. Crop improvement. I. Benkeblia, Noureddine.

S494.5.B563O36 2014
630--dc23
                                                                    2014022714

Visit the Taylor & Francis Web site at
http://www.taylorandfrancis.com

and the CRC Press Web site at
http://www.crcpress.com

Contents

Chapter 1
Masaaki Kobayashi, Hajime Ohyanagi, and Kentaro Yano

Chapter 2
Alberto Valdés, Alejandro Cifuentes, and Virginia García-Cañas

Chapter 3
Lisa W. Alexander and Shaneka S. Lawson

Chapter 4
Shiv S. Verma, Swati Megha, Muhammad H. Rahman, Nat N. V. Kav, and Urmila Basu

Chapter 5
Shoshi Kikuchi

Chapter 6
Ashwani Kumar, Manorma Sharma, Saikat Kumar Basu, Muhammad Asif, Xian Ping Li, and Xiuhua Chen

Chapter 7
Mukhtar Ahmed, Muhammad Asif, Muhammad Kausar Nawaz Shah, Arvind H. Hirani, Muhammad Sajad, Fayyaz-ul-Hassan, and Saikat Kumar Basu

Preface

"Generosity is not what comes from hand, but what comes from heart. Someone can never be better with his knowledge or his wealth, but could be with his value and attitude."

The question that was raised from the end of the twentieth century was how to feed a world population that will likely be ca. 8 billion by 2050, especially considering scientific studies reporting that some yields will probably drop. The main challenge is to address this issue by finding opportunities to improve yields of the main crops of cereals, roots and tubers, and grasses.

In the last three decades, the development of analytical techniques and new emerging technologies such as genomics, transcriptomics, proteomics, metabolomics, and other food-omics has provided promising possibilities for the analysis of crop productivity and possible ways of improving yield and productivity, limiting the land needed to produce this quantity of food, and finally improving the efficiency of resource use, insuring sustainable improvements in crop productivity. These improvements will doubtless lead to improvement in the nutritional, processing, and safety qualities of crops. In addition, these improvements will also have positive impacts on the environment by reducing pollution and the use of fertilizers, pesticides and other chemicals, and water. To achieve these goals, different approaches are suggested such as generating fully sequenced crop genomes or metabolic profiling of these crops.

This book provides additional information on how crops can be improved and how these improvements would be beneficial to humans, environmentally friendly, and socioeconomically sustainable.

Noureddine Benkeblia
University of the West Indies

Editor

Noureddine Benkeblia is a professor of crop science involved in food science, focusing on food-plants biochemistry and physiology. His work is mainly devoted to pre- and postharvest metabolism in crops. A few years ago, he introduced a new concept in systems biology—metabolomics—to investigate the mechanisms of biosynthesis and accumulation of fructans in liliaceous plants. Professor Benkeblia received his BSc, MPhil, and Doctor in Food Sciences from the Institut National Agronomique (Algeria) and Doctor in Agriculture (PhD) from Kagoshima University (Japan). After a few years of teaching in Algeria, he joined the Institut National de la Recherche Agronomique, Avignon, France, as a postdoctoral scientist from 2000. From 2002 to 2007, he worked as a visiting professor at the University of Rakuno Gakuen, Ebetsu, Japan, and research associate at Hokaido University. Professor Benkeblia joined the Department of Life Sciences, the University of the West Indies, Jamaica, in 2008, continuing his work on the physiology, biochemistry, and metabolomics of fructan-containing plants in Jamaica. He also works on the postharvest physiology and biochemistry of local fruits. Professor Benkeblia has published over 150 papers, over 37 books and book chapters, and has been the recipient of many awards, including the University of the West Indies Award for the Most Outstanding Researcher in 2011 and 2013.

Contributors

Ganesh Kumar Agrawal
Research Laboratory for Biotechnology and
  Biochemistry
Kathmandu, Nepal

Mukhtar Ahmed
Department of Agronomy
Arid Agriculture University Rawalpindi
Punjab, Pakistan

and

Department of Biological System
  Engineering
Washington State University
Pullman, Washington

Lisa W. Alexander
Department of Plant Sciences
University of Tennessee
Knoxville, Tennessee

Muhammad Asif
National Agricultural Research Centre
Islamabad, Pakistan

and

Department of Agricultural, Food and
  Nutritional Science
University of Alberta
Edmonton, Alberta, Canada

Atanas Atanassov
Joint Genomic Center
Sofia, Bulgaria

Ilian Badjakov
AgroBioInstitute
Sofia, Bulgaria

Saikat Kumar Basu
Department of Biological Sciences
University of Lethbridge
Lethbridge, Alberta, Canada

Urmila Basu
Department of Agricultural, Food and
  Nutritional Science
University of Alberta
Edmonton, Alberta, Canada

Rossitza Batchvarova
AgroBioInstitute
Sofia, Bulgaria

Noureddine Benkeblia
Department of Crop Science
The University of the West Indies
Kingston, Jamaica

Chumki Bhattacharjee
Department of Biochemistry
Garden City College
Bangalore, India

Luz Castro
Botanical Garden Institute of Mérida
Universidad de Los Andes, Mérida
Mérida, Venezuela

Xiuhua Chen
Institute of Food Crops
Yunan Academy of Agricultural Sciences
Kunming, People's Republic of China

Alejandro Cifuentes
Instituto de Investigación en Ciencias de la
  Alimentación
Consejo Superior de Investigaciones
  Científicas
Madrid, Spain

Prasanta K. Dash
NRC on Plant Biotechnology
Indian Agricultural Research Institute
  Campus
New Delhi, India

Ivayla Dincheva
AgroBioInstitute
Sofia, Bulgaria

Teodora Dzhambazova
AgroBioInstitute
Sofia, Bulgaria

Fayyaz-ul-Hassan
Department of Agronomy
Arid Agriculture University Rawalpindi
Punjab, Pakistan

Gustavo Fermin
Botanical Garden Institute of Mérida
Universidad de Los Andes, Mérida
Mérida, Venezuela

Yoichiro Fukao
Graduate School of Biological Sciences
Nara Institute of Science and Technology
and
Plant Global Educational Project
Nara Institute of Science and Technology
Ikoma, Japan

Kishor Gaikwad
NRC on Plant Biotechnology
Indian Agricultural Research Institute Campus
New Delhi, India

Gad Galili
Department of Plant Science
The Weizmann Institute of Science
Rehovot, Israel

Virginia García-Cañas
Instituto de Investigación en Ciencias de la
  Alimentación
Consejo Superior de Investigaciones Científicas
Madrid, Spain

Maria Georgieva
Research Institute of Mountain Stockbreeding
  and Agriculture
Troyan, Bulgaria

Vijay K. Gupta
Department of Biochemistry
Kurukshetra University
Kurukshetra, India

Arvind H. Hirani
Department of Plant Science
University of Manitoba
Winnipeg, Manitoba, Canada

Muhammad Iqbal
National Institute for Genomics and Advanced
  Biotechnology
National Agricultural Research Centre
Islamabad, Pakistan

Galin Ivanov
Department of Food Preservation and
  Refrigeration Technology
Plovdiv, Bulgaria

Pradeep K. Jain
NRC on Plant Biotechnology
Indian Agricultural Research Institute
  Campus
New Delhi, India

Rekha Kansal
NRC on Plant Biotechnology
Indian Agricultural Research Institute Campus
New Delhi, India

Nat N.V. Kav
Department of Agricultural Food and
  Nutritional Science
University of Alberta
Edmonton, Alberta, Canada

Shoshi Kikuchi
Agrogenomics Research Center
National Institute of Agrobiological Sciences
Tsukuba, Japan

Masaaki Kobayashi
School of Agriculture
Meiji University
Tama-ku, Japan

and

CREST
Japan Science and Technology Agency
Kawasaki, Japan

Violeta Kondakova
AgroBioInstitute
Sofia, Bulgaria

Sarada Krishnan
Department of Horticulture and Center for
  Global Initiatives
Denver Botanic Gardens
Denver, Colorado

Ashwani Kumar
Department of Botany
University of Rajasthan
Jaipur, India

Shaneka S. Lawson
Department of Forestry and Natural Resources
Purdue University
West Lafayette, Indiana

Xian Ping Li
Institute of Industrial Crops
Yunan Academy of Agricultural Sciences
Kunming, People's Republic of China

Andrey Marchev
Institute of Microbiology
Bulgarian Academy of Sciences
Plovdiv, Bulgaria

Swati Megha
Department of Agricultural Food and
  Nutritional Science
University of Alberta
Edmonton, Alberta, Canada

Kiril Mihalev
Department of Food Preservation and
  Refrigeration Technology
University of Food Technologies
Plovdiv, Bulgaria

Muslima Nazir
Department of Botany
Jamia Hamdard University
New Delhi, India

Hajime Ohyanagi
School of Agriculture
Meiji University
Tama-ku, Japan

and

CREST
Japan Science and Technology Agency
Kawasaki, Japan

Atanas Pavlov
Institute of Microbiology
University of Food Technologies
Plovdiv, Bulgaria

Muhammad H. Rahman
Department of Agricultural Food and
  Nutritional Science
University of Alberta
Edmonton, Alberta, Canada

Rashmi Rai
School of Biotechnology
SK University of Agricultural Sciences and
  Technology
Chatha, India

Randeep Rakwal
Research Laboratory for Biotechnology and
  Biochemistry
Kathmandu, Nepal

and

Organization for Educational Initiatives
University of Tsukuba
Tsukuba, Japan

and

Department of Anatomy I
Showa University School of Medicine
Shinagawa, Japan

Muhammad Sajad
Department of Plant Breeding and Genetics
The Islamia University of Bahawalpur
Punjab, Pakistan

Muhammad Kausar Nawaz Shah
Department of Plant Breeding and Genetics
PMAS, Arid Agriculture University
  Rawalpindi
Punjab, Pakistan

Manorma Sharma
Department of Botany
University of Rajasthan
Jaipur, India

Jedrzej Szymanski
Max Planck Institute for Molecular Plant
  Physiology
Potsdam, Germany

and

Department of Plant Science
The Weizmann Institute of Science
Rehovot, Israel

Pooja Choudhary Taxak
NRC on Plant Biotechnology
Indian Agricultural Research Institute
  Campus
New Delhi, India

Paula Tennant
Department of Life Sciences
University of the West Indies
Kingston, Jamaica

Savarni Tripathi
Indian Agricultural Research Institute
Agriculture College Campus
Pune, India

Ivan Tsvetkov
AgroBioInstitute
Sofia, Bulgaria

Alberto Valdés
Instituto de Investigación en Ciencias de la
  Alimentación
Consejo Superior de Investigaciones Científicas
Madrid, Spain

Shiv S. Verma
Department of Agricultural Food and
  Nutritional Science
University of Alberta
Edmonton, Alberta, Canada

Kentaro Yano
School of Agriculture
Meiji University
Tama-ku, Japan

and

CREST
Japan Science and Technology Agency
Kawasaki, Japan

Sajad Majeed Zargar
School of Biotechnology
SK University of Agricultural Sciences and
  Technology
Chatha, India

and

Graduate School of Biological Sciences
Nara Institute of Science and Technology
Ikoma, Japan

CHAPTER 1

Omics Databases and Gene Expression Networks in Plant Sciences

Masaaki Kobayashi, Hajime Ohyanagi, and Kentaro Yano

CONTENTS

1.1 INTRODUCTION

Advances in next-generation sequencing (NGS) technology have opened up a new avenue for quantitative high-throughput sequencing studies. This approach encompasses complete genome sequencing (e.g., Huang et al. 2009; The Tomato Genome Consortium 2012; The Potato Genome Sequencing Consortium 2011), comprehensive analysis of DNA polymorphisms such as single

1

nucleotide polymorphisms (SNPs) (e.g., Asamizu et al. 2012; Austin et al. 2011; Arai-Kichise et al. 2011; Yamamoto et al. 2010), and investigation of whole-genome gene expression profiles (e.g., Schmidt et al. 2011; Suzuki et al. 2013; Tsai et al. 2013). NGS platforms reveal further information on genomic DNA and cDNAs in various crops and other plants, including nonmodel plants.

Large-scale omics data, including sequence data, are very important for providing a better understanding of complex systems in organisms. In particular, the sequence data and gene expression data generated by NGS and microarray platforms, which are rapidly being accumulated as part of the International Nucleotide Sequence Databases Collaboration (INSDC) (e.g., Barrett et al. 2013; Kodama et al. 2012), facilitate large-scale comparative analyses among many experimental samples (e.g., species, organs, developmental stages, and growing conditions) (e.g., Jia et al. 2013; Morris et al. 2013; Novák et al. 2013). As examples, genes specifically expressed in some species and under certain biological conditions, genes with similar expression profiles, gene expression controlled by microRNAs, gene families with a shared ancestry, and speciation in many plant taxa will be elucidated by comparative analysis of large-scale omics data (e.g., Aya et al. 2011; Heyndrickx and Vandepoele 2012; Koenig et al. 2013; de Meaux et al. 2008; Sakurai et al. 2011).

In this chapter, we introduce the current status of web databases for omics data and bioinformatics tools for plant research. In addition, new bioinformatics approaches are discussed. Along with the recent vast increase in experimental data, it is becoming hard to perform analysis efficiently and effectively by conventional analytical approaches. Large-scale analysis with current methods requires extensive computer resources, such as central processing units (CPUs), memory modules, and computational time, even when large-scale computer systems are available to researchers. Novel methods that can treat large sets of omics data using a general laboratory computer system must be developed and utilized to acquire new knowledge efficiently. The significance of the availability of some new approaches (Hamada et al. 2011; Manickavelu et al. 2012; Nishida et al. 2012) with large-scale omics data is also introduced in this chapter.

1.2 INFORMATION ON WEB RESOURCES FOR GENE EXPRESSION DATA

Microarray and NGS technologies have been widely used as the main platforms to survey genome-wide gene expression. Collection of gene expression data under particular biological conditions (such as developmental stages, organs, and biotic and abiotic stresses) allows spatiotemporal expression profiles to be viewed and genes expressed under these conditions to be identified. Lists of specifically expressed genes facilitate further functional analysis of genomes, genes, and gene products.

Microarray platforms can simultaneously quantitate genome-wide gene expression levels. The current microarray platforms generally contain more than 40,000 DNA probes. Custom and commercial microarray chips have been designed for many plants, such as *Arabidopsis*, rice, cotton, maize, wheat, and soybean. The wealth of experimental platforms in various plant species has triggered the accumulation of expression data based on microarray technology. To promote practical usage of microarray data stored in public databases, it is required to deposit information on experimental and statistical methods into those databases according to the minimum information about a microarray experiment (MIAME) guidelines (Brazma et al. 2001).

1.2.1 Databases for Microarray Gene Expression Profiling

Microarray data have been provided from public databases, such as the NCBI Gene Expression Omnibus (GEO) (Barrett et al. 2013). At the time of writing (August 2013), data from more than 10,000 platforms and approximately 1,000,000 samples have been stored in the GEO (http://www.ncbi.nlm.nih.gov/geo/). Entire data sets can be searched for platform, gene expression data, and experiments. The expression profiles of each retrieved gene set are shown graphically on the web page.

Gene expression profiles obtained from microarray experiments are also available from individual web databases. *Arabidopsis* expression data are accessible from The Arabidopsis Information Resource (TAIR) (Lamesch et al. 2011) and the AtGenExpress project (Schmid et al. 2005). Microarray data from rice are provided by databases such as the Rice Oligonucleotide Array Database (ROAD) (Cao et al. 2012) and the Rice Expression Profile Database (RiceXPro) (Sato et al. 2013b). The Plant Expression Database (PLEXdb) (Dash et al. 2012) focuses on gene expression of plants and plant pathogens and provides gene expression profiles and results of gene clustering analysis.

1.2.2 Databases for Gene Expression Profiling by NGS

Recently, NGS technology has also been widely used to obtain genome-wide sequencing data. At present, NGS data are mainly generated by three platforms: the Illumina Genome Analyzer, Roche 454 GS System, and Life Technologies SOLiD. For gene expression analysis, while it is necessary to design DNA probes for microarray experiments, NGS technology requires neither complete genome sequences nor DNA probes. For this reason, NGS technology has been applied for genome-wide gene expression analysis in both model and nonmodel organisms for which microarray platforms have not been designed to date (e.g., Suzuki et al. 2013). For expression analysis using NGS (mRNA-Seq analysis), the procedure consists of sequencing, mapping, and expression profiling (see Section 1.4). Since NGS technology has also been improved, resulting in lower costs, the applications of NGS continue to grow for expression analysis as well as other purposes, including genomic DNA sequencing and detection of small RNAs.

Sequencing data from mRNA-Seq analysis are available from public databases. The members of the International Nucleotide Sequence Database Collaboration (INSDC), namely the Sequence Read Archive (SRA) (Wheeler et al. 2008), the European Nucleotide Archive (ENA) of the EMBL-EBI (Leinonen et al. 2011), and the DDBJ Sequence Read Archive (DRA) (Kodama et al. 2012), store and distribute raw sequencing data produced by NGS. In both DRA and ENA, raw sequence data (read data) are downloadable in FASTQ format, which contains both the read sequences themselves and their base quality. Sequencing data stored in SRA are formatted in a specific SRA format, which can be easily converted into FASTQ format using the SRA toolkit (http://trace.ncbi.nlm.nih.gov/Traces/sra/sra.cgi?view=software). In order to download read data sets more quickly, a web tool called Aspera (http://asperasoft.com/) is available in all three archives.

1.2.3 Reference Genome Sequences

The reference genomes required to map and count the mRNA-Seq reads have been determined through DNA sequencing projects and made available as web resources. For example, The 1001 Genomes Project (http://signal.salk.edu/atg1001/) covering broad accessions in *Arabidopsis*, the Oryza Map Alignment Project (OMAP) covering multiple species in the genus *Oryza* (Wing et al. 2005), and the SOL-100 sequencing project for the Solanaceae (http://solgenomics.net/organism/sol100/view) have been launched to collect comprehensive genome sequence data. The availability of reference genomes assists in accurately measuring gene expression levels with mRNA-Seq analysis.

1.3 WEB DATABASES FOR GENE EXPRESSION NETWORKS IN PLANTS

The idea of gene clustering analysis is based on the rationale that clusters of genes showing similar expression patterns across multiple experimental conditions tend to be functionally related (Usadel et al. 2009). This also means that the biological functions of genes might be

Table 1.1   Major Gene Coexpression Databases

Name of Database	URL	Species	References
ATTED-II	http://atted.jp/	*Arabidopsis*, rice	Obayashi et al. (2011)
BAR	http://bar.utoronto.ca/welcome.htm	*Arabidopsis*, poplar	Toufighi et al. (2005)
CressExpress	http://cressexpress.org	*Arabidopsis*	Srinivasasainagendra et al. (2008)
GeneCAT	http://genecat.mpg.de	*Arabidopsis*, poplar, barley, rice	Mutwil et al. (2008)
OryzaExpress	http://bioinf.mind.meiji.ac.jp/OryzaExpress/	Rice	Hamada et al. (2011)
PlantArrayNet	http://arraynet.mju.ac.kr/arraynet/	Rice, brassica, *Arabidopsis*	Lee et al. (2009)
RiceFREND	http://ricefrend.dna.affrc.go.jp	Rice, *Arabidopsis*	Sato et al. (2013a)

predicted by their gene expression profiles. A bunch of coexpressed genes provides a global view of their network and has the potential to provide biological insight into characteristics of the network. With the proliferation of microarray data in public repositories (see Section 1.2), there has been an upsurge in the number of databases specializing in gene expression profiling of various crops. Similarly, an increasing number of gene coexpression databases for various plant species such as *Arabidopsis* and rice have also been developed (Table 1.1) (Hamada et al. 2011; Lee et al. 2009; Mutwil et al. 2008; Obayashi et al. 2011; Sato et al. 2013a; Srinivasasainagendra et al. 2008; Toufighi et al. 2005). This section will briefly introduce a few of these major coexpression databases.

1.3.1 Gene Expression Networks in *Arabidopsis*

ATTED-II (Obayashi et al. 2011) is a data repository for gene coexpression, mainly for *Arabidopsis*, that can be utilized for functional identification of genes and for studying gene regulatory relationships. ATTED-II was developed with multiple functionalities for constructing gene networks. It introduced a new measure of gene coexpression to retrieve functionally related genes more accurately. It also provides clickable maps with step-by-step navigation for all gene networks, and has a Google Maps application programming interface to create a single map for a large network. It also includes information about protein–protein interactions and conserved gene coexpression patterns.

1.3.2 Gene Expression Networks in Rice

OryzaExpress (Hamada et al. 2011) serves as a data repository for functional annotations of genes, reaction names in metabolic pathways, locus identifiers (IDs) assigned from the Rice Annotation Project (RAP) and Michigan State University (MSU) rice databases, gene expression profiles, and gene expression networks (GENs). It has been maintained with the aim of serving as a single hub for multiple plant (mainly rice) databases. By searching by keyword or ID, it is easy to access genes, the reaction names of metabolic pathways in Kyoto Encyclopedia of Genes and Genomes (KEGG) and RiceCyc, locus IDs in RAP and MSU, and probe names of the Affymetrix Rice Genome Array or Agilent Rice Oligo Microarrays (22K). No matter how the locus IDs and probe names for identical genes differ among these public databases and microarray platforms, OryzaExpress allows simultaneous retrieval of information from independent public databases. In addition, each GEN inferred from the microarray data of the Affymetrix Rice Genome Array is accessible with interactive network viewers.

RiceFREND (Sato et al. 2013a) integrates mutual rice expression profiles of various organs and tissues under different conditions, developmental stages, and other sampling fractions. The database provides a rice gene coexpression database aimed at identifying gene modules with similar expression profiles. It includes coexpression analysis of rice genes performed across more than 800 microarray data points using a single microarray platform. The database provides single-guide gene search and multiple-guide gene search options for efficient retrieval of coexpressed module information. It also provides an interface for visualization and interpretation of gene coexpression networks in HyperTree, Cytoscape Web, and Graphviz formats. Moreover, analysis tools for enriched gene ontology terms and cis-elements facilitate better prediction of biological functions inferred from the coexpressed gene information.

1.4 GENE EXPRESSION ANALYSIS BY NGS

In the early 2000s, microarray analysis was the primary method for obtaining global gene expression profiles. Since the proliferation of NGS technology, while microarray techniques have been predominantly employed for gene expression analyses, particularly for well-annotated model species, one of the major applications of NGS, mRNA-Seq, has been gaining popularity and recognition of its advantages over microarray technologies (Wang et al. 2009). While there are a few experimental considerations regarding the selective extraction of polyadenylated RNA molecules from mixtures of total RNA, the rationale of mRNA-Seq is quite simple: to estimate the absolute gene expression level by counting the number of mRNA molecules for each gene model. In this section, the bioinformatics procedures for mRNA-Seq, particularly with Illumina instruments, will be reviewed, with a few related topics.

1.4.1 Bioinformatics for mRNA-Seq

In terms of bioinformatics, mRNA-Seq analysis comprises multiple steps: (1) preprocessing, (2) mapping, (3) assembly, and (4) expression profiling.

1.4.1.1 Preprocessing

Short reads should be cleaned, to ensure they are free of dubious bases and any contaminating bases other than from mRNA.

First, if the samples have been multiplexed for efficient and economic profiling, they should be categorized into multiple sample records according to the barcode (index) sequences. This step is often called demultiplexing, and is mainly conducted by CASAVA (an Illumina bundled software).

Next, it is good idea to check the general quality of each sample. FastQC (Table 1.2) offers a sophisticated user interface (both a character user interface [CUI] and a graphical user interface [GUI]) to comprehensively check a broad range of quality statistics of short read samples.

After this, persisting adaptor sequences within the reads should be trimmed off. Both cutadapt and FASTXToolkit (Table 1.2) are appropriate programs for this purpose. In addition, poly(A) (or poly(T) on the reverse strands) could also be removed by cutadapt or FASTXToolkit.

Then, the reads derived from rRNA and tRNA should be considered. These molecules might be included in the samples due to incomplete efficiency of poly(A) selection. Conventional homology search programs such as BLAST (Table 1.2) (Altschul et al. 1997) or BLAT (Table 1.2) (Kent 2002) can eliminate them.

Finally, the low-quality bases in each read should be trimmed off, or the reads with low-quality bases might simply be eliminated from each sample. To this end, the FASTXToolkit (Table 1.2) can be used, as can custom-made script programs, which will be more flexible solutions.

Table 1.2   Tools for mRNA-Seq Analysis

Name	URL	Use Application	References
FastQC	http://www.bioinformatics. babraham.ac.uk/projects/ fastqc/	Quality control	
Cutadapt	http://code.google.com/p/ cutadapt/	Adapter trimming	
FASTX-Toolkit	http://hannonlab.cshl.edu/ fastx_toolkit/	General read manipulation	
BLAST	http://blast.ncbi.nlm.nih.gov/ Blast.cgi	rRNA/tRNA elimination	Altschul et al. (1997)
BLAT	http://genome.ucsc.edu/FAQ/ FAQblat.html	rRNA/tRNA elimination	Kent (2002)
TopHat2	http://tophat.cbcb.umd.edu	Reference mapping	Kim et al. (2013)
Bowtie2	http://bowtie-bio.sourceforge. net/bowtie2/index.shtml	Reference mapping	Langmead and Salzberg (2012)
Cufflinks	http://cufflinks.cbcb.umd.edu	Transcript assembly and profiling	Trapnell et al. (2012)
Trans-ABySS	http://www.bcgsc.ca/platform/ bioinfo/software/trans-abyss	*De novo* transcript assembly	Robertson et al. (2010)
Trinity	http://trinityrnaseq. sourceforge.net	*De novo* transcript assembly	Grabherr et al. (2011)
Oases	http://www.ebi.ac.uk/~zerbino/ oases/	*De novo* transcript assembly	Schulz et al. (2012)

1.4.1.2 Mapping

Once the preprocessing step has been completed, the rest of the reads are processed by the following step, mapping on the reference genome sequences. In contrast to the simpler genome DNA sequence samples, the mRNA-Seq reads reflect exon–intron structures (e.g., they might be segmentally mapped on the genome). This sort of mRNA mapping process can be optimally performed by the combination of TopHat2 (Table 1.2) (Kim et al. 2013) and Bowtie2 (Table 1.2) (Langmead and Salzberg 2012).

1.4.1.3 Assembly

After deduction of the transcribed position of each read, gene models should be reconstructed using information from the mapped reads for the following gene expression profiling step. Cufflinks (Table 1.2) combined with TopHat2 (Trapnell et al. 2012) can perform this assembly step. Along with the reference genome DNA sequences, the reference gene annotations can also be transmitted to Cufflinks, though novel gene models can also be predicted according to the assembly of short reads. Competency in novel gene detection is one of the advantages of the mRNA-Seq framework over microarray analyses, which require primers to be designed in advance. Cufflinks also calculates the fragments per kilobase of exon per million mapped reads (FPKM) of each gene model, the normalized intensity of gene expression (see Section 1.4.2).

1.4.1.4 Expression Profiling

Finally, the assembly results of multiple samples are often compared, with the goal of extracting biological significance from expression profiles across tissues, treatments, fractions, time course samples, and so on. Merging gene models and FPKM and detecting differentially expressed genes can be conducted by cuffcompare and cuffdiff, the subprograms of Cufflinks (Table 1.2) (Trapnell et al. 2012).

1.4.2 RPKM and FPKM

In the profiling process, the tag count of each gene should be normalized against the exon length of each gene and the total number of mapped reads, because exon lengths vary among genes, and the total number of tags also differs among samples from multiple conditions. Here, RPKM stands for reads per kilobase of exon model per million mapped reads (Mortazavi et al. 2008). The RPKM value reflects the molar concentration of a transcript in the sample by normalizing for RNA length and for the total read number in the measurement. This facilitates transparent comparison of transcript levels both within and between samples. FPKM (Cufflinks 2009; Trapnell et al. 2012) is almost the same value. While the RPKM simply sums the applicable reads as the numerator and denominator of the formula, FPKM takes care of each of the "paired reads" in order to avoid over-estimation of gene expression, because both members of each pair must be derived from an identical mRNA molecule (Cufflinks 2009; Trapnell et al. 2012).

1.4.3 New Technologies for mRNA-Seq

In past years, mRNA-Seq analysis remained an orientation-free methodology (i.e., knowledge of the strand origin of each mRNA molecule was unavailable). Now, a strand-specific mRNA-Seq procedure is available (Vivancos et al. 2010), revealing the biological significance of dubious anti-sense transcripts of particular genes that had been assumed to be experimental noise.

For nonmodel organisms, *de novo* transcript assembly would reveal the entire transcriptome atlas without knowledge of reference genome sequences. For this purpose some programs designed for *de novo* transcript assembly have been developed (Table 1.2) (Grabherr et al. 2011; Robertson et al. 2010; Schulz et al. 2012), and a number of challenges are currently being addressed.

Challenges in mixed sampling methods have been overcome (Kawahara et al. 2012; Mosquera et al. 2009) in terms of determining the biological interaction between species via a GEN. Such mutual gene network analyses provide an opportunity to unveil the biology of plant infections and symbioses.

While we have focused on simple mRNA-Seq technology, other RNA molecules can be determined and quantified with the latest NGS applications for cell biology. Current technological limitations (e.g., a relatively short read length) will need to be overcome for NGS to be compatible with these RNA-Seq applications, which will likely occur in the coming decade. Overcoming the challenges in RNA-Seq applications with new NGS schemes would further reveal the significance of RNA biology.

1.5 INDICES FOR SIMILARITIES IN GENE EXPRESSION PROFILES

A gene set (gene module) showing similar expression profiles (coexpression) provides a candidate set of genes that play the same functional role or are involved in the same regulatory mechanism controlling gene expression. Therefore, a gene module with similar expression profiles has been employed to understand biological functions, metabolic pathways involving gene products, and mechanisms of gene expression regulation (e.g., Aya et al. 2009; Ewing et al. 1999; Suwabe et al. 2008). For the categorization of genes by expression profiles, hierarchical clustering methods (Eisen et al. 1998) have been widely used with a graphical viewer to display dendrograms and heat maps.

A GEN viewer is a powerful tool to globally grasp gene modules with similar expression profiles from large-scale expression data. The network graph consists of nodes and edges (Figure 1.1). Generally, nodes in a GEN correspond to genes, and edges to similarities in expression profiles between two genes. To construct GENs with expression data from microarray and NGS platforms,

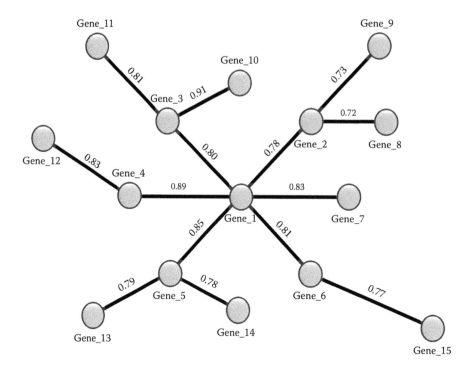

Figure 1.1 An example of a gene expression network. A node refers to a gene; an edge shows genes with similar expression profiles.

calculations of the similarities of expression profiles for all gene pairs are performed. GENs for gene pairs showing significant similarity, which are selected by a threshold value for the similarity index (see below), can be depicted with network graph tools. Many excellent GUI tools to depict and view GENs, such as Cytoscape (Smoot et al. 2011), Graphviz (Ellson 2003), and Pajek (Batagelj and Mrvar 2003), have been developed and distributed. Viewers in most tools can also import annotation data.

With information on the similarities of gene expression profiles, functional annotations for these genes assist in identifying novel attractive candidates. In general, since a great number of gene pairs can show similar expression profiles, other omics information, such as functional annotations, should be investigated. Annotations from Basic Local Alignment Search Tool (BLAST) searches (Altschul et al. 1990), gene ontology terms (The Gene Ontology Consortium 2000) from InterProScan (Quevillon et al. 2005), KEGG pathway names obtained by the KEGG automatic annotation server (KAAS) (Moriya et al. 2007), and other annotation data have been frequently used to detect significant candidate gene pairs.

To detect gene pairs with significantly similar expression profiles, some statistical methods and indices have been widely proposed and used. Among them, the Pearson correlation coefficient (PCC) has been frequently employed as an index to evaluate similarities in expression profiles for gene pairs. The majority of GENs have been constructed based on PCCs or an index modified from PCCs (Obayashi et al. 2011). While PCCs are easily calculated by a variety of computer software packages, false positives are frequently contained among gene pair candidates detected by PCCs. In addition, calculation of PCC values for all gene pairs in a genome takes a long time even if large-scale computers are available. To solve these problems, other methods have been proposed. For example, a distance index with correspondence analysis has been introduced to mine gene modules with similar expression profiles from large-scale expression data (Hamada et al. 2011; Manickavelu et al. 2012; Yano et al. 2006). Regardless of the kind of index, the threshold value for extraction of genes with significantly similar profiles should be decided for each analysis, since the value depends considerably on the expression data. By random sampling of gene pairs around the threshold value, the similarity of expression profiles should be confirmed. The currently favored indices are introduced in the following section.

1.5.1 Pearson Correlation Coefficient (PCC)

The PCC is frequently used to measure the strength of a linear association between two variables, called vectors. Consider vectors x and y, which have elements x_i and y_i ($i = 1 - n$). In expression data, vector x indicates an expression profile through n experimental samples (or conditions) of a given gene X, where each element x_i means an expression level in the ith sample. Similarly, vector y indicates an expression profile of a given gene Y. The PCC value r for the two vectors is given by the following equation:

$$r = \frac{\sum_{i=1}^{n}(x_i - \bar{x})(y_i - \bar{y})}{\sqrt{\sum_{i=1}^{n}\left[(x_i - \bar{x})^2\right]}\sqrt{\sum_{i=1}^{n}\left[(y_i - \bar{y})^2\right]}},$$

where $\bar{x}$ and $\bar{y}$ are the average of elements (gene expression levels) for vectors x and y, respectively.

Gene pairs selected by PCCs sometimes contain false positives. For false positives, the similarity of expression profiles between two vectors (genes) is poor, while the PCC value is moderately high. To solve this problem, a new index called a mutual rank (MR) was proposed to detect gene pairs with significantly similar expression profiles (Obayashi et al. 2011). The MR index has been used to provide information on coexpressed *Arabidopsis* genes in the database ATTED-II.

1.5.2 Distance Index by Correspondence Analysis

Besides false positives, PCC calculations to mine genes with similar expression profiles have another problem: calculation time. When the data set (the number of genes and/or samples) is large, calculation time is long. Taking into account the recent rapid increase in information in the expression data sets employed in research, methods requiring long calculation times are not suitable for routinely analyzing large-scale expression data. For rapid detection of genes with similar expression profiles from large-scale expression data, correspondence analysis (Greenacre 1993; Yano et al. 2006), a multivariate analysis method, is useful. It does not require computer resources such as CPUs, memory, or calculation time.

Correspondence analysis allows an originally high-dimensional data matrix (row [gene] and column [sample]) to be summarized by a low-dimensional projection. The scores (coordinates) in the low-dimensional space are given to genes and samples (phenotypes or experimental conditions). The coordinates of the first three dimensions allow visual assessment of associations between genes and samples in a 3-D subspace. Theoretically, genes with similar expression profiles have almost the same coordinates. Therefore, genes located close together in the 3-D subspace can be selected as candidates with similar expression profiles. For theoretical reasons, the distance between genes indicates the similarity of the expression profiles. This distance index from correspondence analysis has been used to detect genes with similar expression profiles and construct GENs. For example, a GEN in rice, which was constructed using a distance index against an extensive expression data set, is available from the database OryzaExpress (Hamada et al. 2011).

1.6 TOWARD MORE EFFICIENT AND EFFECTIVE APPROACHES WITH LARGE-SCALE EXPRESSION DATA

Here we have introduced various repositories and approaches to handle large-scale gene expression data. Gene expression profiles that can be employed in statistical analyses have proliferated; hence, it is crucial to pursue ever more efficient and effective methods to identify biological significance from large-scale expression data. In particular, large-scale data frequently require large-scale computational resources, such as CPUs, memory modules, and calculation time, while mining genes of interest must be done with limited time and resources. Therefore, websites and journals that provide information on efficient and effective tools will be invaluable. For example, the web server issue published by *Nucleic Acids Research* contains useful tools (Benson 2013).

Because computational analyses in bioinformatics tend to identify too many candidate genes, more evidence is needed to identify the ones that are in fact responsible for particular biological features. While it is difficult to validate the accuracy of each mining result, employing other omics information such as DNA or protein sequences, functional domains, and biological annotations for gene products provides valid collateral evidence that assists in filtering the more appropriate candidates from the original results. Omics information is also provided from many web databases. Lists of useful databases are available in the database issues of *Nucleic Acids Research* (Fernández-Suárez and Galperin 2013) and *Plant and Cell Physiology* (Matsuoka and Yano 2010; Obayashi and Yano 2013).

Although the research environment for bioinformatics in plant science is improving, there still seems to be a large gap between the large-scale data obtained from experiments such as NGS and bioinformatics methods. More efficient and effective analysis methods need to be developed in order for plant biology to advance.

REFERENCES

Altschul, S. F., W. Gish, W. Miller, E. W. Myers, and D. J. Lipman. 1990. Basic local alignment search tool. *J Mol Biol* 215:403–410.

Altschul, S. F., T. L. Madden, A. A. Schäffer et al. 1997. Gapped BLAST and PSI-BLAST: A new generation of protein database search programs. *Nucleic Acids Res* 25:3389–3402.

Arai-Kichise, Y., Y. Shiwa, H. Nagasaki et al. 2011. Discovery of genome-wide DNA polymorphisms in a land-race cultivar of *Japonica* rice by whole-genome sequencing. *Plant Cell Physiol* 52:274–282.

Asamizu, E., K. Shirasawa, H. Hirakawa et al. 2012. Mapping of Micro-Tom BAC-end sequences to the reference tomato genome reveals possible genome rearrangements and polymorphisms. *Int J Plant Genomics* 2012:8.

Austin, R. S., D. Vidaurre, G. Stamatiou et al. 2011. Next-generation mapping of *Arabidopsis* genes. *Plant J* 67:715–725.

Aya, K., G. Suzuki, K. Suwabe et al. 2011. Comprehensive network analysis of anther-expressed genes in rice by the combination of 33 laser microdissection and 143 spatiotemporal microarrays. *PLoS ONE* 6(10):e26162.

Aya, K., M. Ueguchi-Tanaka, M. Kondo et al. 2009. Gibberellin modulates anther development in rice via the transcriptional regulation of GAMYB. *Plant Cell* 21:1453–1472.

Barrett, T., S. E. Wilhite, P. Ledoux et al. 2013. NCBI GEO: Archive for functional genomics data sets—Update. *Nucleic Acids Res* 41:D991–D995.

Batagelj, V. and A. Mrvar. 2003. Pajek—Analysis and visualization of large networks. In M. Jünger and P. Mutzel (eds), *Graph Drawing Software*, pp. 77–103. Dordrecht: Springer-Verlag.

Benson, G. 2013. Editorial. *Nucleic Acids Res* 41:W1–W2.

Brazma, A., P. Hingamp, J. Quackenbush et al. 2001. Minimum information about a microarray experiment (MIAME)-toward standards for microarray data. *Nat Genet* 29:365–371.

Cao, P., K. H. Jung, D. Choi et al. 2012. The rice oligonucleotide array database: An atlas of rice gene expression. *Rice* 5:17–25.

Cufflinks. 2009. Transcript assembly, differential expression, and differential regulation for RNA-Seq. http://cufflinks.cbcb.umd.edu/faq.html#fpkm (accessed: July 11, 2013).

Dash, S., J. V. Hemert, L. Hong et al. 2012. PLEXdb: Gene expression resources for plants and plant pathogens. *Nucleic Acids Res* 40:D1194–D1201.

de Meaux, J., J. Y. Hu, U. Tartler et al. 2008. Structurally different alleles of the ath-*MIR824* microRNA precursor are maintained at high frequency in *Arabidopsis thaliana*. *Proc Natl Acad Sci USA* 105:8994–8999.

Eisen, M. B., P. T. Spellman, P. Q. Brown, and D. Botstein. 1998. Cluster analysis and display of genome-wide expression patterns. *Proc Natl Acad Sci USA* 95:14863–14868.

Ellson, J. 2003. Graphviz and dynagraph—Static and dynamic graph drawing tools. In E. R. Gansner, E. Koutsofios, S. C. North, and G. Woodhull (eds), *Graph Drawing Software*, pp. 127–148. Dordrecht: Springer-Verlag.

Ewing, R. M., A. B. Kahla, P. Olivier et al. 1999. Large-scale statistical analyses of rice ESTs reveal correlated patterns of gene expression. *Genome Res* 9:950–959.

Fernández-Suárez, X. M. and M. Y. Galperin. 2013. The 2013 *Nucleic Acids Research* database issue and the online molecular biology database collection. *Nucleic Acids Res* 41:D1–D7.

Grabherr, M. G., B. J. Haas, M. Yassour et al. 2011. Full-length transcriptome assembly from RNA-Seq data without a reference genome. *Nat Biotechnol* 29:644–652.

Greenacre, M. J. 1993. *Correspondence Analysis in Practice*. London: Academic Press.

Hamada, K., K. Hongo, K. Suwabe et al. 2011. OryzaExpress: An integrated database of gene expression networks and omics annotations in rice. *Plant Cell Physiol* 52:220–229.

Heyndrickx, K. S. and K. Vandepoele. 2012. Systematic identification of functional plant modules through the integration of complementary data sources. *Plant Physiol* 159:884–901.

Huang, S., R. Li, Z. Zhang et al. 2009. The genome of the cucumber, *Cucumis sativus* L. *Nat Genet* 41:1275–1281.

Jia, J., S. Zhao, X. Kong et al. 2013. *Aegilops tauschii* draft genome sequence reveals a gene repertoire for wheat adaptation. *Nature* 496:91–95.

Kawahara, Y., Y. Oono, H. Kanamori et al. 2012. Simultaneous RNA-seq analysis of a mixed transcriptome of rice and blast fungus interaction. *PLoS One* 7:e49423–e49437.

Kent, W. J. 2002. BLAT—The BLAST-like alignment tool. *Genome Res* 12:656–664.

Kim, D., G. Pertea, C. Trapnell et al. 2013. TopHat2: Accurate alignment of transcriptomes in the presence of insertions, deletions and gene fusions. *Genome Biol* 14:R36–R48.

Kodama, Y., M. Shumway, and R. Leinonen. 2012. The sequence read archive: Explosive growth of sequencing data. *Nucleic Acids Res* 40:D54–D56.

Koenig, D., J. M. Jiménez-Gómez, S. Kimura et al. 2013. Comparative transcriptomics reveals patterns of selection in domesticated and wild tomato. *Proc Natl Acad Sci USA* 110:E2655–E2662.

Lamesch, P., T. Z. Berardini, D. Li et al. 2011. The Arabidopsis Information Resource (TAIR): Improved gene annotation and new tools. *Nucleic Acids Res* 40:D1202–D1210.

Langmead, B. and S. L. Salzberg. 2012. Fast gapped-read alignment with Bowtie 2. *Nat Methods* 9:357–359.

Lee, T. H., Y. K. Kim, T. T. Pham et al. 2009. RiceArrayNet: A database for correlating gene expression from transcriptome profiling, and its application to the analysis of coexpressed genes in rice. *Plant Physiol* 151:16–33.

Leinonen, R., R. Akhtar, E. Birney et al. 2011. The European nucleotide archive. *Nucleic Acids Res* 39:D28–D31.

Manickavelu, A., K. Kawaura, K. Oishi et al. 2012. Comprehensive functional analyses of expressed sequence tags in common wheat (*Triticum aestivum*). *DNA Res* 19:165–177.

Matsuoka, M. and K. Yano. 2010. Editorial. *Plant Cell Physiol* 51:1247.

Moriya, Y., M. Itoh, S. Okuda, A. Yoshizawa, and M. Kanehisa. 2007. KAAS: An automatic genome annotation and pathway reconstruction server. *Nucleic Acids Res* 35:W182–W185.

Morris, G. P., P. Ramu, S. P. Deshpande et al. 2013. Population genomic and genome-wide association studies of agroclimatic traits in sorghum. *Proc Natl Acad Sci USA* 110:453–458.

Mortazavi, A., B. A. Williams, K. McCue, L. Schaeffer, and B. Wold. 2008. Mapping and quantifying mammalian transcriptomes by RNA-Seq. *Nat Methods* 5:621–628.

Mosquera, G., M. C. Giraldo, C. H. Khang et al. 2009. Interaction transcriptome analysis identifies *Magnaporthe oryzae* BAS1-4 as biotrophy-associated secreted proteins in rice blast disease. *Plant Cell* 21:1273–1290.

Mutwil, M., J. Obro, W. G. Willats, and S. Persson. 2008. GeneCAT—Novel webtools that combine BLAST and co-expression analyses. *Nucleic Acids Res* 36:W320–W326.

Nishida, H., R. Abe, T. Nagayama, and K. Yano. 2012. Genome signature difference between *Deinococcus radiodurans* and *Thermus thermophilus*. *Int J Evol Biol* 2012:6.

Novák, P., P. Neumann, J. Pech, J. Steinhaisl, and J. Macas. 2013. RepeatExplorer: A galaxy-based web server for genome-wide characterization of eukaryotic repetitive elements from next-generation sequence reads. *Bioinformatics* 29:792–793.

Obayashi, T., K. Nishida, K. Kasahara, and K. Kinoshita. 2011. ATTED-II updates: Condition-specific gene coexpression to extend coexpression analyses and applications to a broad range of flowering plants. *Plant Cell Physiol* 52:213–219.

Obayashi, T. and K. Yano. 2013. Editorial. *Plant Cell Physiol* 54:169–170.

Quevillon, E., V. Silventoinen, S. Pillai et al. 2005. InterProScan: Protein domains identifier. *Nucleic Acids Res* 33:W116–W120.

Robertson, G., J. Schein, R. Chiu et al. 2010. *De novo* assembly and analysis of RNA-seq data. *Nat Methods* 7:909–912.

Sakurai, N., T. Ara, Y. Ogata et al. 2011. KaPPA-View4: A metabolic pathway database for representation and analysis of correlation networks of gene co-expression and metabolite co-accumulation and omics data. *Nucleic Acids Res* 39:D677–D684.

Sato, Y., N. Namiki, H. Takehisa et al. 2013a. RiceFREND: A platform for retrieving coexpressed gene networks in rice. *Nucleic Acids Res* 41:D1214–D1221.

Sato, Y., H. Takehisa, K. Kamatsuki et al. 2013b. RiceXPro Version 3.0: Expanding the informatics resource for rice transcriptome. *Nucleic Acids Res* 41:D1206–D1213.

Schmid, M., T. S. Davison, S. R. Henz et al. 2005. A gene expression map of *Arabidopsis thaliana* development. *Nat Genet* 37:501–506.

Schmidt, M. A., W. B. Barbazuk, M. Sandford et al. 2011. Silencing of soybean seed storage proteins results in a rebalanced protein composition preserving seed protein content without major collateral changes in the metabolome and transcriptome. *Plant Physiol* 156:330–345.

Schulz, M. H., D. R. Zerbino, M. Vingron, and E. Birney. 2012. Oases: Robust *de novo* RNA-seq assembly across the dynamic range of expression levels. *Bioinformatics* 28:1086–1092.

Smoot, M., K. Ono, J. Ruscheinski, P.-L. Wang, and T. Ideker. 2011. Cytoscape 2.8: New features for data integration and network visualization. *Bioinformatics* 27:431–432.

Srinivasasainagendra, V., G. P. Page, T. Mehta, I. Coulibaly, and A. E. Loraine. 2008. CressExpress: A tool for large-scale mining of expression data from Arabidopsis. *Plant Physiol* 147:1004–1016.

Suwabe, K., G. Suzuki, H. Takahashi et al. 2008. Separated transcriptomes of male gametophyte and tapetum in rice: Validity of a laser microdissection (LM) microarray. *Plant Cell Physiol* 49:1407–1416.

Suzuki, T., K. Igarashi, H. Dohra et al. 2013. A new omics data resource of *Pleurocybella porrigens* for gene discovery. *PLoS ONE* 8:e69681.

The Gene Ontology Consortium. 2000. Gene ontology: Tool for the unification of biology. *Nat Genet* 25:25–29.

The Potato Genome Sequencing Consortium. 2011. Genome sequence and analysis of the tuber crop potato. *Nature* 475:189–195.

The Tomato Genome Consortium. 2012. The tomato genome sequence provides insights into fleshy fruit evolution. *Nature* 485:635–641.

Toufighi, K., S. M. Brady, R. Austin, E. Ly, and N. J. Provart. 2005. The botany array resource: e-Northerns, expression angling, and promoter analyses. *Plant J* 43:153–163.

Trapnell, C., A. Roberts, L. Goff et al. 2012. Differential gene and transcript expression analysis of RNA-seq experiments with TopHat and Cufflinks. *Nat Protoc* 7:562–578.

Tsai, W. C., C. H. Fu, Y. Y. Hsiao et al. 2013. OrchidBase 2.0: Comprehensive collection of Orchidaceae floral transcriptomes. *Plant Cell Physiol* 54:e7.

Usadel, B., T. Obayashi, M. Mutwil et al. 2009. Co-expression tools for plant biology: Opportunities for hypothesis generation and caveats. *Plant Cell Environ* 32:1633–1651.

Vivancos, A. P., M. Güell, J. C. Dohm, L. Serrano, and H. Himmelbauer. 2010. Strand-specific deep sequencing of the transcriptome. *Genome Res* 20:989–999.

Wang, Z., M. Gerstein, and M. Snyder. 2009. RNA-Seq: A revolutionary tool for transcriptomics. *Nat Rev Genet* 10:57–63.

Wheeler, D. L., T. Barrett, D. A. Benson et al. 2008. Database resources of the National Center for Biotechnology Information. *Nucleic Acids Res* 36:D13–D21.

Wing, R. A., J. S. Ammiraju, M. Luo et al. 2005. The *Oryza* map alignment project: The golden path to unlocking the genetic potential of wild rice species. *Plant Mol Biol* 59:53–62.

Yamamoto, T., H. Nagasaki, J. Yonemaru et al. 2010. Fine definition of the pedigree haplotypes of closely related rice cultivars by means of genome-wide discovery of single-nucleotide polymorphisms. *BMC Genomics* 11:267.

Yano, K., K. Imai, A. Shimizu, and T. Hanashita. 2006. A new method for gene discovery in large-scale microarray data. *Nucleic Acids Res* 34:1532–1539.

CHAPTER 2

Foodomics Strategies for the Analysis of Genetically Modified Crops

Alberto Valdés, Alejandro Cifuentes, and Virginia García-Cañas

CONTENTS

2.1 INTRODUCTION

Since its early applications in the 1970s, genetic engineering (or recombinant DNA technology) has become one of the principal technological advances in modern biotechnology. Genetic engineering allows selected individual genes to be transferred from one organism into another and also between nonrelated species. Organisms derived from recombinant DNA technology are termed genetically modified organisms (GMOs), and these are defined as those organisms in which the genetic material has been altered in a way that does not occur naturally by mating or natural recombination (WHO 2002). Recombinant DNA technology has become valuable for the experimental investigation of many aspects of plant biochemistry and physiology that cannot be addressed easily by any other experimental means (Wisniewski et al. 2002). This technology offers an extraordinary opportunity to investigate the molecular mechanisms of important processes, such as plant–microbe interactions, development, response to abiotic and biotic stress, and signal transduction pathways, by the analysis of gene function and regulation in transgenic plants (Twyman et al. 2002). Besides, the rapid progress of this technology has opened new opportunities to create genetically modified (GM) plants, which are used to grow GM crops for the production of food, feed, fiber, forest, and so on (Petit et al. 2007). In 2012, the global area of approved GMOs was 170 million hectares in 30 countries, representing an almost hundredfold increase since the first commercialized GM crop in 1996 (James 2012). Since then, the total accumulated land area cultivated with transgenic crops has markedly increased, to 1.5 billion hectares. Until the year 2012, the dominant GM crops, in terms of cultivated area, were soybean, maize, cotton, and rapeseed. However, it is expected that other GM crop species, for example, sugar beet, rice, potato, banana, tomato, and wheat, undergoing field trials worldwide will enter the world markets in the near future. Among the most important traits present in authorized GMOs, tolerance to herbicide and resistance to insects are prevalent worldwide. The transformation events that have received the greatest number of regulatory approvals worldwide are the herbicide-tolerant maize event NK603 and the herbicide-tolerant soybean event GTS-40-3-2, followed by the insect-resistant maize events MON810 and Bt11 and insect-resistant cotton events MON531 and MON1445. Nevertheless, a number of novel traits, such as different micronutrient content, faster ripening, improved feed value, high levels of antioxidants, tolerance to drought, and so on, might be in the pipeline for commercialization in coming years (Cahoon et al. 2003; Robinson 2001; Schubert 2008; Ye et al. 2000).

Since its beginning more than three decades ago, development, release into the environment, and commercialization of GMOs have been widely debated, not only within the public sector but also in the scientific community (Berg et al. 1975). The main controversial issues focus on four areas: environmental concerns (Thomson 2003; Wolfenbarger and Phifer 2000), potential harm to human health (Craig et al. 2008; Domingo 2007; Garza and Stover 2003), concerns related to patent issues (Herring 2008; Vergragt and Brown 2008), and ethical concerns over interference with nature and individual choice (Frewer et al. 2004).

In spite of the predictable accuracy of recombinant DNA technology for genetic modification, possible unintended effects originating during the genetic transformation process might occur. Unintended effects have been defined as those effects that fall beyond the primary expected effects of the genetic modification and represent statistically significant differences in a phenotype compared with an appropriate phenotype control (Cellini et al. 2004). Examples of unintended effects are alterations linked to secondary effects of gene expression (Ali et al. 2008; Kuiper and Kleter 2003), or unexpected transformation-induced mutations such as insertions, rearrangements, deletions, and so on (Hernandez et al. 2003; Latham et al. 2006). As they are unpredictable, unintended effects are considered as a significant source of uncertainty that might have an impact on human health and/or the environment (Ioset et al. 2007). On the other hand, unintended modifications might be noticed if the changes result in a different phenotype, including compositional alterations.

In the European Union and other countries, the mentioned controversial aspects on commercialization of GMOs have led to the enforcement of strict regulations concerning different aspects, including risk assessment, marketing, labeling and traceability of GMOs. The concept typically referred to as "substantial equivalence" or "comparative safety assessment" has been internationally adopted for the safety assessment of GM foods. Substantial equivalence relies upon the assumption that traditional crop-plant varieties currently on the market that have been consumed for decades have gained a history of safe use. Consequently, these traditional varieties can be used as comparators for the safety assessment of new GM crop varieties derived from established plant lines. Although this approach has been generally accepted, there is a lack of international harmonization among regulations that has originated an "asynchronous approval" of these GM crops, an issue of growing concern for its potential economic impact on international trade. Similarly, labeling and traceability requirements for GMO and GMO-derived food ingredients are also quite heterogeneous across national jurisdictions. In this regard, labeling of foodstuffs may be voluntary or mandatory, and the specific thresholds set for labeling vary between countries. For instance, in the European Union, any food containing more than 0.9% GM content has to be labeled as such, provided that the presence of this GM ingredient is technically unavoidable or adventitious, whereas in Australia and Japan, the threshold for labeling has been established at 1% and 5% GM content, respectively.

2.2 CHALLENGES IN GMO ANALYSIS

2.2.1 GMO Detection in the Food Supply Chain

At present, there is a need for suitable strategies to address new challenges in GMO analysis. The enforcement of new regulations implies that analytical tools must be available to verify their compliance in terms of control for nonauthorized GMOs, and also for labeling and traceability of approved GMOs. In this context, appropriate analytical tools should enable the rapid detection, identification, and accurate quantification of approved and unapproved GMOs in a given sample. However, as illustrated above, the number of approved GM crops and the extension areas where they are cultivated are steadily increasing around the globe. More than 90 novel GMOs were in advanced stages of the development, authorization, or commercialization process in 2009 and might enter the market in the near future (Stein and Rodríguez-Cerezo 2009). As the number of GMOs that can be present in the food chain is increasingly growing, the analytical solution aimed at GMO detection becomes more challenging.

For years, the detection, identification, and quantification of GMOs and GM-based materials in foods have been conventionally addressed by target-based analytical strategies, DNA being the chosen target molecule for this type of analysis (Alderborn et al. 2010; Deisingh and Badrie 2005; García-Cañas et al. 2004; Marmiroli et al. 2008; Michelini et al. 2008). Most DNA-based analytical methods for GMOs involve the use of polymerase chain reaction (PCR) to detect, identify, and quantify inserted DNA sequences in food samples. Due to its specificity, its sensitivity, and the fact that it allows a rapid and relatively low-cost analysis, this amplification technique in its different formats has been established as the dominant technique for GMO detection and traceability. In spite of the feasibility of PCR-based methods, assessing a given sample for each and every existing GMO event is not practical, since it would be necessary to carry out as many independent analyses as there are GMOs in the market. In this context, screening analysis has become essential to minimize the analytical effort to rapidly assess whether or not a sample under investigation is likely to contain GMOs. Ideally, screening methods should provide some clues about the transgenic character of a given sample, allowing a further preselection of more specific analysis for identification and quantification. Following this concept, numerous screening methods are based on the amplification and detection of DNA sequences found in as many different GMOs as possible (Akiyama et al. 2009; Bahrdt et al. 2010; Barbau-Piednoir et al. 2010; Chaouachi et al. 2008; Grohmann et al. 2009). The two sequences most frequently used for this purpose are the promoter, P-35S, from cauliflower mosaic virus (CaMV) and the nopaline synthase gene terminator, T-nos, from *Agrobacterium tumefaciens*. However, as the number of new GMOs grows, novel gene encoding sequences and transcription control elements are included in their production; therefore, a screening approach based on a very limited number of targets is no longer effective (Van Götz 2009). Also, current screening approaches present some limitations for the detection of the most recent generations of GMOs, including those that contain transgenic elements that are not common in other known GMOs, or those whose insertions are intragenic constructs derived from the recipient species. Additional challenging issues regarding GMO detection are those associated with so-called stacked GMOs, which are typically hybrid crosses of more than one different GMO. At present, stacked events cannot be easily discriminated from their parental GMOs in complex mixtures.

2.2.2 Safety Evaluation of GMOs

As mentioned before, in several countries, including those of the European Union, any GMO or derived product has to pass an approval system in which safety for humans, animals, and the environment is thoroughly assessed before it can be placed on the market (EFSA 2006). Regarding food safety, the comparative compositional analysis of GM-derived foods with their appropriate comparators already on the market establishes the foundation for the detection of any potential unintended effects in the GM crop variety. Compositional equivalence between GM crops and conventional (non-GM) comparators is considered to provide an equal or increased assurance of the safety of foods derived from GM plants. In recent years, many questions have been raised regarding the selection of the comparator, the need for comparing varieties grown in different areas and seasons, the key components to be analyzed, and so on (Kok et al. 2008). Regarding the latter aspect, comparisons between the GMO and its comparator are typically performed on the basis of targeted analysis of predefined compounds that include macro- and micronutrients, antinutrients, natural toxins, and so on following recommendations in the Organisation of Economic Co-operation and Development (OECD) consensus documents for individual crops (Cellini et al. 2004; Shepherd et al. 2006). Numerous reports have shown that differences due to genetic transformation are negligible, and usually not reproducible, in comparison with the differences associated with the genotype and environmental factors (Harrigan et al. 2010). In this regard, the availability of databases that compile compositional information derived from the analysis of crop species will assist in the determination of the nutritional and toxicological relevance of detected differences between GM and non-GM crops (Kok et al. 2008). Although targeted approach has enabled the identification of unintended effects in some cases (Hashimoto et al. 1999; Shewmaker et al. 1999; Ye et al. 2000), its application to comparing the compositional differences between GMOs and their comparators has raised several concerns. Specifically, it has been pointed out that this targeted strategy is biased and that some unforeseen, unintended effects that could result directly or indirectly from the genetic transformation may remain undetected (Millstone et al. 1999). In contrast, a broader survey of GMO composition would theoretically improve the opportunities for detecting potential unintended effects that might not be noticeable using targeted analysis. Consequently, the European Food Safety Authority has recommended the development and use of profiling technologies with the potential to extend the coverage in comparative analyses of GMOs (EFSA 2006). Also, a panel of experts on risk assessment and management has recently suggested the application of profiling, specifically in those cases where the most scientifically valid isogenic and conventional comparator would not grow, or would not grow as well, under the relevant stress condition (AHTEG 2010).

2.3 FOODOMICS APPLICATIONS IN GMO ANALYSIS

Recently, foodomics has been defined as a new discipline that studies the food and nutrition domains through the application and integration of advanced omics technologies in order to improve consumers' well-being and confidence (Cifuentes 2009; Herrero et al. 2010). This novel discipline could play an important role in the investigation of GM crops that are intended for entry into the human or animal food chain. Thus, foodomics can provide valuable data about molecular profiling, based on genomic, transcriptomic, proteomic, or metabolomic analysis, individually or in combination: information that could be essential for the detection, traceability, and characterization of GMOs. Molecular profiling enables the simultaneous measurement and comparison of thousands of compounds. Thus, molecular profiling might be useful for the study of primary effects of the genetic modification during the different stages of GM crop development. Also, this analytical strategy offers unprecedented opportunities to study the molecular mechanisms leading to a particular phenotype or the mechanisms operating in important cellular processes, such as the response to different stresses (García-Cañas et al. 2011, 2012). As will be discussed in the following sections, the application of molecular profiling tools, such as DNA microarray, might increase the actual coverage of transgenic DNA markers that can be simultaneously detected, reducing the time required for assaying the presence of GMOs in a given sample. Advanced tools in transcriptomics, proteomics, and metabolomics, in combination with bioinformatics and chemometrics, may also provide chemical compositional data complementary to the information obtained in targeted-based analyses, which might be helpful to address any effects (intended or not) derived from the genetic transformation. To illustrate this, the following sections will focus on the recent applications and advances made in the area of molecular profiling for the analysis of GMOs, discussing some of their advantages and drawbacks.

2.3.1 Comprehensive GMO Screening

Due to the growing number of novel GMO releases, screening methods need to be as comprehensive as possible to face the complexity of this topic. Comprehensive GMO screening requires a high degree of parallel tests or improved multiplexing capabilities in order to gain time and cost-effectiveness in GMO detection. To this end, novel schemes for GMO screening have been envisaged in recent years. However, the prevailing strategy is based on the matrix approach (Holst-Jensen et al. 2012). This strategy is based on a stepwise procedure in which a different set of target DNA sequences (called analytical modules) are assayed in each round of analysis (screening). The information associated with target sequences is arranged in a relational matrix. Such sequences are carefully chosen to detect the most ubiquitous transgenic control elements (e.g., P-35S, T-nos, bar, pat, nptII, etc.) in known GMOs in the early rounds of the screening procedure. The results obtained in the first round of screening are then compared with the reference pattern of the matrix to narrow down the list of candidate GMOs that may be present in the sample. This process will help the analyst to select further analytical modules for the next round of analysis. The idea behind this strategy is to reduce the number of analyses needed to confirm the identity of one or more GMOs in the sample, a test that is usually carried out with event-specific analytical modules.

Nowadays, microarray technology has the greatest capabilities for parallel and high-throughput analysis. Therefore, it provides a good opportunity for comprehensive screening, since it improves the chances of detecting a broad range of GMOs compared with conventional PCR and real-time quantitative PCR (RT-qPCR) screening methods. In general, microarrays are collections of oligonucleotides or probes attached to a substrate, usually a glass slide, at predefined locations within a grid pattern. This technique is based on hybridization of specific nucleic acids and it can be used to measure quantities of specific nucleic acid sequences in a given sample for thousands of genes simultaneously. Regardless of the platform used for the analysis, the typical experimental procedure is based on the same analytical steps: nucleic acids are extracted from a source of interest (tissue, cells, or other materials), labeled with a detectable marker (typically, fluorescent dye) and allowed to hybridize to the microarrays, with individual DNA sequences hybridizing to their complementary sequence-specific probes on the microarray. Once hybridization is complete, samples are washed and imaged using a confocal laser scanner. Theoretically, the signal of derivatized nucleic acids bound to any probe is a function of their concentration. Signal intensities are then extracted and transformed to a numeric value (Storhoff et al. 2005). In order to convert images into gene expression data, dedicated image processing software is normally used. Typically, imaging processing involves the following steps: (i) identification and distinction of spot signals from spurious signals; (ii) determination of the spot areas and estimation of background hybridization; and (iii) reporting summary statistics and determination of spot intensity (raw data). The signal intensity raw data are then subjected to transformation and normalization. In general, microarray systems developed for GMO analysis require an amplification step of target DNA sequences prior to labeling and hybridization in the array. In recent years, several microarray methods, most of them based on low-density formats, have been reported for screening analysis with different GMO coverages (Bordoni et al. 2004, 2005; Germini et al. 2005; Kim et al. 2010; Rudi et al. 2003). For instance, the commercial microarray system DualChipGMO V2.0 (Eppendorf Array Technologies, Namur, Belgium) combined three multiplex PCR methods with a colorimetric detection and scanning system (Silverquant system) to detect up to 30 different screening targets, covering more than 80% of all GMOs known at that moment (Hamels et al. 2009).

Tengs et al. (2007, 2010) used a more comprehensive strategy, based on tiling microarrays, to detect transgenic sequences in genomic DNA from plants. In their first work, they designed and fabricated a high-density microarray containing 37,257 overlapping probes (25 base pairs long) covering 235 vector sequences (containing P35S sequences) found in genetic databases. Sample preparation involved whole-genome amplification of the genomic extracts prior to hybridization onto the microarray. After microarray data normalization, a threshold of 30 probes per window was established for data analysis. A vector sequence was scored as positive in a sample when more than 70% of the probes in the window were positive. Following this approach, several transgenic sequences ranging in size from 147 to 325 bases were detected in genomic DNA extracted from GMOs. Interestingly, this method enabled the detection of transgenic sequences in unknown GMOs containing sequences common to specific vector sequences included in the microarray. In a further report, the microarray coverage was extended to include full-length sequence data from the majority of trait genes listed in the AGBIOS database (Tengs et al. 2010). In this case, the applicability of the method was demonstrated by the detection of known transgenic soybean and maize, and an unknown soybean event, demonstrating good performance for the characterization of certain unknown GMOs. However, among the limitations of this approach are the inability to detect organisms transformed with transgenic sequences that are not present in the microarray, and the need for pure genomic material for the assay, which restricts its use for the characterization of GMO mixtures.

Next-generation sequencing is well suited for GMO analysis. Tengs et al. (2009) have proposed an analytical approach based on the combination of high-throughput pyrosequencing with computational subtraction in order to evaluate whether an organism has been genetically modified or not. This sophisticated method relies on the detection of unexpected cDNA sequences by comparing sequence reads from the sample against a set of reference sequences using sequence similarity search algorithms. The main feature of this methodology is the improved capability for comprehensive characterization at the genome level, allowing the search for novel genetic constructions and transgenic elements other than those included in known GMOs. However, it should be noted that the method theoretically would fail in those cases in which the modification is directed to alter the expression of endogenous genes.

2.3.2 Comparative Compositional Analysis of GMOs

As previously mentioned, bias and uncertainties associated with targeted analysis in comparative compositional evaluation of GMOs have motivated several research groups to direct their attention to the development and application of "omics" technologies in order to investigate chemical compositional differences between GM and non-GM crops as part of a comparative safety assessment strategy. Thus, transcriptomics, proteomics, and metabolomics have been applied to obtain profiles of mRNA transcripts, proteins, and metabolites, respectively, from a number of GM varieties.

A debate has recently opened around the usefulness of omics techniques within the framework of safety assessment of GMOs (Doerrer et al. 2010). One of the main arguments against profiling relies on the lack of standardized and validated procedures that allow their routine use among laboratories. Another relevant issue of omics techniques is linked with the poor predictive capacity of the obtained profiles for safety evaluation. Although molecular profiling can effectively measure relative differences in compounds between two varieties with high sensitivity, the biological relevance of such differences cannot be determined without previous knowledge of the natural variability of the crop composition (Ricroch et al. 2011). Furthermore, in most cases where bandwidth variation of a compound is known, it remains difficult to decipher the biological meaning of the detected differences in terms of food safety risk (Doerrer et al. 2010). With respect to the natural variation of crop chemical composition, it should be evaluated in a wide range of situations (plants grown in different locations, climates, seasons, and under different farming practices) to make this overview as complete as possible (Davies 2010). Moreover, the evaluation of natural variability of chemical composition should be accompanied by the development and maintenance of suitable databases containing the information about the natural range of expression for each particular compound. In addition, it has been argued that profiling approaches accumulate false positives, which would provide superfluous data that may distract from the detection of relevant biological information (Chassy 2010).

Despite the criticisms raised about the value of molecular profiling for GMO risk assessment, there are a number of reports based on the use of different profiling approaches for comparative profiling analysis of GMOs and investigation of unintended effects, suggesting good acceptance of these techniques by the scientific community (Heinemann et al. 2011). It is evident that high-throughput profiling approaches are powerful analytical tools in GMO research (García-Cañas et al. 2011; Herrero et al. 2012); however, current omics techniques will need to evolve in order to overcome the mentioned limitations that hamper their use in food safety assessment. It is interesting to note that the majority of the published omics studies report some differences that can be linked to genetic transformation. However, findings consistently observed in omics studies point out that the differences between conventional varieties are, in general, more pronounced than divergences observed between GM and non-GM crops. This observation can be also extended to the differences found when the same variety is cultured in different locations or seasons or both. Representative examples of the application of omics techniques, including gene expression and protein and metabolite profiling, to comparative profiling analysis of GM crops are summarized in Table 2.1. Some of the works carried out in this field will be discussed in the following sections.

2.3.2.1 Gene Expression Profiling

In recent years, due to its extensive optimization and standardization, gene expression microarray has become the leading analytical technology in transcriptomics studies. Generally, gene expression microarrays are used to measure the relative quantities of specific mRNAs in two or more samples for thousands of genes simultaneously. Microarray technology has been shown to be a valuable profiling method to assess possible unintended effects of genetic transformation in crops (potato, rice, wheat, and maize; see Table 2.1).

However, the importance of first investigating the bandwidth of natural variation in gene expression patterns in conventionally bred crop varieties has been widely recognized. To this end, DNA microarray has emerged as a holistic approach to discover changes present in the natural variation of specific genes in different conditions considered to be usual in agricultural practices (Van Dijk et al. 2009). Early studies, focusing on the application of microarray technology to the study of the differences between the transcriptional profiles of untransformed and a derived transgenic wheat line, indicated that the presence of transgenes did not significantly alter gene expression (Baudo et al. 2006). Later, high-density microarray (covering 51,279 transcripts from two rice subspecies) was applied to study the extent of transcriptome modification occurring during improvement through transgenesis versus mutation breeding in rice plants (Batista et al. 2008). In this case, transcriptomic data indicated that altered expression of untargeted genes was more extensive in mutagenized rice than in GM rice. Transcriptome profiling of rice crops has been the object of several recent studies. For instance, Montero et al. (2011) conducted a study on the impact of the insertion site and associated rearrangements on the transcriptomic regulation in transgenic rice.

Table 2.1 Omics Techniques for Comparative Studies in GM Crops

Application	Analysis Technique	GM Crop	Donor Species	Phenotype	Tissue	Genetic Modification	References
Gene expression profiling	Microarray	Wheat	Triticum aestivum	Nutritionally enhanced	Seed and leaf	Glu-A1, Glu-D1	Baudo et al. (2006)
	Microarray	Rice	Hordeum vulgare	Control stress-inducible genes	Seed	BCBF1	Batista et al. (2008)
	Microarray	Soybean	Agrobacterium tumefaciens	Herbicide tolerance	Leaf	CP4 EPSPS	Cheng et al. (2008)
	Microarray	Maize	Bacillus thuringiensis	Insect resistance	Seed	Cry1Ab	Coll et al. (2008)
	Microarray	Maize	B. thuringiensis	Insect resistance	Leaf	Cry1Ab	Coll et al. (2010)
	Microarray	Rice	Aspergillus giganteus	Fungal resistance	Leaf	afp	Montero et al. (2011)
	Microarray	Rice	B. thuringiensis	Insect resistance	Leaf	Cry1Ab	Liu et al. (2012)
Protein profiling	MALDI-TOF	Potato	Solanum tuberosum	Sprouting delay	Tuber	G1-1	Careri et al. (2003)
	2-DGE, MALDI-TOF	Tomato	Tomato spotted wild virus (TSWV)	Virus resistance	Seed	TSWV-N	Corpillo et al. (2004)
	2-DGE, LC-ESI-IT	Potato	Aureobasidium pullulans	Waxy phenotype, modified metabolism	Tuber	W, Mal1, SamDC	Lehesranta et al. (2005)
	2-DGE, MALDI-TOF/TOF	Maize	B. thuringiensis	Insect resistance	Seed	Cry1Ab	Albo et al. (2007)
	CE-ESI-IT, CE-ESI-TOF	Maize	B. thuringiensis	Insect resistance	Grain	Cry1Ab	Erny et al. (2008)
	2-DGE, LC-ESI-IT	Maize	B. thuringiensis	Insect resistance	Seed	Cry1Ab	Zolla et al. (2008)
	2-DGE, MALDI-TOF, LC-ESI-IT	Tomato	Cucumber mosaic virus (CMV)	Virus resistance	Leaf	ScFv (G4)	Di Carli et al. (2009)
	LC-ESI-IT	Maize	B. thuringiensis	Insect resistance	Grain	Cry1Ab	García-López et al. (2009)
	2-DGE, ESI-Q/TOF	Pea	Phaseolus vulgaris	Insect resistance	Seed	αAI1	Islam et al. (2009)
	2-DGE, MALDI-Q/TOF	Soybean	A. tumefaciens	Herbicide tolerance	Seed	CP4 EPSPS	Brandao et al. (2010)
	CE-ESI-TOF	Soybean	A. tumefaciens	Herbicide tolerance	Seed	CP4 EPSPS	Simó et al. (2010)
	2-DGE, MALDI-TOF/TOF	Maize	B. thuringiensis	Insect resistance	Leaf	Cry1Ab	Balsamo et al. (2011)
	2-DGE, LC-ESI-IT	Maize	B. thuringiensis	Insect resistance	Grain	Cry1Ab	Coll et al. (2011)
	2-DGE, LC-ESI-Q/TOF	Soybean	A. tumefaciens	Herbicide tolerance	Seed	CP4 EPSPS	Barbosa et al. (2012)
	LC-ESI-TOF	Maize	B. thuringiensis	Insect resistance	Grain	Cry1Ab	Koc et al. (2012)

Metabolite profiling	GC-EI-Q	Potato	S. tuberosum	Altered starch composition	Tuber	AGPase, StcPGM, StpPGM	Roessner et al. (2001a)
	GC-EI-Q	Tomato	Arabidopsis thaliana	Altered carbohydrate metabolism	Leaf and fruit	AtHXK1	Roessner-Tunali et al. (2003)
	NMR	Pea	Streptomyces hygroscopicus	Herbicide tolerance	Leaf	Bar	Charlton et al. (2004)
	NMR	Potato	A. pullulans, S. tuberosum	Waxy phenotype, modified metabolism	Tuber	W2, Mal1, SamDC	Defernez et al. (2004)
	NMR	Maize	B. thuringiensis	Insect resistance	Grain	Cry1Ab	Manetti et al. (2004)
	GC-EI-TOF, LC-ESI-Q	Potato	Cynara scolymus	Inulin synthesis	Tuber	1-SST, 1-FFT	Catchpole et al. (2005)
	NMR	Wheat	T. aestivum	Nutritionally enhanced	Seed	Glu-A1, Glu-D1	Baker et al. (2006)
	GC-EI-Q	Potato	A. pullulans, S. tuberosum	Starch biosynthesis, leaf morphology, ethylene production	Tuber	W2, FK, Mal1, SamDC	Shepherd et al. (2006)
	LC-ESI-IT	Rice	Zea mays	Flavonoid production	Leaf	C1, R-S	Shin et al. (2006)
	CE-ESI-Q	Rice	Oryza sativa	Stress tolerance	Seed	YK1	Takahashi et al. (2006)
	GC-EI-Q, LC-ESI-IT	Grapevine	Escherichia coli	Abiotic stress	Leaf	Adh	Tesniere et al. (2006)
	LC-ESI-IT	Wheat	H. vulgare, Ustilago maydis	Antifungal activity	Leaf	Glu, Chi, KP4	Ioset et al. (2007)
	LC-ESI-Q	Tomato	Vitis vinifera	Resveratrol synthesis	Fruit	Stilbene synthase	Nicoletti et al. (2007)
	NMR	Lettuce	E. coli	Growth enhanced	Leaf	asn A	Sobolev et al. (2007, 2010)
	GC-EI-Q	Soybean	A. tumefaciens	Herbicide tolerance	Seed	CP4 EPSPS	Bernal et al. (2008)
	CE-ESI-TOF	Soybean	A. tumefaciens	Herbicide tolerance	Seed	CP4 EPSPS	Garcia-Villalba et al. (2008)
	CE-ESI-TOF	Maize	B. thuringiensis	Insect resistance	Grain	Cry1Ab	Levandi et al. (2008)
	CE-ESI-TOF	Soybean	A. tumefaciens	Herbicide tolerance	Seed	CP4 EPSPS	Giuffrida et al. (2009)

(continued)

Table 2.1 (Continued) Omics Techniques for Comparative Studies in GM Crops

Application	Analysis Technique	GM Crop	Donor Species	Phenotype	Tissue	Genetic Modification	References
	GC-EI-Q	Maize	*B. thuringiensis*	Insect resistance	Grain	*Cry1Ab*	Jimenez et al. (2009)
	FT-ICR	Maize	*B. thuringiensis*	Insect resistance	Grain	*Cry1Ab*	León et al. (2009)
	LC-ESI-Q	Rice	*E. coli*	Nutritionally enhanced	Seed	*glgC-TM*	Nagai et al. (2009)
	NMR	Maize	*B. thuringiensis*	Insect resistance	Grain	*Cry1Ab*	Piccioni et al. (2009)
	GC-EI-Q, GC-EI-TOF	Cucumber	*Thaumatococcus daniellii*	Sweet flavor	Fruit	*Thaumatin-II*	Zawirska-Wojtasiak et al. (2009)
	GC-EI-Q	Maize	*B. thuringiensis*	Insect resistance	Grain	*Cry1Ab*	Barros et al. (2010)
	GC-EI-Q	Rice	*O. sativa, H. vulgare, B. thuringiensis*	Antifungal activity, insect resistance	Seed	*RCH10, RAC22, β-1,3-Glu, B-RIP, Cry1Ac, sck*	Jiao et al. (2010a)
	GC-EI-Q, LC-ESI-Q	Papaya	*Carica papaya*	Virus resistance	Pulp and leaf	*rep*	Jiao et al. (2010b)
	LC-ESI-IT	Barley	*Bacillus amyloliquefaciens*	Antifungal activity	Seed	*GluB, ChGP*	Kogel et al. (2010)
	LC-ESI-Q	Rice	*Nicotiana tabacum*	Nutritionally enhanced	Seed	*ASA2*	Matsuda et al. (2010)
	GC-EI-TOF, LC-ESI-Q/ TOF, CE-ESI-Q/TOF	Tomato	*Richadella dulcifica*	Sweet flavor	Fruit	*Miraculin*	Kusano et al. (2011)
	LC-ESI-Q/TOF	Rice	*B. thuringiensis*	Insect resistance	Seed	*Cry1Ac, sck*	Chang et al. (2012)
	GC-EI-Q	Maize	*B. thuringiensis, A. tumefaciens*	Insect resistance, herbicide tolerance	Grain	*Cry1Ab CP4 EPSPS*	Frank et al. (2012)

Their results suggested that 35% of the detected differences can be attributed to the procedure used to obtain GM plants (in vitro culture, cell dedifferentiation, plant regeneration, etc.), whereas around 15% of the changes between the GM and the isogenic line depended on the transformation event, associated with the insertion site and genome rearrangements occurring during transformation. Recently, another transcriptomic study has been carried out with the aim of revealing the molecular basis for unintended effects (susceptibility to rice brown spot mimic lesion and sheath blight diseases) observed in the insect-resistant transgenic KMD rice (Liu et al. 2012). Gene expression profiles revealed significant changes in 1.36% of all 680 analyzed rice transcripts, three of which, associated with the transgenic sequence, were extremely overexpressed. Further functional enrichment analysis using the bioinformatic tool Plant MetGenMAP enabled the identification of those pathways more frequently represented in the list of differentially expressed genes. The results indicated a significant number of genes directly related to plant stress/defense responses, and also to amino acid metabolism in KMD rice. In order to confirm these results, amino-acid profiling analysis of seedlings was performed using isobaric tags for relative and absolute quantitation (iTRAQ), and liquid chromatography (LC) coupled to mass spectrometry (MS) (iTRAQ®-LC-MS/MS). Interestingly, these analyses showed changes in 10 amino acids, and also in γ-amino-n-butyric acid, a typical stress-response amino acid, in the transgenic rice plants relative to the conventional counterparts, which allowed some links to be established between the reported unintended effects and the changes at the transcriptional and metabolite levels.

Coll et al. (2008, 2010) reported the transcriptomic profiles of the transgenic MON810 maize event using gene expression microarray technology in a series of reports. In the first study, authors performed a comparative study of several transgenic maize lines containing the MON810 event with their near-isogenic comparators (Coll et al. 2008). Maize plantlets cultured in vitro under highly controlled experimental conditions in order to minimize transcriptome changes due to environmental factors were analyzed using Affymetrix high-density microarrays. Gene expression profiles revealed different patterns in terms of number and type of genes between two MON810 varieties and their respective counterparts. Gene expression data of 30 genes using RT-qPCR analysis confirmed the microarray results. However, the expression pattern was not confirmed when another four MON810/non-GM pairs were tested. Differences in gene expression patterns were also found between MON810 and non-GM near-isogenic maize plants grown in different field environments and cultural conditions (Coll et al. 2010). Pair-wise comparisons of transgenic and control maize varieties grown under low-nitrogen and control conditions showed differences ranging from 0.07% to 0.2% of maize transcriptome covered by the microarray (corresponding to 7 and 36 sequences, respectively). A total of 13 and 23 genes showed differential expression between the transgenic and nontransgenic maize in control and low-level nitrogen conditions, respectively. The expression of 37 sequences was confirmed by RT-qPCR technique with a 71.1% degree of coincidence between microarray and PCR data. In order to investigate the sources of variation in data sets, transcriptional profiles of all 37 sequences in the different samples and conditions were subjected to principal component analysis (PCA) and unsupervised clustering (Kohonen self-organizing maps) (Figure 2.1). It could be demonstrated that maize variety had the highest impact on gene expression pattern, followed by the nitrogen treatment, whereas the transgenic character of the sample had the lowest effect on gene expression variation.

Cheng et al. (2008) investigated five soybean cultivars (two transgenic and three conventional) that were genotyped using simple sequence repeats marker analysis in order to establish the genetic distances between cultivars. PCA and unsupervised hierarchical clustering of a total of 25 gene expression profiles grouped samples in three major groups. The classification indicated divergences between the oldest variety samples (mandarin cultivar), samples displaying more RNA degradation, and another 15 samples of non-GM and GM soybeans. As the two transgenic cultivars did not cluster independently from the nontransgenic samples, it was indicative that the insertion of the transgene had only marginal impact on global gene expression. When the gene expression

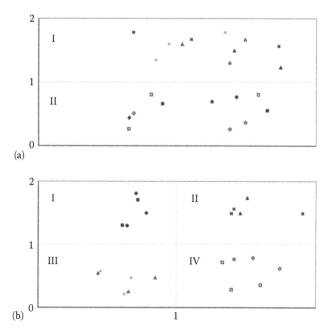

Figure 2.1 Kohonen self-organizing maps (SOM) for samples of Helen Bt (diamond), Helen (squares), Beles Sur (triangles), and Sancia (crosses) plants grown under either control (open figures) or low-N (filled figures) conditions. Maps with two (a) and four (b) cells are shown for classification into two and four groups, respectively. Autoscaled logarithmic expression levels are plotted. (Reprinted from Coll, A., Nadal, A., Collado, R. et al. 2010. Natural variation explains most transcriptomic changes among maize plants of MON810 and comparable non-GM varieties subjected to two N-fertilization farming practices. *Plant Mol Biol* 73:349–62. With kind permission from Springer Science and Business Media.)

profiles of GM cultivars and their non-GM counterparts were compared, the number of differentially expressed genes identified was lower than the number obtained from the comparison between non-GM varieties. Among the differentially expressed genes detected in the pair-wise comparisons between the GM and non-GM cultivars, only two genes (i.e., cysteine protease inhibitor and dihydroflavonol-4-reductase) were downregulated in both transgenic cultivars. However, it could not be ruled out that these changes were an effect of the insertion event, an effect of the transgenic product, or a natural variation of the parent genotype.

2.3.2.2 Protein Profiling

With respect to food safety, proteins are especially interesting because they may act as toxins, antinutrients, or allergens. Proteomics, able to quantify hundreds of proteins simultaneously, has become a key technology in comparative proteomic studies of GM plants and their nonmodified counterparts. Two conceptually and methodologically different strategies can be followed in comparative proteomics. The first, referred to as the bottom-up approach, has been widely applied to investigate substantial equivalence and potential unintended effects in GMOs, while the second, the so-called shotgun approach, has been scarcely used in the field of GMO analysis.

Bottom-up proteomics involves the use of two-dimensional gel electrophoresis (2-DGE), followed by image analysis (Figure 2.2). Identification of individual proteins excised from gel spots is then performed by MS, usually after hydrolysis with trypsin. Typically, matrix-assisted laser desorption/ionization (MALDI), coupled to a time-of-flight mass (TOF) spectrometer, or different

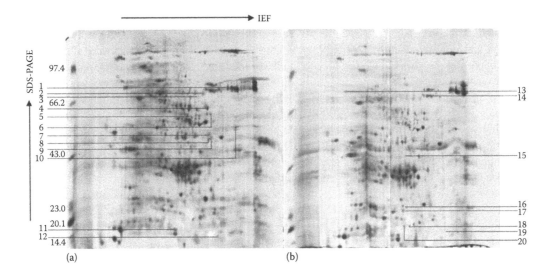

Figure 2.2   2-DGE of mature wheat seeds: (a) Yumai 18; (b) transgenic Yumai 18. (Reprinted from Guo, H., H. Zhang, Y. Li, J. Ren, X. Wang, H. Niu and J. Yin. 2011. Identification of changes in wheat (*Triticum aestivum* L.) seeds proteome in response to anti-trx s gene. *PLoS ONE* 6: e22255.)

variants of LC-MS are used. So far, 2-DGE provides the highest protein-resolution capability with a low instrumentation cost.

However, this technique suffers from some technical limitations that prevent the separation of highly hydrophobic proteins or those with an extreme isoelectric point or a high molecular weight (MW). Also, gel-to-gel variation is one of the major sources of error, which makes an exact match of spots in the image-analysis process difficult. Despite these limitations, the 2-DGE technique has proven to be useful for molecular profiling of GMOs, providing highly resolved proteomic patterns constituting thousands of proteins (Ruebelt et al. 2006a). As a general trend, comparison of protein profiles of GM varieties with those obtained from the non-GM lines often does not reveal more significant differences than those observed between different nontransgenic varieties/genotypes. In this regard, as with comparative transcriptomics, the relevance of knowing the natural bandwidth of variation in the proteome of plants grown under a range of different environments and farming practices, to avoid any over- or mis-interpretation of the results, has been pointed out (Ruebelt et al. 2006b). Some representative examples of the application of this methodology for profiling the proteome of GMOs versus their corresponding non-GM counterparts are the studies on the insect-resistant GM MON810 maize (Albo et al. 2007; Coll et al. 2011; Zolla et al. 2008), herbicide-tolerant soybean (Brandao et al. 2010), tomatoes with a genetically added resistance to virus and insect attacks (Corpillo et al. 2004; Di Carli et al. 2009), and potatoes showing a delayed sprouting process and modified cell wall structure (Careri et al. 2003; Lehesranta et al. 2005).

Coll et al. (2011) investigated potential unintended effects of genetic modification in two different MON810 maize varieties using bottom-up proteomics. In that study, the analysis of grain proteins extracted from plants grown in agricultural fields showed several hundreds of protein spots per gel, although only a small number of spots ($\leq$1.2% analyzed proteins) showed statistical differences in each of the two GM/non-GM comparisons. None of these changes were observed in both comparisons, and the calculated fold changes were minimal. The molecular functions and biological processes predicted for the identified differentially expressed proteins were diverse, and it was not possible to establish any functional correlation between the differentially expressed proteins in the GM maize varieties. Balsamo et al. (2011) used a similar procedure to compare leaf protein profiles of four MON810 maize varieties with their corresponding four isogenic lines cultured in environmentally controlled conditions. After data processing, a total of 12 spots indicated qualitative differences when the GM line was compared with its control variety. As in the previous work on maize grains, the differential protein patterns observed in leaf extracts were variety specific, supporting the idea that the genetic modification does not significantly alter the global protein expression in the plant.

The differential in gel electrophoresis (DIGE) technique has also been applied to comparative analysis of GMOs (Islam et al. 2009). This technique minimizes the gel-to-gel variability in comparative proteomic analysis. To achieve this, different samples labeled with ultrahigh-sensitive fluorescent dyes (typically Cy5 and Cy3) are loaded in the same gel. After separation and image acquisition, specific software is then used to comparatively analyze gel images obtained using two different detection channels, allowing the simultaneous detection of protein spots labeled with the two fluorescent dyes. DIGE methodology has been recently used in a study conducted on the GM soybean proteome (Barbosa et al. 2012). In that work, a total of four spots were detected as differentially expressed when compared with the control soybean variety. Proteomic data could be correlated with results obtained from enzymatic determination of catalase, superoxide dismutase, glutathione reductase, and ascorbate peroxidase activities, which suggested an overall increased level of oxidative stress in the transformed soybean line. In another recent study on GMOs using DIGE methodology, two insect-resistant transgenic rice varieties were investigated for the potential presence of unintended effects (Gong et al. 2012). Apart from the study of two GM indica rice varieties and their corresponding control lines, the study included the evaluation of proteomic profiles of other indica and japonica cultivars. In addition to the effect of the genetic transformation, the experimental design allowed the investigation of the effects of conventional genetic breeding and natural genetic variation on the rice proteome. Multivariate analysis of the protein patterns indicated the highest degree of divergence between indica and japonica cultivars, followed by the differences registered for the three indica varieties. In contrast, GM and control varieties showed high similarity, and, although some differentially expressed proteins were detected, most of the expression changes were associated with the variety and not with the transgenic character. A total of 234 spots corresponding to differentially expressed proteins, found across all the comparisons performed between rice varieties, were successfully identified using MALDI-TOF/TOF MS. Mass spectrometry results indicated a significant number of proteins related to metabolism, in particular the tricarboxylic acid cycle and glycolysis; protein folding and modification; and defense response. The main conclusion is that genetic transformation does not significantly modify the rice seed proteomes as compared with natural genetic variation and conventional genetic breeding.

Bottom-up proteome profiling can also be approached using gel-free separation techniques, namely LC and capillary electrophoresis (CE), which have been shown to be suitable for protein (or peptide) analysis. In comparison with 2-DGE, these techniques require a smaller amount of starting material, enable direct coupling to a mass spectrometer, and provide high-throughput capabilities, full automation, and better reproducibility than 2-DGE in terms of qualitative and quantitative analysis. Based on this concept, García-López et al. (2009) developed a LC-MS method using electrospray ionization (ESI) and an ion trap (IT) mass analyzer to compare the protein profiles of several insect-resistant MON810 maize varieties with those obtained from their corresponding nontransgenic line. The analyses revealed spectral signals that seemed to be very similar between GM and the unmodified lines, whereas several proteins were found to be characteristic for samples from specific regions. More recently, Koc et al. (2012) have reported the results obtained using LC-TOF MS to profile the low-molecular-weight protein fraction in both GM and non-GM maize. Protein extraction from maize flour involved five fractionation steps using different water- and ethanol-based buffer solutions. Interestingly, LC-TOF MS analysis of the resulting fractions (enriched in albumin, globulin, zein, zein-like glutelin, and glutelin proteins, respectively) suggested a higher degree of divergence in protein profiles between globulin fractions from GM and non-GM maize flours.

Erny et al. (2008) investigated the applicability of CE-ESI-MS to profile the insect-resistant transgenic maize proteome. The study also focused on the comparison of the performance of two different mass analyzers, TOF and IT, coupled to CE. For this evaluation, intact zein-protein fractions were extracted from three different MON810 maize varieties and their corresponding isogenic lines. Although results indicated that both instruments showed similar sensitivity and repeatability, CE-ESI-TOF MS provided better results with regard to the number of identified proteins. The comparison of the protein profiles obtained by CE-ESI-TOF MS did not show significant differences between any of the GM/non-GM pairs under study.

Shotgun protein analysis involves the digestion of proteins without any prefractionation or separation of the proteome. The resulting peptides are separated by LC, followed by MS analysis to provide a rapid and automatic identification of proteins in the sample. Although this strategy has already proven to be suitable for proteome profiling (Yates 2013), it has been barely used in GMO analysis. A demonstrative example of the application of shotgun analysis for investigation of unintended effects in GM soybean was reported by Simó et al. (2010). In that work, several parameters affecting the CE-ESI-TOF MS separation and detection of peptides were optimized during the first stages of method development. Optimal CE-MS conditions allowed the detection of 151 peptides in each protein digest derived from soybean flours. Comparisons of peptide profiles showed no statistical differences between GM soybean samples and those from the comparators.

The iTRAQ technique can also be combined with shotgun proteomics for quantitation purposes. This combination has been successfully applied to quantify differences in protein profiles between transgenic and nontransgenic rice (Luo et al. 2009). The procedure involved treatment of four different digested samples with four independent isobaric reagents, designed to react with all primary amines of protein hydrolyzates. Treated samples were subsequently pooled and analyzed by tandem mass spectrometry. The analyses using this technique revealed significant differences between GM and wild-type rice in 103 proteins out of the 1883 proteins identified in rice endosperm samples. Interestingly, some of these changes were associated with downregulation of endogenous storage proteins and carbohydrate metabolism-related proteins, while some others, linked to proteasome-related proteins and chaperones, were overexpressed in the transgenic samples.

2.3.2.3 *Metabolite Profiling*

Metabolomics is aimed at the study of the metabolome by the identification and quantification of all small molecules in biological systems. Recombinant DNA technology is frequently used to modify the metabolism for optimal production of plant metabolites, which may directly benefit human health and plant growth (Okazaki and Saito 2012). An illustrative example of this is the transgenic rice, commonly known as "golden rice," whose genetic modification is being developed to biosynthesize β-carotene in grain (Ye et al. 2000). In this context, metabolomics has the potential of playing an important role within the frame of GMO analysis, assisting in the detection of intended and unintended effects that might occur as a result of the genetic transformation (Shepherd et al. 2006). However, metabolomics faces several challenges, since metabolites encompass a wide range of chemical species with divergent physicochemical properties. In addition, the relative concentration of metabolites in a biological sample can range from the picomolar to the millimolar level. Regarding the technical requirements for comprehensive metabolomics analysis, high sensitivity and resolution are the key parameters to consider for the selection of an appropriate method (Villas-Boas et al. 2005). Metabolomic analysis is usually performed using two analytical platforms: MS- and nuclear magnetic resonance-based systems. These techniques are complementary and, therefore, frequently used in parallel in metabolomic studies to achieve more extensive metabolome coverage (Shulaev 2006). Compared with nuclear magnetic resonance (NMR), MS is more sensitive; also, MS coupled to gas chromatography (GC), LC, or CE allows higher resolution and sensitivity for low-abundance metabolites (Issaq et al. 2009; Xiayan and Legido-Quigley 2008).

High- and ultra-high-resolution analyzers (TOF, Fourier transform ion-cyclotron MS, Orbitrap®) are crucial in those applications where there is a requirement for accurate mass measurements in order to determine metabolites' elemental composition and to perform their tentative identification using databases (Brown et al. 2005; Xu et al. 2010). In addition to this, tandem mass analysis (MS/MS or MSn), especially when product ions are analyzed at high resolution (with Q-TOF, TOF-TOF, or linear trap quadrupole [LTQ]-Orbitrap), provides additional structural information for the identification of metabolites.

The general procedure in metabolic profiling analysis includes: (i) metabolite extraction, which often has to be adapted on a case-by-case basis depending on the type of sample and analytical platform chosen; (ii) sample preparation, which may include partial purification and derivatization steps; (iii) instrumental analysis of samples; (iv) detection and quantification of metabolite signals in raw data to generate a data matrix listing metabolites and their intensity data; (v) statistical analysis of metabolite profiles (Saito and Matsuda 2010). With respect to the analysis of the vast amount of data obtained in these studies, the application of chemometric and bioinformatic approaches (e.g., PCA, cluster analysis, etc.) has become essential to fully exploit the benefits of metabolic profiling for discovering significant differences in order to discriminate among different plant varieties.

Metabolic profiling of GMOs has also been performed by several metabolomics techniques, based on NMR or MS analytical platforms or both. In this context, and compared with MS-based analysis, the application of NMR to investigate metabolic profiles in GMOs has been less frequent. This technique provides qualitative and quantitative data on many metabolites (usually the more abundant constituents) in a given sample. Some remarkable examples of the application of NMR for profiling the metabolome of GMOs versus their corresponding non-GM counterparts are studies on maize (Manetti et al. 2004; Piccioni et al. 2009), pea plant (Charlton et al. 2004), potato (Defernez et al. 2004), wheat (Baker et al. 2006), and lettuce (Sobolev et al. 2007, 2010).

In some studies, the use of NMR has enabled the detection of discriminant metabolites that accumulate differently in GM and non-GM samples. This was the case for the metabolites choline, asparagine, histidine, and trigonelline, levels of which were lower in transgenic samples than in controls (Manetti et al. 2004). Similarly, Piccioni et al. (2009) found that quantitative differences between transgenic and nontransgenic maize samples allowed discrimination of the genotypes by multivariate analysis (PCA). In their work, the use of one- and two-dimensional NMR techniques enabled profiling of 40 water-soluble metabolites, among which ethanol, lactic acid, citric acid, lysine, arginine, glycine-betaine, raffinose, trehalose, R-galactose, and adenine were identified for the first time in the ^{1}H-NMR spectrum of maize seeds. In a study of different metabolic aspects of transgenic lettuce with enhanced growth properties, Sobolev et al. (2007) reported increases (up to 30 times) of short-chain inulin oligosaccharides levels unexpectedly observed in transgenic lettuce leaves when compared with leaves from the wild-type genotype.

In other studies based on the application of NMR, the use of PCA and other multivariate tools failed to provide acceptable classification of transgenic and nontransgenic samples. This can be attributed to the similarity of the NMR profiles derived from the different transgenic lines and the apparent wider divergence of the spectral data obtained from the wild-type samples, suggesting that factors other than the presence of the transgene have a large effect on metabolite profile (Charlton et al. 2004). Similar conclusions were reached in a NMR study on wheat varieties overexpressing high-molecular-weight subunits of glutenin (Baker et al. 2006). In that study, the levels of amino acids also were analyzed in the same samples by GC-MS in order to gain more information on the metabolic composition of the samples. NMR was useful in identifying some differences in the levels of the disaccharides maltose and sucrose between parental and transgenic lines. However, comparative studies of the null transformant with the control line, and the transgenic lines with their respective controls, suggested that neither the transformation process nor the expression of the transgenes had significant and reproducible impact on the grain metabolome. More recently, a NMR study by Sobolev et al. (2010) indicated that NMR spectra of three transgenic and one wild-type lettuce lines were similar, although signal intensity varied with the growth stages and genotype. In that work, metabolic data, including the intensity of 24 selected hydrosoluble metabolites from 180 samples, were subjected to PCA, showing evident separation of the samples at three developmental stages. Some amino acids, sugars, and choline showed the highest differences during maturation (Sobolev et al. 2010).

GC-MS is one of the most reported profiling techniques in the literature to study the metabolome of GMOs. This technique allows the determination of the levels of primary metabolites such as amino acids, organic acids, and sugars by employing chemical derivatization. GC-MS combines high separation efficiency and reproducibility with the stable ionization achieved by electron impact (EI) ionization, which makes this technique well suited to analyzing complex samples. The pioneer work by Roessner et al. (2000) demonstrated the potential of GC-MS to characterize the metabolite profiles of GMOs. In their first study, GC-MS was applied to analyze polar metabolites in extracts obtained from potato tubers with modified starch or sugar metabolism (Roessner et al. 2000). Methoximation and silylation derivatization steps were performed in order to volatilize various classes of compounds. GC-MS results were compared with data compiled in mass spectrometry libraries, resulting in the identification of 77 out of 150 compounds detected by GC-MS. Identified compounds provided valuable information regarding the altered metabolic pathways and unexpected changes in the GM potatoes. A further study by the same group suggested massive elevation of amino-acid levels determined by GC-MS analysis of transgenic potato tubers with altered sucrose catabolism (Roessner et al. 2001b). Later studies on GM potato and tomato varieties demonstrated the benefits of using data-mining tools (e.g., PCA and hierarchical clustering) with GC-MS to discover differences that allow the discrimination of the GM crop lines from the respective nonmodified lines (Roessner et al. 2001a; Roessner-Tunali et al. 2003). Using GC-MS, Catchpole et al. (2005) achieved a broader survey of the potato metabolome. In particular, two different GC-MS techniques were applied to obtain metabolite information in transgenic potato designed to contain high levels of inulin-type fructans and its conventional counterpart. In the first stage of the study, flow-injection analysis (FIA) ESI-MS was used to analyze 600 potato extracts. Data sets were analyzed by PCA to identify top-ranking ions for genotype identification. Then, approximately 2000 tuber samples were also analyzed by GC-TOF MS to obtain complementary data. This methodology enabled the detection of 242 individual metabolites (90 positively identified, 89 assigned to a specific metabolite class, and 73 unknown). Chemometric analysis of the profiles showed that, apart from the expected result of the genetic modification, transgenic potato samples showed a metabolite composition within the range detected in conventional cultivars.

Metabolic profiles of GM rice have been investigated by Jiao et al. (2010a) using GC-MS. In this case, however, GC-MS data were complemented with the results obtained from near-infrared reflectance, inductively coupled plasma atomic emission spectroscopy, and liquid chromatography to investigate the compositional differences between three transgenic rice varieties expressing antifungal proteins. Multivariate analysis of the results revealed several changes in some nutrients, such as proteins, amino acids, fatty acids, and vitamins, that were variety specific.

Some metabolomic studies based on GC-MS analysis of GMOs have paid special attention to the metabolite extraction procedure. For instance, in a comparative study of two different methods for the extraction of volatile fraction from transgenic cucumber, Zawirska-Wojtasiak et al. (2009) observed that solid-phase microextraction yielded a higher number of extracted compounds than microdistillation. Further, the analysis of cucumber extracts using GC-EI-Q MS and GC-EI-TOF MS revealed significant quantitative, rather than qualitative, differences between GM and non-GM cucumber, regardless of the MS-based technique used. Also, Bernal et al. (2008) and Jimenez et al. (2009) combined GC-MS with several selective extraction methods, including supercritical fluids or accelerated solvents, to investigate unintended effects in GMOs. These techniques were applied for the selective extraction of amino acids and fatty acids from soybean and maize for subsequent profiling and quantification. In a recent metabolomic study on insect-resistant and herbicide-tolerant

GM maize varieties by GC-MS, sample preparation involved a complex extraction scheme designed to obtain four fractions containing major lipids; minor lipids; sugars, sugar alcohols and acids; and amino acids and amines (Frank et al. 2012). Fractions obtained from grains from plants grown at distinct locations and in different seasons were independently analyzed using a GC coupled to a quadrupole mass spectrometer with an EI ion source. The results obtained by multivariate statistical analysis for the individual fractions revealed compounds from the two polar factions as the major contributors to the variation observed between samples grown in different locations and seasons (Figure 2.3). However, none of the fractions indicated a separation of GM maize from the near-isogenic variety. Univariate statistical analysis for all pair-wise comparisons performed with ANOVA supported these results, showing a higher number of significant differences (10%–43% of the number of covered peaks) between maize grown at different locations and/or seasons than the number of differences (2%–3% of the number of covered peaks) found between the GM lines and their comparators.

The application of LC-MS to metabolic profiling of GM crops has demonstrated several benefits, which include wide dynamic range and reproducible quantitative analysis; the ability to separate and analyze extremely complex samples; and high versatility for metabolite analysis. LC-MS is commonly used to profile compounds that are polar/nonvolatile, large, thermolabile, or any combination of these, demonstrating good performance in the profiling of secondary metabolites and complex lipids. For instance, LC-MS has been applied to the investigation of flavonoids in GM wheat and rice (Ioset et al. 2007; Shin et al. 2006). Also, LC-MS was useful for detecting differences between

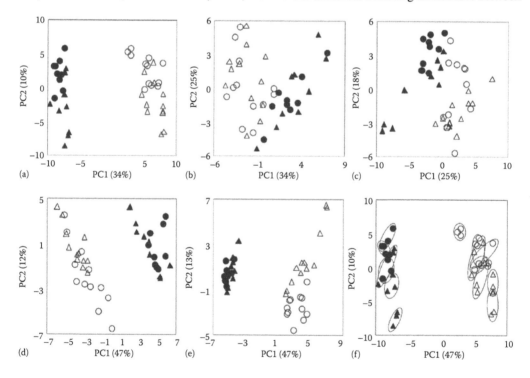

Figure 2.3 Principal component analysis of GC-MS metabolite profiling data from combined fractions I–IV (a and f) and of fractions I (b), II (c), III (d), and IV (e) of Bt-maize (▲, △) and its isogenic counterpart (●, ○) grown at Pfaffenhofen (▲, ●) and Neuhof (△, ○). Three and four field replicates at Pfaffenhofen and Neuhof, respectively, were analyzed in triplicate. The circles in plot F indicate the data from triplicate analysis of field replicates. (Reprinted with permission from Frank, T., R. M. Röhlig, H. V. Davies, E. Barros, and K. Engel. Metabolite profiling of maize kernels: Genetic modification versus environmental influence. *J Agric Food Chem* 60: 3005–12. Copyright 2012 American Chemical Society.)

the polyphenol profiles obtained from control and transgenic tomato, overexpressing a grapevine gene that encoded the enzyme stilbene synthase (Nicoletti et al. 2007). Typically, reversed-phase is the most frequent mode used in LC-MS metabolite profiling in GMO analysis; however, other modes, such as hydrophilic interaction liquid chromatography (HILIC), have been demonstrated to be useful as well (Nagai et al. 2009).

Other representative examples of LC-MS application to metabolic profiling of GMOs are the studies performed on rice. Thus, Matsuda et al. (2010) used this technique to investigate the effect of the α-subunit of anthranilate synthase expression on GM rice with increased tryptophan content. The metabolic profiles of different plant tissues revealed that tryptophan concentration changes in a time-dependent manner, showing a tissue-dependent profile of accumulation. In a recent report, a LC-MS method has been developed to explore the unintended effects of the genetic transformation in three varieties of insect-resistant transgenic rice (Chang et al. 2012). In this work, a comparative study with different solvent compositions was performed in order to improve metabolite extraction efficiency. PCA and partial least squares discriminant analysis (PLS-DA) of metabolic profiles enabled the classification of samples. Metabolites having a significant influence on the classification of the different groups were identified by targeted MS/MS analysis. Using this methodology, 15 metabolites were found that significantly contributed to the separation of the groups harvested at different times and locations. Interestingly, the transgenic rice grains were clearly separated from the unmodified rice grains after discriminant analysis. In this case, another 15 metabolites were found to be responsible for this separation. However, the fold changes were comparatively lower than those obtained from the comparison among different sowing times and locations, suggesting that the effects of the environmental factors were higher than those derived from the genetic transformation.

CE-MS can be useful for the analysis of a wide range of analytes, including ionic and polar thermolabile compounds. Among the main features of CE-MS are high efficiency, analysis speed, and resolution, requiring, moreover, little sample pretreatment. On the other hand, moderate sensitivity is frequently achieved due to the minute sample volumes injected in CE-MS. Currently, this technique is considered complementary to LC-MS and GC-MS; its applicability for metabolic profiling of GMOs has already been proven (Garcia-Villalba et al. 2008; Levandi et al. 2008). Thus, CE-ESI-TOF MS analysis has been combined with multivariate statistical analysis to detect relevant significant differences in metabolic profiles of varieties of conventional and insect-resistant GM maize (Levandi et al. 2008). Data obtained by this methodology indicated some relevant differences in the levels of L-carnitine and stachydrine between conventional and insect-resistant GM maize. In addition to this method, Garcia-Villalba et al. (2008) developed a novel CE-ESI-TOF MS methodology for the comparative analysis of metabolic profiles of transgenic herbicide-tolerant soybean and its corresponding unmodified parental line. In this case, over 45 different metabolites, including carboxylic acids, isoflavones, and amino acids, were tentatively identified. Differences in metabolic profiles between the two lines were more evident in the concentration of three free amino acids, proline, histidine, and asparagine, while the level of a metabolite tentatively identified as 4-hydroxy-L-threonine decreased in the transgenic soybean compared with the control line. In a later report, Giuffrida et al. (2009) demonstrated the usefulness of modified cyclodextrins in CE-ESI-TOF MS analysis of chiral amino acid profiles in herbicide-tolerant GM soybean. Results revealed some quantitative divergences in the chiral amino acid profile between the GM soybean and the untransformed genotype.

Fourier transform ion-cyclotron MS (FT-ICR MS), as well as other ultrahigh-resolution mass spectrometers, enables accurate estimates of chemical formulae of the detected peaks, which facilitates annotation procedures for unknown compounds. However, factors including equipment cost, the difficulties in hardware handling, and the extremely large amount of data generated hamper the application of this technique. Despite these limitations, FT-ICR MS has already demonstrated good potential for GMO analysis (León et al. 2009). FT-ICR MS has also been used in combination with

CE-TOF MS for the metabolic profiling of six varieties of maize, three GM insect-resistant lines, and their corresponding near-isogenic lines. The spectral data obtained in both positive and negative ESI mode with FT-ICR MS were processed using bioinformatic tools and databases in order to identify maize-specific metabolites (Suhre and Schmitt-Kopplin 2008). Then, PLS-DA of the data indicated the most discriminative masses that contributed to differentiating the GM samples from the non-GM controls. Interestingly, electrophoretic mobilities and m/z values provided by CE-TOF MS were helpful in the identification of those compounds that could not be unequivocally identified by FT-ICR MS, such as isomers having the same molecular formula.

The combined use of more than one analytical platform for metabolic profiling has been demonstrated to improve the description of the metabolome status of GMOs. For instance, Tesniere et al. (2006) used LC-MS and GC-MS for the comparative analysis of GM grapevine varieties with their unmodified counterpart. Also, Jiao et al. (2010b) used the same techniques to investigate the compositional differences between two transgenic ringspot virus-resistant papaya varieties and their nonmodified counterparts. In that study, 80 papaya samples were extracted using a prefractionation strategy with subsequent parallel analysis for selective and sensitive detection of metabolites. Thus, profiles of volatile organic compounds and sugar/polyols were obtained using GC-MS, whereas profiles of organic acids, carotenoids, and alkaloids were detected using LC-MS. When the compositional profiles were analyzed by PCA, data points from transgenic and unmodified papaya genotypes formed a tight cluster, showing great similarity in composition, while papaya samples from different harvesting times were separated.

In a different report, Kusano et al. (2011) have proposed a broader strategy based on GC-TOF MS, LC-QTOF MS, and CE-TOF MS analysis to assess the substantial equivalence between GMOs and the unmodified counterparts. In their study, two GM tomato varieties overexpressing miraculin glycoprotein and a panel of traditional tomato cultivars were selected to prove the proposed methodology. Multivariate analysis was performed on single consensus data sets containing summarized data from the three analytical platforms. This methodology allowed the tentative identification of over 175 unique metabolites. Using this analytical setup, the achieved metabolite coverage was approximately 85% of the chemical diversity found in the LycoCyc database (Figure 2.4). Also, application of orthogonal PLS-DA (OPLS-DA) to look for changes attributable to genetic modification indicated a higher overall proximity between the transgenic lines and the controls than the proximity observed among ripening stages and traditional cultivars.

There are few studies regarding the use of different high-throughput omics platforms for GMO analysis. This strategy allows as much information as possible to be recorded in order to obtain a broader picture of GMO composition that might increase the opportunities to identify potential unintended effects in a GMO under specific conditions. Also importantly, the possibility of analyzing diverse types of biomolecules implies that interrelationships between the different levels of information can be explored using different omics platforms.

In studies where several omics technologies are applied, the general trend is to perform statistical analysis on each independent omics data set. For instance, Kogel et al. (2010) carried out a study on transcriptome and metabolome profiling analyses of transgenic barley cultivars exhibiting different disease-resistance and nutritional traits. Analyses were performed on leaves from barley plants (GM and non-GM genotypes) grown in fields with and without amendment of soil with mycorrhizal fungi. Interestingly, gene expression microarray data were not as sensitive as metabolome analysis with LC-ESI-MS for revealing minor differences in three of the four varieties in response to mycorrhizal fungi treatment. Furthermore, consistently with other profiling studies, statistical analysis of omics data from both platforms indicated more significant differences between the unmodified cultivars than the divergences found between GM and non-GM barley plants. Also, Barros et al. (2010) have studied the transcriptome, proteome, and metabolome of two transgenic insect-resistant and herbicide-tolerant maize varieties. Gene expression microarray and 2-DGE analysis were applied for transcriptomics and proteomics analyses, respectively, whereas two metabolomics platforms,

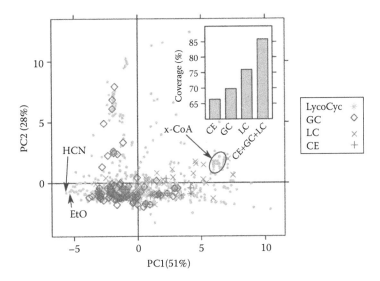

Figure 2.4   **(See color insert)** Evaluation of the achieved coverage. PCA was performed on the predicted physicochemical properties of the detected metabolites and the metabolites in the LycoCyc database. The score plots show that the distribution of the detected metabolites occupies a similar space as the reference metabolites. (Reprinted from Kusano, M., H. Redestig, T. Hirai et al. 2011. Covering chemical diversity of genetically-modified tomatoes using metabolomics for objective substantial equivalence assessment. *PLoS ONE* 16: e16989.)

^{1}H-NMR and GC-MS, were used for metabolome profiling of maize grain samples. Evaluation of the effect of individual variables (e.g., year of harvest, agricultural practices, and location) for the factor genotype (i.e., GM and non-GM maize) was systematically performed with one-way ANOVA analysis of individual omics data sets. Also, PCA was applied to each omics data set in order to classify samples according to genotype and growing season or location. The results indicated that growing seasons, as well as locations, had a stronger overall influence than the genetic modification on the three levels of information of the three maize genotypes. Minor changes registered between non-GM and GM maize varieties included the downregulation of maize allergen Zea m14 in both transgenic maize lines and the accumulation of glucose and fructose in the insect-resistant maize and the decreased level of γ-tocopherol and inositol in the herbicide-tolerant line.

2.3.3 Investigation of Primary Effects of Genetic Modification

Apart from the study of potential unintended effects, the investigation of the primary intended effects induced by the genetic modification at the different molecular levels can be approached by omics techniques. For instance, proteomics can be useful to investigate the molecular mechanisms leading to a particular phenotypic characteristic. Thus, Guo et al. (2011) studied the resistance of transgenic wheat seeds to preharvest sprouting. In this work, the authors applied 2-DGE and MALDI-TOF MS techniques to decipher proteome differences between wild-type and transgenic wheat seeds expressing a heterologous version of the antisense thioredoxin gene. The electrophoretic analysis provided about 600 matched protein spots in control and transgenic samples, but the expression of only 16 proteins was considered significantly altered. The identification of the expressed proteins and their functions was the foundation for proposing a molecular regulation model involving thioredoxin protein in wheat seed germination. The combined use of 2-DGE and MS is also the leading analytical strategy to study the mechanisms involved in the response of GMOs under a variety of abiotic (chemicals, drought, salinity, etc.) and biotic (pathogens, parasites, etc.) stresses. For instance, Horvath-Szanics et al. (2006) studied the effect of drought stress on the proteomic expression of a herbicide-tolerant transgenic wheat using 2-DGE followed by MALDI-TOF MS. In another study, 2-DGE and quadrupole TOF (QTOF) MS were used to study the effect of the over-expression of calcium-dependent protein kinase 13 and calreticulin-interacting protein 1 involved in the cold-stress response in GM rice (Komatsu et al. 2007). The proteomic study of GMOs might also focus on the technological benefits of a particular transformed event. For instance, Di Luccia et al. (2005) carried out a study on two GM durum wheat cultivars (Svevo B730 1-1 and Ofanto B688 1-2) with modified functional performance of the grain in order to investigate the technological properties of gluten. Also, Scossa et al. (2008) investigated, at the transcriptome and proteome level, bread wheat showing an unusually high expression of low-molecular-weight glutenin subunit (LMW-GS). In their work, wheat cDNA microchips covering transcripts from 7835 genes were used to profile gene expression in transgenic and wild-type wheat harvested at different time points. For comparative proteomic analyses, a procedure involving different protein fractionation steps was adopted for sample preparation prior to 2-DGE and LC-MS/MS analysis. In general, results from microarray experiments were in good agreement with 2-DGE data. Data from both platforms revealed that overexpression of LMW-GS was coincident with a massive downregulation of other classes of storage proteins. Their findings opened new insight into the compensatory responses during wheat seed maturation.

2.4 CONCLUSIONS AND FUTURE OUTLOOK

The market for novel foods and food products containing GMOs is still emerging. In view of the evolution of GMO development and commercialization worldwide, and in order to provide appropriate tools for the control of organisms, future research in this field will continue to concentrate on methods able to reliably detect, quantify, and characterize GMOs. In this context, foodomics strategies are expected to continue to play a relevant role in GMO research laboratories. On the other hand, omics technologies need to face important challenges in order to achieve implementation in enforcement and commercial laboratories. Currently, lack of establishment of common standardized experimental protocols is a major limitation for omics techniques. The unification and validation of analytical platforms and protocols will allow the comparison of experimental data among different laboratories. High-throughput technologies are still costly and require highly specialized personnel. It can be anticipated that in future years implementation of these techniques in control laboratories will be more affordable than now, due to the availability of less expensive consumables and the development of new compact and portable instruments.

Regarding global screening of GMOs, the requirement for a DNA amplification step prior to hybridization onto a microarray chip for the most common microarray platforms limits the real high-throughput and quantitative capabilities of this technology (Dobnik et al. 2010). Further developments in this area will include more efficient and robust preamplification strategies, such as unbiased whole-genome amplification (Tengs et al. 2007), the padlock probe ligation technique (Prins et al. 2008), and nucleic acid sequence-based amplification, NASBA (Morisset et al. 2008), which might provide microarray platforms with true multiplexing and quantitative capabilities for GMO detection. With respect to gene expression profiling, microarray is currently the dominant technique; however, next-generation sequencers can be considered as potential tools for genome characterization and transcriptome profiling of GMOs. Thus, the novel techniques termed RNA-Seq methods might represent good alternatives for the comprehensive study of GMOs at transcriptome level. Regarding proteomics methods, it may be expected that some limitations of the current technologies, especially the resolution of peptides, will be improved to provide increased protein coverage and more reliable identification. Also, advanced technologies focused on the analysis of posttranslational modifications in proteins will gain more attention in the future. It can also be anticipated that improvements and innovations in proteomic technology with regard to gel-to-gel variability will help proteomic profiling to become standard practice in GMO analysis. Similarly, recent advances in the metabolomics field will find application in the area of GMO analysis in the near future. Among these advances, comprehensive multidimensional techniques, such as GC×GC or LC×LC coupled to MS, will have the potential of improving metabolic profiling of GMOs by achieving enhanced resolution, selectivity, and sensitivity in comparison with conventional separation techniques. Also, new instrumental development will be important, such as the novel MS interfaces with minimum requirements for sample preparation and novel analytical platforms, including NMR instruments with higher sensitivity and with compatibility for online MS coupling. Moreover, the issues regarding the current low rate of analyte identification must be improved, for instance, by the development and maintenance of the metabolite databases providing structural information. Also, the availability of bioinformatic tools that allow faster annotation of the peaks recorded in the profiles (Saito and Matsuda 2010) and tools able to summarize multiplatform data sets will be crucial for efficient analysis of metabolomics data (Redestig et al. 2010). Although several public databases focused on integrating metabolite information with data on genes and proteins are available, much effort is still needed to link identified plant metabolites to metabolic pathways and other relevant biological information. Such complementary information, frequently termed metadata, provides the opportunity to improve the biological interpretation of the results and to realize the full potential of metabolomics in GMO research. Also, regarding integration, there is a need for novel and suitable integrative strategies that enable dealing with the high level of complexity in the different omics areas (Martín et al. 2010). In this context, appropriate holistic systems biology is proposed as a promising strategy to integrate, summarize, and visualize data from different omics domains (Gehlenborg et al. 2010). As opposed to the traditional reductionist approach, systems biology aims to study biological processes using network modeling with an integrated approach. However, before data integration procedures can be practically applied to GMO analysis, much effort will have to be made to develop appropriate computational frameworks for describing molecular systems and connecting omics databases. This will require much collaborative work to compare, share, and validate data within the scientific community.

ACKNOWLEDGMENTS

This work was supported by AGL2011-29857-C03-01 project (Ministerio de Economía y Competitividad, Spain), and CSD2007-00063 FUN-C-FOOD (Programa CONSOLIDER, Ministerio de Educación y Ciencia, Spain). A.V. thanks the Ministerio de Economía y Competitividad for his FPI predoctoral fellowship.

REFERENCES

AHTEG (Ad Hoc Technical Expert Group). 2010. Guidance Document on risk assessment of living modified organisms. Available at: https://www.cbd.int/doc/meetings/bs/bsrarm-02/official/bsrarm-02-05-en.pdf. United Nations Environment Programme Convention for Biodiversity.

Akiyama, H., F. Nakamura, C. Yamada et al. 2009. A screening method for the detection of the 35S promoter and the nopaline synthase terminator in genetically modified organisms in a real-time multiplex polymerase chain reaction using high-resolution melting-curve analysis. *Biol Pharm Bull* 32:1824–1829.

Albo, A. G., S. Mila, G. Digilio, M. Motto, S. Aime, and D. Corpillo. 2007. Proteomic analysis of a genetically modified maize flour carrying Cry1Ab gene and comparison to the corresponding wild-type. *Maydica* 52:443–455.

Alderborn, A., J. Sundstrom, D. Soeria-Atmadja, M. Sandberg, H. C. Andersson, and U. Hammerling. 2010. Genetically modified plants for non-food or non-feed purposes: Straightforward screening for their appearance in food and feed. *Food Chem Toxicol* 48:453–464.

Ali, S., Y. Zafar, Z. Xianyin, G. M. Ali, and T. Jumin. 2008. Transgenic crops: Current challenges and future perspectives. *Afr J Biotechnol* 7:4667–4676.

Bahrdt, C., A. B. Krech, A. Wurz, and D. Wulff. 2010. Validation of a newly developed hexaplex real-time PCR assay for screening for presence of GMOs in food, feed and seed. *Anal Bioanal Chem* 396:2103–2112.

Baker, J. M., N. D. Hawkins, J. L. Ward et al. 2006. A metabolomic study of substantial equivalence of field-grown genetically modified wheat. *Plant Biotechnol J* 4:381–392.

Balsamo, G. M., G. C. Cangahuala-Inocente, J. B. Bertoldo, H. Terenzi, and A. C. M. Arisi. 2011. Proteomic analysis of four Brazilian MON810 maize varieties and their four non-genetically-modified isogenic varieties. *J Agric Food Chem* 59:11553–11559.

Barbau-Piednoir, E., A. Lievens, G. Mbongolo-Mbella et al. 2010. SYBR® Green qPCR screening methods for the presence of "35S promoter" and "NOS terminator" elements in food and feed products. *Eur Food Res Technol* 230:383–393.

Barbosa, H. S., S. C. C. Arruda, R. A. Azevedo, and M. A. Z. Arruda. 2012. New insights on proteomics of transgenic soybean seeds: Evaluation of differential expressions of enzymes and proteins. *Anal Bioanal Chem* 402:299–314.

Barros, E., S. Lezar, M. J. Anttonen et al. 2010. Comparison of two GM maize varieties with a nearisogenic non-GM variety using transcriptomics, proteomics and metabolomics. *Plant Biotechnol J* 8:436–451.

Batista, R., N. Saibo, T. Lourenc, and M. M. Oliveira. 2008. Microarray analyses reveal that plant mutagenesis may induce more transcriptomic changes than transgene insertion. *Proc Natl Acad Sci USA* 105:3640–3645.

Baudo, M. M., R. Lyons, S. Powers et al. 2006. Transgenesis has less impact on the transcriptome of wheat grain than conventional breeding. *Plant Biotechnol J* 4:369–380.

Berg, P., D. Baltimore, S. Brenner, R. O. Roblin, and M. F. Singer. 1975. Asilomar conference on recombinant DNA molecules. *Science* 188:991–994.

Bernal, J. L., M. J. Nozal, L. Toribio, C. Diego, R. Mayo, and R. Maestre. 2008. Use of supercritical fluid extraction and gas chromatography–mass spectrometry to obtain amino acid profiles from several genetically modified varieties of maize and soybean. *J Chromatogr A* 1192:266–272.

Bordoni, R., A. Mezzelani, C. Consolandi et al. 2004. Detection and quantitation of genetically modified maize (Bt-176 transgenic maize) by applying ligation detection reaction and universal array technology. *J Agric Food Chem* 52:1049–1054.

Brandao, A. R., H. S. Barbosa, and M. A. Z. Arruda. 2010. Image analysis of two-dimensional gel electrophoresis for comparative proteomics of transgenic and non-transgenic soybean seeds. *J Proteomics* 73:1433–1440.

Brown, S. C., G. Kruppa, and J. L. Dasseux. 2005. Metabolomics applications of FT-ICR mass spectrometry. *Mass Spectrom Rev* 24:223–231.

Cahoon, E. B., S. E. Hall, K. G. Ripp, T. S. Ganzke, W. D. Hitz, and S. J. Coughland. 2003. Metabolic redesign of vitamin E biosynthesis in plants for tocotrienol production and increased antioxidant content. *Nat Biotechnol* 21:1082–1087.

Careri, M., L. Elviri, A. Mangia, I. Zagnoni, C. Agrimonti, G. Visioli, and N. Marmiroli. 2003. Analysis of protein profiles of genetically modified potato tubers by matrix-assisted laser desorption/ionization time-of-flight mass spectrometry. *Rapid Commun Mass Sp* 17:479–483.

Catchpole, G. S., M. Beckmann, D. P. Enot et al. 2005. Hierarchical metabolomics demonstrates substantial compositional similarity between genetically modified and conventional potato crops. *Proc Natl Acad Sci USA* 102:14458–14462.

Cellini, F., A. Chesson, I. Colquhoun et al. 2004. Unintended effects and their detection in genetically modified crops. *Food Chem Toxicol* 42:1089–1125.

Chang, Y., C. Zhao, Z. Zhu et al. 2012. Metabolic profiling based on LC/MS to evaluate unintended effects of transgenic rice with cry1Ac and sck genes. *Plant Mol Biol* 78:477–487.

Chaouachi, M., M. N. Fortabat, A. Geldreich et al. 2008. An accurate real-time PCR test for the detection and quantification of cauliflower Mosaïc virus (CaMV): Applicable in GMO screening. *Eur Food Res Technol* 227:789–798.

Charlton, A., T. Allnutt, S. Holmes et al. 2004. NMR profiling of transgenic peas. *Plant Biotechnol J* 2:27–35.

Chassy, B. M. 2010. Can -omics inform a food safety assessment? *Regul Toxicol Pharm* 58:S62–S70.

Cheng, K. C., J. Beaulieu, E. Iquira, F. J. Belzile, M. G. Fortin, and M. V. Stromvik. 2008. Effect of transgenes on global gene expression in soybean is within the natural range of variation of conventional cultivars. *J Agric Food Chem* 56:3057–3067.

Cifuentes, A. 2009. Food analysis and foodomics. *J Chromatogr A* 1216:7109–7110.

Coll, A., A. Nadal, R. Collado et al. 2010. Natural variation explains most transcriptomic changes among maize plants of MON810 and comparable non-GM varieties subjected to two N-fertilization farming practices. *Plant Mol Biol* 73:349–362.

Coll, A., A. Nadal, M. Palaudelmas et al. 2008. Lack of repeatable differential expression patterns between MON810 and comparable commercial varieties of maize. *Plant Mol Biol* 68:105–117.

Coll, A., A. Nadal, M. Rossignol, P. Puigdomenech, and M. Pla. 2011. Proteomic analysis of MON810 and comparable non-GM maize varieties grown in agricultural fields. *Transgenic Res* 20:939–949.

Corpillo, D., G. Gardini, A. M. Vaira et al. 2004. Proteomics as a tool to improve investigation of substantial equivalence in genetically modified organisms: The case of a virus-resistant tomato. *Proteomics* 4:193–200.

Craig, W., M. Tepfer, G. Degrassi, and D. Ripandelli. 2008. An overview of general features of risk assessments of genetically modified crops. *Euphytica* 164:853–880.

Davies, H. 2010. The role of "omics" technologies in food safety assessment. *Food Control* 21:1601–1610.

Defernez, M., Y. M. Gunning, A. J. Parr, L. V. T. Shepherd, H. V. Davies, and I. J. Colquhoun. 2004. NMR and HPLC-UV profiling of potatoes with genetic modifications to metabolic pathways. *J Agric Food Chem* 52:6075–6085.

Deisingh, A. K. and N. Badrie. 2005. Detection approaches for genetically modified organisms in foods. *Food Res Int* 38:639–649.

Di Carli, M., M. E. Villani, G. Renzone et al. 2009. Leaf proteome analysis of transgenic plants expressing antiviral antibodies. *J Proteome Res* 8:838–848.

Di Luccia, A., C. Lamacchia, C. Fares et al. 2005. A proteomic approach to study protein variation in GM durum wheat in relation to technological properties of semolina. *Ann Chim* 95:405–414.

Dobnik, D., B. Morisset, and K. Gruden. 2010. NAIMA as a solution for future GMO diagnostics challenges. *Anal Bioanal Chem* 396:2229–2233.

Doerrer, N., G. Ladics, S. McClain et al. 2010. Evaluating biological variation in non-transgenic crops: Executive summary from the ILSI Health and Environmental Sciences Institute workshop. *Regul Toxicol Pharm* 58:S2–S7.

Domingo, J. L. 2007. Toxicity studies of genetically modified plants: A review of the published literature. *Crit Rev Food Sci* 47:721–733.

EFSA (European Food Safety Authority). 2006. Guidance document for the risk assessment of genetically modified plants and derived food and feed by the scientific panel on genetically modified organisms (GMO). Parma, Italy: EFSA Communications Department.

Erny, G. L., C. León, M. L. Marina, and A. Cifuentes. 2008. Time of flight versus ion trap MS coupled to CE to analyse intact proteins. *J Sep Sci* 31:1810–1818.

Frank, T., R. M. Röhlig, H. V. Davies, E. Barros, and K. Engel. 2012. Metabolite profiling of maize kernels-genetic modification versus environmental influence. *J Agric Food Chem* 60:3005–3012.

Frewer, L., J. Lassen, B. Kettlitz, J. Scholderer, V. Beekman, and K. G. Berdal. 2004. Societal aspects of genetically modified foods. *Food Chem Toxicol* 42:1181–1193.

García-Cañas, V., A. Cifuentes, and R. González. 2004. Detection of genetically modified organisms in foods by DNA amplification techniques. *Crit Rev Food Sci* 44:425–436.

García-Cañas, V., C. Simó, M. Herrero, E. Ibáñez, and A. Cifuentes. 2012. Present and future challenges in food analysis: Foodomics. *Anal Chem* 84:10150–10159.

García-Cañas, V., C. Simó, C. León, E. Ibáñez, and A. Cifuentes. 2011. MS-based analytical methodologies to characterize genetically modified crops. *Mass Spectrom Rev* 30:396–416.

García-López, M. C., V. García-Cañas, and M. L. Marina. 2009. Reversed-phase high-performance liquid chromatography-electrospray mass spectrometry profiling of transgenic and non-transgenic maize for cultivar characterization. *J Chromatogr A* 1216:7222–7228.

Garcia-Villalba, R., C. León, G. Dinelli et al. 2008. Comparative metabolomic study of transgenic versus conventional soybean using capillary electrophoresis–time-of-flight mass spectrometry. *J Chromatogr A* 1195:164–173.

Garza, C. and P. Stover. 2003. General introduction: The role of science in identifying common ground in the debate on genetic modification of foods. *Trends Food Sci Tech* 14:182–190.

Gehlenborg, N., S. I. O'Donoghue, N. S. Baliga et al. 2010. Visualization of omics data for systems biology. *Nat Methods* 7:S56–S68.

Germini, A., S. Rossi, A. Zanetti, R. Corradini, C. Fogher, and R. Marchelli. 2005. Development of a peptide nucleic acid array platform for the detection of genetically modified organisms in food. *J Agric Food Chem* 53:3958–3962.

Giuffrida, A., C. León, V. García-Cañas, V. Cucinotta, and A. Cifuentes. 2009. Modified cyclodextrins for fast and sensitive chiral-capillary electrophoresis-mass spectrometry. *Electrophoresis* 30:1734–1742.

Gong, C. Y., Q. Li, H. T. Yu, Z. Wang, and T. Wang. 2012. Proteomics insight into the biological safety of transgenic modification of rice as compared with conventional genetic breeding and spontaneous genotypic variation. *J. Proteome Res* 11:3019–3029.

Grohmann, L., C. Brünen-Nieweler, A. Nemeth, and H. U. Waiblinger. 2009. Collaborative trial validation studies of real-time PCR-based GMO screening methods for detection of the bar gene and the ctp2-cp4epsps construct. *J Agric Food Chem* 57:8913–8920.

Guo, H., H. Zhang, Y. Li, J. Ren, X. Wang, H. Niu, and J. Yin. 2011. Identification of changes in wheat (*Triticum aestivum* L.) seeds proteome in response to anti–trx s gene. *PLoS ONE* 6:e22255.

Hamels, S., T. Glouden, K. Gillard et al. 2009. A PCR-microarray method for the screening of genetically modified organisms. *Eur Food Res Technol* 228:531–551.

Harrigan, G. G., D. Lundry, S. Drury et al. 2010. Natural variation in crop composition and the impact of transgenesis. *Nat Biotechnol* 28:402–404.

Hashimoto, W., K. Momma, T. Katsube et al. 1999. Safety assessment of genetically engineered potatoes with designed soybean glycinin: Compositional analyses of the potato tubers and digestibility of the newly expressed protein in transgenic potatoes. *J Sci Food Agr* 79:1607–1612.

Heinemann, J. A., B. Kurenbach, and D. Quist. 2011. Molecular profiling—A tool for addressing emerging gaps in the comparative risk assessment of GMOs. *Environ Int* 37:1285–1293.

Hernandez, M., M. Pla, T. Esteve, S. Prat, P. Puigdomenech, and A. Ferrando. 2003. A specific real-time quantitative PCR detection system for event MON810 in maize YieldGard based on the 3′-transgene integration sequence. *Transgenic Res* 12:179–218.

Herrero, M., V. García-Cañas, C. Simo, and A. Cifuentes. 2010. Recent advances in the application of CE methods for food analysis and foodomics. *Electrophoresis* 31:205–228.

Herrero, M., C. Simó, V. García-Cañas, E. Ibáñez, and A. Cifuentes. 2012. Foodomics: MS-based strategies in modern food science and nutrition. *Mass Spectrom Rev* 31:49–69.

Herring, R. J. 2008. Opposition to transgenic technologies: Ideology, interests and collective action frames. *Nat Rev Genet* 9:458–463.

Holst-Jensen, A., Y. Bertheau, M. de Loose et al. 2012. Detecting un-authorized genetically modified organisms (GMOs) and derived materials. *Biotechnol Adv* 30:1318–1335.

Horvath-Szanics, E., Z. Szabo, T. Janaky, J. Pauk, and G. Hajos. 2006. Proteomics as an emergent tool for identification of stress-induced proteins in control and genetically modified wheat lines. *Chromatographia* 63:S143–S147.

Ioset, J. R., B. Urbaniak, K. Ndjoko-Ioset et al. 2007. Flavonoid profiling among wild type and related GM wheat varieties. *Plant Mol Biol* 65:645–654.

Islam, N., P. M. Campbell, T. J. V. Higgins, H. Hirano, and R. J. Akhurst. 2009. Transgenic peas expressing an α-amylase inhibitor gene from beans show altered expression and modification of endogenous proteins. *Electrophoresis* 30:1863–1868.

Issaq, H. J., Q. N. Van, T. J. Waybright, G. M. Muschik, and T. D. Veenstra. 2009. Analytical and statistical approaches to metabolomics research. *J Sep Sci* 32:2183–2199.

James, C. 2012. Global Status of Commercialized Biotech/GM Crops: 2012. ISAAA Brief No. 44. Ithaca, NY: ISAAA.

Jiao, Z., J. C. Deng, G. K. Li, Z. M. Zhang, and Z. W. Cai. 2010b. Study on the compositional differences between transgenic and non-transgenic papaya (*Carica papaya* L.). *J Food Compos Anal* 23:640–647.

Jiao, Z., X. Si, G. Li, Z. Zhang, and X. Xu. 2010a. Unintended compositional changes in transgenic rice seeds (*Oryza sativa* L.) studied by spectral and chromatographic analysis coupled with chemometrics methods. *J Agric Food Chem* 58:1746–1754.

Jimenez, J. J., J. L. Bernal, M. J. Nozal, L. Toribio, and J. Bernal. 2009. Profile and relative concentrations of fatty acids in corn and soybean seeds from transgenic and isogenic crops. *J Chromatogr A*
1216:7288–7295.

Kim, H., J. Kim, and M. Oh. 2010. Regulation and detection methods for genetically modified foods in Korea.
Pure Appl Chem 82:129–137.

Koc, A., A. Cañuelo, J. F. Garcia-Reyes, A. Molina-Diaz, and M. Trojanowicz. 2012. Low-molecular weight
protein profiling of genetically modified maize using fast liquid chromatography electrospray ionization
and time-of-flight mass spectrometry. *J Sep Sci* 35:1447–1461.

Kogel, K. H., L. M. Voll, P. Schafer et al. 2010. Transcriptome and metabolome profiling of fieldgrown
transgenic barley lack induced differences but show cultivar-specific variances. *Proc Natl Acad Sci USA*
107:6198–6203.

Kok, E. J., J. Keijer, G. A. Kleter et al. 2008. Comparative safety assessment of plant-derived foods. *Regul
Toxicol Pharm* 50:439–444.

Komatsu, S., G. Yang, M. Khan, H. Onodera, S. Toki, and M. Yamaguchi. 2007. Over-expression of calcium-
dependent protein kinase 13 and calreticulin interacting protein 1 confers cold tolerance on rice plants.
Mol Genet Genomics 277:713–723.

Kuiper, H. A. and G. A. Kleter. 2003. The scientific basis for risk assessment and regulation of genetically
modified foods. *Trends Food Sci Tech* 14:277–293.

Kusano, M., H. Redestig, T. Hirai et al. 2011. Covering chemical diversity of genetically-modified tomatoes
using metabolomics for objective substantial equivalence assessment. *PLoS ONE* 16:e16989.

Latham, J. R., A. K. Wilson, and R. A. Steinbrecher. 2006. The mutational consequences of plant transformation. *J Biomed Biotechnol* 25376:1–7.

Lehesranta, S. J., H. V. Davies, L. V. Shepherd et al. 2005. Comparison of tuber proteomes of potato varieties,
landraces, and genetically modified lines. *Plant Physiol* 138:1690–1699.

León, C., I. Rodriguez-Meizoso, M. Lucio et al. 2009. Metabolomics of transgenic maize combining Fourier
transform-ion cyclotron resonance-mass spectrometry, capillary electrophoresis-mass spectrometry and
pressurized liquid extraction. *J Chromatogr A* 1216:7314–7323.

Levandi, T., C. Leon, M. Kaljurand, V. García-Cañas, and A. Cifuentes. 2008. Capillary electrophoresis-time
of flight-mass spectrometry for comparative metabolomics of transgenic vs. conventional maize. *Anal
Chem* 80:6329–6335.

Liu, Z., Y. Li, J. Zhao, X. Chen, G. Jian, Y. Peng, and F. Qi. 2012. Differentially expressed genes distributed
over chromosomes and implicated in certain biological processes for site insertion genetically modified
rice kemingdao. *Int J Biol Sci* 8:953–963.

Luo, J., T. Ning, Y. Sun, J. Zhu, Y. Zhu, Q. Lin, and D. Yang. 2009. Proteomic analysis of rice endosperm cells
in response to expression of hGM-CSF. *J Proteome Res* 8:829–837.

Manetti, C., C. Bianchetti, M. Bizzarri et al. 2004. NMR-based metabonomics study of transgenic maize.
Phytochemistry 65:3187–3198.

Marmiroli, N., E. Maestri, M. Gulli et al. 2008. Methods for detection of GMOs in food and feed. *Anal Bioanal
Chem* 392:369–384.

Martín, L., A. Anguita, V. Maojo, and J. Crespo. 2010. Integration of omics data for cancer research. In W. C.
S. Cho (ed.), *An Omics Perspective on Cancer Research*, pp. 249–266. New York: Springer.

Matsuda, F., A. Ishihara, K. Takanashi et al. 2010. Metabolic profiling analysis of genetically modified rice
seedlings that overproduce tryptophan reveals the occurrence of its inter-tissue translocation. *Plant
Biotechnol* 27:17–27.

Michelini, E., P. Simoni, L. Cevenini, L. Mezzanotte, and A. Roda. 2008. New trends in bioanalytical tools for
the detection of genetically modified organisms: An update. *Anal Bioanal Chem* 392:355–367.

Millstone, E., E. Brunner, and S. Mayer. 1999. Beyond substantial equivalence. *Nature* 401:525–526.

Montero, M., A. Coll, A. Nadal, J. Messeguer, and M. Pla. 2011. Only half the transcriptomic differences
between resistant genetically modified and conventional rice are associated with the transgene. *Plant
Biotechnol J* 9:693–702.

Morisset, D., D. Dobnik, S. Hamels, J. Žel, and K. Gruden. 2008. NAIMA: Target amplification strategy
allowing quantitative on-chip detection of GMOs. *Nucleic Acids Res* 36:e118.

Nagai, Y. S., C. Sakulsinghroj, G. E. Edwards et al. 2009. Control of starch synthesis in cereals: Metabolite
analysis of transgenic rice expressing an up-regulated cytoplasmic ADP-glucose pyrophosphorylase in
developing seeds. *Plant Cell Physiol* 50:635–643.

Nicoletti, I., A. De Rossi, G. Giovinazzo, and D. Corradini. 2007. Identification and quantification of stilbenes in fruits of transgenic tomato plants (*Lycopersicon esculentum* Mill.) by reversed phase HPLC with photodiode array and mass spectrometry detection. *J Agric Food Chem* 55:3304–3311.

Okazaki, Y. and K. Saito. 2012. Recent advances of metabolomics in plant biotechnology. *Plant Biotechnol Rep* 6:1–15.

Petit, L., G. Pagny, F. Baraige, A. C. Nignol, and D. Zhang. 2007. Characterization of genetically modified maize in weakly contaminated seed batches and identification of the origin of the adventitious contamination. *J AOAC Int* 90:1098–1106.

Piccioni, F., D. Capitani, L. Zolla, and L. Mannina. 2009. NMR metabolic profiling of transgenic maize with the Cry1A(b) gene. *J Agric Food Chem* 57:6041–6049.

Prins, T. W., J. P. van Dijk, H. G. Beenen et al. 2008. Optimised padlock probe ligation and microarray detection of multiple (non-authorised) GMOs in a single reaction. *BMC Genomics* 9:584.

Redestig, H., M. Kusano, A. Fukushima, F. Matsuda, K. Saito, and M. Arita. 2010. Consolidating metabolite identifiers to enable contextual and multi-platform metabolomics. *BMC Bioinformatics* 11:214.

Ricroch, A. E., J. B. Berge, and M. Kuntz. 2011. Evaluation of genetically engineered crops using transcriptomic, proteomic, and metabolomic profiling techniques. *Plant Physiol* 155:1752–1761.

Robinson, C. 2001. Genetic modification technology and food. In W. Hammer (ed.), *Consumer Health and Safety*. ILSI Europe Concise Monograph Series.

Roessner, U., A. Luedemann, D. Brust et al. 2001a. Metabolic profiling allows comprehensive phenotyping of genetically or environmentally modified plant systems. *Plant Cell* 13:11–29.

Roessner, U., C. Wagner, J. Kopka, R. N. Trethewney, and L. Willmitzer. 2000. Simultaneous analysis of metabolites in potato tuber by gas chromatography-mass spectrometry. *Plant J* 23:131–142.

Roessner, U., L. Willmitzer, and A. R. Fernie. 2001b. High-resolution metabolic phenotyping of genetically and environmentally diverse potato tuber systems. Identification of phenocopies. *Plant Physiol* 127:749–764.

Roessner-Tunali, U., B. Hegemann, A. Lytovchenko et al. 2003. Metabolic profiling of transgenic tomato plants overexpressing hexokinase reveals that the influence of hexose phosphorylation diminishes during fruit development. *Plant Physiol* 133:84–99.

Rudi, K., I. Rud, and A. Holck. 2003. A novel multiplex quantitative DNA array based PCR (MQDA-PCR) for quantification of transgenic maize in food and feed. *Nucleic Acids Res* 31:e62.

Ruebelt, M. C., N. K. Leimgruber, M. Lipp et al. 2006a. Application of two-dimensional gel electrophoresis to interrogate alterations in the proteome of genetically modified crops. 1. Assessing analytical validation. *J Agric Food Chem* 54:2154–2161.

Ruebelt, M. C., M. Lipp, T. L. Reynolds, J. D. Astwood, K. H. Engel, and K. D. Jany. 2006b. Application of two-dimensional gel electrophoresis to interrogate alterations in the proteome of genetically modified crops. 2. Assessing natural variability. *J Agric Food Chem* 54:2162–2168.

Saito, K. and F. Matsuda. 2010. Metabolomics for functional genomics, systems biology, and biotechnology. *Annu Rev Plant Biol* 61:463–489.

Schubert, D. R. 2008. The problem with nutritionally enhanced plants. *J Med Food* 11:601–605.

Scossa, F., D. Laudencia-Chingcuanco, O. D. Anderson et al. 2008. Comparative proteomic and transcriptional profiling of a bread wheat cultivar and its derived transgenic line overexpressing a low molecular weight glutenin subunit gene in the endosperm. *Proteomics* 8:2948–2966.

Shepherd, L. V. T., J. W. McNicol, R. Razzo, M. A. Taylor, and H. V. Davies. 2006. Assesing the potential for unintended effects in genetically modified potatoes perturbed in metabolic and developmental processes. Targeted analysis of key nutrients and anti-nutrients. *Transgenic Res* 15:409–425.

Shewmaker, C. K., J. A. Sheehy, M. Daley, S. Colburn, and D. Y. Ke. 1999. Seed-specific overexpression of phytoene synthase: Increase in carotenoids and other metabolic effects. *Plant J* 20:401–412.

Shin, Y. M., H. J. Park, S. D. Yim et al. 2006. Transgenic rice lines expressing maize C1 and R-S regulatory genes produce various flavonoids in the endosperm. *Plant Biology* 4:303–315.

Shulaev, V. 2006. Metabolomics technology and bioinformatics. *Brief Bioinform* 7:128–139.

Simó, C., E. Domínguez-Vega, M. L. Marina, M. C. García, G. Dinelli, and A. Cifuentes. 2010. CE-TOF MS analysis of complex protein hydrolyzates from genetically modified soybeans-A tool for foodomics. *Electrophoresis* 31:1175–1183.

Sobolev, A. P., A. L. Segre, D. Giannino et al. 2007. Strong increase of foliar inulin occurs in transgenic lettuce plants (*Lactuca sativa* L.) overexpressing the asparagine synthetase A gene from *Escherichia coli*. *J Agric Food Chem* 55:10827–10831.

Sobolev, A. P., G. Testone, F. Santoro et al. 2010. Quality traits of conventional and transgenic lettuce (*Lactuca sativa* L.) at harvesting by NMR metabolic profiling. *J Agric Food Chem* 58:6928–6936.

Stein, A. J. and E. Rodríguez-Cerezo. 2009. The global pipeline of new GM crops. Implications of asynchroneous approval for international trade. EU23846-EN, European Communities, Luxemburg.

Storhoff, J. J., S. S. Marla, V. Garimella, and C. A. Mirkin. 2005. Labels and detection methods. In U. R. Müller and D. V. Nicolau (eds), *Microarray Technology and Its Applications*, pp. 147–180. Berlin: Springer-Verlag.

Suhre, K. and P. Schmitt-Kopplin. 2008. MassTRIX: Mass translator into pathways. *Nucleic Acids Res* 36:481–484.

Takahashi, H., M. Hayashi, F. Goto et al. 2006. Evaluation of metabolic alteration in transgenic rice overexpressing dihydroflavonol-4-reductase. *Ann Bot* 98:819–825.

Tengs, T., A. B. Kristoffersen, K. G. Berdal et al. 2007. Microarray-based method for detection of unknown genetic modifications. *BMC Biotechnol* 7:91.

Tengs, T., A. B. Kristoffersen, H. Zhang, K. G. Berdal, M. Løvoll, and A. Holst-Jensen. 2010. Non-prejudiced detection and characterization of genetic modifications. *Food Anal Methods* 3:120–128.

Tengs, T., H. Zhang, A. Holst-Jensen et al. 2009. Characterization of unknown genetic modifications using high throughput sequencing and computational subtraction. *BMC Biotechnol* 9:87.

Tesniere, C., L. Torregrosa, M. Pradal et al. 2006. Effects of genetic manipulation of alcohol dehydrogenase levels on the response to stress and the synthesis of secondary metabolites in grapevine leaves. *J Exp Bot* 57:91–99.

Thomson, J. 2003. Genetically modified food crops for improving agricultural practice and their effects on human health. *Trends Food Sci Tech* 14:210–228.

Twyman, R. M., P. Christou, and E. Stöger. 2002. Genetic transformation of plants and their cells. In K. M. Oksman-Kandentey and W. H. Barz (eds), *Plant Biotechnology and Transgenic Plants*, pp. 114–144. New York: Marcel Dekker.

Van Dijk, J. P., K. Cankar, S. J. Scheffer et al. 2009. Transcriptome analysis of potato tubers: Effects of different agricultural practices. *J Agric Food Chem* 57:1612–1623.

Van Götz, F. 2009. See what you eat: Broad GMO screening with microarrays. *Anal Bioanal Chem* 396:1961–1967.

Vergragt, P. J. and H. S. Brown. 2008. Genetic engineering in agriculture: New approaches for risk management through sustainability reporting. *Technol Forecast Soc* 75:783–798.

Villas-Boas, S. G., S. Mas, M. Akeson, J. Smedsgaard, and J. Nielsen. 2005. Mass spectrometry in metabolome analysis. *Mass Spectrom Rev* 24:613–646.

Wisniewski, J. P., N. Frangne, A. Massonneau, and C. Dumas. 2002. Between myth and reality: Genetically modified maize, an example of a sizeable scientific controversy. *Biochimie* 84:1095–1103.

Wolfenbarger, L. L. and P. R. Phifer. 2000. The ecological risk and benefits of genetically engineered plants. *Science* 290:2088–2093.

World Health Organization (WHO). 2002. Foods derived from modern technology: 20 questions on genetically modified foods. Available at: htpp://www.who.int/fsf/GMfood/, accessed: October 13, 2013.

Xiayan, L. and C. Legido-Quigley. 2008. Advances in separation science applied to metabonomics. *Electrophoresis* 29:3724–3736.

Xu, Y., J. F. Heilier, G. Madalinski, E. Genin, E. Ezan, J. C. Tabet, and C. Junot. 2010. Evaluation of accurate mass and relative isotopic abundance measurements in the LTQ-orbitrap mass spectrometer for further metabolomics database building. *Anal Chem* 82:5490–5501.

Yates, J. R. 2013. The revolution and evolution of shotgun proteomics for large-scale proteome analysis. *J Am Chem Soc* 135:1629–1640.

Ye, X., S. Al-Babili, A. Kloti et al. 2000. Engineering the provitamin A (β-carotene) biosynthetic pathway into (carotenoid-free) rice endosperm. *Science* 287:303–305.

Zawirska-Wojtasiak, R., M. Goslinski, J. Gajc-Wolska, and S. Mildner-Szkudlarz. 2009. Aroma evaluation of transgenic, thaumatin II-producing cucumber fruits. *J Food Sci* 74:204–210.

Zolla, L., S. Rinalducci, P. Antonioli, and P. G. Righetti. 2008. Proteomics as a complementary tool for identifying unintended side effects occurring in transgenic maize seeds as a result of genetic modifications. *J Proteome Res* 7:1850–1861.

CHAPTER 3

Genomics in Hardwood Tree Improvement
Applications of a Growing Resource

Lisa W. Alexander and Shaneka S. Lawson

CONTENTS

3.1 INTRODUCTION

Forest tree genomics has recently entered an exciting phase of rapid development and discovery. Genomics tools are being developed and applied to understand natural populations of forest trees and to accelerate the domestication and improvement of important forest tree species for human benefit. For many years, there was a perception that forests would continue as an unlimited source of hardwood trees for ecological and economic services. However, pressure on important hardwood species due to overharvesting, lack of harvestable land, pests and pathogens, and imminent climate change has forced a change in dogma from forest management to species management. Individual tree species are now the subjects of in-depth genomic investigations driven by two separate needs, resulting in two branches of forest genomics research. The first need is for support of genetic improvement programs for short-rotation species such as many conifers, eucalypts, and poplars. Genomics technologies will accelerate the development and deployment of breeding populations that satisfy major economic needs for wood products, pulp, fiber, and biofuels. The second need is to develop diagnostic tools for the conservation and management of natural populations. Research toward this goal seeks to decipher the distribution and evolution of adaptive diversity in an ecological and biogeographical context. Research goals of the two branches often overlap, and results from studies in natural populations often provide tools for applied tree breeding programs.

Most traditional crop breeding strategies are not applicable to forest trees because of their outcrossing nature, large size, long generation times, and the advanced age at which many traits (i.e., stem form, biomass, wood quality) are evaluated. Tree breeding generally involves recurrent selection of phenotypically superior "plus" trees and population improvement in each cycle of breeding, where each cycle takes many years (White et al. 2007). The success of selection based on plus trees has been shown to be moderate at best (Finkeldey and Hattemer 2007), and the most widely planted species of pines and eucalypts have not moved beyond the third cycle of breeding (Gailing et al. 2009). Thus, the vast majority of forest trees have never been subjected to systematic breeding. Forest tree breeders are literally banking on the promise of genomics to speed selection and increase its efficiency. Advances in genotyping and gene discovery in the genera *Pinus*, *Picea*, and *Pseudotsuga* aimed at dissecting the architecture of important adaptive traits have provided the most advanced genomic platforms and analysis methods available for forest trees (see Neale and Kremer 2011 for a thorough review). These proven techniques, along with new advances arising directly from hardwood genomic research, have provided unprecedented access to the structure and function of hardwood tree genomes.

3.1.1 Genomic Properties of Forest Trees

Forest trees are large, long-lived, woody, perennial plants. Hardwood trees (angiosperms) are predominantly outcrossing, with efficient means of seed and pollen dispersal, such that high levels of diversity are harbored within populations (Hamrick et al. 1992). Pre- and postzygotic incompatibility systems prevent self-pollination of many species (Brewbaker 1957; McKay 1942; Petit and Hampe 2006). Often, species with a huge native range appear (based on neutral genetic markers) to be a single, randomly mating population (Petit and Hampe 2006). Individuals have high levels of heterozygosity and carry large genetic loads that can cause severe inbreeding depression (Hamrick 2004; Petit and Hampe 2006; Vos et al. 1995). Hardwood trees are generally very fecund, such that strong selection takes place at the seedling stage in natural populations (Poorter 2007). An important feature for breeders is the lack of linkage disequilibrium (LD) in natural populations due to high rates of outcrossing. Resequencing projects in conifers have shown that r^2 (a common measure of LD) among single-nucleotide polymorphisms (SNPs) within candidate genes declines rapidly to <0.20 within 1–2 kb (Burdon and Wilcox 2007; Neale and Savolainen 2004; Zhu et al. 2008). The lack of LD influences every aspect of forest tree genomics including mapping, gene discovery, and the potential for marker-assisted selection (MAS).

Forest trees have large genomes compared to other plants, with conifer genomes being exceptionally large. For example, the haploid genome sizes of *Pinus*, *Quercus*, *Populus*, and *Arabidopsis* are roughly 30 Gb, 1.5 Gb, 4.8 Mb, and 1.2 Mb, respectively (Arabidopsis Genome Initiative 2000; Neale and Kremer 2011; Tuskan et al. 2006), meaning that the *Populus* genome is 40 times larger than *Arabidopsis* and 40 times smaller than pine. The genomes of poplar and oak appear to have gone through a whole-genome duplication followed by chromosome fusion that led to a loss of some genes and extensive duplication of others (Bodenes et al. 2012; Tuskan et al. 2006). Most hardwood trees are diploid, though polyploidy has been observed in several species, including some poplars (Yin et al. 2008). Resequencing in conifers has revealed high levels of genome diversity, with an estimated 1 SNP in every 100 base pairs for species that have been surveyed (Neale 2007; Savolainen and Pyhajarvi 2007). Although estimates of mutation rates remain rare in trees, evidence has now accumulated that shows forest trees evolve more slowly than annual plants at the DNA sequence level for chloroplast, mitochondrial, and nuclear genes, particularly at silent sites (Petit and Hampe 2006).

3.1.2 Mapping and Gene Discovery in Forest Trees

Genetic maps (also known as linkage maps) are important tools that provide insights into the genome structure of organisms. DNA-based genetic markers, known in the parents, are assessed in a large group of progeny for recombination rates and placed on a genetic map in linkage groups (LGs) comprised of relatively closely spaced markers. In practice, each LG represents one chromosome, such that the number of LGs for a species is equal to its haploid number. Adding DNA markers to a genetic map increases its utility in identifying genomic regions influencing quantitative trait loci (QTLs), facilitating positional cloning of genes from bacterial artificial chromosome-based physical maps, and providing markers for applied breeding programs. Genetic maps are often improved as new marker types and mapping algorithms become available; the advent of high-throughput short-read sequencing has led to the addition of many simple sequence repeats (SSRs) and SNPs to genetic maps originally generated by amplified fragment length polymorphism (AFLP) and random amplified DNA (RAPD) markers. The high-density genetic maps necessary for association genetics and physical map anchoring are available or in development for several species of *Populus*, *Quercus*, and *Castanea* (Bodenes et al. 2012; Fang et al. 2012).

Several life-history traits shared by forest trees make linkage mapping difficult and imprecise. They are long-lived with long generation times, such that even generating and phenotyping a single offspring generation takes many years. An F_2 mapping population, used in genetic mapping of many crop species, is virtually unattainable. Further, the obligate outcrossing nature of forest trees precludes the development of inbred lines and leads to high individual heterozygosity. A strategy commonly used to overcome these biological hurdles is known as the two-way pseudo-testcross (Grattapaglia and Sederoff 1994). In this method, any two unrelated trees, preferably from wide genetic backgrounds, are crossed to form an F_1 population. In a cross between highly heterozygous parents, many dominant markers will be heterozygous in one parent and null in the other, resulting in a 1:1 segregation of progeny as in a testcross. (The term pseudo-testcross is applied because homozygosity at marker loci is not known or assumed before the testcross; it is inferred by segregation of the marker in the progeny.) Markers present in the male parent are placed on a "male" map, and those from the female parent are placed on the "female" map. The male and female linkage maps can be aligned using codominant markers (i.e., SSRs) segregating in both parents (e.g., Grattapaglia and Sederoff 1994; Jermstad et al. 2001; Krutovsky and Neale 2005; Saintagne et al. 2004).

Genetic mapping is the first step in gene discovery. Placing markers in order along chromosomes allows for statistical comparison of marker–trait relationships (i.e., QTL analysis). Detection of a QTL indicates that the trait of interest is influenced by a gene or genes within the genomic region covered by the confidence interval of the QTL. QTL analysis has successfully located genomic regions underlying important traits in hardwoods related to growth, bud phenology, wood properties, and resistance to biotic and abiotic stresses (Howe et al. 2003; Kremer et al. 2007; Neale and Kremer 2011). While some QTL have been corroborated by comparative or expression analysis (e.g., Casasoli et al. 2006; Chagne et al. 2003), QTL analysis has distinct limitations for gene discovery in forest trees. Creating a progeny population, as mentioned above, is laborious and time- and land-consuming. Most importantly, rapid decay of LD within populations precludes the utility of a QTL outside of a single pedigree; that is, marker–trait associations are pedigree specific. The confidence interval covered can be as large as an entire LG and contain over 1000 genes. Precise pinpointing of a causative polymorphism is not possible, especially with the small number of individuals (often ~100) commonly used, and the effect of many QTLs are likely overestimated. Finally, the detection of QTL is limited by the allelic variation in the crossing parents (Figure 3.1a). Because of the large size of forest tree genomes and the expense of mapping, positional cloning of QTLs has not been attempted. Thus, QTL analysis has not led to the widespread discovery of the genes that underlie important traits, as has happened in many crop species (Mauricio 2001).

Association genetics appears to be a better method than traditional QTL analysis for resolving gene function in forest trees (Burdon and Wilcox 2007; Neale and Savolainen 2004; Zhu et al. 2008). No pedigreed population is required; rather, a large, unstructured population is phenotyped for a trait of interest and genotyped at many marker loci. Populations represent generations of recombination so many, many gene combinations occur, allowing the detection of any true marker–trait association (Figure 3.1b). Genome-wide association studies require extremely dense genetic maps, as the lack of LD requires that a marker be extremely close to the causative polymorphism (probably within the gene itself) for an association to be found. Once an association is found, it is likely to remain through many recombination events and be valid in all pedigrees. Thus, markers found through association studies are good candidates for MAS. With the development of high-throughput SNP discovery and genotyping platforms (such as the Illumina Golden Gate assay, www.illumina.com; Neale 2007), large SNP haplotypes will eventually be used to scan the whole genome for marker–trait associations. Until then, the candidate gene association studies, where certain genomic regions are preselected for marker–trait analysis, will continue to be the preferred method of gene discovery in forest trees. Complementary methods of gene discovery, such as comparative genomics and functional genomics/proteomics/metabolomics, will be used increasingly as reference sequences and computational platforms become available for nonmodel species (Devey 2004; Galas 2002; Gonzalez-Martinez et al. 2006).

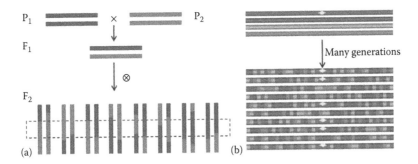

Figure 3.1 **(See color insert)** A schematic representation of traditional QTL mapping (a) and association mapping (b). QTL mapping can suffer from low resolution as single markers are typically in LD with large genomic regions. In association mapping, higher resolution is obtained because markers are in LD with much smaller genomic regions, primarily due to many generations of recombination and high genetic diversity in large, unstructured populations. A mutated allele (gold diamond) will be in LD only with closely located markers. (Reprinted with permission from Zhu, C., Gore, M., Buckler, E.S., and Yu, J., *Plant Genome*, 1, 5–20, 2008.)

This chapter describes genomics tools as they are used to understand and conserve natural populations and inform breeding programs aiming to manage and deploy value-added tree crops. The first section summarizes insights gained from *Populus*, a short-rotation wood fiber crop and the first tree species with a fully sequenced genome. The second section focuses on adaptive trait variation in oaks, with an emphasis on tools and techniques for gene discovery in forest trees. The third section comprises an in-depth examination of genomic applications for American chestnut disease-resistance breeding and reintroduction. Both current and potential genomics applications are presented in a comprehensive framework of chestnut ecology, life history, breeding objectives, and reintroduction goals. Finally, a compendium of forest tree genomics databases and resources is presented along with user information and potential applications.

Multitudes of genomic data in the form of genetic maps, sequences, and functional analyses are often produced from large-scale genomics projects. We have placed an emphasis on the uses, location, and accessing of these resources; as a result, many links to web-based instruments appear throughout the chapter. This chapter will provide the student with an overview of genomics applications and progress in forest tree improvement and provide the professional a comprehensive resource guide to genomics tools for *Populus*, *Quercus*, and *Castanea* species.

3.2 *POPULUS*: THE MODEL TREE

The *Populus* genus is composed of nearly 30 species within 5 distinct sections: Aigeiros (cottonwoods), Tacamahaca (balsam poplars), Leuce Duby. (white poplars), Turanga (subtropical poplars), and Leucoides (bigleaf poplars). Poplar species are found around the world in a myriad of habitats but reside primarily in temperate regions of the Northern Hemisphere (Yadav et al. 2010). Drought is one of the primary factors affecting productivity and survival of the poplar throughout its native and introduced ranges (Jiang et al. 2012; van Mantgem et al. 2009; Yang et al. 2010). The ability of *Populus* spp. to adapt to diverse conditions is indicative of variability within the transcriptome.

Early in the twenty-first century, the Department of Energy and the Joint Genome Institute initiated a genome sequencing project for *P. trichocarpa*, one of a number of different poplar varieties (Table 3.1). By 2004, the sequencing project had neared completion and the results were freely available in 2006 (Tuskan et al. 2006). This "model" species quickly became a popular choice for research study around the world because of its many uses. Poplar trees of many varieties are important for plantation farmers, lumber, pulp and biofuels production, and ecological diversity and as noninvasive methods of

Table 3.1 *Populus* spp. Common Names and Related Research Articles

Family	Species	Common Names	Reference
Aigeiros[a]	*P. deltoides* L.	Eastern cottonwood	Justin et al. (2010), Meirmans et al. (2010), Induri et al. (2012)
	P. d. deltoids[f]	Southern cottonwood, Carolina poplar,	Saša et al. (2009), Justin et al. (2010)
		Eastern poplar, necklace poplar	
	P. d. monilifera (syn. *P. sargentii*)[f]	Plains cottonwood	Not referenced[g]
	P. d. wislizeni[f]	Rio Grande cottonwood	Not referenced[g]
	P. fremontii	Fremont cottonwood	Not referenced[g]
	P. nigra L.	Black poplar	Regier and Frey (2010), Chenault et al. (2011), Fabbrini et al. (2012)
	P. canadensis–(*P. deltoides* × *P. nigra*)	Canadian poplar	Cohen et al. (2010), Gourcilleau et al. (2010), Wang and Jia (2010)
Populus[b]	*Populus* spp.	Aspen	Al-Masri et al. (2010), Yadav et al. (2010), Marmiroli et al. (2011b), Monclus et al. (2012)
	P. alba	White poplar, silver poplar, silverleaf poplar	Berta et al. (2010), Paolucci et al. (2010), Wang et al. (2010a), Schroeder et al. (2012)
	P. adenopoda	Chinese aspen	Wang et al. (2010a)
	P. davidiana	Korean aspen	Lee et al. (2011)
	P. grandidentata	Bigtooth aspen, American aspen, white poplar	Isabel et al. (2013)
	P. hopeiensis	N/A	Song et al. (2012b)
	P. tomentosa	Chinese white poplar	Du et al. (2012), Song et al. (2012a), Du et al. (2013)
	P. tremula	Aspen, common aspen, Eurasian aspen, European aspen, quaking aspen	Kieffer et al. (2009a,b), Grisel et al. (2010), Ismail et al. (2012), Schroeder et al. (2012)
	P. tremuloides	Quaking aspen, trembling aspen, American aspen, quakies, mountain or golden aspen, white poplar	Kelleher et al. (2012), Schroeder et al. (2012)
Leucoides[c]	*P. heterophylla*	Swamp or river cottonwood, downy poplar	Not referenced[h]
		Swamp poplar	Not referenced[h]
	P. lasiocarpa	Chinese necklace poplar	Not referenced[h]
	P. wilsonii	Wilson's poplar	Not referenced[h]
Tacamahaca[d]	*P. angustifolia*	Willow-leaved poplar, narrow leaf cottonwood	Coble and Kolb 2012
	P. balsamifera	Balsam poplar, eastern Balsam poplar, Hackmatack	Breen et al. (2009), Almeida-Rodriguez et al. (2010), Hamanishi et al. (2010), Meirmans et al. (2010), Ismail et al. (2012)
	P. cathayana	N/A	Chen and Peng (2010), Zhang et al. (2010a), Yang et al. (2010), Zhang et al. (2012)

Table 3.1 (Continued) *Populus* **spp. Common Names and Related Research Articles**

Family	Species	Common Names	Reference
	P. simonii	Simon's poplar	Almeida-Rodriguez et al. (2010)
	P. simonii × P. nigra		Yuan et al. (2011), Chen et al. (2012)
	P. simonii × P. balsamifera		Almeida-Rodriguez et al. (2010)
	P. trichocarpa	Black cottonwood, western balsam Poplar, California Poplar	Geraldes et al. (2011), Hamelin (2012), Induri et al. (2012), Ismail et al. (2012)
	P. trichocarpa × P. deltoids	N/A	Caron et al. (2010)
	P. yunnanensis	Yunnan poplar	Jiang et al. (2012)
Turanga[e]	*P. euphratica*	Euphrates poplar	Li et al. (2009), Brinker et al. (2010), Janz et al. (2010), Qiu et al. (2011), Wang et al. (2011a,b), Yan et al. (2012), Xu et al. (2013)
	P. pruinosa	Desert poplar	Wang et al. (2011b)

[a] Black poplars, some of the cottonwoods: North America, Europe, Western Asia; temperate.
[b] Aspens and white poplar: circumpolar subarctic and cool temperate, and mountains farther south (white poplar warm temperate).
[c] Necklace poplars or bigleaf poplars: Eastern North America, Eastern Asia; warm temperate.
[d] Balsam poplars: North America, Asia; cool temperate.
[e] Subtropical poplars: Southwest Asia, East Africa; subtropical to tropical.
[f] Subspecies of *P. deltoides* L.
[g] Not referenced in the chapter but included as examples of the members of the Aigeiros family.
[h] Not referenced in the chapter but included as examples of the members of the Leucoides family.

phytoremediation (Al-Masri et al. 2010; Caron et al. 2010; Grisel et al. 2010; Induri et al. 2012; Justin et al. 2010; Kieffer et al. 2009a,b). Black cottonwood was the first woody plant species chosen for genome sequencing (Tuskan et al. 2006), and since then many authors have continued to report the discovery of genes and proteins involved in a number of stress responses. A plethora of data existed illustrating drought-stress responses among poplar plants; however, additional research was needed to link these data to what was known about the gene expression changes that occurred during drought (Yang et al. 2009a,b; Yin et al. 2004). Caruso et al. (2008) looked at genes differentially expressed during the drought response in *P. canadensis* leaves with the aid of polyethylene glycol. Using differential display, they were able to identify variability in gene expression during the early portion of the stress period (10 days), before leaf growth was altered. Caruso et al. (2008) were able to identify genes involved in cellular protection (serine proteases, peroxidases), results similar to those found in *Arabidopsis* and *P. euphratica* studies (Brosché et al. 2005; Zimmermann et al. 2004). Another study with poplar involving water availability stresses indicated that similar genes were active in *P. deltoides* and *P. trichocarpa* (Street et al. 2006).

Ogata et al. (2009) reported on the compilation of a poplar gene coexpression database based on information obtained from nine publicly available DNA microarray datasets from the NCBI's Gene Expression Omnibus database (http://www.ncbi.nlm.nih.gov/geo). Their database contained vital information on the genome-wide prediction of poplar gene function and stress responses and was posted at http://webs2.kazusa.or.jp/kagiana/cop/. During an in-depth analysis of the *P. balsamifera* L. drought transcriptome, Hamanishi et al. (2010) concluded that a lack of a pair-wise alignment between single polymorphisms and geographic origin between and among genotypes indicated local adaptation of some genotypic responses, while others were more widespread. Altogether these findings provided insight into the distribution of various poplar genotypes across many different terrains.

3.2.1 Sequence Data

Yang et al. (2009) wrote a comprehensive review of advances in the field of *Populus* research genetics and stated that additional efforts were needed in the areas of genome annotation and in-depth sequence analysis. The response from the research community was tremendous and many different experiments encompassing multiple subject areas were initiated. Grönlund et al. (2009) used a multinetwork algorithm on microarray data of poplar gene coexpression to reveal gene-to-gene communication and show how multiple tissues and genes are involved in stress responses. Ismail et al. (2012) studied SNPs in two balsam poplar (*P. trichocarpa* and *P. balsamifera*) and one Eurasian aspen (*P. tremula*) species across 4.5, 4.1, and 3.8 kb of sequence. Nine loci for stress response, freezing tolerance, and defense genes were analyzed, with variation among Eurasian aspen being highest.

Rathmacher et al. (2009) indicated in studies of *P. canadensis* that correct identification of hybrid poplar is vital to both breeding programs and genetic data acquisition. These data illustrate how genetic diversity between and among the various poplar species requires use of multiple loci to make inferences related to population history (Breen et al. 2009). Whole-genome coverage is likely to reveal a plethora of data regarding loci and genetic diversity. Other researchers have used SNP screening of nuclear genes to identify specific clones, hybrids, and individual species within *Populus* (Fladung and Buschbom 2009; Ingvarsson 2008; Isabel et al. 2013; Schroeder and Fladung 2010), a feat difficult to attain with SSRs if more than two species are being compared (Fossati et al. 2005; Fussi et al. 2010; Liesebach et al. 2010; Rathmacher et al. 2009).

Progressive research exposed the limitations of isozyme research and additional methods such as the use of nuclear microsatellites (Lee et al. 2011; Li et al. 2012; Liesebach et al. 2010), AFLPs (Chen and Peng 2010; Gao et al. 2009a), and SSRs (Du et al. 2012; Saša et al. 2009; Schroeder and Fladung 2010; Wang et al. 2010a; Zhang et al. 2009b) were developed. Using SSRs and AFLPs in poplar species differentiation, while useful, is more labor intensive and cost prohibitive than using SNPs and fails to distinguish between homozygous genotypes, null alleles, and homologous recombinations (Liesebach et al. 2010; Rathmacher et al. 2009). Schroeder et al. (2012) used SNP screening combined with PCR–RFLPs and length polymorphisms of >20 chloroplast regions (intergenic and coding), 23 published barcoding primer combinations, and 17 newly designed combinations to accurately differentiate between *Populus* species. Reliability of previously published primers was roughly 50%, while the new primers based on the sequined cpDNA genome of *P. trichocarpa* were more successful, and these data revealed 21 species-specific SNPs and 5 species-specific indels. Yuan et al. (2011) used the shotgun method to identify 119 proteins involved in photosynthesis, transport and signaling, primary and secondary metabolism, or redox reactions. Eighty-five of these were chloroplast proteins, with the remainder being nuclear proteins, and nearly half were associated with the thylakoid, thus providing a valuable starting point for further investigations into photosynthesis. Wang et al. (2011b) used chloroplast DNA (cpDNA), nuclear ITS sequences, and eight SSR loci to differentiate between *P. euphratica* and *P. pruinosa* species. These authors also stated that cpDNA is less useful for differentiation between species than nuclear DNA (Liesebach et al. 2010).

3.2.2 Mapping Populations

Genetic analysis has revealed *Populus* species have high reproductive capabilities and have interspecific breeding tendencies (Meirmans et al. 2010). Typically, the most used species within breeding programs are *P. balsamifera*, *P. deltoides*, *P. maximowiczii*, *P. nigra*, and *P. trichocarpa*. Chen and Peng (2010) used AFLPs to analyze and genotype six *P. cathayana* Rehd populations and noted highly elevated levels of genetic diversity and differentiation within populations. Worthy of note was the fact that stress responses were proposed to be sex-linked, thus hybrids must be screened carefully to reproduce previous finding for a particular cross. Zhang et al. (2012) noted that male and female poplar trees differed in response to chilling stress, drought stress (Zhang et al. 2010a), and salt stress (Chen et al. 2011; Jiang et al. 2012). Song et al. (2012a) identified 27 sex-specific methylation sites in the first screen for sex-specific DNA methylation and gene expression in poplar. Paolucci et al. (2010) used comparative analysis of *P. alba* L. genetic linkage maps for sex-linked markers with the 19 haploid chromosomes of the *Populus* genome sequence. Species comparisons indicated location differences for the sex locus and a skew in the sex segregation ratio toward males in the *P. alba*, *P. tremula* × *P. tremuloides*, and *P. deltoides* × *P. nigra* pedigrees, but not in *P. trichocarpa*.

Evaluation of breeding potential among poplar species should start as early as possible. Zhang et al. (2009b) looked at the mechanism of pollen formation in the hybrid poplars *P.* × *euramericana* and *P.* × *popularis* using SSRs. The authors proposed the use of *2n* gametes in the production of polyploid clones for breeding studies. Their discovery has many end uses but one of the primary benefits is the acceleration of early selection for high-value poplar hybrids. Wang et al. (2010a) used two economically and ecologically valuable poplar species, *P. adenopoda* and *P. alba*, to generate the first SSR and sequence-related amplified polymorphism (SRAP) genetic linkage maps as a means for comparing poplar genotypes. These data are imperative for subsequent identification of QTL and regions of similarity within the *Populus* genus. They used 1142 SSR primer pairs and 163 SRAP primer combinations in their study, an indication of the complexity and advancement of this field of research. While Yin et al. (2001) generated parental maps using RAPDs, Wang et al. (2010a) used Joinmap software to map 192 SSRs and 50 SRAPs to a new consensus map.

3.2.3 Linkage Mapping and QTL Discovery

Generation of biomass and conversion to biofuels is often the rationale for developing extensive plantations of poplar hybrids. Identification of the most abiotic stress–tolerant, biomass-producing clones is the desired result of plantation owners worldwide. After the completion of the *P. trichocarpa* genome sequence (Tuskan et al. 2006), research into genetic improvements and familial histories of poplar were accelerated. Wang et al. (2010b) stated that current *Populus* resources can be favorably compared with resources related to crop species and future studies focused on identification of particular QTLs are needed for improving strategies for MAS. Rae et al. (2009) used QTL mapping to identify biomass-related "hotspots" within short-rotation coppiced poplar. Of the >200 QTLs mapped, roughly 5.6 were mapped per trait and 9.4 per LG. The results of this study can be used to identify correlations of physiological traits with overall biomass and limit losses when growing poplar for biomass-related end products. These data were very promising; however, failure to state the heritage of the poplar used is a considerable drawback as many poplar species and hybrid crosses have been used for research study, and proper identification of clones is essential for experiment repeatability. In a study of nitrogen levels related to wood quality and growth, Novaes et al. (2009) used QTL analysis and discovered an average QTL contained 426 genes and nearly half had unknown functions. QTLs related to growth and root formation were mapped by Zhang et al. (2009a) using a variety of marker systems including SSRs, AFLPs, RAPDs, ISSRs, and SNPs on a full-sib family of 93 *P. deltoides* and *P. euramericana* hybrids. These authors mapped 67 maternal markers and 65 paternal markers and noted a clustered distribution for each set across four LGs. Highly heritable traits such as root length and number were observed, suggesting the identified QTLs influenced root capacity. These results are important for obtaining superior stock material for subsequent vegetative propagation.

Pakull et al. (2009) used genetic data obtained from SSR markers and AFLP of 61 *P. tremula* L. × *P. tremuloides* Michx. hybrids to compile the first genetic linkage maps for aspen, while Drost et al. (2009) genotyped 154 pseudo-backcross progeny of *P. trichocarpa* and *P. deltoides* using microarray analysis and SSRs to produce a high-density map of gene-based markers. Results of the study are likely to be used in genetic mapping efforts for the remainder of the unassigned genes within the *Populus* genome. Gao et al. (2009b) used nine unique primer pairs and AFLP to aid in genetic identification of 22 elite Chinese *P. deltoides* and *P.* × *canadensis* accessions. Analysis of 461 AFLP bands indicated strong consistency between morphologic traits and phylogeny.

Chenault et al. (2011) used SSRs to genotype a mature population of 413 *P. nigra* L. at 11 loci. These authors were able to genotype 379 individuals and used subtractive analysis to assign 401 of these trees to one of 37 multilocus lineages (MLLs). No geographic pattern was observed as clonality within the site was evenly distributed. Data obtained in this study indicated that Lombardy poplar (considered a *P. nigra* species) possessed a few rare *P. nigra* alleles. Notably, of the 13 Lombardy poplar sampled, 11 were similar to the reference genotype, while the remaining two different at only one or two loci. Wang et al. (2011a) initiated a study of *P. euphratica* populations to determine the degree of diversity within long- and short-range groups. Attempts to find correlations between distance and genetic variation were unsuccessful; however, diversity within the species was exceptional with >12 alleles per locus. Anthropogenic changes were targeted as the cause of widespread habitat fragmentation and contributed to the decline of this now endangered desert species rather than lack of diversity. Xu et al. (2013) constructed several polymorphic SSR multiplexes for *P. euphratica* based on *P. trichocarpa* sequence data. These authors developed two "marker kits," one with genomic SSRs (gSSRs) and another with expressed sequence tag (EST) SSRs (eSSRs). Successful amplicons were obtained from 88 of the 499 primers developed using *P. trichocarpa* sequences; only 52 fragments contained microsatellite repeats for submission to GenBank. These data can be used as added references when genotyping additional *Populus* species.

Identification of gene divergence within a population is important for understanding gene variations and how genes are segregated among species. Saša et al. (2009) used a combination of AFLP and SSRs to identify DNA variation within 13 poplar genotypes. These authors identified three clusters using SSR and four using AFLP markers. Both SSR and AFLP data indicated *P. nigra* and *P. deltoids* segregated into different groups. Interestingly, two clones of *P. trichocarpa* were found to be genetically different, thus questioning the backgrounds of *P. trichocarpa* individuals used for other comparison tests. These data indicate the usefulness of using SSRs and AFLP markers to identify relationships and genetic backgrounds among numerous *Populus* spp. Du et al. (2012) used 20 species-specific microsatellite markers to genotype a group of 460 unrelated Chinese white poplar (*P. tomentosa*) individuals and noted genetic variation followed a discernible pattern. The greatest levels of variation were found in the southern region, while the northern regions were much less varied. In this study, no correlations between distance and genetic variation were found. Nearly 100 alleles with almost 5 alleles were found per locus in this species using SSRs, a low number when compared with several other species.

QTL analysis can be used to find candidate genes for a number of important morphological or physiological traits. Monclus et al. (2012) examined 330 F_1 *P. deltoides* × *P. trichocarpa* progeny using QTL analysis for identification of candidate genes for productivity, architecture, and water-use efficiency (WUE). These authors tested 110 SSRs and were able to identify 77 QTLs matching 11 distinct gene ontology (GO) terms such as branch numbers and adventitious root development. This work represented the first study using GO terms and QTL analysis to examine gene function. Fabbrini et al. (2012) used *P. trichocarpa* genome sequences and QTL to identify 13 gene models and 67 expressed and 6 functional candidate genes (PhyA, Co-L2, FHY1.2, PAT1, ZGT1like, and ZGT2like), all in the light signaling pathway, related to bud set in poplar. Several identified QTL that contained no identified candidate genes will be used to identify potential new bud set genes. These data were compared to *P. nigra* for similarity and will likely serve as important breeding targets. LD is an important consideration when determining if a candidate gene or a genome-wide approach to genetic differentiation is necessary (Du et al. 2013). Earlier studies have indicated low LD, and Kelleher et al. (2012) noted incidents of decay within tree species. They also selected 35 gene regions of *P. tremuloides* involved primarily in lignin biosynthesis and wood formation and genes involved in disease resistance and light responses and developed primers using ESTs from various *Populus* spp. A second study of wood formation and growth traits used SSRs to show strong correlations between the wood and growth traits of 1200 F_1 poplar progeny of a cross between a *P. alba* × *P. glandulosa* female and *P. tomentosa* male. Negative correlations were seen between lignin content and holocellulose, α-cellulose, and diameter (Du et al. 2013).

3.2.4 Gene Discovery

Despite the many advances made in the field, there appear to be several key areas that have been overlooked. Geraldes et al. (2011) examined 0.5 million SNPs in 26,595 genes expressed in developing xylem tissue of *P. trichocarpa*. Attempts at mapping revealed that >96% of the SNPs were unmapped. This indicated a shocking lack of annotation within the *Populus* genome and an example of one area of poplar research requiring additional research. Genes with differing expression patterns in response to abiotic stress are likely to be prime targets for further research into breeding tree varieties with the greatest abiotic tolerances using advanced genetic techniques (Chen et al. 2011; Hamelin 2012; Kilian et al. 2012; Li et al. 2009; Pakull et al. 2009). Therefore, closer inspection of the transcriptome may be warranted if additional information regarding gene expression is wanted. These data indicated that numerous methods for data acquisition and analysis are available; however, the different genetic and molecular approaches are fickle in their reliability of quantification and analysis. Regier and Frey (2010) reported that because of this phenomenon, results computed from the same dataset can lead to considerably different conclusions.

Drought tolerance has been studied extensively in many species, and understanding the mechanisms, genes, and pathways involved in the drought response will allow for ease in determining the best candidates for biomass studies, plantations, and reclamation in water-limited environments. Hamanishi et al. (2010) noted variations within the transcriptome of six *P. balsamifera* clones in response to drought conditions. Almeida-Rodriguez et al. (2010) compared *P. balsamifera*, a drought-tolerant clone with higher rates of gas exchange and lower vulnerability, to *P. simonii* × *balsamifera*, a drought-avoidant clone characterized by rapid stomatal closure, using sequence analysis and qRT-PCR. Cohen et al. (2010) used comparative transcriptomics to look at drought responses of two hybrid poplar transcriptomes and revealed several potential gene candidates for additional drought tolerance research, while Almeida-Rodriguez et al. (2010) also found a number of candidate genes for further study of drought tolerance within poplar species. It is anticipated that a plethora of candidate genes will be discovered in the coming years and a suite of stress tolerance aspects will be identified and examined for use in high-throughput screening activities.

3.2.5 Case Study: Breeding for Drought Resistance in *Populus* spp.

Black cottonwood was the first woody plant species to have its genome sequenced (Tuskan et al. 2006), and since then many authors have continued to report the discovery of genes and proteins involved in a number of stress responses. Tschaplinski et al. (2006) examined hybrid poplar clones of black cottonwood (*P. trichocarpa*) and eastern cottonwood (*P. deltoides*) for phenotypic variation and identified heritable QTLs that were presumed to be drought tolerance indicators. Using 33 genotypes from F_1 progeny of a *P. deltoides* × *P. trichocarpa* cross, Monclus et al. (2009) presented data to support their claims, and noted leaf area increases and reductions were closely associated with the observed variability in WUE.

WUE has been studied extensively in poplar species. Understanding the mechanisms, genes, and pathways involved in the poplar drought response will allow for ease in determining the best candidates for biomass studies, plantations, and reclamation in water-limited environments. Tschaplinski et al. (2006) examined hybrid poplar clones of black cottonwood (*P. trichocarpa*) and eastern cottonwood (*P. deltoides*) for phenotypic variation and QTLs that may allow for ease in identifying clones prone to drought tolerance. The authors were able to reliably identify QTLs across sites and across drought-stress treatments in both parents, thereby presumably serving as heritable indicators of drought tolerance in the ensuing offspring. Monclus et al. (2009) presented information to support their claims that tolerance to water stress could be correlated across 33 genotypes from the F_1 progeny of a *P. deltoides* $\times$ *P. trichocarpa* cross. Leaf area increases and reductions were closely associated with the observed variability in WUE, prompting them to suggest that future studies focus on a QTL-based approach to better understand these interactions.

Drought conditions are becoming more and more frequent worldwide; therefore, trees with greater resistances to abiotic stressors are likely to be more beneficial for plantation owners (Chen and Polle 2010). Future climate scenarios have indicated that water stress and drought will be major obstacles facing proper growth and development of many forest and plantation tree species. The majority of poplar species are very sensitive to drought conditions. In response to water deficit, declines in productivity have been seen. Poplar species differ in tolerance levels and stress responses (Bonhomme et al. 2009a; Coble and Kolb 2012). Lack of water, specifically freshwater, can prove challenging; therefore, trees with a greater tolerance to salt are essential if pure water is not plentiful. Brinker et al. (2010) used *P. euphratica*–specific microarrays to identify clusters of coexpressed genes involved in salt-stress tolerance. Qiu et al. (2011) analyzed 86,777 stressed and unstressed *P. euphratica* ESTs using Solexa sequencing data. Greater than 25% of the ESTs were differentially expressed in response to salt stress. Janz et al. (2010) indicated that these differential responses were linked to evolutionary adaptations in mechanisms of stress tolerance.

Transcriptome data has been collected from a myriad of poplar species under various abiotic stress conditions (Berta et al. 2010; Chen et al. 2012; Jiang et al. 2012; Song et al. 2012b), and their tolerance to stressful environments varied considerably among species, populations, and clones because of their genetic diversity (Xiao et al. 2009). Bonhomme et al. (2009a) noted differential responses of *P. deltoides* $\times$ *P. nigra*, cv. Agathe_F and Cima to drought stress. The cultivar "Agathe_F" responded with increased photosynthetic enzymes, while "Cima" increased photorespiration and oxidative stress enzymes. Bonhomme et al. (2009b) noted that eight *P.* $\times$ *euramericana* genotypes with varied leaf carbon isotope discriminations (∂) also varied in chloroplastic carbon fixation proteins and stomatal conductance or CO_2 assimilation proteins. Cseke et al. (2009) used transcriptome comparisons of leaves from two trembling aspen (*P. tremuloides* Michx.) to show that individual clones respond differently to elevated CO_2. One clone, "271," responded by devoting a greater proportion of energy into growth, while clone "216" relocated carbon to secondary compounds rather than increased growth. Yan et al. (2012) exposed *P. euphratica* to long-term drought conditions and noted increased responses in heat shock proteins using microarray and qRT-PCR. These data indicated *P. euphratica* protected proteins from oxidative damage with a robust drought-stress response and emphasize the importance of genetic approaches and our ability to work toward understanding tree responses to climate change. Physiological and genetic analysis has been used in agriculture for many years and is currently being applied to many tree species (Khan et al. 2010; Pijut et al. 2011; Tester and Langridge 2010; Witcombe et al. 2008).

Development of specialized responses to stress by different poplar species provides an apt topic for continued advancement within genetic research studies. Gourcilleau et al. (2010) reported results from DNA methylation studies of six genotypes of *P. deltoides* $\times$ *P. nigra* subjected to artificial water stress conditions. Positive correlations between productivity and DNA methylation were observed, and irregular responses were recorded among hybrids—an indication of variability in response mechanisms. Yang et al. (2010) used proteomics approaches such as protein extraction and enzyme assays to show *P. kangdingensis* and *P. cathayana* populations exhibited differential responses to artificial drought conditions, while Song et al. (2012b) observed differential responses in *P. hopeiensis* exposed to water stress using 256 primer combinations for cDNA-AFLP analysis.

Bogeat-Triboulot et al. (2007) used *P. euphratica* plants to observe shoot and root responses to soil water deficit. The physiological responses and transcriptional changes, when examined along with proteomic analysis, provided a blueprint of drought acclimatization in *P. euphratica*. Caruso et al. (2008) looked at genes differentially expressed during the drought response in *P. canadensis* leaves with the aid of polyethylene glycol and differential display. The authors identified variability in gene expression during the early portion of the stress period (10 days), before altered leaf growth and genes involved in cellular protection (serine proteases, peroxidases) were upregulated, results similar to those found in *Arabidopsis, P. euphratica, P. deltoides,* and *P. trichocarpa* studies (Brosché et al. 2005; Street et al. 2006; Zimmermann et al. 2004). A plethora of data exist illustrating drought-stress responses among poplar plants; however, additional research was needed to link these data to what was known about gene expression changes during drought (Yang et al. 2009a,b; Yin et al. 2004). Ogata et al. (2009) compiled a poplar gene coexpression database based on information obtained from nine publicly available DNA microarray data sets from the NCBI's Gene Expression Omnibus database (http://www.ncbi.nlm.nih.gov/geo) that contained vital information on the genome-wide prediction of poplar gene function and stress responses and was posted at http://webs2.kazusa.or.jp/kagiana/cop/.

Other advanced techniques for studying evolutionary changes in plants have since been implemented in poplar. Bonhomme et al. (2009a) used two-dimensional electrophoresis to compare the genetic variation and drought response of two *P. × euramericana* genotypes (Agathe_F and Cima) and showed that, regardless of ontogeny, genotypic differences in response to water deficit experiments, whether in the field or in the greenhouse, could be reliably repeated. Bonhomme et al. (2009a) indicated different combinations of genes were responsible for drought tolerance in the two genotypes and noted drought responses among eight *P. euramericana* genotypes were not distinguishable by drought tolerance levels, yet significant differences were observed in the chloroplast proteins involved in carbon fixation and in photosynthesis-related proteins (Bonhomme et al. 2009b). This study was the first to examine genetic variation within the poplar leaf proteome. During an in-depth analysis of the *P. balsamifera* L. drought transcriptome, Hamanishi et al. (2010) concluded that a lack of a pair-wise alignment between single polymorphisms and geographic origin between and among genotypes indicated local adaptation of some genotypic responses, while others were more widespread. All together, these findings provided insight into the distribution of various poplar genotypes across many different terrains.

3.2.6 Case Study: Leaves as Engineering Targets

As climates change, water shortages in many areas increase the likelihood that many tree species will succumb to drought. Gortan et al. (2009) noted that leaf mass areas (LMAs) were identical for plants growing in water-stressed and nonstressed environments, and leaf hydraulic conductance was lower in trees accustomed to drought. This study was one of the first to focus on the leaves as possible targets for studying drought tolerance. Additional research groups have focused their efforts on the leaf surface, as stomatal pores are responsible for the majority of water loss in a plant (Figure 3.2). Muthuri et al. (2009) proposed that leaf phenology played a greater role in foliar gas exchange, while Hallik et al. (2009) postulated whether a species' shade- and drought tolerance capability can be found in leaf-level analysis. Their data indicated that the correlations discovered between drought and shade tolerances could only partially be explained during analysis, and more in-depth research projects were needed. In this manner, Damour et al. (2010) suggested stomata, which are key in plant adaptation (Berry et al. 2010), be closely examined along with aquaporins (Johansson et al. 2000), as these features may provide valuable targets for future modification. In examination of leaves for engineering targets, Sandquist and Ehleringer (2003) considered analysis of trichomes, which had been proven to decrease solar radiation, while altering WUE and drought tolerance.

Agrawal et al. (2008) speculated that heritability of ecophysiological traits was variable and that adaptive responses were likely to vary as well. Gouveia and Freitas (2009) examined the

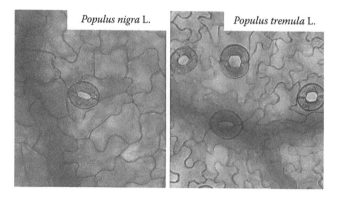

Figure 3.2   Examination of the adaxial surface of a leaf of *Populus nigra* L. compared with one of *P. tremuloides* L. indicated clear differences in adaxial stomatal density.

relationships between leaf plasticity and drought responses. Specific leaf area (SLA), a ratio of leaf area to dry biomass, and leaf thickness were shown to positively correlate with rainfall along environmental gradients. Their data indicated that areas with decreased rainfall supported trees exhibiting decreased SLA and leaf thickness, while the opposite was true for those areas with increased rainfall. It was surmised that cork oak (*Quercus suber*), and perhaps other varieties of oak, modulates gas exchange in relation to water availability, thereby increasing WUE. An alternate theory was proposed by Fichot et al. (2009), who examined xylem anatomy in response to changes in artificially maintained water availabilities. It was reported that in response to limiting water supplies, vessel densities increased while cell sizes decreased. De Micco et al. (2008) focused on differences in anatomical features between the juvenile and mature stages for a total of eight woody plant species from both mesic and xeric habitats and showed that variations between immature and mature forms were large in mesic species and slight in xeric species. De Micco et al. (2008) proposed that observed differences corresponded well to water availability and stated that mesic species possessed additional protective mechanisms to negate drought effects as seasons change, while xeric species were less inclined to experience highly variable water regimes. Nitrogen content of an average leaf was generally between 2.0% and 2.5% (Isaakidis et al. 2004), although increased leaf nitrogen could indicate increased photosynthetic processes. However, this was only likely to occur when carbon assimilation was operating at full capacity (Meziane and Shipley 2001). Additional studies addressing leaf physiology and its relation to growth performance were needed as Monclus et al. (2006) stated that there was no clear evidence that WUE was related to growth performance or whether or not some leaf traits could be used as predictors for productivity in *P. deltoides* × *P. nigra* hybrids.

Bogeat-Triboulot et al. (2007) used *P. euphratica* plants to observe shoot and root responses to soil water deficit. The physiological responses and transcriptional changes, when examined along with proteomic analysis, provided a blueprint of drought acclimatization in *P. euphratica*. Other advanced techniques for studying evolutionary changes in plants have since been implemented in poplar. Bonhomme et al. (2009a) used two-dimensional electrophoresis to compare the genetic variation and drought response of two *P. × euramericana* genotypes (Agathe_F and Cima). This first attempt at examination of the leaf proteome indicated, regardless of ontogeny, genotypic differences in response to water deficit experiments whether in the field or in the greenhouse that could be reliably repeated. Their findings also indicated different combinations of the genes responsible for drought tolerance in the two genotypes. In a second study using poplar, Bonhomme et al. (2009b) looked at drought responses among eight *P. euramericana* genotypes using measurements of WUE and carbon isotope discrimination. The genotypes were not distinguishable by drought tolerance levels, yet significant differences were observed in the chloroplast proteins involved in carbon fixation and in photosynthesis-related proteins. This study was the first to examine genetic variation within the poplar leaf proteome.

Efforts to find links between biomass and productivity have shown that a correlation between $\delta^{13}C$ and productivity was highly variable in woody and herbaceous species, with both positive and negative relationships being reported in the literature (Marron et al. 2005). Marron et al. (2005) looked for correlations between leaf characteristics and biomass production in 31 genotypes of poplar. Clones that exhibited the fastest leaf growth rates had greater total leaf area, but lower SLA than less vigorous clones and indicated that vigor does not necessarily correlate to leaf traits. They also discovered that $\delta^{13}C$ did not correlate to biomass production and indicated the possibility of selecting clones with greater biomass production combined with higher WUE.

3.3 GENOMICS OF ADAPTIVE TRAITS IN *QUERCUS*

Understanding and characterizing genes responsible for adaptive variation is a primary task for population and evolutionary genomics. Forest geneticists and forest resource managers have traditionally used genetic information such as population differentiation of desirable traits and neutral marker genetic diversity to implement and monitor tree breeding and tree improvement programs and to monitor and delineate conservation zones in natural populations. However, these traditional methods provide no information concerning the genomic location, spatial distribution, and variation of genes that control the traits of interest. Estimates of genetic diversity based on neutral markers may be high relative to diversity of genes controlling adaptive traits such that estimates of genetic diversity and population structure used by resource managers to develop breeding or *in situ* conservation programs may be misinformed (Howe et al. 2003). Further, trees growing today and in the future will face relatively high rates of climate change compared to the warming that occurred after the last glaciations (Intergovernmental Panel on Climate Change 2007). This section examines genomic resources and techniques currently used to dissect the diversity of important adaptive traits in *Quercus*, with an emphasis on data and results that are directly transferable to tree improvement programs.

The most commonly studied oak species are the European white oaks *Quercus robur* (pedunculate oak) and *Q. petraea* (sessile oak) and the northern red oak, *Quercus rubra*. *Q. robur* and *Q. petraea* are large deciduous trees that are widely distributed in Europe from northern Spain to southern Scandinavia and from Ireland to Eastern Europe. The natural range of *Q. petraea* is included in that of *Q. robur*, except that its eastern limit is the Ukraine, whereas *Q. robur* reaches the Ural Mountains. Like most oaks, they are obligate outcrossers and predominately wind-pollinated. Pedunculate oak prefers fertile and well-watered soils, and mature trees will tolerate flooding. Sessile oak is more tolerant to drought and poor soil than pedunculate oak but more sensitive to airless soil conditions (Ducousso and Bordacs 2004). The species freely hybridize across their range. Pure species can be differentiated by the presence of a short leaf stalk (3–8 mm) and stalked acorn in *Q. robur*, as these features are not present in *Q. petraea*; hybrids are intermediate in appearance. Northern red oak (*Q. rubra* L. section *Lobatae*) is an economically and ecologically important forest tree of North America. It is a major-dominant hardwood species with a wide native range extending east from Nebraska to the Atlantic coast (60 –96 W longitude) and from northern Ontario to southern Alabama (32 –47 N latitude). Northern red oak has adapted to wide ranges of mean annual temperature and rainfall. Often it is the dominant oak species on lower slopes and north-facing slopes (Abrams 2002), and it usually co-occurs with several other red oak species. Intraspecific variation for adaptive traits such as height growth, phenology, drought resistance, and cold-hardiness has been reported on altitudinal, latitudinal, and longitudinal clines across the species' ranges (Gall and Taft 1973; Kriebel 1993; McGee 1974; Schlarbaum and Bagley 1981).

3.3.1 Gene Sequences

Within the genus *Quercus*, *Q. petraea* and *Q. robur* have the most developed genomic resources as they have long been subjects of investigations in hybridization, postglacial migration, and distribution of adaptive variation (Table 3.2, Figure 3.3). ESTs are sequences of DNA originating from coding regions of the genome. ESTs are generated in transcriptome studies, where cDNA from individuals contrasting for a trait of interest is sequenced, annotated, and compared to find genes underlying trait variation. Because adaptive variation and its underlying genes are the most common genomic studies of oak, ESTs comprise the largest sequence type currently available. Twenty-five ESTs (GenBank accessions CF369265–CF369290) were first generated by Porth et al. (2005) from cDNA libraries of *Q. petraea* callus cells undergoing osmotic stress. Derory et al. (2006) developed cDNA libraries for six stages of *Q. robur* apical bud development. Transcriptome analysis of these libraries led to the generation of 801 ESTs deposited in the EMBL nucleotide database (accessions CR627501–CR628310). In the first large-scale exploration of the oak transcriptome, Ueno et al. (2010) produced and analyzed ESTs from Sanger and 454 sequencing of 34 cDNA libraries. Sanger reads were produced from 20 cDNA libraries (10 from *Q. petraea* and 10 from *Q. robur*) and subjected to quality standards including the removal of library-specific cloning-vector and adaptor sequences, low-quality sequences, and contaminants and the masking of low complexity sequences. The resulting Sanger catalog contained 125,886 high-quality sequences from 111,631 cDNA clones. Sequence chromatograms are available via SURF from the INRA website: http://mulcyber.toulouse.inra.fr/projects/surf/. A total of 1,948,579 reads were generated from 454 sequencing of 14 libraries (9 from *Q. petraea* and 5 from *Q. robur*). Clustering and assembly of the Sanger and 454 reads together produced 1,704,117 EST sequences that reduced to 222,671 unigene elements. The unigene elements were comprised of 69,154 (31%) tentative oak consensus sequences (those common to both Sanger and 454 reads) and 153,517 (69%) singletons. Examination of the EST catalog for homologs of genes implicated in bud break, cuticle formation,

Table 3.2   Genomic Resources for *Quercus* and *Castanea* Include Markers Developed for Linkage Mapping and Markers, Short Sequence Reads, and Read Assemblies from Sequenced cDNA and BAC Libraries

Species (Pedigree)	EST	RAPD, AFLP	SSR[a]	Isozyme	SNP
Q. robur (3P × A4)	19,113	217,074	54,513	1,794	18,653
Q. robur × *Q. robur* subsp. *slavonica* (EF03 × SL03)	–	21,504	21,120	–	–
Q. robur individual tree	81,671	1,564	569	12	132
Q. petraea (Qs28 × Qs21)	372	–	46,591	–	93
Q. robur × *Q. petraea* (11P × Qs29)	486	–	5,922	–	–
Q. petrae individual tree	8	–	500	–	2
Q. alba individual tree	203,206	–	4,350	–	8,589
Q. rubra individual tree	277,154	–	6,187	–	12,152
C. sativa (Bursa × Hopa)	5,406	78,864	14,780	930	–
C. sativa individual tree	102	848	246	10	–
C. mollissima (Mahogany) × *C. dentata*	–	521	29	8	447
C. mollissima (Mahogany × Nanking) (Vanuxem × Nanking)	–	–	250	–	906
C. mollissima individual tree	847,952	–	24,655	–	41,584
C. dentata individual tree	688,198	–	6,605	–	11,924

Sources: Quercus Map Database, http://w3.pierroton.inra.fr/CartoChene/index.php; Fagaceae Genomics Web, www.fagaceae.org; National Center for Biotechnology Information NBCI, www.ncbi.nlm.nih.gov.
Notes: A dash (–) indicates no data or unavailable data.
[a] SSRs include neutral markers (gSSRs) and EST-derived markers (eSSRs).

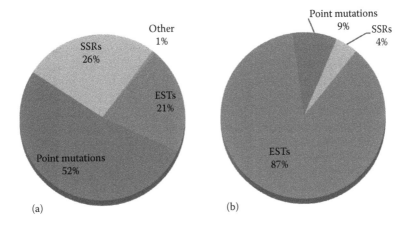

Figure 3.3 Relative proportion of genomic data types available for (a) European oaks (*Q. robur* and *Q. petraea*) and (b) North American oaks (*Q. alba* and *Q. rubra*). "Point mutations" include RAPDs, AFLPs, and SNPs; "other" includes isozymes, SCARs, and 5sRNA markers. (From Fagaceae Genomics Web and Quercus Map Database, 2013.)

and cell wall formation indicated a high coverage of genes involved in those traits. These resources are publically available from the oak contig browser at http://genotoul-contigbrowser.toulouse.inra. fr:9092/Quercus_robur/index.html.

EST catalogs also represent the largest type of sequence in *Q. rubra* and *Q. alba*, two of the most highly valued hardwood species in North America. There are over 277,000 ESTs from *Q. rubra* that have been assembled into a unigene set containing 28,041 contigs. For *Q. alba*, 203,206 ESTs were assembled into 22,102 contigs. GO and KEGG analyses are available for the most current EST assemblies of both species (see Section 5.3 for a description of these analyses). Several thousand SSR and SNP markers are also available for these species.

A complete chloroplast genome sequence is available for 14 *Quercus* species (Kremer et al. 2012). The full chloroplast genome of *Q. robur* was generated using traditional Sanger sequencing of DNA amplified by conserved chloroplast markers. Sequence gaps were filled in by designing internal primers and sequencing across the gaps. Twelve other European white oaks have full chloroplast sequences generated by high-throughput resequencing using the Illumina platform. The complete chloroplast of *Q. rubra* (GenBank accession JX97037) was sequenced using the GS-FLX 454 platform and the pseudoreference mapping method of Cronn et al. (2008), where *de novo* contigs are compared to a reference mapping assembly. Pseudoreference mapping helps alleviate alignment and base calling bias when the reference sequence and reads to be aligned are not from the same species. Read alignments and base calls from the *de novo* contigs are favored so that the new sequence is not artificially conformed to the reference. Because *Q. rubra* is a red oak in the section *Rubrae*, while all other oaks with a sequenced chloroplast are from the white oak section *Quercus*, the pseudoreference mapping method will reveal any differences in chloroplast structure between red and white oaks. All *Quercus* chloroplasts are comprised of between 161,295 and 161,482 bp. Oak chloroplast genomes were annotated using the Dual Organellar GenoMe Annotator package (http://www.dogma.ccbb.utexas.edu, Wyman et al. 2004) and refined manually. As expected, gene content and synteny are highly conserved among *Quercus* species; small structural differences between species or individuals were observed for two genes, *accD* and *petA* (Kremer et al. 2012). Comparison of all Fagaceae chloroplasts (the *Quercus* chloroplasts plus those of *Castanea sativa* and *Fagus sylvatica*) revealed 1252 polymorphic sites, including SSRs, SNPs, and insertion/deletions, which allowed discrimination of all *Quercus* species haplotypes (Kremer et al. 2012).

3.3.2 Mapping Populations

Most oak genetic mapping has taken place in Europe focusing on two major species, *Q. robur* (pedunculate oak) and *Q. petraea* (sessile oak). Genetic mapping in oak began with the goals of detecting genomic regions involved in species differentiation, detecting QTLs controlling traits of adaptive significance, and analyzing Fagaceae genome evolution through comparative mapping. Three mapping populations have been created from full-sibling families of *Q. robur* (pedigree 3P×A4, Barreneche et al. 1998; pedigree A1×A2, Bakker et al. 2001; and pedigree SL03×EF03, Bodenes et al. 2012). Other mapping populations include a full-sibling *Q. petraea* family (QS28×QS21, Kremer et al. 2007) and one interspecific F_1 full-sibling family *Q. robur×Q. petraea* (11P×QS29, Kremer et al. 2007). Mapping populations (except A1×A2) were generated by carrying out controlled crosses according to the techniques described by Zanetto et al. (1996).

Generating pedigreed populations of oak can be difficult, time consuming, and expensive due to the logistics of controlled pollinations and harvest. An alternative to controlled pollination uses molecular markers to reconstruct full-sibling families from a half-sibling family. This method was used in Bakker's (2001) *Q. robur* (A1×A2) map. The maternal tree was located in an urban area surrounded by three other oak trees; 397 acorns were collected in 1998. One nearby tree was selected to be the male parent. Paternity analysis revealed that 26% of the collected seeds were sired by the male parent, leading to a mapping population of 101 full-siblings. The same method was employed to generate a *Q. rubra* mapping population (Romero-Severson et al. 2003). Open-pollinated seeds were collected from a maternal tree in 2001, 2003, and 2007. Microsatellite-based exclusion methods (Aldrich et al. 2002, 2004) were used to identify 97, 438, and 21 full-siblings from the 2001, 2003, and 2007 cohorts, respectively. This offspring population is being genotyped and phenotyped for genetic map construction and discovery of genes responsible for disease (e.g., sudden oak death, *Phytophthera ramorum*) resistance and wood quality.

3.3.3 Linkage Mapping and QTL Analysis

Mapping efforts to date have relied primarily on the *Q. robur* 3P×A4 full-sibling family (hereafter, reference population; Table 3.3). A sample of 94 full-siblings was used to create a linkage map comprised of 307 markers (271 RAPDs, 10 SCARs, 18 SSRs, 1 minisatellite, 6 isozymes, and 1 5s rDNA; Barreneche et al. 1998). A map for the female parent (893.2 cM, 12 LGs) and a map for the male parent (921.7 cM, 12 LGs) were created based on the pseudo-testcross mapping strategy of Grattapaglia and Sederoff (1994). Both maps provided >85% genome coverage and homologies between the female and male maps were identified based on 74 intercross markers segregating in expected Mendelian ratios. In each map, 18% of the markers showed segregation distortion; over 60% of the skewed markers were due to higher than expected heterozygosity in the parents (Barreneche et al. 1998). Saintagne et al. (2004) genotyped an additional 184 full-siblings from the *Q. robur* reference population for a total of 278 individuals mapped. They used 128 markers (84 AFLPs, 34 SSRs, 1 SCAR, and 9 RAPDs) to map regions of the genome involved in species differentiation. Nineteen of the SSRs and all 9 RAPDs were common to both maps. The female and male maps of Saintagne et al. (2004) were comprised of 75 and 92 markers, respectively, covering about 75% of the genome. Casasoli et al. (2006) used EST-derived markers to genotype a subset of 57 individuals from the *Q. robur* reference population. They added 51 EST-derived markers to the framework map, merged data from all previous mapping experiments, and merged the female and male maps. The *Q. robur* consensus map is available from http://www.pierroton.inra.fr/biogeco/genetique/recherches/Oak-map/index.html.

Recently, studies examining ESTs for the presence of SSRs and SNPs have added these markers to the frameworks of previous linkage maps. Durand et al. (2010) assembled a catalog of 103,000

Table 3.3 The Majority of Genetic and QTL Mapping Resources for *Quercus* Are Based on F_1 Pedigrees of Three Species

Map Type	Genetic			QTL			
Species (Pedigree)	Number of Linkage Groups	Total Genetic Distance (% Genome Saturation)	Number of Marker Loci[a]	Trait	Number of QTLs[b]	QTL Range, Mean (cM)	References
Q. robur (3P×A4)	12	950 (80)	295,024	CO₂ response	22	12.3–81.1, 42.7	Barreneche et al. (2004), Saintagne et al. (2004), Scotti-Saintagne et al. (2004a,b), Porth et al. (2005a), Kremer et al. (2007) and Brendel et al. (2008)
				Leaf morphology	12–14[c]	18.0–80.0, 46.0	
				Bud break	12	4.0–71.0, 23.9	
				Osmotic stress	7	0.98–65.9, 13.3	
				Height growth	1–4	–	
				Delta	5	–	
				Senescence	20	–	
Q. robur (A1×A2)	13	659 (64)	361	Height growth	1	–	Bakker (2001) and Kremer et al. (2007)
				Bud break	2	–	
Q. robur × Q. robur subsp. *slavonica* (EF03×SL03)	12	1135/939 (95/78)	42,624	Leaf morphology	6–20	1.0–103.0, 20.0	Gailing (2008)
Q. petraea (Qs28×Qs21)	12	767 (70)	46,684	–	–	–	–
Q. robur × Q. petraea (11P×Qs29)	12	756/584 (69/53)	5,922	Osmotic stress	6	–	Porth et al. (2005a)

Notes: Data reflect the most recent map build in cases of pedigrees with multiple genetic maps.
a Number of marker loci reflects all markers available in the Quercus Map Database for each pedigree.
b Number of QTLs is based on a consensus map if available; otherwise, QTLs from female and male maps were combined.
c A range of numbers indicates QTLs found for multiple metrics of a single trait. A dash (–) indicates no data or unavailable data. References are listed in order of year. Unpublished data were gathered from the Quercus Map Database (http://w3.pierroton.inra.fr/CartoChene/index.php) and the National Center for Biotechnology Information NBCI (www.ncbi.nlm.nih.gov).

ESTs into 28,024 unigenes from which over 5,000 presented one or more SSR motif. Primers were designed for 748 unigenes; 283 primer pairs (37.7%) amplified a single polymorphic locus in the *Q. robur* full-sibling family 3P×A4. A bin mapping approach (Vision et al. 2000) was then used to assign 256 EST-SSRs to the female and male framework maps of Saintagne et al. (2004). Bodenes et al. (2012) advanced the establishment of a dense EST-SSR map by including individuals from all five *Quercus* mapping populations in a single experiment. Populations included three *Q. robur* full-sibling families, one *Q. petraea* full-sibling family, and a full-sibling family from an interspecific cross of *Q. robur*×*Q. petraea*. They added 397 EST-SSRs to the *Q. robur* map and generated a *Q. petraea* consensus map. These represent the first gene-based maps for *Quercus*, where all markers were derived from coding regions. Because expressed genes tend to be conserved among related species, gene-based maps will be useful in comparative mapping and dissection of QTLs affecting complex traits of ecological importance (Derory et al. 2010). These maps are available through the Quercus Portal at https://w3.pierroton.inra.fr/QuercusPortal/.

QTLs have been detected in *Q. robur* for bud break phenology and height growth through linkage and comparative mapping experiments. A sample of 278 full-siblings from the *Q. robur* reference population were vegetatively propagated and established in three field tests. Over the three tests, 32 independently detected QTLs representing at least 12 unique chromosomal regions explained 3%–11% of the variance in bud break among clones. Nine QTLs were detected in at least two field sites and five QTLs were detected at all three sites. Two to four QTLs explained 4%–19% of the variance in height growth among clones (Scotti-Saintagne et al. 2004a). Their position on the map indicated that one QTL was common to all four growth traits and that all growth traits shared some common QTLs. In a comparative QTL study, the 32 independently observed QTL for bud break were collapsed into 19 unique QTL, 9 of which were colocalized in *C. sativa* (Casasoli et al. 2006). Two QTL for height growth collocated in *Q. robur* and *C. sativa*. Many of the EST-based markers used in the study were developed from cDNA libraries from bud tissues at different developmental stages. Thus, a QTL for bud break appearing over time, space, and between species combined with its mapping to the same region as an EST developed from a bud break cDNA library is a strong indicator that the QTL is real and the region is a good candidate gene for further investigation of that trait. Examples of ESTs located within conserved QTLs for bud break include *cons58* (auxin-repressed protein) and *08C11* (metal nicotianamine transporter; Casasoli et al. 2006). Derory et al. (2010) identified 19 bud break QTLs in the *Q. robur* reference population, which contributed to >12% of the variation in bud break. These 19 QTL were similar to those of Scotti-Saintagne et al. (2004a) and Casasoli et al. (2006), except standard deviations of QTL positions were greatly reduced due to multiyear and multisite observations.

QTLs for species differentiation and osmotic stress have also been detected in oak. Saintagne et al. (2004) detected one to three QTL for each of 13 leaf morphological traits in the *Q. robur* reference population. QTLs controlling the five most morphologically divergent traits were clustered on six LGs with two clusters containing QTLs for at least two traits. Porth et al. (2005) investigated osmotic stress in the *Q. robur* reference population and the *Q. robur*×*Q. petraea* (11P×QS29) interspecific cross. Seven and six QTLs were detected and mapped in the reference population and interspecific cross, respectively, for a total of thirteen unique QTLs. Two of the 13 QTLs collocated on the *C. sativa* linkage map (Casasoli et al. 2006).

3.3.4 BAC Libraries and Physical Mapping

Bacterial artificial chromosomes (BACs) are DNA constructs based on a plasmid molecule naturally occurring in some bacteria (Shizuya et al. 1992). The circular molecule, known as the F′ plasmid, is able to accept DNA inserts up to 200 kb—about 20 times longer than DNA inserts accepted by traditional phage plasmids. Genomic (or other target) DNA is cut with restriction enzymes and each piece of sheared DNA is ligated to an F′ plasmid. The plasmid is taken up by *Escherichia coli* via electroporation and the *E. coli* is transferred to media favorable for its growth. Each bacterium containing a unique piece of DNA—called a clone—replicates many thousands of times so that each unique piece of DNA is represented in many copies. All clones generated from the same genomic DNA in the same experiment are collectively called a library. It is possible for a BAC library with a large enough number of clones to contain all the DNA of an organism's genome. With the recently introduced strategy of combining BAC and Sanger sequences with next-generation read data (Rampant et al. 2011), sequencing the oak genome is now well within reach.

One of the key goals in oak genomics is to uncover genes controlling traits of adaptive variation for forest conservation and management and to provide resources to tree improvement programs. Deep-coverage, large-insert genomic libraries are important tools for the marker development, physical mapping, and genome sequencing necessary for attaining that goal. To that end, Rampant et al. (2011) reported the generation of an *Eco*R1-based BAC library of *Q. robur* genomic DNA. The female parent (3P) of the *Q. robur* reference population was used as the source of genomic DNA. The library consisted of 92,160 clones, of which 7% had no DNA insert. Mean insert length was estimated at 135 kb and <3% and <1% contamination was reported from chloroplast and mitochondrial DNA, respectively. Based on mean insert size, number of clones, and contamination frequency, they estimated 12× coverage of the *Q. robur* haploid genome. The probability of recovering any sequence of interest from the library was estimated at more than 99%. A second BAC library was constructed for the same *Q. robur* genotype using *Hind*III as the restriction enzyme. Both BAC libraries are available from The French Plant Genomic Resource Center (CNRGV) at http://cnrgv.toulouse.inra.fr/ and The Platform for Integrated Clone Management (PICME) at http://www.picme.at/.

Approximately 15,000 clones from the *Q. robur Eco*R1 BAC library were sequenced from both ends. After quality trimming of reads, 20,056 BAC ends comprising 12,018,238 bp were retained for further analysis (GenBank: HN154083–HN174138). Protein-coding regions were represented by 2712 (13.5%) of the BAC end sequences. A total of 1823 BAC end sequences significantly aligned with at least one *A. thaliana* sequence in the protein database Swissprot. Of these, 799 were assigned a functional category. Based on the number of BAC end sequences aligning to *A. thaliana* in the Swissprot database (1,823), the mean sequence length of each BAC end (599 kb), the estimated size of the oak genome (740 Mb), the mean size of a gene (about 2 kb), and previous bioinformatics analysis of the oak unigene set (Ueno et al. 2010), the authors estimated the *Q. robur* genome to contain at least 32,467 genes. Of the approximately 20,000 BAC end sequences generated, only 176 and 81 aligned to the grape and poplar genome sequence using a MEGABLAST alignment. Thus, the sequences generated in this experiment may be unique, species-specific genes with the potential to enhance basic understanding of the organization and function of plant genomes. The large amount of well-covered sequence makes this BAC library an attractive resource for *Q. robur* physical mapping, map-based cloning, marker development, and genome sequencing (Rampant et al. 2011).

3.3.5 Case Studies

Large-scale genomics projects produce wide arrays of data for use in diverse applications. In the next two sections, we present case studies in specific areas of active oak research that highlight the use of genomic data to solve grand challenges in oak ecology, evolutionary biology, and conservation. The first case study focuses on the development of association genetics methods to study the architecture of bud break, an important adaptive trait in forest trees. The second case study details the genomic tools available for, and current progress in, population differentiation and management of northern red oak.

3.3.5.1 *Adaptive Trait Variation: From Linkage Mapping To Association Genetics*

Phenological traits—those related to the timing of growth, flowering, and senescence—are important in natural forest tree populations as drivers of population spatial distribution and abundance. Chuine et al. (2000) found that accurate spatial distribution models could be generated using a mechanistic approach driven by spring phenology for five of the six European tree species they studied. As the date of bud break is largely driven by temperature (winter chilling requirements and/or spring heat sum) in temperate species (Chuine and Cour 1999), warming climate patterns are expected to have a major impact on the phenology of trees. Climate-based species distribution models predict drastic redistributions of tree species in the next century in response to changing climate (Malcolm et al. 2002), yet few studies have considered the biological and genetic ability of species to do so. Predicting the fate of forest tree populations in a changing climate requires the integration of knowledge across biological scales from individual genes and genomes to ecosystems and across temporal scales from annual phenology traits to glacial and interglacial periods (Aitken et al. 2008). Bud break timing is also important to production plant breeders, as growth initiation and cessation are directly related to the amount of biomass a tree can produce and its relationship to insect and fungal pests. Thus, genomic tools and information resulting from the study of bud break in natural populations can be directly applied to populations in the form of markers for characterization and selection and guidelines for diversity maintenance and deployment zones for certain genotypes.

Aitken et al. (2008) suggested that studies in the phenotypic, neutral marker, and nucleotide variation of genes involved in adaptation to climate in areas where populations span steep climatic gradients may yield considerable knowledge about genes and traits involved in rapid adaptation to climate. Indeed, the strongest patterns of local adaptation in temperate and boreal forest trees reflect the critical synchronization of the annual growth and dormancy cycle (phenology) of populations with their local seasonal temperature regimes (Savolainen et al. 2007). Many studies of phenological traits have shown population differentiation in bud break or bud set dates across latitudinal and altitudinal gradients (reviewed in Morgenstern 1996). In several studies, the heritability of bud phenology was higher than that of any other measured trait (Bradshaw and Stettler 1995). Local adaptation, inferred from phenotypic associations to steep environmental gradients and often seen in forest tree populations despite high gene flow, has been convincingly shown to be a major source of variation in phenological traits for many forest tree species such as *Q. rubra* (McGee 1974), *Q. petraea* (Kremer et al. 2009), *P. tremula* (Ingvarsson et al. 2006), *Pinus taeda* (Brown et al. 2001), and *Pseudosteuga menziezii* (Krutovsky and Neale 2005). Indeed, *Q. petraea* accessions from southern and maritime sources were injured during a spring frost event in an inland common garden in France (Ducousso et al. 1996). Conversely, northern and high-altitude accessions began growth later and ended growth sooner than other *Q. petraea* accessions in the same French common garden.

The lack of LD in natural forest tree populations coupled with the difficulty of generating large, pedigreed populations precludes the efficiency of QTL analysis and genome-wide scans for detecting key genes controlling traits. Thus, association mapping using candidate genes appears to be the most promising method for discovering genes underlying adaptive traits in forest trees (Krutovsky and Neale 2005; Neale and Savolainen 2004). *Q. rubra* and *Q. petraea* are convenient study organisms for association genomics because they have abundant genetic and phenotypic variation, are open-pollinated, and show low to moderate LD (making it more probable that a marker associated with a phenotype is linked to or within a gene controlling a complex trait), and the existing infrastructure of American and European tree improvement programs provides replicated genetic tests across multiple environments that facilitate the evaluation of complex trait variation (Krutovsky 2006; Neale and Savolainen 2004).

A candidate gene is a gene believed to have an important functional role in a trait based on indirect evidence from expression studies, a putative role in a different species, or both. Candidate genes for bud burst in *Q. petraea* have been identified by QTL analysis (positional candidates), by analysis of transcriptomes from nascent spring buds (expression candidates), and from orthologous genes in *Arabidopsis* (Derory et al. 2010). A total of nine candidate genes have been examined in *Q. petraea*, with genes from expression analysis that collocated with a bud break QTL being of particular importance. SNPs are potentially the best type of genetic marker for candidate gene–based association genetics because of their abundance in the genome and their potential association with disease and adaptive traits (Gonzalez-Martinez et al. 2006). SNPs have been reported in candidate genes controlling adaptive traits in *Eucalyptus* (Thumma et al. 2005), Douglas fir (Neale and Ingvarsson 2008), *Populus* (Ingvarsson et al. 2006), *Quercus* (Derory et al. 2010), and *Pinus* (Brown et al. 2004; Gonzalez-Martinez et al. 2006). Hall et al. (2007) confirmed the presence of three SNPs within a phenology candidate gene in *P. tremula* that showed latitudinal clines similar in frequency to the observed latitudinal cline of phenology differentiation. Derory et al. (2010) found no associations between 125 SNPs within 9 candidate genes and phenological or geographical populations of *Q. petraea*. However, five of the candidate genes collocated with QTLs, and four of them were located within the three QTLs explaining the largest the amounts of variance in bud break. This contrast illustrates the need for experiments using larger numbers of individuals and SNPs to increase the probability of revealing associations between SNP frequencies and phenotypes.

The large number of individuals and SNPs required to identify and validate a causal variant within a candidate gene requires a genotyping platform capable of producing multilocus genotypes in a large panel of individuals. The next-generation Illumina BeadArray combined with GoldenGate assay (http://www.illumina.com) is capable of genotyping up to 1536 polymorphic sites in 384 individuals in a single reaction using allele-specific oligonucleotides (ASOs) to discriminate between the allelic states at a SNP locus (Fan et al. 2003; Oliphant et al. 2002). Throughput can be high (500 samples/day for all SNPs) and cost is relatively low (approximately US$0.10 per data point; Neale 2007). Comparing Illumina genotype data with traditional Sanger sequencing for 56 SNPs revealed 99% and 100% genotyping accuracy rates in two lines of wheat (Akhunov et al. 2009). The Illumina platform is expected to generate a large number of SNPs in the amplicons assayed as the average frequency of SNPs in the conifer genome is on the order of 1 in 50 nucleotides (Neale 2007), and Derory et al. (2010) estimated an insertion/deletion mutation or SNP every 31 nucleotides in *Q. petraea*.

It is becoming increasingly apparent that epigenetics may play an important and possibly pivotal role in the adaptive response of forest trees (Aitken et al. 2008). Tree populations are often highly differentiated for quantitative traits relating to the timing of growth and dormancy, while the same populations show no to weak differentiation among populations for selectively neutral genetic markers. Derory et al. (2010) found that the mean population differentiation within candidate genes of *Q. petraea* was at the same level as differentiation of neutral markers but far less than differentiation of the phenological trait of interest (bud break). In a seminal publication, LeCorre and Kremer (2003) showed that phenotypic differentiation between populations (Q_{ST}, Spitze 1993) was driven partly by variation in allelic effects at loci controlling a trait and partly by covariances among allelic effects at the different loci. The former is dependent on the population differentiation at the loci controlling the trait (F_{ST}, Wright 1951), whereas the latter is generated by allelic associations between loci. Thus, lack of genetic variation within candidate genes may mean that allelic associations rather than a specific gene variant (haplotype) are responsible for phenotypic differentiation and local adaptation.

Recent resources aimed at understanding bud break phenology in oak include over 145,000 Sanger ESTs from 20 cDNA libraries made from bud (collected at various phenological stages), leaf, differentiating xylem, and root tissues. Also generated was 2M of 454 reads from candidate genes of early and late flushing populations of *Q. petraea*. Analysis of these data is aimed at the discovery of genes differentially expressed between early and late flushing populations. A combined Sanger-454 unigene set is under construction that will include a detailed catalog of the oak transcriptome and provide a source of SNP and SSR markers for linkage and association mapping (Kremer et al. 2012). These resources, available through the Quercus Portal (https://w3.pierroton.inra.fr/QuercusPortal/) offer a powerful tool set for investigating the genomic architecture of important adaptive traits in *Quercus* and other forest tree species.

3.3.5.2 *Population Differentiation in Northern Red Oak: Chloroplast Genomes*

Chloroplast intergenic markers have been used extensively to examine genetic differentiation of European white oak populations (*Quercus* section *Quercus*; Petit et al. 2002), to map the northward colonization of white oaks in Europe following the last glacial maximum approximately 21,000–18,000 years ago (Kremer et al. 2012; Petit et al. 2001, 2004), and to elucidate the invasion-by-hybridization model of gene flow in the *Q. robur–Q. petraea* forests of southwestern Europe (Petit 2004). However, compared to other oak species, the neutral-marker population differentiation (G_{ST}) and chloroplast haplotype diversity of northern red oak is remarkably low (Feng et al. 2008; Hokanson et al. 1993; Magni et al. 2005; Sork et al. 1993). White oaks from the eastern United States had a mean chloroplast G_{ST} value of 0.87 ± 0.07 (Whittemore and Schaal 1991), while six species of European white oaks exhibited coefficients of chloroplast differentiation ranging from 0.78 in *Q. robur* to 0.96 in *Q. pyrenaica* (Petit et al. 2002). Magni et al. (2005) found the value of cpDNA differentiation in *Q. rubra* ($G_{ST} = 0.46$) was lower than the mean value of 22 European forest trees and shrubs ($G_{ST} = 0.54$, Petit et al. 2005) and the overall mean of angiosperms ($G_{ST} = 0.76$, Petit et al. 2003), though Tovar-Sanchez et al. (2008) reported an R_{ST} (similar to G_{ST}) value of 0.398 for *Quercus crassipes* in an introgression zone.

The strikingly low differentiation of *Q. rubra* chloroplasts among populations and low haplotype diversity has been attributed to short postglacial migration routes and a particularly high seed-mediated gene flow (Birchenko et al. 2009; Magni et al. 2005). Whatever the historical factors causing the phenomena, the low neutral marker diversity and conservative nature of the chloroplast genome evolution often necessitates large amounts of cpDNA sequence in order to detect statistically robust population differentiation (Small et al. 1998). Large amounts of sequence have traditionally been generated by cloning and Sanger sequencing or primer walking approaches (Jansen et al. 2005; Petit and Vendramin 2006), but these approaches can be laborious and time-consuming. Recently introduced next-generation sequencing technologies provide the possibility of acquiring entire genomes—or many simple genomes—at a fraction of the cost and time of traditional approaches. Multiplex tagging methods have the potential to spread the capacity of high-capacity sequencers across many genomes and strike an acceptable balance between coverage, throughput, and cost (Cronn et al. 2008). Second-generation sequencing has been proven useful in sequencing angiosperm chloroplast genomes (Moore et al. 2006, 2007) and has facilitated the Angiosperm Tree of Life (AToL) project. As a consequence, there has been a rapid increase in the number of chloroplast genomes added to public databases in recent years.

Given the low intraspecific chloroplast diversity detected in northern red oak, more powerful tools are necessary to more accurately characterize *Q. rubra* chloroplast diversity. A fully sequenced and annotated chloroplast genome containing locations of interspecific and intraspecific polymorphisms is essential for studying population differentiation, phylogeography, and evolutionary history of this species. Tools for detecting intraspecific variation are also needed for management goals such as monitoring reintroduced populations, tracking wood products, and certifying seed lots and forests (Deguilloux et al. 2003). To this end, we report the sequencing, assembly, and annotation of the chloroplast genome of northern red oak. The *Q. rubra* chloroplast genome (GenBank accession JX97037) contains 161,304 bp, which is at the small end of the range of oak chloroplast sizes (Kremer et al. 2012). The 138 chloroplast genes code for 32 tRNA genes, 4 rRNA genes, and 82 protein-coding genes, a total of 118 proteins, higher than the average of 90 for green plants (Ravi et al. 2008) and 113 for *Nandina* and American sycamore (Moore et al. 2006). SNPs were detected using standard and high-quality (HQ) settings and each SNP and DIP was visually examined for accuracy. A total of 737 (99 HQ) SNPs and 214 DIPs were detected between *Q. robur* and *Q. rubra*; 23 (5 HQ) SNPs and 163 DIPs were detected between *Q. nigra* and *Q. rubra*; and 8 (6 HQ) SNP and 45 DIPs were detected within the genomes of the 4 sequenced *Q. rubra* individuals. All polymorphisms were in the LSC and SSC; no polymorphisms were found in the inverted repeat regions. Of the 23 SNPs detected between *Q. nigra* and *Q. rubra*, 9 were located within genes. Of the 6 HQ SNPs detected within *Q. rubra*, 3 and 3 were located in coding and noncoding regions respectively; 10 of 45 DIPs were found in coding regions. The *Q. rubra* chloroplast genes *rpoC2* and *ycf1* showed the most variability. The availability of this chloroplast sequence with known polymorphic sites should provide the resolution necessary to accurately characterize *Q. rubra* chloroplast diversity.

3.3.6 Summary

Genomics resources for hardwood trees are growing, but remain underdeveloped. The majority of genomic resources are within the genera *Populus*, *Pinus*, and *Eucalyptus*, although *Quercus* genomic resources now include over 1 million ESTs assembled into 2 separate unigene sets, over 1500 SSRs, and 16 genetic or QTL map sets on 5 intra- and interspecific pedigrees (Table 3.3). Development of deeper EST databases for *Quercus* and related species, more densely populating genetic maps with SNP markers, identifying SNPs that explain variation in complex traits and monitoring their frequencies in a larger population in order to assess intergenic allelic associations among populations, and the comparison of these data between species will help to account for the overall population differentiation among candidate genes. Identifying the genes responsible for adaptive trait variation, surveying their frequencies and distribution in natural populations, and applying those results to forest management practices are essential for conserving oak species and ecosystems under serious threat of rapid climate change.

3.4 GENOMICS OF DISEASE RESISTANCE IN *CASTANEA*

While most hardwood tree improvement programs are in the first stages of germplasm collection and breeding, the largest American chestnut improvement programs have reached the sixth generation of selection and breeding for chestnut blight resistance. The American chestnut, *Castanea dentata*, is expected to be reintroduced and managed in its former forest habitat after the incorporation of sufficient chestnut blight resistance. Breeders face unusual challenges when seeking to reintroduce populations to their former habitats or to supplement wild populations that may no longer be viable. Conservation breeders may use traditional breeding methods, but unlike crop breeders, they must release the products of their program into environments marked by dynamic change and nonuniformity. In effect, conservation breeders are seeking to reverse the typical work flow of breeding. Crop breeders typically incorporate wild germplasm into cultivated or domesticated lines and seek to improve the predictability of yield by manipulating the genetics and culture of the crop. American chestnut breeders introgress genes from a cultivated species into a wild relative with the hope that their selections will thrive in unmanaged or minimally managed stands across its entire former native range.

The following sections examine how outcomes of historical and current American chestnut breeding may affect reintroduced chestnut populations. What levels of genetic diversity and family structure will characterize reintroduced populations? How will reintroduced populations survive, mate, and spread? What will be the fate of native American chestnut genes? Genomics tools and resources will be essential for managing breeding populations, identifying and characterizing genes important for local adaptation and fitness, assessing the stability of chestnut blight resistance in reintroduced populations, and monitoring variation in the success and spread of hybrid chestnut families (Worthen et al. 2010a).

3.4.1 Chestnut Population Genetics

The genus *Castanea* is comprised of three sections and seven species. Four species, *Castanea mollissima* Blume (Chinese chestnut), *C. seguinii* Dode. (Seguin chestnut), *C. crenata* Sieb. and Zucc. (Japanese chestnut), and *C. henryi* Skan (Chinese chinquapin) occur in eastern Asia. The European chestnut, *C. sativa* Mill., is native to western Asia and southern Europe, and *C. dentata* (Marsh.) Borkh. (American chestnut) and *C. pumila* Mill. (American chinquapin, often divided into subspecies Ozark and Alleghany chinquapin) occur in eastern North America.

Genetic variation, population structure, and phylogenetic relationships within and among *Castanea* species have been analyzed using a variety of marker systems (Dane et al. 1999, 2003; Han et al. 2007; Huang et al. 1994, 1998; Kubisiak and Roberds 2003; Lang et al. 2006, 2007; Lang and Huang 1999; Tanaka et al. 2005; Villani et al. 1991; Wang et al. 2006). Sequence-based phylogenetic analysis of the variable *trn*T-L-F region of *Castanea* cpDNA showed *C. crenata* is the most basal species and placed the Chinese species in a monophyletic clade with the North American and European species as a sister group (Lang et al. 2006). In general, *Castanea* species show high levels of genetic diversity at neutral loci and little population structure, and they partition most genetic diversity within populations—population genetic traits shared by many other long-lived, outcrossing forest trees (Petit and Hampe 2006). In North America, *Castanea* species were driven southward to one or two southern glacial refugia during the Wisconsin glacial maximum 18,000–20,000 years ago. *C. dentata* greatly expanded its range after the glacial retreat about 10,000 years ago, spreading northward along the Appalachian ridge (Delcourt 2002; Delcourt and Delcourt 1984; Huang et al. 1998) and establishing itself as a dominant species in 800,000 km² of eastern forest (Braun 1950). *C. dentata*, like all *Castanea* species, is diploid ($2n=2x=24$) and will easily hybridize with its congeners, although offspring from some combinations suffer from low vigor or male sterility (Jaynes 1975).

The preblight population genetics of the American chestnut must be inferred from knowledge of postblight American chestnut populations and other forest trees with similar life history characteristics. Analyses using allozymes have shown that *C. mollissima* has the highest population mean genetic variability of all the members of the genus (total genetic variability (H_T)=0.321, H_{exp}=0.311; Lang and Huang 1999) and *C. dentata* has the lowest mean variability at isozyme loci (H_T=0.214, H_{exp}=0.167; Huang et al. 1998). Pigliucci et al. (1991) reported an intermediate mean expected heterozygosity for *C. sativa* (H_{exp}=0.24–0.27). Kubisiak and Roberds (2003) employed 6 microsatellite and 19 RAPD loci to examine 17 populations of root crown collar sprouts of *C. dentata*. They used Nei's (1975) gene diversity (*h*) and calculated a mean genetic diversity across RAPD loci (*h*=0.226) that, barring differences in marker systems, is similar to the diversity for American chestnut reported by Huang et al. (1998). Both reports are well within genetic diversity limits of similar forest tree species (Hamrick et al. 1992). Higher than average heterozygosities have been found in postblight American chestnut populations, leading some to speculate on the role of heterozygote advantage in the species' remarkable growth (Stilwell et al. 2003).

The spatial genetic structure characteristic of wind-pollinated, self-incompatible forest tree species is most often negligible or weakly present at only very close distances (Berg and Hamrick 1995; Streiff et al. 1998). Pierson et al. (2007) used minisatellite band sharing to calculate the differentiation between sampling plots in a large woodland of naturalized American chestnut 600 km west of the species' native range. They determined that no differentiation among sampling plots had occurred ($F_{ST}<0.031$), and that gene flow was not restricted within the woodland. Kubisiak and Roberds (2003) reported a similar differentiation value ($G_{ST}=0.036$) and concluded that preblight populations of *C. dentata* followed a pattern consistent with the hypothesis of a single metapopulation where genetic drift was a dominant force in structuring populations. Conversely, Huang et al. (1998) reported a threefold higher measure of genetic differentiation among American chestnut populations ($G_{ST}=0.110$) and determined that four large metapopulations existed within the range of American chestnut. Both the study of Huang et al. (1998) and that of Kubisiak and Roberds (2003) reported a cline in allele frequencies along an axis from the southwestern to the northeastern limits of the species' natural range. Differences in their findings may have been the result of differences in sampling strategies. Kubisiak and Roberds (2003) excluded individuals with a *C. pumila* chloroplast type, whereas Huang et al. (1998) did not attempt to exclude cryptic hybridization and thus may have inflated between-population differences. Also, Kubisiak and Roberds (2003) sampled continuously down the range of *C. dentata*, whereas Huang and colleagues sampled fewer and more outlying populations. In *C. sativa* and *C. dentata*, allele frequencies showed differentiation over altitudinal and edaphic clines, suggesting the possibility of local adaptation across the wide range of the two species (Kubisiak and Roberds 2003; Villani et al. 1991).

3.4.2 Chestnut Blight Disease

Chestnut blight disease is incited by the fungus *Cryphonectria parasitica* (Murr.) Barr. The fungus causes a girdling canker on stems and branches of American chestnut that is lethal to the trees (Roane et al. 1986). It was introduced to North America on Japanese chestnut trees (Anagnostakis 1992; Cunningham 1984) and was discovered at the New York Zoological Gardens in 1904 (Merkel 1906). By 1950, the disease had spread across the entire range of the American chestnut, eliminating it as an overstory tree in eastern forests (Newhouse 1990). An estimated 4 billion trees were killed by chestnut blight disease, resulting in significant changes in forest communities and communities of people that depended on the tree for food, timber, and livestock feed (Russell 1987). Blight-killed trees often sprout from the root crown collar. It has been shown that root systems are weakened by the disease and produce fewer sprouts with each cycle of resprouting and dying (Griffin 2000). The prolific root crown collar sprouting of American chestnut provides a continuous supply of susceptible host tissue, a continuing reservoir of fungal inoculum in the forest, and a source of genetic diversity for chestnut conservation.

Chestnut blight is generally not lethal to the Asiatic species of chestnut, although a range of resistance among *C. mollissima* cultivars has been reported (Huang et al. 1996), and *C. sativa* and *C. pumila* will both contract and succumb to chestnut blight infection (Jaynes 1975). American chestnuts are generally accepted to be fully susceptible to chestnut blight fungus, although the presence of trees within the native range that have survived blight, and evidence that grafts from blight survivors show relatively high levels of resistance when inoculated with *C. parasitica*, have led some to hypothesize some level of blight resistance in American chestnuts (Griffin et al. 1983; Jaynes 1975). Hybrids between susceptible American chestnuts and resistant chestnut species have intermediate levels of resistance (Clapper 1952).

In Europe, *C. sativa* suffered severe damage in the 1930s as the result of infection by the chestnut blight fungus, but populations have regenerated in recent decades. This recovery was due to the emergence and spread of naturally occurring hypovirulent strains of *C. parasitica*, along with the deployment to chestnut growers of hypovirulent strains propagated *in vitro* (Grente and Berthelay-Sauret 1978; Grente and Sauret 1969). Hypovirulence is widely used in Europe to reduce mortality of *C. sativa*; however, hypovirulent strains of *C. parasitica* have not shown any degree of success at the population level in the recovery of American chestnut (Milgroom and Cortesi 2004), probably because there are more mycelial compatibility groups in North America than in Europe (Anagnostakis et al. 1986; Double and Macdonald 2002).

3.4.3 Genomic Tools for Blight Resistance Breeding and Improvement

Early breeding results demonstrated that F_1 American × Chinese hybrids of above-average resistance backcrossed to Chinese chestnut produced a high (>75%) percentage of resistant progeny, leading to the conclusion that inheritance of blight resistance may be under oligogenic control, perhaps conditioned by as few as two partially dominant genes (Clapper 1952). Standard deviations of canker size in F_2 populations were compatible with models for one or two incompletely dominant genes controlling blight resistance, based on Wright's method for estimating the number of factors controlling a segregating trait (Falconer 1960; Hebard 2006). Linkage mapping and breeding efforts have been aimed at discovering the areas in the Chinese chestnut genome that confer resistance to chestnut blight fungus. In 2006, a multidisciplinary NSF-funded project called Genomic Tool Development for the Fagaceae was initiated with the goal of developing sequence and mapping resources for chestnut, oak, and beech (www.fagaceae.org). Chinese chestnut was selected for whole-genome sequencing and analysis as the model Fagaceae species because it is resistant to the chestnut blight fungus. Genomic analysis of chestnut is expected to provide the location and annotation of genes, allow for complex trait dissection (e.g., for blight resistance, wood properties, and bud break), and provide a base resource for genome-wide association studies (Wheeler and Sederoff 2009). Used together, gene and genome sequences, genetic and physical maps, and *in situ* phenotype-to-genotype experiments represent a powerful resource for American chestnut improvement (Table 3.2, Figure 3.4). They will allow researchers to identify and potentially clone specific blight resistance genes in breeding programs, making it possible to eliminate blight-susceptible progeny in the seedling stage. Genotyping young seedlings would save large amounts of time, effort, and money traditionally spent on growing trees to the appropriate age for disease-resistance screening. Further, the genome of the chestnut blight fungus, *C. parasitica*, is now available from the Joint Genome Institute (http://genome.jgi-psf.org/euk_home.html). Studies of host and parasite genome interaction will shed even more light on pathways involved in blight resistance and candidate genes in the response to resistance that will allow pyramiding of multiple resistance factors into reintroduced American chestnut hybrids.

3.4.3.1 Sequence Data

The *C. mollissima* cultivar Vanuxem was chosen as the reference genotype for the family Fagaceae due to its high level of chestnut blight resistance and key role in the American Chestnut Foundation's breeding program. Whole-genome sequencing is being completed by the Forest Health Initiative Genome Resources and Tools Project (www.foresthealthinitiative.org). Libraries of genomic DNA from healthy Vanuxem stem tissues were prepared for sequencing, resulting in 55.5 Gb of data. The ends of over 43,000 BAC clones were also sequenced to help guide assembly.

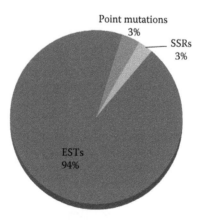

Figure 3.4 Relative proportion of genomic data types available for chestnut (*C. sativa*, *C. dentata*, and *C. mollissima*). "Point mutations" include RAPDs, AFLPs, and SNPs. (Data from Fagaceae Genomics Web, 2013.)

The genomic DNA libraries and BAC libraries (generated for physical mapping and described below) used for *C. mollissima* genome sequencing can be downloaded from the Fagaceae Genomics Web (http://www.fagaceae.org/libraries). The reference genome is currently being assembled *de novo* from 454 sequence data. The most recent assembly placed 925 Mb into 1,147,939 contigs with a relatively short mean read size of 1,369 bp. The BAC end data provided a loose framework for contig assembly, and contigs were merged into 51,766 scaffolds covering over 500 Mb of sequence.

The current assembly appears to be a good representation of the genes found in Chinese chestnut. One hundred percent of Chinese chestnut ESTs aligned to the genome scaffolds with >97% sequence identity. The largest gene in *Arabidopsis*, an ATPase over 43 kb in length, was found intact in one scaffold. The set of predicted genes contained >550 genes related to disease resistance. However, the current assembly remains too fragmented to serve as a reference sequence for other Fagaceae species. Orienting all 51,766 scaffolds is not possible with the relatively small number of markers (1,156) on the genetic map. The next steps include correcting and extending the assembly using Illumina sequence data, adding markers to the genetic map, and building pseudochromosomes by integrating the genome sequence with the genetic and physical map. A complete genome sequence of the *C. mollissima* chloroplast is available from NCBI (GenBank accession HQ336404, Jansen et al. 2010).

The three genomic regions (QTL) coferring resistance to chestnut blight have been completely sequenced in Chinese chestnut. This was possible because the three QTL were completely spanned in the physical map of Chinese chestnut by four large contigs of BAC clones. Each QTL was sequenced as a pool of between 40 and 97 BAC clones. The QTL on LG B assembled into 6.6 Mb with 994 genes, the LG F QTL contains 3.99 Mb and 548 genes, and the LG "G" QTL includes 2.94 Mb representing 410 genes. Candidate genes are currently the target of functional studies by the Forest Health Institute. The sequence of each QTL, complete with annotated features, can be viewed and downloaded through the Fagaceae Genomics Web CMAP application (www.fagaceae.org/genetic_maps).

Several cDNA libraries for both *C. mollissima* and *C. dentata* have been generated for large-scale transcriptomic studies aimed at discovering genes for chestnut blight resistance (Barakat et al. 2009). RNA was isolated from various tissues of *C. mollissima* infected with *C. parasitica*, *C. mollissima* noninfected, *C. dentata* infected with *C. parasitica*, and *C. dentata* noninfected. The tissues sampled varied from species to species. For infected and noninfected samples from Chinese and American chestnut, living tissue from the region at the potential or actual infection site was collected and immediately frozen for shipment to a laboratory. For all species, other tissues such as flowers, leaves, buds, xylem, phloem, seeds, and roots were quick-frozen for RNA extraction. cDNA sequences were generated in two ways: by pyrosequencing on Roche Biosystems FLX® Genome Sequencer and by conventional Sanger sequencing of 5′ directionally cloned cDNAs and 3′ poly-T primed sequences (Barakat et al. 2009).

Barakat et al. (2009) sequenced the transcriptomes of *C. mollissima* and *C. dentata* with the objective of comparing the transcriptomes for genes differentially expressed between the two species. Differentially expressed genes may have a role in conferring blight resistance to *C. mollissima* and genes identified through expression studies are important corroborations for QTL identified during mapping. In total, for all libraries, they generated 317,842 and 856,618 sequence reads from American and Chinese chestnut, respectively. Two unigene sets (i.e., EST assemblies) are available for *C. dentata*. The first (Version 1 or v1) contains 28,890 contigs assembled by GS-FLX assembly software, and the second (v2) contains 34,800 contigs assembled using SeqManPro. Both unigene sets are based on sequences from the same five cDNA libraries. Two unigene sets for *C. mollissima* also represent two different assemblies of sequence from the same library; the second assembly version contains over 838,000 reads assembled into 48,501 contigs. A third *C. mollissima* unigene set (CCall) contains sequences generated from 14 cDNA libraries representing various tissue types and disease status. 838,472 GS-FLX reads and 9,480 Sanger ESTs were assembled into 48,355 contigs. GO analysis showed that, overall, canker tissue from both species presented a similar transcriptome. Gene function categories associated with metabolic processes, especially those associated with resistance to biotic and abiotic stresses, were highly represented in both transcriptomes. A small number of genes with known roles in plant pathogen response were found to be preferentially expressed in Chinese chestnut canker tissue (Barakat et al. 2009). Roughly 50% of reads could be annotated using either the *Arabidopsis* or *Populus* proteome. The half that could not be annotated may correspond to potential chestnut-specific genes (Barakat et al. 2009). Combined with physical mapping data indicating low levels of macrosynteny between the chestnut and poplar genomes (Fang et al. 2012), the moderate level of proteome similarity between chestnut and poplar reinforces the complexity of forest tree genomes and the need for species-specific genomic data. The genome and transcriptome sequence of *C. mollissima* will undoubtedly be a valuable tool for investigating plant functional and comparative genomics and genome evolution. Detailed descriptions of cDNA libraries, raw cDNA sequence data, and assembled contigs are available to download from the Fagaceae Genomics Web (http://www.fagaceae.org/sequences).

3.4.3.2 Mapping Populations

A European chestnut mapping population was generated through controlled crosses between individual trees from opposite sides of the Anatolian peninsula, an ecologically diverse area known to be rich in chestnut genetic diversity. The female parent Bursa was from a drought-adapted population in western Turkey and the male parent Hopa was from a wetter climate in eastern Turkey. A total of 96 F_1 full-sibs were used to construct a female and male linkage map with the goal of identifying genomic regions affecting WUE, bud phenology, and growth rate (Villani et al. 1999b).

Mapping efforts using Chinese and American chestnuts have been largely aimed at discovering genomic areas conferring blight resistance. A mapping population of F_2 progeny was constructed by crossing two *C. mollissima* × *C. dentata* F_1 hybrids. In each case, the female parent of the F_1 hybrid was the *C. mollissima* cultivar Nanking. Two different *C. dentata* trees were used as males. A total of 102 F_2 progeny were used for genetic map construction and QTL detection.

Two crosses were made to create a genetic map for Chinese chestnut. The first was a cross between the Chinese chestnut cultivars Vanuxem and Nanking, and the second was between the Chinese chestnut cultivars Mahogany and Nanking. Mahogany and Nanking are widely planted cultivars (Anagnostakis 1992; Burnham 1986), and Vanuxem is a highly blight-resistant Chinese chestnut clone originating from American Chestnut Foundation research orchards. Progeny trees were subjected to paternity analysis and 192 confirmed trees from each of the two mapping populations were selected for genetic mapping (Fang et al. 2012).

3.4.3.3 Linkage Mapping and QTL Discovery

The first chestnut genetic map was published in 1997 (Kubisiak et al. 1997). Based on the F_2 offspring of two *C. mollissima* × *C. dentata* F_1 hybrids, the map contained 241 markers, including 216 RAPDs, 17 RFLPs, and 8 isozymes. Twelve LGs were identified corresponding to the 12 chromosomes of *Castanea*. Initially, seven genomic regions (QTLs) were found to affect blight resistance; however, only three of the QTL held up after further investigation (Kubisiak et al. 1997). Clark et al. (2001) added 275 AFLPs to the map and Sisco et al. (2005) added 30 STS markers. After reanalysis of the Chinese × American F_2 mapping population with the additional markers, Sisco et al. (2005) generated a new map that confirmed the presence of three loci (*Cbr1*, *Cbr2*, and *Cbr3*) conditioning resistance to chestnut blight. The current *C. mollissima* × *C. dentata* genetic map contains over 1000 marker loci (447 SNP; 521 RAPD, AFLP, ISSR; 29 SSR; and 8 isozyme) covering a total genetic distance of 686 cM (Table 3.4).

The genetic map of *C. sativa* (Casasoli et al. 2001) was generated using the two-way pseudo-testcross method where markers were placed on the map of either the female (*C. sativa* Bursa) or male (*C. sativa* Hopa) parent. A total of 381 RAPD and ISSR markers segregating in the F_1 Bursa × Hopa

Table 3.4 Genetic and QTL Mapping Resources for *Castanea* Are Based on F$_1$ or F$_2$ Pedigrees of Three Species

Map Type	Genetic				QTL		
Species (Pedigree)	Number of Linkage Groups	Total Genetic Distance (% Genome Saturation)	Number of Marker Loci[a]	Trait	Number of QTLs[b]	QTL Range, Mean (cM)[c]	References
C. sativa (Bursa×Hopa)	12	865 (82)	94,574	Bud break	12	6.0–112.0, 39.6	Casasoli et al. (2001, 2004) and Kremer et al. (2007)
				Height	6	19.6–32.0, 25.7	
				Water use	7	47.3–112.8, 63.7	
				Diameter growth	4		
C. mollissima (Mahogany)×*C. dentata*	12	686 (86)	1,006	Disease resistance	3	–	Kubisiak et al. (1997, 2012), Clark et al. (2001) and Sisco et al. (2005)
C. mollissima (Mahogany×Nanking) (Vanuxem×Nanking)	12	742 (93)	1,156	Disease resistance	3	3.8–9.5, 7.3	Kubisiak et al. (2012)

Notes: Data reflect the most recent map build in cases of pedigrees with multiple genetic maps.

[a] Number of marker loci reflects all markers published or available in the Quercus Map Database for each pedigree.

[b] Number of QTLs is based on a consensus map if available; otherwise, QTLs from female and male maps were combined.

[c] A range of numbers indicates QTLs found for multiple metrics of a single trait. A dash (–) indicates no data or unavailable data. References are listed in order of year.

full-sib family were placed on the genetic map, which was estimated to cover 70% of the *C. sativa* genome. At present, 517 markers (427 RAPD, AFLP, ISSR; 39 SSR; 46 STS; and 5 isozymes) covering 865 cM have been mapped in *C. sativa*, representing approximately 80% coverage of its genome. The LGs were aligned from the female and male maps to create consensus LGs. These 12 *C. sativa* consensus LGs, as well as maps of the 12 QTL found to influence bud break phenology, are available to view and download from https://w3.pierroton.inra.fr/QuercusPortal/index.php?p=cmap.

A Chinese chestnut genetic map was recently completed by Kubisiak et al. (2012). They used 384 offspring from two related Chinese chestnut crosses to construct a map based on 1156 SNPs and SSRs. This map was used to anchor the physical map. This Chinese chestnut genetic map, as well as the Chinese chestnut physical map and maps of each of the three QTL conferring blight resistance, can be viewed and downloaded from the Fagaceae Genomics Web (www.fagaceae.org/genetic_maps).

3.4.3.4 BAC Libraries and Physical Mapping

Three large-insert BAC libraries were created from the *C. mollissima* cultivar Vanuxem (Fang et al. 2012). Two *Hind*III libraries contain 73,728 and 110,592 clones, respectively, for an estimated combined 24× genome coverage. An *Eco*RI library includes 92,160 clones with a mean insert length of 115 kb, for an estimated 12× genome coverage. These three Chinese chestnut BAC libraries, totaling 36 genome equivalents, were generated by and may be purchased from the Genomics Institute at Clemson University (https://www.genome.clemson.edu/cgibin/orders). A BAC library was also generated from American chestnut genomic DNA digested with the HindII restriction enzyme. The library contains over 73,000 clones with a mean insert of 140 kb, for an estimated 13× genome coverage. The library has not yet been fingerprinted (i.e., probed with sequences derived from markers on the genetic map), as genetic maps and markers specific to American chestnut are still being developed.

BAC end sequences were analyzed for overlap so that contiguous sets of BACs (contigs) could be constructed. The physical map build consisted of 1,377 contigs with a mean length of 951 kb and 12,919 singletons. The map covered approximately 1311 Mb—a physical span about 1.6× the estimated genome size of chestnut (Fang et al. 2012). Over 1000 hybridization probes were derived from DNA markers that had been positioned on the *C. mollissima* genetic map (Kubisiak et al. 2012). These overgo probes were then used to locate 691 corresponding to 376 BAC contigs, allowing correlation of the physical and genetic maps of Chinese chestnut (Fang et al. 2012; Gardiner et al. 2004). Overall, about 50% of the consensus genetic map markers were placed on the physical map and 47% of the length of the physical map was placed on the genetic map. Probes designed for the three blight resistance QTL (*Cbr1*, *Cbr2*, and *Cbr3*) identified BAC clones spanning the region of each locus, which led to the complete sequencing of each blight resistance QTL and the identification of putative candidate genes for blight resistance (Fang et al. 2012). This genetically anchored physical map for Chinese chestnut is expected to facilitate map-based cloning and identification of genes involved in blight resistance. These genes can then be introduced through introgressive backcross breeding strategies (Burnham 1982; Hebard 2006) or transformation directly into American chestnut (Merkle et al. 2006).

3.4.4 Blight Resistance Breeding

The most common method for managing plant species threatened by new pests or diseases is backcross breeding. Resistance genes are transferred from a donor species into germplasm of a recipient species through a process of repetitive intermating and selection among the progeny for resistant individuals, which are then deployed to replace or mate with the remaining native population (Figure 3.5). Breeding of American chestnuts began in the late 1920s, when ecological

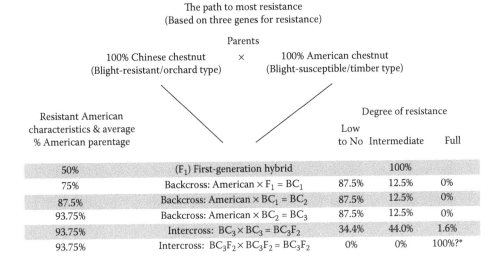

The path to most resistance
(Based on three genes for resistance)

Parents

100% Chinese chestnut   ×   100% American chestnut
(Blight-resistant/orchard type)   (Blight-susceptible/timber type)

Resistant American characteristics & average % American parentage		Degree of resistance		
		Low to No	Intermediate	Full
50%	(F$_1$) First-generation hybrid		100%	
75%	Backcross: American × F$_1$ = BC$_1$	87.5%	12.5%	0%
87.5%	Backcross: American × BC$_1$ = BC$_2$	87.5%	12.5%	0%
93.75%	Backcross: American × BC$_2$ = BC$_3$	87.5%	12.5%	0%
93.75%	Intercross: BC$_3$ × BC$_3$ = BC$_3$F$_2$	34.4%	44.0%	1.6%
93.75%	Intercross: BC$_3$F$_2$ × BC$_3$F$_2$ = BC$_3$F$_2$	0%	0%	100%?*

Figure 3.5   The breeding strategy adopted by The American Chestnut Foundation (TACF) seeks to introgress genes for resistance to chestnut blight (*C. parasitica*) from Chinese chestnut (*C. mollissima*) into American chestnut (*C. dentata*) in order to reintroduce American chestnut into its former habitat in eastern North America. Blight-resistant Chinese chestnut trees are crossed with blight-susceptible American chestnut trees, followed by three generations of backcrossing to American chestnut parents and two generations of intercrossing. In each generation, trees are screened with chestnut blight disease, and only blight-resistant trees are advanced in the breeding population. Each generation takes a minimum of 5 years to complete. *Recovery of full resistance is a theoretical assumption, and has not been proven at this time. (Courtesy of The American Chestnut Foundation.)

extinction of the species was believed to be imminent, but was abandoned when blight resistance in the F$_1$ hybrids proved insufficient (Clapper 1952; Jaynes 1974). Burnham (1982) suggested that if blight resistance was under oligogenic control, as suggested by Clapper (1952), introgressing genes from resistant Asiatic species (*C. mollissima* and *C. crenata*) should be possible by making hybrids and backcrossing to the recurrent American parent only the most resistant individuals as determined by a disease challenge (Figure 3.6). The American Chestnut Foundation (TACF), a nonprofit organization, took up this task upon its inception in 1983. University and government partners have since cooperated with TACF in research, funding, and personnel to create the largest hardwood breeding program is U.S. history. As of 2012, TACF's Meadowview Research Farms had over 55,000 trees at various stages of breeding planted on 65 ha of land (Smith 2012). Most of these trees are descendants of two BC$_1$ hybrids, such that two sources of resistance (from the *C. mollissima* grandparent of each hybrid) provide the foundation for most of the blight resistance in the breeding program. In addition, over 30 Chinese or Japanese chestnut trees have been used in smaller numbers as sources of resistance, with some sources as far advanced as the BC$_3$ generation (Hebard 2006).

Soon after its inception, the TACF organized into state chapters that sought out living American chestnut trees with the goal of introducing locally adapted genotypes into each state's breeding population (Alexander et al. 2005; Burnham et al. 1986). State chapters pollinated trees that were surviving with or had escaped chestnut blight using pollen from advanced backcrosses from one of two TACF resistance sources. State chapters were also encouraged to find unique sources of resistance in Chinese and other chestnut species and make F$_1$ crosses. There are currently more than 300 breeding orchards in 21 states; TACF hopes to have 500 maternal lines in the breeding program representing locally adapted germplasm from throughout the native range.

Management of and selection in the *C. dentata* hybrid breeding program is primarily targeted toward the recovery of American chestnut form and ecological behavior following the initial

(a)   (b)   (c)

Figure 3.6 **(See color insert)** (a) Chestnut trees are cork-borer inoculated with three different strains of chestnut blight: a mildly virulent strain at the top, a moderately virulent strain in the middle, and a highly
virulent strain at the bottom. (b) Blight infections on a resistant tree are characterized by fungal
cankers having defined margins of swollen calluses and living tissue. (c) Blight-susceptible trees
show sunken cankers with necrotic tissue at the margins, which quickly girdle and kill the infected
stem. (Photo by J. Hill Craddock.)

hybridization to Chinese chestnut. On average, 94% of the genome of all third-backcross hybrids
and their descendants comes from American chestnut; 6% is inherited from the Chinese chestnut
ancestor. The *C. dentata* genomic complement in the BC_3 generation could be smaller, however,
if resistance genes are inherited in large blocks, or larger if selection is practiced against those
C. mollissima alleles that are not responsible for conferring resistance to chestnut blight. Diskin
et al. (2006), using an index compiled from 24 leaf, stipule, twig, and bud morphological traits,
determined that the complement of American genome in the BC_3 generation was not significantly
different from expected under additive gene action. They further showed that species assignment
based on measurements of diagnostic morphological traits would classify BC_3 hybrids as American
chestnuts. In other words, most BC_3s in their study were morphologically indistinguishable from
C. dentata. Liu and Carlson (2006) developed a dot blot technique to directly select against Chinese
chestnut DNA in the BC_3 generation. As expected, they detected progressively more American
chestnut DNA in successive hybrid generations. However, it was not possible to predict the morphometric differences between trees with 100% success, likely due to the close relationship between
the American and Chinese chestnut genomes. There was significant individual variability for the
amount of Chinese chestnut genome remaining in the BC_2 and BC_3 generations, indicating that this
technique may be useful in screening individuals within families for advancement in the breeding
program (Liu and Carlson 2006). Although these results are promising, many traits with ecological
and economic significance, such as growth rate and form, were not investigated and bear heavily on
the success of reintroduced American chestnut hybrids.

Since 2005, TACF has harvested increasing numbers of seeds every year from BC_3F_2 trees.
Seeds of this BC_3F_3 generation are called Restoration Chestnuts 1.0. The 1.0 signifies that they are
the first in a series of potentially blight-resistant trees. In 2009, the first Restoration Chestnuts 1.0
were planted in forest environments. The three main objectives of forest trials are to determine
(1) to what degree BC_3F_2 trees resemble American chestnut in a forest setting and to what degree
C. mollissima characters (other than blight resistance) may remain, (2) to what degree BC_3F_2 trees
are resistant to blight; and (3) how long resistance persists in BC_3F_2 plantations (Steiner et al. 2004).

As the chestnut genome sequence is annotated, and markers that are tightly linked to QTL
conferring blight resistance are located and confirmed, high-throughput genotyping may replace

disease inoculation and screening as the method of choice for eliminating moderately susceptible individuals from late-stage breeding populations. Markers found in regions that influence growth rate, form, and wood quality (i.e., American chestnut characters) will be able to discriminate introgressed sections of the Chinese chestnut genome, allowing for removal of those trees with less favorable silvicultural characteristics from forest test plantations before reproductive age. Neutral markers distributed evenly through the genome will allow researchers to monitor and manage genetic diversity of reintroduced populations. Used together, the suite of genomics tools currently under development for chestnut offers the unique ability to test hypotheses related to allelic variation and disease resistance and to quantitatively evaluate reintroduction success.

3.4.5 Monitoring Hybrid Chestnut Reintroduction

The goal of TACF is not only to produce a blight-resistant American chestnut hybrid, but also "to bring blight resistance into wild, naturally regenerating populations of *Castanea dentata* in Appalachian forests and, by doing so, restore the species to its former role" (Steiner et al. 2004). Chestnut's "former role" includes its important economic uses, which should be complementary to its ecological and conservation values (Rajora and Mosseler 2001). As chestnut wood is highly valued for beauty and durability, large numbers of blight-resistant trees will likely be planted in agricultural environments outside of its native range and intensively managed in crop plantations (Jacobs 2007). However, natural regeneration from trees reintroduced to forests in the native range, with accompanying natural selection and the potential for hybridization with native chestnut, are necessary for achieving the goal of ecological restoration (Irwin 2003). Population genetic concerns will likely center on the presence of founder effects, the effects of inbreeding within populations, and rates of population expansion (Worthen et al. 2010b). Genomics tools are expected to play a large role in addressing these concerns in Restoration 1.0 trees and beyond.

3.4.5.1 Markers for Population Management

Reintroduced populations of blight-resistant American chestnut trees will undoubtedly be monitored for many attributes. Molecular tools offer conservation geneticists the opportunity to monitor the ways in which exotic genes for resistance to chestnut blight will interact with the genome of chestnut over space and time. The outcome of ecological interactions between blight-resistant and native chestnuts, as well as between blight-resistant chestnut trees and the chestnut blight fungus, will be an active area of investigation. Genetic markers and sequences from both the nuclear and chloroplast genomes of *Castanea* species can be used to address many questions of genetic diversity, structure, and mating within populations, and paint a picture of how populations expand and interact. Work by various research initiatives will ensure even more molecular tools become available for chestnut reintroduction.

3.4.5.1.1 Chloroplast Markers

cpDNA markers located on the maternally inherited chloroplast genome have proved useful in understanding aspects of chestnut population expansion, colonization, and interpopulation interactions where the diversity, evenness, and aggressiveness of maternal lineages is in question (Dane et al. 1999; Lang et al. 2007). A central question in chestnut reintroduction concerns the ability of newly planted populations to colonize new territories by long-distance seed dispersal. It can be inferred that species with broader ranges are potentially good colonists since they managed to spread in the past (Petit et al. 2004). Indeed, broad-ranging species have distinct genetic features that include relatively high gene diversity and less differentiation among populations at isozyme markers than species with narrower ranges (Hamrick et al. 1992). The American chestnut expanded its range rapidly after the recession of the Pleistocene glaciers (Delcourt and Delcourt 1984; Huang et al. 1994) and has been shown to outgrow other hardwood species in plantings (Jacobs and Severeid 2004; Paillet and Rutter 1989), both of which further support the high colonizing potential of the species. Because colonization of new territories relies on seed dispersal, the level of differentiation among populations at maternally inherited markers may shed some light on the rate of seed dispersal, the relative dispersal ability of maternal lines, and the diversity of cpDNA haplotypes in newly established colonies. Maternal markers can also be used to study the strength and longevity of patches of family structure that develop as the result of a single seed founding a new colony, where each patch represents a maternal clone of large size. In simulation studies with oaks, for example, patches of tens of square kilometers were found to be nearly fixed for a single cpDNA haplotype (LeCorre et al. 1997).

When multiple introductions are employed, genetic studies of zones of secondary contact (where two reintroduced populations meet) can provide information regarding the mode of expansion of populations, depending on whether the expanding lineages form abrupt "suture zones," intermingle, or form gradual clines (Hewitt 1988). In ponderosa pine, analysis of maternally inherited markers revealed that populations expanded in a diffuse "moving front," resulting in abrupt contact, whereas paternally inherited cpDNA markers showed a progressive cline (Johansen and Latta 2003; Latta and Mitton 1999).

Chloroplast markers have been employed in several studies of *Castanea* genetic diversity and phylogeny in Asia, Europe, and North America (e.g., Lang et al. 2006, 2007). American chestnut cpDNA can be distinguished from that of other chestnut species by a 75 bp deletion in the *trn*T-L intergenic spacer region (Kubisiak and Roberds 2003; Lang et al. 2006). Nucleotide variation for noncoding cpDNA intergenic spacer regions is generally low in interspecific comparisons of *Castanea*, and sequence divergence among and within genera is low (Fineschi et al. 2000; Lang et al. 2006). Patterns of the variation in the cpDNA of *C. dentata* and *C. pumila* were geographically structured and not congruent with the current phylogeny based on bur and cupule characteristics, leading the authors to suggest chloroplast capture via introgression in North American *Castanea* species (Lang et al. 2006).

Genomics tools are increasing the likelihood that cpDNA will be used to study the reintroduction of blight-resistant American chestnut. In the genetics era, markers would be needed that distinguish maternal lines within a species, and to date no intraspecific variation has been detected within the *trn*T-L-F intergenic regions of the chloroplast genome of American chestnut (Lang et al. 2007). In the genomics era, complete chloroplast genome sequences are being developed or are available that will allow resequencing of many individuals at thousands of markers. The complete chloroplast genomes of *C. sativa* (Sebastiani 2003) and *C. mollissima* (Jansen et al. 2010) have been sequenced, leading to the development of cpSSRs (Sebastiani et al. 2004) and thousands of cpSNPs (Jansen et al. 2010) that could be used in monitoring lineages of American chestnut. The newly developed chloroplast markers may reveal polymorphisms necessary for monitoring large numbers of American chestnut lineages; alternatively, falling costs of high-throughput second-generation sequencing technologies offer the possibility of completely sequencing the chloroplast genomes of a sample of individuals for more fine-scale monitoring of dispersal patterns, mutations, and introgressions. Once the American chestnut chloroplast genome is sequenced, populations can be monitored for directional introgression of chloroplast haplotypes (Li and Dane 2012; Petit and Vendramin 2006; Shaw et al. 2005).

3.4.5.1.2 Nuclear Markers

Nuclear markers have been used to examine genetic diversity, heterozygosity, and population differentiation between founder populations and their descendants, and between orchard populations and nearby wild or coppiced populations. Allozyme markers were used to detect a zone of hybridization between two distinct races of *C. sativa*, an area of gene pool intergradation that the authors attribute to secondary contact between races that have been differentiated in allopatry (Villani et al. 1999a). Eriksson et al. (2005) found that gene flow and genetic differentiation among wild and coppiced populations of *C. sativa* were higher than among orchard populations, though that was not the case in Japan, where orchard populations showed similar heterozygosities, allele size ranges, and genetic differentiation to wild populations of *C. crenata* (Tanaka et al. 2005). The incongruence in these results may have to do with orchard establishment: in Europe, allochthanous chestnuts were often introduced into cultivated populations, whereas in Japan cultivars seem to have originated from local gene pools (Tanaka et al. 2005).

Nuclear markers may also reveal zones of population admixture, levels of gene flow, and ranges and variances of effective number of males in reintroduced populations. The careful characterization of Turkish chestnut populations by Villani and colleagues illustrates the variety of ways in which population structure and admixture can be examined using nuclear markers. Populations of *C. sativa* in Turkey span an east–west range characterized by a low-rainfall Mediterranean climate in the west and a much higher-rainfall Eurosiberian climate in northeastern Anatolia near the Black Sea. Eastern and western populations were shown to be genetically divergent at allozyme loci and for morphological and physiological traits, with populations of intermediate genetic makeup existing between them that are thought to be the result of introgression (Villani et al. 1991, 1992, 1994). Villani et al. (1999a) used 12 allozyme systems to survey 34 populations of *C. sativa* spanning the eastern, intermediate, and western genetic zones in Turkey and estimated genetic structure and gene flow both within and between zones. They found that most of the overall genetic variance was due to heterogeneity among populations ($F_{ST} = 0.184$) and that allele frequencies were relatively homogenous across regions eastward and westward of the introgression zone where a sharp transition in allele frequencies and an increase in genetic variability (spanning a cline of 324 km) were clearly detected. The number of migrants between populations (*Nm*) decreased with geographic distance; populations <30 km apart exchanged between 4 and 32 (median of 12) migrants per generation, whereas *Nm* flattened out at values of 2–4 for populations separated by 60–90 km. Interestingly, they found that gene flow was significantly higher within genetic zones than between them and that there was a significant decrease of gene flow at the boundaries between the three groups of populations, forming a "partial genetic barrier" between them (Villani et al. 1999a). The authors suggested that this barrier was caused by a limitation in the number of populations in the region of introgression that could behave as genetic links between the eastern and western populations.

3.4.5.1.3 Morphological, Physiological, and Phenological Traits

In addition to molecular markers, the genetic diversity of chestnut can be characterized using a variety of morphological, physiological, and phenological traits. Although variability in vegetative morphological traits may be used to distinguish among races or ecotypes of a species, their connection to ecological processes is not obvious, and for that reason most researchers focus on traits such as bud burst, bud set, and timing of flowering that are closely associated with adaptation. Flowering phenology, the dates when male and female flowering begins and ends, may be the single most important phenological variable because it affects whether or not individuals and populations will intermate (Slavov et al. 2005).

Villani et al. (1999b) suggested that WUE was an adequate physiological indicator of satisfactory growth in water-limited environments. They made controlled crosses of trees within and between two genetically divergent *C. sativa* populations in Turkey, measured 351 seeds for 12 morphometric traits, and measured WUE in a sample of 6-month-old seedlings. The progeny of within-population crosses showed significantly more homogeneity in all variables than the between-population crosses. Diskin et al. (2006) characterized American chestnuts, Chinese chestnuts, and four hybrid chestnut types using an index of species identity (ISI) calculated from 26 morphometric variables based on leaf, stipule, twig, and bud traits. Although the traits were not necessarily adaptive, they represented consistent morphological differences between American and Chinese chestnuts and were well described, and baseline data including trait means and standard errors were given for each species and hybrid type. These traits might be useful in determining differentiation between populations and may also reveal differences in the retention of Chinese chestnut genomic regions between populations.

Analysis of morphological traits can be coupled with analysis of neutral genetic variation to determine whether local adaptation has occurred in natural populations, without the need for large and expensive common garden plantings. The proportion of variation between populations at neutral loci (F_{ST}) can be compared to Q_{ST}, an analogous measure of differentiation at quantitative loci based on the ratio of between-population genetic variance to total genetic variance for the trait under study. If morphological or physiological traits show significantly higher differentiation than neutral markers ($Q_{ST} > F_{ST}$), it can be concluded that divergent selection is shaping adaptive traits (McKay and Latta 2002).

3.4.5.2 Interactions between Resistant and Native Chestnuts

Some blight-resistant American chestnuts will be reintroduced to areas where chestnut formerly grew or still grows, offering a potential genetic connection between native populations of American chestnuts and reintroduced hybrids. The potential for gene flow out of the introduced populations and into the forest is limited, however, because stump sprouts of autochthonous trees now grow in the forest understory, where conditions of low light inhibit flowering (Griffin 1989). Native trees and their moderately susceptible offspring may flower rarely or not at all, and they may not cross with the BC_3F_3s in nearby plantations. Gametes from moderately susceptible trees that do flower and flow back into the plantation will be swamped, and any resulting offspring will be, on average, much less viable than offspring that result from intraplantation crosses. Crosses between moderately resistant trees and wild trees will be at a strong selective disadvantage to new moderately resistant seedlings produced by plantation × wild crosses. Finally, parent–offspring matings between moderately resistant trees and nearby wild trees are expected to result in reduced seed set, reduced fitness, or both. Thus, moderately resistant trees may not have many viable options for "rescuing" wild American chestnut genes, and their ability to contribute to the blending of blight-resistant hybrids with wild trees could be minimal unless they are managed in a way that promotes the production of flowers and the survival of moderately resistant offspring.

3.4.5.3 Linkage Disequilibrium

Under the plan outlined by the TACF, the third backcross generation, which will be used to reconstitute a population for reintroduction, will be on average 15/16 American chestnut and 1/16 Chinese chestnut (Diskin et al. 2006; Liu and Carlson 2006; Rutter and Burnham 1982). If the introgressed portion of the Chinese chestnut genome confers no fitness advantage, chromosomal recombination in subsequent meioses would effectively "break up" the remaining portions of the Chinese genome. However, when selection favors and rapidly increases the frequency of an advantage-conferring locus, sites linked to the favorable locus "hitchhike" along for the ride (Maynard Smith and Haigh 1974). This phenomenon, often called linkage drag, is commonly encountered in plant breeding when artificial selection favors an introgressed region and drags along unfavorable or neutral linked genes (Brinkman and Frey 1977). In the same way, when natural selection favors a particular allele, sites linked to that locus are dragged along to fixation, resulting in reduced genetic variation in a region around the gene under selection relative to the rest of the genome, a process known as a selective sweep. The more rapid the gene fixation, the more reduced the level of variation around the favored site and the larger the size of the region influenced by the sweep. Likewise, as recombination decreases, the length of the sweep-influenced region increases. Specifically, Kaplan et al. (1989) showed that the approximate distance d at which a neutral site can be influenced by a sweep is a function of the strength of selection s and the recombination fraction c,

$$d \cong 0.01 \frac{s}{c}$$

High effective levels of recombination result in a shorter window of influence around the selective site, resulting in shorter regions of disequilibrium (Walsh 2008). Chromosomal regions with reduced recombination, such as those caused by chromosome inversions or translocations in hybrids, may lead to the production of large linkage blocks that will resist reduction and potentially contribute to evolutionary change (Hoffman and Rieseberg 2008).

LD has important evolutionary consequences, as it decreases levels of genetic variation and therefore decreases the efficiency of selection on linked genes within the region influenced by the sweep. These effects occur because (1) LD reduces the number of independent loci and (2) LD introduces a negative component of variation that is subtracted from the total genetic variance (the Bulmer effect; Bulmer 1971). Furthermore, within the region influenced by the sweep, deleterious alleles have a higher probability of fixation, while favorable alleles have a reduced probability of fixation compared to sites outside of the sweep (Walsh 2008). Once the favored site has become fixed, the signal for the sweep starts to decay through recombination and mutation, but LD generated by a selective sweep may persist for at least N_e (effective population size) generations (Przeworski 2002). Although the length of a hitchhiking chromosomal segment is expected to be large in early backcross generations (Figure 3.7), by screening segregating populations for markers from the recipient genome, the proportion of donor genome retained can be minimized (Hospital 2001). In the near term, the size of the chromosomal segment expected to hitchhike into American chestnut depends primarily on the selection pressure (s, the relative advantage of an allele from Chinese chestnut that confers resistance to $C.\ parasitica$ over the alleles found in American chestnut) and the recombination rate (r), such that hitchhiking is efficient when $r < s$ (Maynard Smith and Haigh 1974). The extent to which genomic

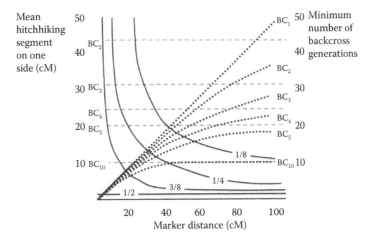

Figure 3.7 Expected mean length (cM) of intact donor chromosome (hitchhiking) on one side of an introgressed locus without MAS (*gray*) or with selection for a recipient genome marker at the indicated distance (cM) for backcross 1–10 (BC_{1-10}; *black dotted lines*). Number of backcross generations required to reduce a donor segment length by the indicated proportion (*black solid lines*) if a recipient genome marker at the indicated distance (cM) is used to assist selection. (Reprinted with permission from DeWoody, J.A., Bickham, J., Michler, C.H. et al., *Molecular Approaches in Natural Resource Management and Conservation*, Cambridge University Press, New York, 2010.)

regions from Chinese chestnut (*C. mollissima*, the donor species) will be retained in American chestnut over the long term are difficult to predict. The models for LD during hitchhiking are complex, even for single, beneficial alleles (Eriksson et al. 2008), and the breeding program for American chestnut includes the introgression of at least three resistance genes, potentially with incomplete dominance and epistatic interactions. Even if the generation time for American chestnut is only 10 years (it is likely to be much longer in the forest) and if the N_e is >500 (which seems probable), the LD caused by selective sweeps around the loci conferring blight resistance could be expected to last for thousands of years in reintroduced American chestnut populations. It remains to be seen if gene flow from reintroduced populations of American chestnut will rescue native American chestnut or initiate a separate and evolutionarily distinct lineage.

3.4.5.4 Local Adaptation and Long-Term Fitness of Reintroduced Populations

As a general rule in forest genetics, populations of tree species adapted to cold or dry climates may perform favorably almost anywhere (i.e., in climates that are warmer or colder, wetter or drier), whereas populations from warmer and wetter climates perform poorly when planted in areas colder or drier than their original provenance (White et al. 2007). Local adaptation for timing of Spring bud burst, a phenological measurement associated with cold injury, has been observed along latitudinal or altitudinal clines in natural populations of sessile oak (*Q. petraea*; Ducousso et al. 1996), northern red oak (McGee 1974) and pitch pine (Steiner and Berrang 1990), to name a few. An analysis of several adaptive traits in Douglas fir (*Pseudotsuga menziesii*), a species often found in mountainous areas, showed complex patterns of variation that were accounted for by a combination of latitude, longitude, altitude, and slope (Rehfeldt et al. 1999). Many more examples of local adaptation in forest trees have been reported (see Savolainen et al. 2007 for a thorough review).

Chestnut breeders are using several methods to reduce the possibility of reintroducing maladapted trees into the landscape. The primary method is to include in the breeding program as many locally adapted sources of American chestnut germplasm as possible. Ideally, hybrids will be introduced back into the area from which their American chestnut parent originated. Under the TACF breeding model, each state chapter is striving to find 20 or more American chestnuts to use in the backcross breeding program for that state (Alexander et al. 2005; Hebard 2005). Resistant selections are expected to be planted and allowed to open-pollinate in three regional seed orchards representing the southern, central, and northern areas of the species' range within the United States. If many relatively small populations are introduced (rather than a few large populations), the loss of one population will not have a dramatic effect on the overall reintroduction effort. A strategy that employs many small reintroductions increases the likelihood that some of the populations will either be on sites to which they are adapted, or they will be able to disperse to suitable sites.

The TACF breeding program was designed to capture as much genetic diversity as possible from American chestnuts, with the goal of maximizing local adaptation. It is reasonable to assume that alleles conferring adaptation to a range of environments were present in preblight populations of American chestnut and that high gene flow and high outcrossing rates perpetually created new combinations of alleles that were more favorable in some environments than others. If the TACF breeding program has captured a large portion of existing alleles, then outcrossing within and between reintroduced populations should again lead to genotypes that are locally adapted to specific sites.

The problem of small populations inhibiting local adaptation can be ameliorated by continual supplementation with new genotypes. Assuming a nonzero survival rate, continual addition of new individuals should ensure that at least some trees will survive to adulthood and reproduce (Ryman and Laikre 1991). One possible scheme for reintroduction that lends itself to population supplementation is the reintroduction of populations in cycles. According to this scheme, populations are reintroduced at specific geographic points such that gene flow can occur between them. Some years later, more populations are introduced to fill the gaps between existing populations and to add more alleles to the total population. In this model, cost, planting, and management effort are spread out over a number of years, and improvements in the TACF breeding population (i.e., more sources of resistance and more American chestnut genes) have a well-defined mechanism to enter the landscape.

3.4.6 Summary

Based on the preceding sections and lessons from other captive-breeding programs (e.g., the salmonid reintroduction program of Fraser 2008), the following general conclusions regarding chestnut reintroduction can be made. It is critical that the effective population size of the breeding population be maintained at a high level by incorporating as many American chestnut alleles as possible and by managing breeding populations such that equal variance in family sizes characterizes the final generation for reintroduction. Many locally adapted genotypes should be included in the program, and genotypes should be deployed to their provenance of origin to preserve both genetic diversity and local adaptation. In order to increase chances of seed germination and seedling survival, factors that favor local regeneration should be identified and sites for reintroduction prepared accordingly. This may involve determining the best types of planting stock, simulating disturbance, controlling competition, and protecting reintroduced populations from herbivory. As many unrelated individuals as possible should be deployed in small populations such that within- and between-population matings are possible. Methods that increase N_e in a breeding population may include the maintenance of a high census population (N), starting the breeding population with as many genetically diverse founders as possible and minimizing the family variance of reintroduced populations (Ryman and Laikre 1991) by equalizing family sizes in the final stage of the breeding program. Reintroduction sites should be managed, and the long-term progress of the trees monitored. Monitoring may include measuring survival, fecundity, variance in reproductive success, the retention of Chinese chestnut genome, the extent of gene flow into and out of introduced populations, and any changes in growth or development that may indicate inbreeding depression. Special attention should be given to population genetic parameters and the fitness of the first generation of trees that regenerates following introduction, as these trees will provide information that will characterize subsequent generations. To close the loop, chestnut breeders must incorporate lessons learned from early reintroductions into future breeding and deployment decisions. The plans of the TACF already embody many of these suggestions. Genomics data (Section 3.4.3) will play a large role in evaluating the success of the TACF reintroduction plan.

Incorporating many sources of blight resistance into reintroduced populations will help increase the variability of resistance and protect the trees from genetic changes in the pathogen. Because interactions between a plant host with a hybrid genome and a fungal pathogen are complex and difficult to predict, reintroduced American chestnut populations will require careful monitoring for incidence of chestnut blight. Careful disease phenotyping combined with resequencing of many individuals will allow for the identification of allelic variants and their relative magnitudes; markers that are in tight linkage with blight resistance QTL will be invaluable in pyramiding these resistance genes. In addition, attacks by other known pests and pathogens of chestnut, such as root rot fungi (*Phytophthera* spp.) and chestnut gall wasp (*Dryocosmus kuriphilus*), are certain. The threat posed by gypsy moth (*Lymantria dispar*), twolined chestnut borer (*Agrilus bilineatus*), ambrosia beetles (Curculionidae), and other potential chestnut pests and pathogens remains to be seen. Genomics technologies are being used to locate genes conferring resistance to *Phytophthera* (Costa et al. 2011; Olukolu et al. 2012) and identify polymorphisms within the genes, with the ultimate goal of incorporating *Phytophthera* resistance into elite lines of blight-resistant American chestnut.

Social forces such as the public's acceptance of new trees, their perception of the reintroduction, and their feelings about changes in the forest as a result of the reintroduction will influence when and where reintroduced American chestnut populations will be successful. While these social forces, as well as the biotic and abiotic considerations enumerated above, are challenges that can be planned for and met, they also represent a large amount of uncertainty that will individually or collectively have a large effect on where American chestnut hybrids will grow in the future. American chestnut will be the first major tree species to be reintroduced throughout its former range (or, to be exact, to have its populations significantly supplemented) by a conservation breeding program. Chestnut could set the stage for the reintroduction or supplementation of other tree species facing similar threats, including hemlock (*Tsuga canadensis*), butternut (*Juglans cinera*), ash, and black walnut (*Juglans nigra*). Further, if transgenic chestnut hybrids produced by *in vitro* propagation truly do increase the speed and efficiency of chestnut reintroduction in a readily apparent way (Merkle et al. 2006), the strong social desire for chestnut restoration could overcome social unease related to transgenic plants. The positive presence of transgenic American chestnuts in the landscape could pave the way toward social acceptance of transgenic technologies in other species.

3.5 FOREST TREE GENOMICS RESOURCES

The tree breeder's reliance on sequence, marker, and map data will only increase as genomic tools become standard for use in improvement of trees for wood, fiber, and energy. Genomic databases are an essential resource for applied biologists. Access to markers and sequences generated from functional and comparative genomics studies provides breeders with tools they need to implement MAS and other time-saving breeding strategies. Genomics resources are contained in generalized repositories, like the Genbank/DDJB/EMBL databases of nucleic acids sequences and the PIR/SwissProt/PDB protein sequence databases, as well as specialized repositories centered on specific organisms, protein families, or tissue types. The next few sections are devoted to describing the specialized genomics databases developed for and used by the forest tree genomics community. Important forest genomics initiatives are also mentioned, with an emphasis on their objectives and access to their results. This chapter serves as a comprehensive guide to publically available hardwood tree genomics resources.

3.5.1 Dendrome Project

The Dendrome Project is a community-focused outreach of the University of California at Davis that provides an integrated research hub for the forest tree genomics community (Wegrzyn et al. 2008). The Dendrome website is the access point to the TreeGenes database (Section 3.5.2), user forums, bioinformatics tools, and information about forest tree genomics community projects, upcoming events, and research opportunities (http://dendrome.ucdavis.edu/resources). The Dendrome page incorporates custom BLAST and FASTA services that directly access all publicly available sequence data as well as custom EST or organism databases. The Tools pages provide information on open-source and freeware software packages related to the processing and analysis of genetic, phenotypic, and genomic data. The Links pages direct users to other online resources and projects relevant to the research community. The Dendrome Plone (http://dendrome.ucdavis.edu/TGPlone) is a controlled-access forum that provides a common environment for researchers to share and obtain data, expertise, and information on a wide variety of forest genomics projects.

3.5.2 TreeGenes

The TreeGenes Database (http://dendrome.ucdavis.edu/treegenes/) is the worldwide home to genomic data on forest tree species (Wegrzyn et al. 2008). While the database primarily focuses on conifer trees, it contains some genomic or transcriptomic data for over 80 genera of forest trees. Genera with the most resources are *Pinus*, *Picea*, *Populus*, and *Eucalyptus*. Also represented by genomic or transcriptomic data are the hardwood genera *Betula* (birches), *Castanea* (chestnuts), *Fagus* (beeches), *Juglans* (butternut and walnuts), and *Quercus* (oaks). The TreeGenes database serves as the primary community hub for genetic maps and data from resequencing, genotyping, and phenotyping forest trees. Custom informatics tools for high-throughput genomics projects help users manage data from sample collection to downstream analyses. Location of the database within the Dendrome pages ensures users' access to an expert community for support and collaborative interpretation.

The TreeGenes Database is organized into highly connected modules that support the storage of data and provide the foundation for web-based searches and visualization tools (Wegrzyn et al. 2008). Database modules include the Sample Tracking module, which holds tree source and DNA extraction information; the Sequencing module, which contains sequences and data describing sequence generation, alignment, and analysis; the Species module, which holds taxonomy information; the Colleague module, which contains information on laboratories and individuals; the Comparative Map module, which stores and visualizes genetic maps and molecular markers; the Literature module, which houses relevant publications; and the EST module, which stores sequence and annotation information. The EST module contains an automated pipeline for sequence processing, functional analysis, and GenBank submission. Interconnection between the modules provides the user with genotype-to-phenotype information about an organism, population, or individual tree in one searchable environment.

Users retrieve data from the TreeGenes database through the desktop-style interface DiversiTree (http://dendrome.ucdavis.edu/DiversiTree/). The main screen menu options allow users to access and visualize ESTs, primers, trace files, SNPs, individual tree data, sequencing chromatograms, genotypes, and phenotypes contained within the TreeGenes database. A sample tracking system and intuitive window design allow users to track individual tree samples through DNA extraction, phenotyping, sequencing, and SNP detection (Wegrzyn et al. 2009). DiversiTree also supports bulk retrieval of sequences (such as GenBank accessions, contigs, or SNPs) with functional annotations via a user-friendly bulk uploader window. Users can connect to the Forest Tree Genetic Stock Center (http://dendrome.ucdavis.edu/ftgsc/) and request specific DNA based on sample, sequence, or marker data (Wegrzyn et al. 2011). Thus, a single interface can be used to retrieve and visualize data and obtain DNA of interest, providing a dynamic center for genotype-to-phenotype investigations in forest trees.

3.5.3 Fagaceae Genomics Web

Unlike Dendrome, which is an information hub for a wide-ranging genomics and bioinformatics community, the Fagaceae Genomics Web was initiated by a single, multi-institutional project dedicated to developing genomic tools for the Fagaceae (www.fagaceae.org). Genomic data from American beech (*Fagus grandifolia*), Chinese chestnut (*C. mollissima*), American chestnut (*C. dentata*), Northern red oak (*Q. rubra*), and white oak (*Q. alba*) have been generated through the project and are publicly available. The Fagaceae Genomics Web page allows users to browse and download project data by type or species. The DNA Libraries tab contains sequences for the BAC and cDNA libraries used in transcriptomic studies and physical mapping. The Sequences tab houses a searchable EST database, and the Markers tab contains candidate SSRs and SNPs predicted from ESTs. The Genetic Maps tab allows viewing of genetic maps using CMap software. The Physical Map tab allows visualization of contigs used in physical mapping and contains mapping statistics. A BLAST tab allows users to search nucleotide and protein databases with project data grouped by species and type.

The Fagaceae Genomics Web page contains an Organism panel where genomic data can be viewed separately for each species. Selecting an organism leads to a species description and overview of genomic features in the database. Tabs available for each species include EST assemblies, GO analysis, KEGG, and Libraries. The EST Assemblies tab contains raw data files and assembled contigs as well as links to the sequences to individual libraries used in each assembly project. New assemblies are built after a significant amount of new markers are added to the project or new software is made available. The GO Analysis and KEGG tabs show functional group assignments to assembled contigs, and links to all libraries associated with an organism are located in the Libraries tab.

3.5.4 EVOLTREE

EVOLTREE, a short form of the name "Evolution of trees as drivers of terrestrial biodiversity," is a network of excellence established by the European Union to study the evolution of trees in response to environmental change. Objectives of the network, which currently includes 23 research groups in 13 European countries, include identifying genes controlling adaptive traits in forest trees and examining patterns of adaptive genetic diversity across Europe. Most project resources were directed toward the development of genomic and genetic resources, the creation of common databases, and the establishment of common infrastructures and protocols. Outcomes of the project, including databases and other community resources, are actively maintained and accessible through the EVOLTREE web page (http://www.evoltree.eu/). The EVOLTREE page contains information on network-related activities, projects, workshops, meetings, and job opportunities and is the entry point to genomic databases and electronic resources. Resources available through the page include a customizable BLAST similarity search, an EST finder, and the eLab, which allows users to place a query against all of the databases available from the site.

The EVOLTREE web page houses several databases that can be searched individually or queried all at once through the eLab service. Data-specific databases include Libraries, which contains cDNA, gDNA, and BAC libraries generated for mapping and sequencing; Sequence Analysis, which provides details and visualization for sequence generation and assembly; and TreePop, which contains genotype, phenotype, and location information for mapping and association populations. Databases for Candidate Genes, SNPs, SSRs, and Microarray Experiments are also maintained. The SNP database currently houses 9118 markers for 3 species (*Q. petraea*, *Q. robur*, and *Populus nigra*). A list of traits allows users to view SNPs in genomic areas putatively associated with adaptive variation. The SSR database contains over 62,000 markers for 22 species and is searchable by species, assembly version, repeat class, and motif class. Species-specific databases include Quercus Phenotypes/Genotypes, Quercus Maps, Salicaceae Phenotypes/Genotypes, and Salicaceae Maps. The Quercus Map database contains detailed genotypic and phenotypic data of over 9000 offspring from 11 oak map builds as well as visualization and comparison of oak maps through the CMap viewer (Section 3.5.6).

3.5.5 Quercus Portal

The Quercus Portal is a comprehensive European genetic and genomic web resource for *Quercus* (https://w3.pierroton.inra.fr/QuercusPortal/). Part of the EVOLTREE network, the Quercus Portal provides basic information on oak trees as well as in-depth information on integrated research projects dedicated to European and American oaks such as GENOAK (https://w3.pierroton.inra.fr/QuercusPortal/index.php?p=OAK_GENOME_SEQUENCING) and Oaks of the Americas (http://quercus.lifedesks.org/). The Quercus Portal provides access to the same genomic databases as the EVOLTREE page and contains additional databases including Oak Provenance, which contains geographic data of source stands used in *Q. robur* and *Q. petraea* provenance tests, and FossilMap, which holds records of macrofossil remains of Fagaceae species. The $(GD)^2$ database (Georeferenced Database of Genetic Diversity) contains genetic, geographic, and ecological information for populations and single trees in natural populations. The primary function of the database is to view data (allele/haplotype frequencies or aggregated diversity statistics) on a Google Earth interface map. The database is also designed for bulk export of genetic and geographic data. The integration of genetic mapping resources, genomics databases, and geographic information makes the Quercus Portal a unique and invaluable tool for understanding genetic mechanisms underlying adaptive variation in oak. This comprehensive database portal is also the largest species-specific source of marker and mapping resources for oak breeding and improvement.

3.5.6 CMap

CMap is a web-based application for viewing and comparing genetic, physical, and QTL maps (Youens-Clark et al. 2009). It was originally written as a comparative mapping resource for grass species and is now part of the Generic Model Organism Database (GMOD), a collection of open-source software tools for creating and managing genome-scale biological databases (http://gmod. org/wiki/CMap). Using CMap in concert with other GMOD tools allows users to create a browser that queries and visualizes genetic map data within or across species. Genetic and QTL mapping data resulting from EVOLTREE and other European hardwood research projects can be queried and viewed in a dedicated CMap database (https://w3.pierroton.inra.fr/QuercusPortal/index. php?p=cmap). The software allows users to view, compare, and search maps for features such as AFLP and SNP markers, QTLs, and ESTs. Multiple genetic maps exist for many species as maps are redrawn based on additional marker sets or new analysis software and added to the database. Each QTL map is based on analysis of a single trait such that many QTL maps exist for well-studied species. Current mapping resources are most developed for *Q. robur* (6 genetic maps and 11 QTL maps), followed by *Q. petraea*, and then *C. sativa*, for which one genetic map and one QTL map are housed in the database. Map and feature data can be downloaded from each map or individual LG. A detailed view of each LG shows an overview of its features and a detailed list containing each feature and all maps on which it appears. The utility of this CMap database will continue to grow with the addition of new maps and species. It is, and likely will continue to be, the most comprehensive source of detailed molecular markers for population management, cultivar identification, marker- and genome-assisted breeding, and other tree improvement applications in the Fagaceae.

3.5.7 Other Genomics Initiatives

The Hardwood Genomics Project, formally entitled "Comparative genomics of environmental stress responses in North American hardwoods," is a multi-institutional collaboration aimed at developing genomic resources for the eight most economically and phylogenetically important hardwood species in North America (http://www.hardwoodgenomics.org/). Species included in the project are sweetgum (*Liquidambar styraciflua*), honeylocust (*Gleditsia triacanthos*), black walnut (*Juglans nigra*), northern red oak (*Q. rubra*), sugar maple (*Acer saccharum*), green ash (*Fraxinus pennsylvanica*), and tulip poplar (*Liriodendron tulipifera*). Genomic data produced by the project will include EST databases for six species, BAC libraries for six species, framework genetic maps for three species, and high-density QTL maps for two species. Project data will be made freely available through the website. Information about project participants, as well as links to community resources such as upcoming meetings and recent publications, is also available from this important community resource.

The Forest Health Initiative is a collaborative effort between government and industry to advance the understanding and role of biotechnology in the United States to address some of today's most pressing forest health challenges (www.foresthealthinitiative.org). The near-term objective of the initiative is to support efforts to safely and effectively use biotechnology to develop American chestnut that is resistant to chestnut blight and *Phytophthera* root disease. The Forest Health Initiative is providing the full genome sequence and assembly for *C. mollissima* (Chinese chestnut) to help meet the initiative's first objective of transgenic chestnut acceptance and restoration. The Chinese chestnut genome sequence will be available from general and species-specific genomic databases upon its completion. While working to restore American chestnut as the test tree, the program will explore new, holistic approaches to enhancing the health and vitality of other trees and forest ecosystems by assessing scientific, social, and regulatory issues.

3.6 FUTURE DIRECTIONS OF HARDWOOD GENOMICS

The genomics era offers new promises of accelerated rates of gene discovery and insights into molecular mechanisms of forest trees, a group of species that, collectively, drives terrestrial biodiversity and provides some of the most valuable commodities in the world economy. Genomics technologies are currently being used to investigate adaptive traits like bud break in large, natural populations of hardwood trees in order to conserve species and ecosystems in the face of climate change and other biotic and abiotic threats. American chestnut exemplifies the integrated use of transcriptome profiling, genetic mapping, genome sequencing, and linkage analyses both for discovering genomic regions conferring resistance to a devastating disease and for increasing the efficiency of introgressing resistance genes into breeding populations. Integrated genomics technologies are also important components of short-rotation poplar breeding and have led to the discovery and characterization of genes in metabolic pathways, such as drought resistance and lignin production, that may contribute to sustainable production of these species in intensively managed plantations. These efforts will continue, and their ultimate outcomes will certainly influence the direction of forest genomics.

Hardwood genomics resources are available primarily for two families, the Fagaceae (beeches, oaks, and chestnuts) and the Salicaceae (willows and poplars). While there are initiatives aimed at extending this list to include members of several other North American hardwoods (Section 5.10), the diversity of genera that have been investigated worldwide is still relatively low. Tropical hardwood trees make up a large proportion of all hardwoods and are economically and ecologically invaluable where they occur, yet there are almost no resources devoted to tropical hardwood genera. Genomic information on these species, which have adapted under different climatic and seasonal regimes than temperate trees, would provide unique insights on the genetic architecture of traits such as wood formation and bud break that respond to environmental cues.

An important step for forest tree genomics will be to show empirical gain—ultimately, of profit—from the use of genomics techniques in tree breeding. Thus far, the efficiency of MAS has been shown primarily through simulation studies, and returns on investments into MAS in hardwoods have yet to be realized in the way of reduced breeding time or increased efficiency of selection. For funding of MAS to continue or to be initiated in new species, hardwood breeding programs need to show real and meaningful gain for MAS versus the best recurrent phenotypic selection methods. The numbers of markers available for MAS and known marker–trait associations will certainly increase as "model" species like pine and chestnut are subjected to genome-wide association studies and genomic selection. To ensure the future of genomics funding, it will be important for these programs to clearly and quickly prove real-world gains from investments in genomics technologies.

Finally, genomics technologies are expected to play a large role in monitoring forest ecosystem health, just as they have revolutionized diagnosis and treatment of human health. Analysis of the human genome, aided by in-depth association studies, has improved human health by estimating risk factors for specific diseases and recommending drug or lifestyle therapies based on a patient's genetic constitution. Similarly, if the genes underlying adaptive disease-resistance traits in forest trees were known, along with allelic variants and their trait values, natural populations could be assessed for risk to biotic or abiotic threats based on their allele frequencies (Neale 2007).

Though prescribing forest management based on population allele frequencies is still in the future, pressure on forest resources due to habitat loss and the threat of global climate change is a current reality. Global priorities centered on the two branches of forest genetics—conserving natural hardwood tree populations and sustainably producing wood products—will continue to drive research priorities in the field of hardwood genomics.

REFERENCES

Abrams, M. D. 2002. The postglacial history of oak forests in eastern North America. In W. J. H. McShea and W. M. Healey (eds), *Oak Forest Ecosystems: Ecology and Management for Wildlife*, pp. 34–45. Baltimore: Johns Hopkins University Press.

Agrawal, A. A., A. C. Erwin, and S. C. Cook. 2008. Natural selection on and predicted responses of ecophysiological traits of swamp milkweed (*Asclepias incarnata*). *J Ecol* 96:536–542.

Aitken, S. N., S. Yeaman, J. A. Holliday, T. Wang, and S. Curtis-McLane. 2008. Adaptation, migration or extirpation: Climate change outcome for tree populations. *Evol Appl* 1:95–111.

Akhunov, E., C. Nicolet, and J. Dvorak. 2009. Single nucleotide polymorphism genotyping in polyploidy wheat with the Illumina GoldenGate assay. *Theor Appl Genet* 119:507–517.

Aldrich, P. R., M. Jagtap, C. H. Michler, and J. Romero-Severson. 2004. Amplification of North American red oak microsatellite markers in European white oak and Chinese chestnut. *Silvae Genet* 52:176–179.

Aldrich, P. R., C. H. Michler, W. Sun, and J. Romero-Severson. 2002. Microsatellite markers for northern red oak (Fagaceae: *Quercus rubra*). *Mol Ecol Notes* 2:472–474.

Alexander, M. T., L. M. Worthen, and J. H. Craddock. 2005. Conservation of *Castanea dentata* germplasm in the southeastern United States. *Acta Hort* 693:485–490.

Al-Masri, M. S., Y. Amin, B. Al-Akel, and T. Al-Naama. 2010. Biosorption of cadmium, lead, and uranium by powder of poplar leaves and branches. *Appl Biochem Biotech* 160:976–987.

Almeida-Rodriguez, A. M., J. E. K. Cooke, F. Yeh, and J. J. Zwiazek. 2010. Functional characterization of drought-responsive aquaporins in *Populus balsamifera* and *Populus simonii* × *balsamifera* clones with different drought resistance strategies. *Physiol Plantarum* 140:321–333.

Anagnostakis, S. L. 1992. Chestnuts and the introduction of chestnut blight. *Annu Rep Northern Nut Growers Assoc* 83:23–37.

Anagnostakis, S. L., B. Hau, and J. Kranz. 1986. Diversity of vegetative compatibility groups of *Cryphonectria parasitica* in Connecticut and Europe. *Plant Dis* 70:536–538.

Arabidopsis Genome Initiative. 2000. Analysis of the genome sequence of the flowering plant *Arabidopsis thaliana*. *Nature* 408:796–814.

Bakker, E. G. 2001. Towards molecular tools for management of oak forest. PhD Thesis, Department of Plant Breeding, University of Wageningen, Alterra, The Netherlands.

Barakat, A., D. S. DiLoreto, Y. Zhang et al. 2009. Comparison of the transcriptomes of American chestnut (*Castanea dentata*) and Chinese chestnut (*Castanea mollissima*) in response to the chestnut blight infection. *BMC Plant Biol* 9:51–62.

Barreneche, T., C. Bodenes, C. Lexer et al. 1998. A genetic linkage map of *Quercus robur* L. (pedunculate oak) based on RAPD, SCAR, microsatellite, minisatellite, isozyme and 5S rDNA markers. *Theor Appl Genet* 97:1090–1103.

Barreneche, T., M. Casasoli, K. Russell et al. 2004. Comparative mapping between *Quercus* and *Castanea* using simple-sequence repeats (SSRs). *Theor Appl Genet* 108:558–566.

Berg, E. E. and J. L. Hamrick. 1995. Fine-scale genetic structure of a turkey oak forest. *Evolution* 49:110–120.

Berry, J. A., D. J. Beerling, and P. J. Franks. 2010. Stomata: Key players in the earth system, past and present. *Curr Opin Plant Biol* 13:233–240.

Berta, M., A. Giovannelli, F. Sebastiani, A. Camussi, and M. L. Racchi. 2010. Transcriptome changes in the cambial region of poplar (*Populus alba* L.) in response to water deficit. *Plant Biol* 12:341–354.

Birchenko, I., Y. Feng, and J. Romero-Severson. 2009. Biogeographical distribution of chloroplast diversity in northern red oak (*Quercus rubra* L.). *Am Midl Nat* 161:134–145.

Bodenes, C., E. Chancerel, O. Gailing et al. 2012. Comparative mapping in the Fagaceae and beyond with EST-SSRs. *BMC Plant Biol* 12:153–171.

Bogeat-Triboulot, M. B., M. Brosché, J. Renaut et al. 2007. Gradual soil water depletion results in reversible changes of gene expression, protein profiles, ecophysiology, and growth performance in *Populus euphratica*, a poplar growing in arid regions. *Plant Physiol* 143:876–892.

Bonhomme, L., R. Monclus, D. Vincent et al. 2009a. Genetic variation and drought response in two *Populus × euramericana* genotypes through 2-DE proteomic analysis of leaves from field and glasshouse cultivated plants. *Phytochemistry* 70:988–1002.

Bonhomme, L., R. Monclus, D. Vincent et al. 2009b. Leaf proteome analysis of eight *Populus × euramericana* genotypes: Genetic variation in drought response and in water-use efficiency involves photosynthesis-related proteins. *Proteomics* 9:4121–4142.

Bradshaw, H. D. and R. F. Stettler. 1995. Molecular genetics of growth and development in *Populus*. IV. Mapping QTLs with large effects on growth, form, and phenology traits in a forest tree. *Genetics* 139:963–973.

Braun, E. 1950. *Deciduous Forests of Eastern North America*. New York: MacMillan.

Breen, A. L., E. Glenn, A. Yeager, and M. S. Olson. 2009. Nucleotide diversity among natural populations of a North American poplar (*Populus balsamifera*, Salicaceae). *New Phytol* 182:763–773.

Brendel, O., D. Le Thiec, C. Scotti-Saintagne, C. Bodénès, A. Kremer, and J. M. Guehl. 2008. Quantitative trait loci controlling water use efficiency and related traits in *Quercus robur* L. *Tree Genet Genome* 4:263–278.

Brewbaker, J. L. 1957. Pollen cytology and self-incompatibility in plants. *J Hered* 48:271–277.

Brinker, M., M. Brosché, B. Vinocur et al. 2010. Linking the salt transcriptome with physiological responses of a salt-resistant *Populus* species as a strategy to identify genes important for stress acclimation. *Plant Physiol* 154:1697–1709.

Brinkman, M. A. and K. J. Frey. 1977. Yield component analysis of oat isolines that produce different grain yields. *Crop Sci* 17:165–168.

Brosché, M., B. Vinocur, E. Alatalo et al. 2005. Gene expression and metabolite profiling of *Populus euphratica* growing in the Negev desert. *Genome Biol* 6:6–12.

Brown, G. R., G. P. Gill, R. J. Kuntz, C. H. Langely, and D. B. Neale. 2004. Nucleotide diversity and linkage disequilibrium in loblolly pine. *Proc Natl Acad Sci USA* 101:15225–15260.

Brown, G. R., E. E. Kadel III, D. L. Bassoni et al. 2001. Anchored reference loci in loblolly pine (*Pinus taeda* L.) for integrating pine genomics. *Genetics* 159:799–809.

Bulmer, M. G. 1971. The effect of selection on genetic variability. *Am Nat* 105:201–211.

Burdon, R. D. and P. L. Wilcox. 2007. Population management: Potential impacts of advances in genomics. *New Forests* 34:187–206.

Burnham, C. R. 1982. Breeding for chestnut blight resistance. *Nutshell* 35:8–9.

Burnham, C. R. 1986. Chestnut hybrids from USDA breeding programs. *J Am Chestnut Found* 1(2):8–12.

Burnham, C. R., P. A. Rutter, and D. W. French. 1986. Breeding blight-resistant chestnuts. *Plant Breed Rev* 4:347–397.

Caron, E., P. Lafrance, J. C. Auclair, and M. Duchemin. 2010. Impact of grass and grass with poplar buffer strips on atrazine and metolachlor losses in surface runoff and subsurface infiltration from agricultural plots. *J Environ Quality* 39:617–629.

Caruso, A., F. Chefdor, S. Carpin et al. 2008. Physiological characterization and identification of genes differentially expressed in response to drought induced by PEG 6000 in *Populus canadensis* leaves. *J Plant Physiol* 165:932–941.

Casasoli, M., J. Derory, C. Morera-Dutrey et al. 2006. Comparison of quantitative trait loci for adaptive traits between oak and chestnut based on an expressed sequence tag consensus map. *Genetics* 172:533–546.

Casasoli, M., C. Mattioni, M. Cherubini, and F. Villani. 2001. A genetic linkage map of European chestnut (*Castanea sativa* Mill.) based on RAPD, ISSR, and isozyme markers. *Theor Appl Genet* 102:1190–1199.

Chagne, D., G. Brown, C. Lalanne et al. 2003. Comparative genome and QTL mapping between maritime and loblolly pines. *Mol Breed* 12:185–195.

Chen, F., S. Zhang, H. Jiang, W. Ma, H. Korpelainen, and C. Li. 2011. Comparative proteomics analysis of salt response reveals sex-related photosynthetic inhibition by salinity in *Populus cathayana* cuttings. *J Proteome Res* 10:3944–3958.

Chen, K. and Y. Peng. 2010. AFLP analysis of genetic diversity in *Populus cathayana* Rehd originating from southeastern Qinghai-Tibetan plateau of China. *Pakistan J Bot* 42:117–127.

Chen, S., J. Jiang, H. Li, and G. Liu. 2012. The salt-responsive transcriptome of *Populus simonii × Populus nigra* via DGE. *Gene* 504:203–212.

Chen, S. and A. Polle. 2010. Salinity tolerance of *Populus*. *Plant Biol* 12:317–333.

Chenault, N., S. Arnaud-Haond, M. Juteau et al. 2011. SSR-based analysis of clonality, spatial genetic structure and introgression from the Lombardy poplar into a natural population of *Populus nigra* L. along the Loire River. *Tree Genet Genome* 7:1249–1262.

Chuine, I., J. Belmonte, and A. Mignot. 2000. A modeling analysis of the genetic variation of phenology between tree populations. *J Ecol* 88:561–570.

Chuine, I. and P. Cour. 1999. Climatic determinants of budburst seasonality in four temperate zone tree species. *New Phytol* 143:339–349.

Clapper, R. B. 1952. Relative blight resistance of some chestnut species and hybrids. *J For* 50:453–455.

Clark, C., T. Kubisiak, B.-C. Lee et al. 2001. AFLPs—Towards a saturated genetic map for *Castanea*. In Plant and Animal Genome IX Conference, San Diego, 13–17 January.

Coble, A. P. and T. E. Kolb. 2012. Riparian tree growth response to drought and altered streamflow along the Dolores River, Colorado. *West J Appl For* 27:205–211.

Cohen, D., M. B. Bogeat-Triboulot, E. Tisserant et al. 2010. Comparative transcriptomics of drought responses in *Populus*: A meta-analysis of genome-wide expression profiling in mature leaves and root apices across two genotypes. *BMC Genomics* 11:630.

Costa, R., C. Santos, F. Tavares et al. 2011. Mapping and transcriptomic approaches implemented for understanding disease resistance to *Phytophthora cinammomi* in *Castanea* sp. *BMC Proc* 5(Suppl 7):O18.

Cronn, R., A. Liston, M. Parks, D. S. Gernandt, R. Shen, and T. Mockler. 2008. Multiplex sequencing of plant chloroplast genomes using Solexa sequencing-by-synthesis technology. *Nucleic Acid Res* 36:19.

Cseke, L. J., C. J. Tsai, A. Rogers et al. 2009. Transcriptomic comparison in the leaves of two aspen genotypes having similar carbon assimilation rates but different partitioning patterns under elevated [CO_2]. *New Phytol* 182:891–911.

Cunningham, I. S. 1984. *Frank N. Meyer, Plant Hunter in Asia*. Ames: Iowa State University Press.

Damour, G., T. Simonneau, H. Cochard, and L. Urban. 2010. An overview of models of stomatal conductance at the leaf level. *Plant Cell Environ* 33:1419–1438.

Dane, F., L. K. Hawkins, and H. Huang. 1999. Genetic variation and population structure of *Castanea pumila* var. ozarkensis. *J Am Soc Hort Sci* 124:666–670.

Dane, F., P. Lang, H. Huang, and Y. Fu. 2003. Intercontinental genetic divergence of *Castanea* species in eastern Asia and eastern North America. *Heredity* 91:314–321.

Deguilloux, M. F., S. Dumolin-Lapegue, L. Gielly, D. Grivet, and R. J. Petit. 2003. A set of microsatellite primers for the amplification of chloroplast microsatellites in *Quercus*. *Mol Ecol Notes* 3:24–27.

Delcourt, H. R. 2002. *Forests in Peril: Tracking Deciduous Trees from Ice-Age Refuges into the Greenhouse World*. Granville: McDonald and Woodward.

Delcourt, H. R. and P. A. Delcourt. 1984. Ice age haven for hardwoods. *Nat Hist* 93:22–28.

De Micco, V., G. Aronne, and P. Baas. 2008. Wood anatomy and hydraulic architecture of stems and twigs of some Mediterranean trees and shrubs along a mesic-xeric gradient. *Trees* 22:643–655.

Derory, J., P. Leger, V. Garcia et al. 2006. Transcriptome analysis of bud burst in sessile oak (*Quercus petraea*). *New Phytol* 170:723–738.

Derory, J., C. Scotti-Saintagne, E. Bertocchi et al. 2010. Contrasting relations between the diversity of candidate genes and variation of bud burst in natural and segregating populations of European oaks. *Heredity* 105(4):401–411.

Devey, M. 2004. Genomics and gene discovery in forest trees. In C. Walter and M. Carson (eds), *Plantation Forest Biotechnology for the 21st Century*. Trivandrum, India: Research Signpost.

Diskin, M., K. C. Steiner, and F. V. Hebard. 2006. Recovery of American chestnut characteristics following hybridization and backcross breeding to restore blight ravaged *Castanea dentata*. *Forest Ecol Manag* 223:439–447.

Double, M. L. and W. L. Macdonald. 2002. Hypovirus deployment, establishment, and spread: Results after six years of canker treatment. *Phytopathology* 92:S94.

Drost, D. R., E. Novaes, C. Boaventura-Novaes et al. 2009. A microarray-based genotyping and genetic mapping approach for highly heterozygous outcrossing species enables localization of a large fraction of the unassembled *Populus trichocarpa* genome sequence. *Plant J* 58:1054–1067.

Du, Q., W. Pan, B. Xu, B. Li, and D. Zhang. 2013. Polymorphic simple sequence repeat (SSR) loci within cellulose synthase (*PtoCesA*) genes are associated with growth and wood properties in *Populus tomentosa*. *New Phytol* 197:763–776.

Du, Q., B. Wang, Z. Wei, D. Zhang, and B. Li. 2012. Genetic diversity and population structure of Chinese white poplar (*Populus tomentosa*) revealed by SSR markers. *J Hered* 103:853–862.

Ducousso, A. and S. Bordacs. 2004. *EUFORGEN Technical Guidelines for Genetic Conservation and Use for Pedunculate and Sessile Oaks (*Quercus robur *and* Q. petraea*).* Rome, Italy: International Plant Genetic Resources Institute.

Ducousso, A., J. P. Guyon, and A. Kremer. 1996. Latitudinal and altitudinal variation of bud burst in western populations of sessile oak (*Quercus petraea* (Matt) Liebl). *Ann Sci For* 53:775–782.

Durand, J., C. Bodenes, E. Chancerel et al. 2010. A fast and cost-effective approach to develop and map EST-SSR markers: Oak as a case study. *BMC Genomics* 11:570–583.

Eriksson, A., P. Fernstrom, B. Mehlig, and S. Sagitov. 2008. An accurate model for genetic hitch-hiking. *Genetics* 178:439–451.

Eriksson, G., A. Pliura, J. Fernandez-Lopez et al. 2005. Management of genetic resources of the multi-purpose tree species *Castanea sativa* Mill. *Acta Hort* 693:373–386.

Fabbrini, F., M. Gaudet, C. Bastien et al. 2012. Phenotypic plasticity, QTL mapping and genomic characterization of bud set in black poplar. *BMC Plant Biol* 12:47.

Falconer, D. S. 1960. *Introduction to Quantitative Genetics.* New York: Ronald.

Fan, J. B., A. Oliphant, R. Shen et al. 2003. Highly parallel SNP genotyping. *Cold Spring Harb Symp Quant Biol* 68:69–78.

Fang, G., B. P. Blackmon, M. E. Staton et al. 2012. A physical map of the Chinese chestnut (*Castanea mollissima*) genome and its integration with the genetic map. *Tree Genet Genome* 9:525–537.

Feng, Y., Y. L. Sun, and J. Romero-Severson. 2008. Heterogeneity and spatial autocorrelation for chloroplast haplotypes in three old growth populations of northern red oak. *Silvae Genet* 57(4–5):212–220.

Fichot, R., F. Laurans, R. Monclus, A. Moreau, G. Pilate, and F. Brignolas. 2009. Xylem anatomy correlates with gas exchange, water-use efficiency and growth performance under contrasting water regimes: Evidence from *Populus deltoides* × *Populus nigra* hybrids. *Tree Phys* 29:1537–1549.

Fineschi, S., D. Taurchini, F. Villani, and G. G. Vendramin. 2000. Chloroplast DNA polymorphism reveals little geographical structure in *Castanea sativa* Mill. (Fagaceae) throughout southern European countries. *Mol Ecol* 9:1495–1503.

Finkeldey, R. and H. H. Hattemer. 2007. *Tropical Forest Genetics.* Berlin: Springer.

Fladung, M. and J. Buschbom. 2009. Identification of single nucleotide polymorphisms in different *Populus* species. *Trees* 23:1199–1212.

Fossati, T., I. Zapelli, S. Bisoffi et al. 2005. Genetic relationships and clonal identity in a collection of commercially relevant poplar cultivars assessed by AFLP and SSR. *Tree Genet Genome* 1:11–20.

Fraser, D. J. 2008. How well can captive breeding programs conserve biodiversity? A review of salmonids. *Evol Appl* 1:535–586.

Fussi, B., C. Lexer, and B. Heinze. 2010. Phylogeography of *Populus alba* (L.) and *P. tremula* (L.) in Central Europe: Secondary contact and hybridisation during recolonisation from disconnected refugia. *Tree Genet Genome* 6:439–450.

Gailing, O. 2008. QTL analysis of leaf morphological characters in a *Quercus robur* full-sib family (*Q. robur* × *Q. robur* subsp. *slavonica*). *Plant Biol* 10:624–634.

Gailing, O., B. Vornam, L. Leinemann, and R. Finkeldey. 2009. Genetic and genomic approaches to assess adaptive genetic variation in plants: Forest trees as a model. *Physiol Plantarum* 137:509–519.

Galas, D. J. (ed.). 2002. *Genomic Technologies: Present and Future.* Wymondham, UK: Horizon Scientific Press.

Gall, W. and K. Taft. 1973. Variation in height growth and flushing of northern red oak (*Quercus rubra* L.). In *Proceedings of the 12th Southern Forest Tree Improvement Conference*, pp. 190–199. Baton Rouge, LA, 12–13 June.

Gao, D., Q. Gao, H.-Y. Xu et al. 2009a. Physiological responses to gradual drought stress in the diploid hybrid *Pinus densata* and its two parental species. *Trees* 23:717–728.

Gao, J., Y. Zhang, C. Wang, S. Zhang, L. Qi, and W. Song. 2009b. AFLP fingerprinting of *Populus deltoids* and *Populus* × *canadensis* elite accessions. *New Forests* 37:333–344.

Gardiner, J., S. Schroeder, M. L. Polacco et al. 2004. Anchoring 9,371 maize expressed sequence tagged unigenes to the bacterial artificial chromosome contig map by two-dimensional overgo hybridization. *Plant Physiol* 134:1317–1326.

Geraldes, A., J. Pang, N. Thiessen et al. 2011. SNP discovery in black cottonwood (*Populus trichocarpa*) by population transcriptome resequencing. *Mol Ecol Res* 11(Suppl. 1):81–92.

Gonzalez-Martinez, S. C., K. V. Krutovsky, and D. B. Neale. 2006. Forest-tree population genomics and adaptive evolution. *New Phytol* 170:227–238.

Gortan, E., A. Nardin, A. Gascó, and S. Salleo. 2009. The hydraulic conductance of *Fraxinus ornus* leaves is constrained by soil water availability and coordinated with gas exchange rates. *Tree Phys* 29:529–539.

Gourcilleau, D., M.-B. Bogeat-Triboulot, D. le Thiec et al. 2010. DNA methylation and histone acetylation: Genotypic variations in hybrid poplars, impact of water deficit and relationships with productivity. *Ann For Sci* 67:208.

Gouveia, A. C. and H. Freitas. 2009. Modulation of leaf attributes and water use efficiency in *Quercus suber* along a rainfall gradient. *Trees* 23:267–275.

Grattapaglia, D. and R. Sederoff. 1994. Genetic linkage maps of *Eucalyptus grandis* and *Eucalyptus urophylla* using a pseudo-testcross: Mapping strategy and RAPD markers. *Genetics* 137:1121–1137.

Grente, J. and S. Berthelay-Sauret. 1978. Biological control of chestnut blight in France. In W. L. MacDonald, F. C. Chech, J. Luchok, and H. C. Smith (eds), *Proceedings of the American Chestnut Symposium*, pp. 4–5. Morgantown: West Virginia University Books.

Grente, J. and S. Sauret. 1969. L'hypovirulence exclusive, phénomène original en pathologie végétale. *C.R. Acad Sci* 268:3173–3176.

Griffin, G. J. 1989. Incidence of chestnut blight and survival of American chestnut in forest clearcut and neighboring understory sites. *Plant Dis* 73:123–127.

Griffin, G. J. 2000. Blight control and restoration of the American chestnut. *J For* 98:22–27.

Griffin, G. J., F. V. Hebard, R. W. Wendt, and J. R. Elkins. 1983. Survival of American chestnut trees: Evaluation of blight resistance and virulence in *Endothia parasitica*. *Phytopathology* 73:1084–1092.

Grisel, N., S. Zoller, M. Künzli-Gontarczyk et al. 2010. Transcriptome responses to aluminum stress in roots of aspen (*Populus tremula*). *BMC Plant Biol* 10:185.

Grönlund, A., R. P. Bhalerao, and J. Karlsson. 2009. Modular gene expression in poplar: A multilayer network approach. *New Phytol* 181:315–322.

Hall, D., V. Luquez, V. M. Garcia, K. R. St Onge, S. Jansson, and P. K. Ingvarsson. 2007. Adaptive population differentiation in phenology across a latitudinal gradient in European Aspen (*Populus tremula* L.): A comparison of neutral markers, candidate genes, and phenotypic traits. *Evolution* 61:2849–2860.

Hallik, L., U. Niinemets, and I. J. Wright. 2009. Are species shade and drought tolerance reflected in leaf-level structural and functional differentiation in Northern Hemisphere temperate woody flora? *New Phytol* 184:257–274.

Hamanishi, E. T., S. Raj, O. Wilkins et al. 2010. Intraspecific variation in the *Populus balsamifera* drought transcriptome. *Plant Cell Environ* 33:1742–1755.

Hamelin, R. C. 2012. Contributions of genomics to forest pathology. *Can J Plant Path* 34:20–28.

Hamrick, J. L. 2004. Response of forest trees to global environmental changes. *For Ecol Manag* 197:323–335.

Hamrick, J. L., M. J. W. Godt, and S. L. Sherman-Broyles. 1992. Factors influencing levels of genetic diversity in woody plant species. *New Forests* 6:95–124.

Han, J. C., G. P. Wang, D. J. Kong, Q. X. Liu, and X. Y. Zhang. 2007. Genetic diversity of Chinese chestnut (*Castanea mollissima*) in Hebei. *Acta Hort* 760:573–577.

Hebard, F. V. 2005. Notes from Meadowview Research Farms 2004–2005. *J Am Chestnut Found* 19:27–39.

Hebard, F. V. 2006. The backcross breeding program of the American Chestnut Foundation. In K. C. Steiner and J. E. Carlson (eds), *Restoration of the American Chestnut Tree to Forest Lands—Proceedings of a Conference and Workshop*, pp. 61–77, Natural Resources Rep NPS/NCR/CUE/NRR—2006/001. Washington, DC: National Park Service.

Hewitt, G. M. 1988. Hybrid zones—Natural laboratories for evolutionary studies. *Trends Ecol Evol* 3:158–167.

Hoffman, A. A. and L. H. Rieseberg. 2008. Revising the impact of inversions in evolution: From population genetic markers to drivers of adaptive shifts and speciation? *Annu Rev Ecol Evol Syst* 39:21–42.

Hokanson, S. C., J. G. Isebrands, R. J. Jensen, and J. F. Hancock. 1993. Isozyme variation in oaks of the Apostle Islands in Wisconsin: Genetic structure and levels of inbreeding in *Quercus rubra* and *Q. ellipsoidalis* (Fagaceae). *Am J Bot* 80:1349–1357.

Hospital, F. 2001. Size of donor chromosome segments around introgressed loci and reduction of linkage drag in marker-assisted backcross programs. *Genetics* 158:1363–1379.

Howe, G. T., S. N. Aitken, D. B. Neale, K. D. Jermstad, N. C. Wheeler, and T. H. H. Chen. 2003. From genotype to phenotype: Unraveling the complexities of cold adaptation in forest trees. *Can J Bot* 81(12):1247–1266.

Huang, H., W. A. Carey, F. Dane, and J. D. Norton. 1996. Evaluation of Chinese chestnut cultivars for resistance to *Cryphonectria parasitica. Plant Dis* 80:45–47.

Huang, H., F. Dane, and T. L. Kubisiak. 1998. Allozyme and RAPD analysis of the genetic diversity and geographic variation in wild populations of the American chestnut (Fagaceae). *Am J Bot* 85:1013–1021.

Huang, H., F. Dane, and J. D. Norton. 1994. Allozyme diversity in Chinese, Seguin, and American chestnut (*Castanea* spp.). *Theor Appl Genet* 88:981–985.

Induri, B. R., D. R. Ellis, G. T. Slavov et al. 2012. Identification of quantitative trait loci and candidate genes for cadmium tolerance in *Populus. Tree Phys* 32:626–638.

Ingvarsson, P. K. 2008. Multilocus patterns of nucleotide polymorphism and the demographic history of *Populus tremula. Genetics* 180:329–340.

Ingvarsson, P. K., M. V. Garcia, D. Hall, V. Luquez, and S. Jansson. 2006. Clinal variation in *phyB2*, a candidate gene for day-length-induced growth cessation and bud set, across a latitudinal gradient in European aspen (*Populus tremula*). *Genetics* 172:1845–1853.

Intergovernmental Panel on Climate Change. 2007. Climate change 2007: Impacts, adaptation and vulnerability. Contribution of Working Group II to the 4th Assessment Report of the Intergovernmental Panel on Climate Change. Cambridge: Cambridge University Press.

Irwin, H. 2003. The road to American chestnut restoration. *J Am Chestnut Found* 16:6–13.

Isaakidis, A., T. Sotiropoulos, D. Almaliotis, I. Therios, and D. Stylianidis. 2004. Response to severe water stress of the almond *Prunus amygdalus*, "Ferragnès" grafted on eight rootstocks. *New Zeal J Crop Hort Sci* 32:355–362.

Isabel, N., M. Lamothe, and S. L. Thompson. 2013. A second-generation diagnostic single nucleotide polymorphism (SNP)-based assay, optimized to distinguish among eight poplar (*Populus* L.) species and their early hybrids. *Tree Genet Genomes* 9:621–626.

Ismail, M., R. Y. Soolanayakanahally, P. K. Ingvarsson et al. 2012. Comparative nucleotide diversity across North American and European *Populus* species. *J Mol Evol* 74:257–272.

Jacobs, D. F. 2007. Toward development of silvicultural strategies for forest restoration of American chestnut (*Castanea dentata*) using blight-resistant hybrids. *Biol Conserv* 137:497–506.

Jacobs, D. F. and L. R. Severeid. 2004. Dominance of interplanted American chestnut (*Castanea dentata*) in southwestern Wisconsin, USA. *Forest Ecol Manag* 191:111–120.

Jansen, R. K., L. A. Raubeson, J. L. Boore et al. 2005. Methods for obtaining and analyzing whole chloroplast genome sequences. *Method Enzymol* 395:348–384.

Jansen, R. K., C. Saski, S. B. Lee, A. K. Hansen, and H. Daniell. 2010. Complete plastid genome sequences of three rosids (*Castanea, Prunus, Theobroma*): Evidence for at least two independent transfers of *rpl22* to the nucleus. *Mol Biol Evol* 28:835–847.

Janz, D., K. Behnke, J.-P. Schnitzler, B. Kanawat, P. Schmitt-Kopplin, and A. Polle. 2010. Pathway analysis of the transcriptome and metabolome of salt sensitive and tolerant poplar species reveals evolutionary adaption of stress tolerance mechanisms. *BMC Plant Biol* 10:150.

Jaynes, R. A. 1974. *Genetics of Chestnut.* Washington, DC: Forest Service, USDA.

Jaynes, R. A. 1975. Chestnut. In J. Janick and J. Moore (eds), *Advances in Fruit Breeding*, pp. 590–603, West Lafayette: Purdue University Press.

Jermstad, K. D., D. L. Bassoni, K. S. Jech, N. C. Wheeler, and D. B. Neale. 2001. Mapping of quantitative trait loci controlling adaptive traits in coastal Douglas-fir. 1. Timing of vegetative bud flush. *Theor Appl Genet* 102:1142–1151.

Jiang, H., S. Peng, S. Zhang, X. Li, H. Korpelainen, and C. Li. 2012. Transcriptional profiling analysis in *Populus yunnanensis* provides insight into molecular mechanisms of sexual differences in salinity tolerance. *J Exp Bot* 63:3709–3726.

Johansen, A. D. and R. T. Latta. 2003. Mitochondrial haplotype, distribution, seed dispersal, and patterns of postglacial expansion of ponderosa pine. *Molec Ecol* 12:293–298.

Johansson, I., M. Karlsson, U. Johanson, C. Larsson, and P. Kjellbom. 2000. The role of aquaporins in cellular and whole plant water balance. *Biomembranes* 1465:324–342.

Justin, M. Z., N. Pajk, V. Zupanc, and V. Zupančič. 2010. Phytoremediation of landfill leachate and compost wastewater by irrigation of *Populus* and *Salix*: Biomass and growth response. *Waste Manag* 30:1032–1042.

Kaplan, N. L., R. R. Hudson, and C. H. Langely. 1989. The "hitchhiking effect" revisited. *Genetics* 123:887–889.

Kelleher, C. T., J. Wilkin, Z. Zhuang et al. 2012. SNP discovery, gene diversity, and linkage disequilibrium in wild populations of *Populus tremuloides. Tree Genet Genome* 8:821–829.

Khan, H. R., J. G. Paull, K. H. M. Siddique, and F. L. Stoddard. 2010. Faba bean breeding for drought-affected environments: A physiological and agronomic perspective. *Field Crop Res* 115:279–286.

Kieffer, P., S. Planchon, O. Oufir et al. 2009a. Combining proteomics and metabolite analyses to unravel cadmium stress-response in poplar leaves. *J Proteome Res* 8:400–417.

Kieffer, P., P. Schröder, J. Dommes, L. Hoffmann, J. Renaut, and J.-F. Hausmana. 2009b. Proteomic and enzymatic response of poplar to cadmium stress. *J Proteomics* 72:379–396.

Kilian, J., F. Peschke, K. W. Berendzen, K. Harter, and D. Wanke. 2012. Prerequisites, performance and profits of transcriptional profiling the abiotic stress response. *Biochimica et Biophysica Acta* 1819:166–175.

Kremer, A., A. G. Abbott, J. E. Carlson et al. 2012. Genomics of Fagaceae. *Tree Genet Genome* 8:583–610.

Kremer, A., M. Casaoli, T. Barreneche et al. 2007. Genome mapping and molecular breeding in plants. In C. Kole (ed.), *Forest Trees*, vol. 7, pp. 161–187. Heidelberg: Springer-Verlag Berlin.

Kremer, A., V. LeCorre, R. J. Petit, and A. Ducousso. 2009. Historical and contemporary dynamics of adaptive differentiation in European oaks. In J. A. DeWoody, J. Bickham, C. H. Michler, K. Nichols, G. Rhodes, and K. Woeste (eds), *Molecular Approaches in Natural Resource Conservation*, pp. 101–122. New York: Cambridge University Press.

Kriebel, H. B. 1993. Intraspecific variation of growth and adaptive traits in North American oak species. *Ann For Sci* 50:153s–165s.

Krutovsky, K. V. 2006. From population genetics to population genomics of forest trees: Integrated population genomics approach. *Russ J Genet* 42:1088–1100.

Krutovsky, K. V. and D. B. Neale. 2005. Nucleotide diversity and linkage disequilibrium in cold-hardiness- and wood-quality-related candidate genes in Douglas fir. *Genetics* 171:2029–2041.

Kubisiak, T. L., F. V. Hebard, C. D. Nelson et al. 1997. Molecular mapping of resistance to blight in an interspecific cross in the genus *Castanea*. *Phytopathology* 87:751–759.

Kubisiak, T. L., C. D. Nelson, M. E. Staton et al. 2012. A transcriptome-based genetic map of Chinese chestnut (*Castanea mollissima*) and identification of regions of segmental homology with peach (*Prunus persica*). *Tree Genet Genome* 9:557–571.

Kubisiak, T. L. and J. H. Roberds. 2003. Genetic variation in natural populations of American chestnut. *J Am Chestnut Found* 16:42–48.

Lang, P., F. Dane, and T. L. Kubisiak. 2006. Phylogeny of *Castanea* (Fagaceae) based on chloroplast trnT-L-F sequence data. *Tree Genet Genome* 2:132–139.

Lang, P., F. Dane, T. L. Kubisiak, and H. Huang. 2007. Molecular evidence for an Asian origin and a unique westward migration of species in the genus *Castanea* via Europe to North America. *Mol Phylogenet Evol* 43:49–59.

Lang, P. and H. Huang. 1999. Genetic diversity and geographic variation in natural populations of the endemic *Castanea* species in China. *Acta Bot Sin* 41:651–657.

Latta, R. G. and J. B. Mitton. 1999. Historical separation and present gene flow through a zone of secondary contact in ponderosa pine. *Evolution* 53:769–776.

LeCorre, V. and A. Kremer. 2003. Genetic variability at neutral markers, quantitative trait loci, and trait in a subdivided population under selection. *Genetics* 164:1205–1219.

LeCorre, V., N. Machon, R. J. Petit, and A. Kremer. 1997. Colonization with long-distance seed dispersal and distribution of maternally inherited diversity in forest trees: A simulation study. *Genet Res* 69:117–125.

Lee, K. M., Y. Y. Kim, and J. O. Hyun. 2011. Genetic variation in populations of *Populus davidiana* Dode based on microsatellite marker analysis. *Genes Genomics* 33:163–171.

Li, B., W. Yin, and X. Xia. 2009. Identification of microRNAs and their targets from *Populus euphratica*. *Biochem Biophys Res Commun* 388:272–277.

Li, T., J. Chen, S. Qiu et al. 2012. Deep sequencing and microarray hybridization identify conserved and species-specific microRNAs during somatic embryogenesis in hybrid yellow poplar. *PLoS ONE* 7:e43451.

Li, X. and F. Dane. 2012. Comparative chloroplast and nuclear DNA analysis of *Castanea* species in the southern region of the USA. *Tree Genet Genome* 9:107–116.

Liesebach, H., V. Schneck, and E. Ewald. 2010. Clonal fingerprinting in the genus *Populus* L. by nuclear microsatellite loci regarding differences between sections, species and hybrids. *Tree Genet Genome* 6:259–269.

Liu, S. and J. E. Carlson. 2006. Selection for Chinese versus American genetic material in blight resistant backcross progeny using genomic DNA. In K. C. Steiner and J. E. Carlson (eds), *Restoration of the American Chestnut Tree to Forest Lands—Proceedings of a Conference and Workshop*, Natural Resources Report NPS/NCR/CUE/NRR—2006/001, Washington, DC: National Park Service.

Magni, C. R., A. Ducousso, H. Caron, R. J. Petit, and A. Kremer. 2005. Chloroplast DNA variation of *Quercus rubra* L. in North America and comparison with other Fagaceae. *Molec Ecol* 14:513–524.

Malcolm, J. R., A. Markham, R. P. Nelson, and M. Garaci. 2002. Estimated migration rates under scenarios of global climate change. *J Biogeog* 29:835–849.

Marmiroli, M., G. Visioli, E. Maestri, and N. Marmiroli. 2011. Correlating SNP genotype with the phenotypic response to exposure to Cadmium in *Populus* spp. *Envir Sci Tech* 45:4497–4505.

Marron, N., M. Villar, M. Dreyer et al. 2005. Diversity of leaf traits related to productivity in 31 *Populus deltoides* × *Populus nigra* clones. *Tree Phys* 25:425–435.

Mauricio, R. 2001. Mapping quantitative trait loci in plants: Uses and caveats for evolutionary biology. *Genetics* 2:370–382.

Maynard Smith, J. and J. Haigh. 1974. The hitch-hiking effect of a favorable gene. *Genet Res* 23:23–35.

McGee, C. E. 1974. Elevation of seed sources and planting sites affects phenology and development of red oak seedlings. *For Sci* 20:160–164.

McKay, H. K. and R. G. Latta. 2002. Adaptive population divergence: Markers, QTL, and traits. *Trends Ecol Evol* 17:285–291.

McKay, J. W. 1942. Self-sterility in the Chinese chestnut (*Castanea mollissima*). *Acta Hort* 41:156–160.

Meirmans, P. G., M. Lamothe, M.-C. Gros-Louis et al. 2010. Complex patterns of hybridization between exotic and native North American poplar species. *Am J Bot* 97:1688–1697.

Merkel, H. W. 1906. A deadly fungus on the American chestnut. *Annu Rep New York Zool Soc* 10:97–103.

Merkle, S. A., G. M. Andrade, C. J. Nairn, W. A. Powell, and C. A. Maynard. 2006. Restoration of threatened species: A noble cause for transgenic trees. *Tree Genet Genome* 3:111–118.

Meziane, M. and B. Shipley. 2001. Direct and indirect relationships between specific leaf area, leaf nitrogen and leaf gas exchange effects of irradiance and nutrient supply. *Ann Bot* 88:915–927.

Milgroom, M. G. and P. Cortesi. 2004. Biological control of chestnut blight with hypovirulence: A critical analysis. *Annu Rev Phytopathol* 42:311–338.

Monclus, R., E. Dreyer, M. Villar et al. 2006. Impact of drought on productivity and water use efficiency in 29 genotypes of *Populus deltoides* × *Populus nigra*. *New Phytol* 169:765–777.

Monclus, R., J.-C. Leplé, C. Bastien et al. 2012. Integrating genome annotation and QTL position to identify candidate genes for productivity, architecture and water-use efficiency in *Populus* spp. *BMC Plant Biol* 12:173.

Monclus, R., M. Villar, C. Barbaroux et al. 2009. Productivity, water-use efficiency and tolerance to moderate water deficit correlate in 33 poplar genotypes from a *Populus deltoides* × *Populus trichocarpa* F_1 progeny. *Tree Phys* 29:1329–1339.

Moore, M. J., C. D. Bell, P. S. Soltis, and D. E. Soltis. 2007. Using plastid genome-scale data to resolve enigmatic relationships among basal angiosperms. *Proc Nat Acad Sci USA* 104:19363–19368.

Moore, M. J., A. Dhingra, P. S. Soltis et al. 2006. Rapid and accurate pyrosequencing of angiosperm plastid genomes. *BMC Plant Biol* 6:17.

Morgenstern, E. K. 1996. *Geographic Variation in Forest Trees: Genetic Basis and Application of Knowledge in Silviculture*. Vancouver: University of British Columbia Press.

Muthuri, C. W., C. K. Ong, J. Craigon, B. M. Mati, V. W. Ngumi, and C. R. Black. 2009. Gas exchange and water use efficiency of trees and maize in agroforestry systems in semi-arid Kenya. *Agr Ecosyst Environ* 129:497–507.

Neale, D. B. 2007. Genomics to tree breeding and forest health. *Curr Opin Genet Dev* 17:539–544.

Neale, D. B. and P. K. Ingvarsson. 2008. Population, quantitative, and comparative genomics of adaptation in forest trees. *Curr Opin Plant Biol* 11:149–155.

Neale, D. B. and A. Kremer. 2011. Forest tree genomics: Growing resources and applications. *Nat Rev Genet* 12:111–122.

Neale, D. B. and O. Savolainen. 2004. Association genetics of complex traits in conifers. *Trends Plant Sci* 9:325–330.

Nei, M., T. Maruyama, and R. Chakraborty. 1975. The bottleneck effect and genetic variability in populations. *Evolution* 29:1–10.

Newhouse, J. R. 1990. Chestnut blight. *Sci Am* 263:106–111.

Novaes, E., L. Osorio, L. Drost et al. 2009. Quantitative genetic analysis of biomass and wood chemistry of *Populus* under different nitrogen levels. *New Phytol* 182:878–890.

Ogata, Y., H. Suzuki, and D. Shibata. 2009. A database for poplar gene co-expression analysis for systematic understanding of biological processes, including stress responses. *J Wood Sci* 55:395–400.

Oliphant, A., D. L. Barker, J. R. Stuelpnagel, and M. S. Chee. 2002. BeadArray technology: Enabling an accurate, cost-effective approach to high-throughput genotyping. *Biotech Suppl* 5:6–58.

Olukolu, B. A., C. D. Nelson, and A. G. Abbott. 2012. Genomics assisted breeding for resistance to *Phytophthera cinnamomi* in chestnut (*Castanea* sp.). In R. A. Sniezko, A. D. Yanchuk, J. T. Kliejunas et al. (eds), *Proceedings of the Fourth International Workshop on the Genetics of Host-Parasite Interactions in Forestry: Disease and Insect Resistance in Forest Trees*. Gen. Tech. Rep. PSW-GTR-240. Albany, CA: U.S. Department of Agriculture.

Paillet, F. L. and P. A. Rutter. 1989. Replacement of native oak and hickory tree species by the introduced American chestnut (*Castanea dentata*) in southwestern Wisconsin. *Can J Bot* 67:3457–3469.

Pakull, B., K. Groppe, M. Meyer, T. Markussen, and M. Fladung. 2009. Genetic linkage mapping in aspen (*Populus tremula* L. and *Populus tremuloides* Michx.) *Tree Genet Genome* 5:505–515.

Paolucci, I., M. Gaudet, V. Jorge et al. 2010. Genetic linkage maps of *Populus alba* L. and comparative mapping analysis of sex determination across *Populus* species. *Tree Genet Genome* 6:863–875.

Petit, R. J. 2004. Biological invasions at the gene level. *Divers Distrib* 10:159–165.

Petit, R. J., I. Aguinagalde, J.-L. de Beaulieu et al. 2003. Glacial refugia: Hotspots but not melting pots of genetic diversity. *Science* 300:1563–1565.

Petit, R. J., R. Bialozyt, S. Brewer, R. Cheddadi, and B. Comps. 2001. From spatial patterns of genetic diversity to postglacial migration processes in forest trees. In J. Silvertown and J. Antonovics (eds), *Integrating Ecology and Evolution in a Spatial Context*, pp. 295–318. Oxford: Blackwell Science.

Petit, R. J., R. Bialozyt, P. Garnier-Gere, and A. Hampe. 2004. Ecology and genetics of tree invasions: From recent introductions to Quaternary migrations. *Forest Ecol Manag* 197:117–137.

Petit, R. J., J. Duminil, S. Fineschi, A. Hampe, D. Salvini, and G. G. Vendramin. 2005. Comparative organization of chloroplast, mitochondrial and nuclear diversity in plant populations. *Molec Ecol* 14:689–701.

Petit, R. J. and A. Hampe. 2006. Some evolutionary consequences of being a tree. *Annu Rev Ecol Evol S* 37:187–214.

Petit, R. J., C. Latouche-Hall, M. H. Pemonge, and A. Kremer. 2002. Chloroplast DNA variation in oaks in France and the influence of forest fragmentation on genetic diversity. *For Ecol Manag* 156:115–129.

Petit, R. J. and G. G. Vendramin. 2006. Phylogeography of organelle DNA in plants: An introduction. In S. Weiss and N. Ferrand (eds), *Phylogeography of Southern European Refugia*. Dordrecht: Springer.

Pierson, S. A. M., C. H. Keiffer, B. C. McCarthy, and S. H. Rogstad. 2007. Limited reintroduction does not always lead to rapid loss of genetic diversity: An example from the American chestnut (*Castanea dentata*; Fagaceae). *Restor Ecol* 15:420–429.

Pigliucci, M., C. Paoletti, S. Fineschi, and M. E. Malvolti. 1991. Phenotypic integration in chestnut (*Castanea sativa* Mill.): Leaves versus fruits. *Bot Gaz* (Chicago) 152:514–521.

Pijut, P. M., S. S. Lawson, and C. H. Michler. 2011. Biotechnological efforts for preserving and enhancing temperate hardwood tree biodiversity, health, and productivity. *In Vitro Cell Dev-Plant* 47(S1):123–147.

Poorter, L. 2007. Are species adapted to their regeneration niche, adult niche, or both? *Am Nat* 169:433–442.

Porth, I., C. Scotti-Saintagne, T. Barreneche, A. Kremer, and K. Burg. 2005. Linkage mapping of osmotic stress induced genes of oak. *Tree Genet Genome* 1:31–40.

Przeworski, M. 2002. The signature of positive selection at randomly chosen loci. *Genetics* 160:1179–1189.

Qiu, Q., T. Ma, Q. Hu et al. 2011. Genome-scale transcriptome analysis of the desert poplar, *Populus euphratica*. *Tree Phys* 31:452–461.

Rae, A. M., N. R. Street, K. M. Robinson, N. Harris, and G. Taylor. 2009. Five QTL hotspots for yield in short rotation coppice bioenergy poplar: The poplar biomass loci. *BMC Plant Biol* 9:23.

Rajora, O. P. and A. Mosseler. 2001. Challenges and opportunities for conservation of forest genetic resources. *Euphytica* 118:197–212.

Rampant, P. F., I. Lesur, C. Boussardon et al. 2011. Analysis of BAC end sequences in oak, a keystone forest tree species, providing insight into the composition of its genome. *BMC Genomics* 12:292–305.

Rathmacher, G., M. Niggemann, H. Wypukol, K. Gebhardt, B. Ziegenhagen, and R. Bialozyt. 2009. Allelic ladders and reference genotypes for a rigorous standardization of poplar microsatellite data. *Trees* 23:573–583.

Ravi, V., J. P. Khurana, A. K. Tyagi, and P. Khurana. 2008. An update on chloroplast genomes. *Plant Syst Evol* 271:101–122.

Regier, N. and B. Frey. 2010. Experimental comparison of relative RT-qPCR quantification approaches for gene expression studies in poplar. *BMC Molec Biol* 11:57.

Rehfeldt, G. E., C. C. Ying, D. L. Spittlehouse, and D. A. Hamilton. 1999. Genetic responses to climate change in *Pinus contorta*: Niche breadth, climate change, and reforestation. *Ecol Monogr* 69:375–407.

Roane, M. K., G. J. Griffin, and J. R. Elkins. 1986. *Chestnut Blight, Other Endothia Diseases, and the Genus Endothia*. St. Paul: APS Press.

Romero-Severson, J., P. Aldrich, Y. Feng, W. Sun, and C. Michler. 2003. Chloroplast DNA variation of northern red oak (*Quercus rubra* L.) in Indiana. *New Forest* 26:43–49.

Russell, E. W. B. 1987. Pre-blight distribution of *Castanea dentata* (Marsh.) Borkh. *Bull Torrey Bot Club* 114:183–190.

Rutter, P. J. and R. B. Burnham. 1982. The Minnesota chestnut program—New promise for breeding a blight-resistant American chestnut. *Annu Rep Northern Nut Growers Assoc* 73:81–90.

Ryman, N. and L. Laikre. 1991. Effects of supportive breeding on the genetically effective population size. *Conserv Biol* 5:325–329.

Saintagne, C., C. Bodenes, T. Barreneche, D. Pot, C. Plomion, and A. Kremer. 2004. Distribution of genomic regions differentiating oak species assessed by QTL detection. *Heredity* 92:20–30.

Sandquist, D. R. and J. R. Ehleringer. 2003. Population-and family-level variation of brittlebush (*Encelia farinosa*, Asteraceae) pubescence: Its relation to drought and implications for selection in variable environments. *Am J Bot* 90:1481–1486.

Saša, O., G. Vladislava, Z. Miroslav, K. Branislav, P. Andrej, and Z. Galic. 2009. Evaluation of interspecific DNA variability in poplars using AFLP and SSR markers. *Afr J Biotech* 8:5241–5247.

Savolainen, O. and T. Pyhajarvi. 2007. Genomic diversity in forest trees. *Curr Opin Plant Biol* 10:162–167.

Savolainen, O., T. Pyhajarvi, and T. Knurr. 2007. Gene flow and local adaptation in trees. *Annu Rev Ecol Evol S* 38:595–619.

Schlarbaum, S. E. and W. T. Bagley. 1981. Intraspecific genetic variation of *Quercus rubra* L., northern red oak. *Silvae Genet* 30:50–56.

Schroeder, H. and M. Fladung. 2010. SSR and SNP markers for the identification of clones, hybrids, and species within the genus *Populus*. *Silvae Genet* 59:257–263.

Schroeder, H., A. M. Hoeltken, and M. Fladung. 2012. Differentiation of *Populus* species using chloroplast single nucleotide polymorphism (SNP) markers—Essential for comprehensible and reliable poplar breeding. *Plant Biol* 14:374–381.

Scotti-Saintagne, C., C. Bodenes, T. Barreneche, E. Bertocchi, C. Plomion, and A. Kremer. 2004a. Detection of quantitative trait loci controlling bud burst and height growth in *Quercus robur* L. *Theor Appl Genet* 109:1648–1659.

Scotti-Saintagne, C., S. Mariette, I. Porth et al. 2004b. Genome scanning for interspecific differentiation between two closely related oak species (*Quercus robur* L. and *Q. petraea* (Matt.) Liebl.). *Genetics* 168:1615–1626.

Sebastiani, F. 2003. Sequenziamento del genoma del cloroplasto del castagno, *Castanea sativa* Mill. PhD Thesis, University of Florence, Italy.

Sebastiani, F., S. Carnevale, and G. G. Vendramin. 2004. A new set of mono-and dinucleotide chloroplast microsatellite markers in Fagaceae. *Mol Ecol Notes* 4:259–261.

Shaw, J., E. B. Lickey, E. E. Schilling, and R. L. Small. 2005. The tortoise and the hare II: Relative utility of 21 noncoding chloroplast DNA sequences for phylogenetic analysis. *Am J Bot* 92:142–166.

Shizuya, H., B. Birren, U. J. Kim et al. 1992. Cloning and stable maintenance of 300-kilobase-pair fragments of human DNA in *Escherichia coli* using an F-factor-based vector. *Proc Natl Acad Sci USA* 89:8794–8797.

Sisco, P. H., T. L. Kubisiak, M. Casasoli et al. 2005. An improved genetic map for *Castanea mollissima/Castanea dentata* and its relationship to the genetic map of *Castanea sativa*. *Acta Hort* 693:491–495.

Slavov, G. T., G. T. Howe, and W. T. Adams. 2005. Pollen contamination and mating patterns in a Douglas-fir seed orchard as measured by simple sequence repeat markers. *Can J For Res* 35:1592–1603.

Small, R. L., J. A. Ryburn, R. C. Cronn, T. Seelanan, and J. F. Wendal. 1998. The tortoise and the hare: Choosing between noncoding plastome and nuclear *Adh* sequences for phylogeny reconstruction in a recently diverged plant group. *Am J Bot* 85:1301–1315.

Smith, A. H. 2012. Breeding for resistance: TACF and the Burnham hypothesis. *J Am Chestnut Found* 26:11–15.

Song, Y., K. Ma, W. Bo, Z. Zhang, and D. Zhang. 2012a. Sex-specific DNA methylation and gene expression in andromonoecious poplar. *Plant Cell Report* 31:1393–1405.

Song, Y., Z. Wang, W. Bo, Y. Ren, Z. Zhang, and D. Zhang. 2012b. Transcriptional profiling by cDNA-AFLP analysis showed differential transcript abundance in response to water stress in *Populus hopeiensis. BMC Genomics* 13:286.

Sork, V. L., J. Bramble, and O. Sexton. 1993. Ecology of mast fruiting in 3 species of North American deciduous oaks. *Ecology* 74:528–541.

Spitze, K. 1993. Population genetic structure in *Daphnia obtusa*: Quantitative genetic and allozyme variation. *Genetics* 135:367–374.

Steiner, K., A. Ellingboe, S. Friedman et al. 2004. TACF adopts guidelines for testing blight-resistant American chestnuts. *J Am Chestnut Found* 18:7.

Steiner, K. C. and P. C. Berrang. 1990. Microgeographic adaptation to temperature in pitch pine progenies. *Am Midl Nat* 123:292–300.

Stilwell, K. L., H. M. Wilbur, C. R. Werth, and D. R. Taylor. 2003. Heterozygote advantage in the American chestnut, *Castanea dentata* (Fagaceae). *Am J Bot* 90:207–213.

Street, N. R., O. Skogström, S. Ojödin et al. 2006. The genetics and genomics of the drought response in *Populus. Plant J* 48:321–341.

Streiff, R., T. Labbe, R. Bacilieri, H. Steinkellner, J. Glossl, and A. Kremer. 1998. Within-population genetic structure in *Quercus robur* L. and *Quercus petraea* (Matt.) Liebl. assessed with isozymes and microsatellites. *Mol Ecol* 7:317–328.

Tanaka, T., T. Yamamoto, and M. Suzuki. 2005. Genetic diversity of *Castanea crenata* in Northern Japan assessed by SSR markers. *Breed Sci* 55:271–277.

Tester, M. and P. Langridge. 2010. Breeding technologies to increase crop production in a changing world. *Science* 327:818–822.

Thumma, B. R., M. F. Nolan, R. Evans, and G. F. Moran. 2005. Polymorphisms in cinnamoyl CoA reductase (CCR) are associated with variation in microfibril angle in *Eucalyptus* species. *Genetics* 171:1257–1265.

Tovar-Sanchez, E., P. Mussali-Galante, R. Esteban-Jimenez et al. 2008. Chloroplast DNA polymorphism reveals geographic structure and introgression in the *Quercus crassifolia*×*Quercus crassipes* hybrid complex in Mexico. *Botany* 86:228–239.

Tschaplinski, T. J., G. A. Tuskan, M. M. Sewell, G. M. Gebre, D. E. Todd, and C. D. Pendley. 2006. Phenotypic variation and quantitative trait locus identification for osmotic potential in an interspecific hybrid inbred F_2 poplar pedigree grown in contrasting environments. *Tree Phys* 26:595–604.

Tuskan, G. A., S. DiFazio, S. Jansson et al. 2006. The genome of western black cottonwood, *Populus trichocarpa* (Torr. & Gray ex Brayshaw). *Science* 313:1596–1604.

Ueno, S., G. Le Provost, V. Leger et al. 2010. Bioinformatic analysis of ESTs collected by Sanger and pyrosequencing methods for a keystone forest tree species: Oak. *BMC Genomics* 11:650–674.

van Mantgem, P. J., N. L. Stephenson, J. C. Byrne et al. 2009. Widespread increase of tree mortality rates in the Western United States. *Science* 323:521–524.

Villani, F., M. Lauteri, A. Sansotta et al. 1999a. Genetic structure and quantitative traits variation in F_1 full-sib progenies of *Castanea sativa* Mill. *Acta Hort* 494:395–405.

Villani, F., M. Pigliucci, S. Benedettelli, and M. Cherubina. 1991. Genetic differentiation among Turkish chestnut (*Castanea sativa* Mill.) populations. *Heredity* 66:131–136.

Villani, F., M. Pigliucci, and M. Cherubini. 1994. Evolution of *Castanea sativa* Mill, in Turkey and Europe. *Genet Res Camb* 63:109–116.

Villani, F., M. Pigliucci, M. Lauteri, and M. Cherubini. 1992. Congruence between genetic, morphometric, and physiological data on differentiation of Turkish chestnut (*Castanea sativa*). *Genome* 35:251–256.

Villani, F., A. Sansotta, M. Cherubini, D. Cesaroni, and V. Sbordoni. 1999b. Genetic structure of natural populations of *Castanea sativa* in Turkey: Evidence of a hybrid zone. *J Evol Biol* 12:233–244.

Vision, T. J., D. G. Brown, D. B. Shmoys, R. T. Durrett, and S. D. Tanksley. 2000. Selective mapping: A strategy for optimizing the construction of high-density linkage maps. *Genetics* 155:407–420.

Vos, P., R. Hogers, M. Bleeker et al. 1995. AFLP: A new technique for DNA fingerprinting. *Nucleic Acids Res* 23:4407–4414.

Walsh, B. 2008. Using molecular markers for detecting domestication, improvement, and adaptation genes. *Euphytica* 161:1–17.

Wang, X. and Y. Jia. 2010. Study on adsorption and remediation of heavy metals by poplar and larch in contaminated soil. *Environ Sci Pollut Res Int* 17:1331–1338.

Wang, J., Z. Li, Q. Guo, G. Ren, and Y. Wu. 2011a. Genetic variation within and between populations of a desert poplar (*Populus euphratica*) revealed by SSR markers. *Ann For Sci* 68:1143–1149.

Wang, J., Y. Wu, G. Ren, Q. Guo, J. Liu, and M. Lascoux. 2011b. Genetic differentiation and delimitation between ecologically diverged *Populus euphratica* and *P. pruinosa*. *PLoS ONE* 6:e26530.

Wang, Y., M. Kang, and H. Huang. 2006. Subpopulation genetic structure in a panmicitc population as revealed by molecular markers: A case study of *Castanea seguinii* using SSR markers. *J Plant Ecol* 30:147–156.

Wang, Y., X. Sun, and B. Tan. 2010a. A genetic linkage map of *Populus adenopoda* Maxim.×*P. alba* L. hybrid based on SSR and SRAP markers. *Euphytica* 173:193–205.

Wang, Y. C., G. Z. Qu, H. Y. Li et al. 2010b. Enhanced salt tolerance of transgenic poplar plants expressing a manganese superoxide dismutase from *Tamarix androssowii*. *Molec Biol Reps* 37:1119–1124.

Wegrzyn, J. L., J. M. Lee, J. Liechty, and D. B. Neale. 2009. PineSAP—Sequence alignment and SNP identification pipeline. *Bioinformatics* 25:2609–2610.

Wegrzyn, J. L., J. M. Lee, B. R. Tearse, and D. B. Neale. 2008. TreeGenes: A forest tree genome database. *Int J Plant Genomics* 2008:412875.

Wegrzyn, J. L., D. Main, B. Figueroa et al. 2011. Uniform standards for genome databases in forest and fruit trees. *Tree Genet Genome* 8:549–557.

Wheeler, N. and R. Sederoff. 2009. Role of genomics in the potential restoration of the American chestnut. *Tree Genet Genome* 5:181–187.

White, T. L., W. T. Adams, and D. B. Neale. 2007. *Forest Genetics*. Cambridge: CAB International.

Whittemore, A. T. and B. A. Schaal. 1991. Interspecific gene flow in sympatric oaks. *Proc Natl Acad Sci USA* 88:2540–2544.

Witcombe, J. R., P. A. Hollington, C. J. Howarth, S. Reader, and K. A. Steele. 2008. Breeding for abiotic stresses for sustainable agriculture. *Philos T Roy Soc B* 363:703–716.

Worthen, L., C. H. Michler, and K. Woeste. 2010a. Genetic ramifications of restoration of blight-resistant American chestnut. In J. A. DeWoody, J. Bickham, C. H. Michler, K. Nichols, G. Rhodes, and K. Woeste (eds), *Molecular Approaches in Natural Resource Conservation and Management*, pp. 307–309. New York: Cambridge University Press.

Worthen, L., K. Woeste, and C. Michler. 2010b. Breeding American chestnut for blight resistance. *Plant Breed Rev* 33:305–339.

Wright, S. 1951. The genetical structure of populations. *Ann Eugen* 15:323–354.

Wyman, S. K., R. K. Jansen, and J. L. Boore. 2004. Automatic annotation of organellar genomes with DOGMA. *Bioinformatics* 20:3252–255.

Xiao, X., F. Yang, S. Zhang, H. Korpelainen, and C. Li. 2009. Physiological and proteomic responses of two contrasting *Populus cathayana* populations to drought stress. *Physiol Plantarum* 136:150–168.

Xu, F., S. Feng, R. Wu, and F. K. Du. 2013. Two highly validated SSR multiplexes (8-plex) for Euphrates' poplar, *Populus euphratica* (Salicaceae). *Molec Ecol Res* 13:144–153.

Yadav, R., P. Arora, S. Kumar, and A. Chaudhury. 2010. Perspectives for genetic engineering of poplars for enhanced phytoremediation abilities. *Ecotoxicology* 19:1574–1588.

Yan, D.-H., T. Fenning, S. Tang, X. Xia, and W. Yin. 2012. Genome-wide transcriptional response of *Populus euphratica* to long-term drought stress. *Plant Sci* 195:24–35.

Yang, F., Y. Wang, and L.-F. Miao. 2010. Comparative physiological and proteomic responses to drought stress in two poplar species originating from different altitudes. *Physiol Plantarum* 139:388–400.

Yang, F., X. Xiao, S. Zhang, H. Korpelainen, and C. Li. 2009a. Salt stress responses in *Populus cathayana* Rehder. *Plant Sci* 176:669–677.

Yang, F., X. Xu, X. Xiao, and C. Li. 2009b. Responses to drought stress in two poplar species originating from different altitudes. *Biol Plantarum* 53:511–516.

Yang, X., U. C. Kalluri, S. P. DiFazio et al. 2009. Poplar genomics: State of the science. *Crit Rev Plant Sci* 28:285–308.

Yin, C., B. Duan, B. Wang, and C. Li. 2004. Morphological and physiological responses of two contrasting poplar species to drought stress and exogenous abscisic acid application. *Plant Sci* 167:1091–1097.

Yin, T., S. P. Difazio, L. E. Gunter et al. 2008. Genome structure and emerging evidence of an incipient sex chromosome in *Populus*. *Genome Res* 18:422–430.

Yin, T. M., M. R. Huang, M. X. Wang, L. H. Zhu, Z. B. Zeng, and R. L. Wu. 2001. Preliminary interspecific genetic maps of the *Populus* genome constructed from RAPD markers. *Genome* 44:602–609.

Youens-Clark, K., B. Faga, I. V. Yap, L. Stein, and D. Ware. 2009. CMap 1.01: A comparative mapping application for the Internet. *Bioinformatics* 25:3040–3042.

Yuan, H.-M., K.-L. Li, R.-J. Ni et al. 2011. A systemic proteomic analysis of *Populus* chloroplast by using shotgun method. *Molec Biol Rep* 38:3045–3054.

Zanetto, A., A. Kremer, G. Muller-Starck, and H. Hattemer. 1996. Inheritance of isozymes in pedunculate oak (*Quercus robur* L.). *J Heredity* 87:364–470.

Zhang, B., C. Tong, T. Yin et al. 2009a. Detection of quantitative trait loci influencing growth trajectories of adventitious roots in *Populus* using functional mapping. *Tree Genet Genome* 5:539–552.

Zhang, J.-F., Z.-Z. Wei, D. Li, and B. Li. 2009b. Using SSR markers to study the mechanism of 2n pollen formation in *Populus* × *euramericana* (Dode) Guinier and *P.* × *popularis*. *Ann Forest Sci* 66:506.

Zhang, S., F. Chen, S. Peng, W. Ma, H. Korpelainen, and C. Li. 2010. Comparative physiological, ultrastructural and proteomic analyses reveal sexual differences in the responses of *Populus cathayana* under drought stress. *Proteomics* 10:2661–2677.

Zhang, S., L. Feng, H. Jiang, W. Ma, H. Korpelainen, and C. Li. 2012. Biochemical and proteomic analyses reveal that *Populus cathayana* males and females have different metabolic activities under chilling stress. *J Proteome Res* 11:5815–5826.

Zhu, C., M. Gore, E. S. Buckler, and J. Yu. 2008. Status and prospects of association mapping in plants. *Plant Genome* 1:5–20.

Zimmermann, P., M. Hirsch-Hoffmann, L. Hennig, and W. Gruissem. 2004. GENEVESTIGATOR. *Arabidopsis* microarray database and analysis toolbox. *Plant Phys* 136:2621–2632.

CHAPTER 4

MicroRNA Omics Approaches to Investigate Abiotic and Biotic Stress Responses in Plants

Shiv S. Verma, Swati Megha, Muhammad H. Rahman, Nat N. V. Kav, and Urmila Basu

CONTENTS

4.1 INTRODUCTION

MicroRNAs (miRNAs) are endogenous, noncoding single-stranded RNA molecules of ~22 nucleotides (nt) in length, which regulate gene expression in both plants and animals by targeting mRNAs through cleavage or translational repression. The first miRNA to be discovered was *lin4*, known to control the timing of larval development in *Caenorhabditis elegans* (Lee et al. 1993). As part of the regulatory pathway, *lin4* RNAs are paired to the 3′ untranslated region (UTR) of target mRNA, leading to its translational repression, which, in turn, activates the transition from cell divisions of the first stage to the second stage of larval development (Lee et al. 1993; Wightman et al. 1993). In *Arabidopsis*, 16 differentially expressed plant miRNAs were first identified during development, and the plant miRNAs are generally 20–24 nt long, transcribed from independent genes, and evolutionarily conserved (Reinhart et al. 2002).

Plants, being sessile in nature, are exposed to a plethora of environmental stresses—both biotic and abiotic. Abiotic stress is marked by conditions such as extremes in temperatures, salinity, water deficit, excess water resulting in flooding and anaerobic stress, oxidative stress, heavy metals, radiation, and mechanical stress. A large number of tissue-specific stress-induced miRNAs and biotic or abiotic stress-regulated miRNAs have been identified using expressed sequence tag (EST)-based studies in different plant species (Zhang et al. 2005). Of the total, 26% of miRNA-ESTs were found to be from stress-induced tissues, as shown in Figure 4.1a and b (Zhang et al. 2005), which can adversely affect the productivity of crops (Gao et al. 2007). Induction of stress-responsive genes alters physiological or biochemical pathways, causing accumulation of metabolites, which may in turn result in adaptation to stress conditions. As normal cellular functions are disrupted during stressed conditions, a quick and wide reprogramming by transcriptional, posttranscriptional, and translational regulation of stress responsive genes is required. In plants, posttranscriptional regulation is most common under stress conditions (Jagadeeswaran et al. 2009; Lu et al. 2008; Sunkar and Zhu 2004; Zhu 2002), and manipulation of miRNA-guided gene regulation can help in the engineering of stress-resistant plants (Ivashuta et al. 2011; Sablok et al. 2011). There are numerous reports in the literature that miRNAs are involved in the regulation of many of these processes, which may, ultimately, determine whether or not the plant is able to survive the imposed stress. Beginning with an overview of miRNA biogenesis, the following sections discuss in detail the role of miRNAs in mediating plant responses to stress.

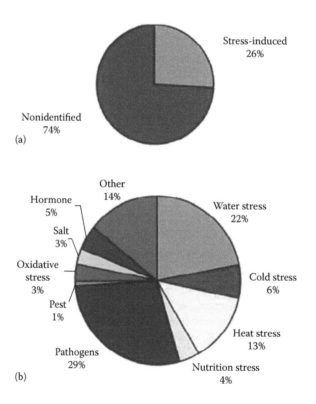

Figure 4.1 miRNAs are induced and regulated by environmental biotic and abiotic stresses. (a) 26% of ESTs were obtained from stress-induced tissues. (b) ESTs containing miRNAs were obtained from tissue by various biotic and abiotic environmental stresses. (Reprinted by permission from Macmillan Publishers Ltd. Zhang et al. Identification and characterization of new plant microRNA s using ES T analysis. *Cell Res.* 15: 336–360, copyright 2005.)

4.2 miRNA BIOGENESIS PATHWAY: AN OVERVIEW

miRNAs in plants arise from defined genetic loci, which are independent, protein-coding transcription units (Allen et al. 2004). There have been reports of transcripts harboring tandem precursors in plants (Boualem et al. 2008; Chuck et al. 2007) located in the untranslated regions of mRNA (Rajagopalan et al. 2006). Transcripts of ~1 kb in size, termed primary-miRNA or pri-miRNA, are encoded by the miRNA loci by RNA polymerase II (pol II). Although the precise mechanism of routing pri-miRNA into the miRNA pathway, instead of the translational machinery, is yet to be elucidated fully, the framework for miRNA biogenesis in plants has emerged, based mainly on genetic and biochemical studies in *Arabidopsis* (Xie et al. 2010). There are two lines of evidence which suggest that pol II is actually involved in the transcription of miRNA genes. First, primary transcripts are capped and polyadenylated, characteristic of a pol II transcript (Xie et al. 2010), and, second, the core promoter elements (TATA box and transcription initiator elements) typically found in pol II genes have been identified in miRNA loci of *A. thaliana* by sequence analysis (Xie et al. 2005; Zhou et al. 2007). The pri-miRNA transcript is characterized by its ability to form a characteristic hairpin-like imperfect loop structure (miRNA precursor), subsequently resulting in only one unique mature miRNA species. However, polycistronic miRNA precursors encoding functional nonhomologous miRNAs have also been reported (Merchan et al. 2009).

The pri-miRNAs are cleaved within the nucleus by miRNA processing machinery to produce miRNA:miRNA duplexes. Dicer-Like 1 protein (DCL1) is the core component of miRNA processing machinery and is a RNAse type III endonuclease. DCL1 is assisted by the dsRNA binding protein Hyponastic Leaves 1 (HYL1) (Han et al. 2004; Vazquez et al. 2004) and the C_2H_2 zinc-finger protein Serrate (SE) (Grigg et al. 2005; Lobbes et al. 2006). DCL1, HYL1, and SE co-localize in subnuclear bodies termed Dicing bodies or D-bodies (Fang and Spector 2007; Song et al. 2007). The first cleavage by DCL1 removes the flanking sequence of pri-miRNA, liberating the precursor transcripts with characteristic stem loop structure, the "pre-miRNA." The miRNA, duplexed with its near reverse complement, the miRNA*, is separated from the stem loop structure by a second cleavage mediated by DCL1. DCL1 action results in 2 nt overhang at the 3′ end of 20–22 nt miRNA:miRNA* duplex (Figure 4.2). End methylation of miRNA by nuclear protein Hua Enhancer 1 (HEN1) protects the 3′ ends from uridylation and subsequent degradation (Boutet et al. 2003; Yu et al. 2005).

The last step in miRNA processing is the transport of methylated miRNA duplex into the cytoplasm and association with silencing complex. The transport of mature miRNA is mediated by Hasty 1 (HST1), an exportin protein (Park et al. 2005). Maturation of miRNAs involves incorporation of one strand of the duplex into an Argonaute (AGO) protein, which is the catalytic component of RNA induced silencing complex (RISC) (Figure 4.2). The miRNA strand of the duplex acts as a guide strand and is selectively loaded onto the AGO complex, whereas the miRNA* strand is excluded from the complex (Xie et al. 2010). miRNAs synthesized in this way are involved in the regulation of various processes, as discussed in Sections 4 and 5. The next section discusses various techniques employed to profile the miRNAs.

4.3 miRNA PROFILING TECHNOLOGIES

As mentioned previously, plant miRNAs play an essential role in various biological and metabolic processes by modulating the expression of genes, either through their effects on mRNA stability or by repressing translational efficiency (Chen and Rajewsky 2007). Comprehensive information about miRNA transcriptomes and their targets, obtained through the application of genome-wide approaches, can be useful for system-level studies in developmental biology, novel miRNA discovery, and miRNA–mRNA interactions and their role in gene regulatory

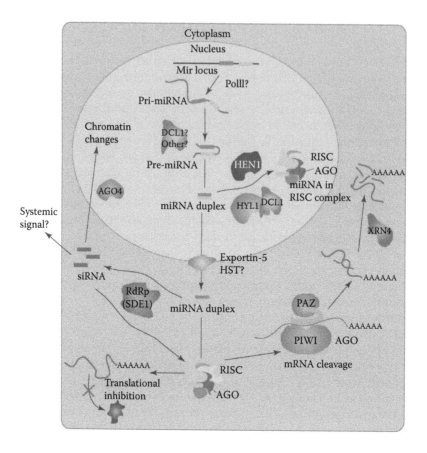

Figure 4.2 **(See color insert)** Model for miRNA biogenesis and activity in plants. (Reprinted from *Curr Opin Plant Bio*, 8, Kidner, C. A. and R. A. Martienssen. The developmental role of microRNA in plants, 38–44. Copyright 2005, with permission from Elsevier.)

networks (Pritchard et al. 2012). Accurate quantification of miRNA molecules is challenging, mainly due to small size, lack of a common sequence such as poly(A) tails, sequence length heterogeneity, end polymorphism, and large dynamic range in gene expression (Newman et al. 2011; Pritchard et al. 2012). The most commonly employed techniques for profiling miRNAs are deep sequencing, microarrays, and quantitative real-time PCR (qRT-PCR), each of which has its own advantages and disadvantages (Baker 2010; Git et al. 2010; Pritchard et al. 2012). The choice of platform for miRNA analysis is further dictated by the precision and accuracy required, cost, and sample availability (Aldridge and Hadfield 2012). Some of these factors are discussed below.

qRT-PCR uses miRNA-specific stem loop for reverse transcription, followed by the amplification and monitoring of the reverse-transcribed product in real time using a miRNA-specific forward primer and the universal reverse primer (Chen et al. 2005; Mestdagh et al. 2008; Varkonyi-Gasic and Hellens 2011). Several medium–high-throughput platforms, such as microfluidic cards containing preplated qRT-PCR primers, offer experimental advantages for robust workflow in profiling miRNAs (Pritchard et al. 2012). However, to perform hundreds of miRNA assays simultaneously for high-throughput studies, standardizing optimal miRNA primer hybridization conditions due to sequence-specific differences in primer annealing, remains a challenge (Balcells et al. 2011). Furthermore, qRT-PCR is also ineffective for novel miRNA discovery, since prior sequence information is a prerequisite for this technique.

Another approach, miRNA microarray, uses DNA-based probes to capture fluorescently labeled miRNAs, followed by scanning of slides and quantification of fluorescence. Using miRCURY Locked-Nucleic-Acid array technology and TaqMan miRNA assays, 45 miRNAs were found to be expressed in *Arabidopsis* mature pollen; however, neither technology was able to detect low-abundance miRNAs (Chambers and Shuai 2009). In another study, miRNA microarrays were used to profile the expression patterns of 19 annotated cadmium-responsive miRNAs in rice, of which only 6 were further validated experimentally by qRT-PCR (Ding et al. 2011). Overall, this approach relies on genome sequence information, and different isoforms of very similar miRNAs cannot be differentiated because of potential cross-hybridization problems (Liu et al. 2008; Baker 2010). Computational prediction approaches (Tables 4.1 and 4.2) can only be used for identification of highly conserved miRNAs (Adai et al. 2005; Bonnet et al. 2004; Breakfield et al. 2012), and not miRNAs which are species specific or are expressed in different developmental stages.

4.3.1 Next-Generation Sequencing for miRNA Profiling

Next-generation sequencing (NGS) technologies provide a rapid and high-throughput tool to measure the abundance of miRNAs with high sensitivity and specificity in multiple species, and to identify *de novo* miRNA involved in specific processes, without any previous sequence information.

In this approach, small RNAs (~18–35 nt) are isolated from the sample of interest, ligated to the 5′ and 3′ adapters and reverse transcribed into cDNA using a primer specific to the 5′ adapter. The cDNA product is amplified by PCR with two primers specific to the adapters followed by the massively parallel sequencing of millions of individual cDNA molecules from the library (Metzker 2010). The NGS approach was used in the identification of miRNAs in *Arabidopsis* (Lu et al. 2005; Rajagopalan et al. 2006). Using the 454 sequencing platform, a total of 48 nonconserved miRNA families were identified in *Arabidopsis* from wild type and mutants defective in miRNA biogenesis (Fahlgren et al. 2007). Since then, high-throughput sequencing technology has been successfully used in many plant species, including rice (Sunkar et al. 2008), soybean (Subramanian et al. 2008), *Phaseolus vulgaris* (Arenas-Huertero et al. 2009), *Medicago truncatula* (Lelandais-Briere et al. 2009), and sweet orange (Xu et al. 2010). Recently, Solexa sequencing identified 321 conserved miRNA families and 228 novel miRNAs in the Chinese cabbage, with 2308 and 736 potential target genes for 221 conserved and 125 novel miRNAs, respectively (Wang et al. 2012a,b). In *Panax*

Table 4.1   List of Available miRNA Databases

Name	Description	Link	References
miRBase	Searchable database of published miRNA sequences and annotation	http://www.mirbase.org/	Griffiths-Jones et al. (2006)
deepBase	Database for annotating and discovering small and long noncoding (nc) RNA from high-throughput deep sequencing data	http://deepbase.sysu.edu.cn/	Yang et al. (2010)
miRanda-microRNA.org	Database for experimentally observed miRNA expression patterns and predicted miRNA targets and target downregulating scores	http://www.microrna.org/microrna/home.do	Betel et al. (2008)
miRGen 2.0	Database of miRNA genomic information and regulation	http://www.microrna.gr/mirgen	Alexiou et al. (2010)
miRNAMap	Genomic maps of miRNA genes and their target genes in mammalian genomes	http://mirnamap.mbc.nctu.edu.tw/	Hsu et al. (2008)
PMRD	Plant miRNA database	http://bioinformatics.cau.edu.cn/PMRD/	Zhang et al. (2010)

Table 4.2   List of Available miRNA Target Gene Databases

Name	Description	Link	References
starBase	Database for exploring miRNA-target interaction maps from Argonaute (AGO) CHIP-seq (HITS-CLIP) and degardome sequencing, PARE data	http://starbase.sysu.edu.cn/	Yang et al. (2011)
miRwalk	Database and Webserver: aggregates and compares results from different miRNA databases	http://www.umm.uni-heidelberg.de/apps/zmf/mirwalk/	Dweep et al. (2011)
targetScan	Database and Webserver for predicted miRNA targets in animals	http://www.targetscan.org/	Lewis et al. (2005)
TarBase	Database of animal miRNA targets	http://www.diana.pcbi.upenn.edu/tarbase	Sethupathy et al. (2006)
Diana-microT	Webserver combines consereved and nonconserved miRNA recognition elements into a final prediction score	www.microrna.gr/microT	Maragkakis et al. (2009)
miRecords	Database for miRNA–target interactions	mirecords.biolead.org/	Xiao et al. (2009)
picTar	Webserver for miRNA target predictions	http://pictar.mdc-berlin.de/	Krek et al. (2005)
PITA	Webserver for miRNA target predictions as determined by base-pairing interactions within the mRNA	http://genie.weizmann.ac.il/pubs/mir07/mir07_data.html	Kertesz et al. (2007)
RepTar	Database of inverse miRNA target predictions independent of evolutionary conservation considerations, not limited to seed pairing sites	http://bioinformatics.ekmd.huji.ac.il/reptar/	Elefant et al. (2011)
RNA22	Webserver: predictions for all transcripts, visualize the predictions within a cDNA map and find multiple miRNAs of the target of interest	http://cbcsrv.watson.ibm.com/rna22.html	Miranda et al. (2006)
micTarBase	Webserver for experimentally verified miRNA targets	http://mirtarbase.mbc.nctu.edu.tw/	Hsu et al. (2011)
miRGator	Database of miRNA expression and targets	http://mirgator.kobic.re.kr	Cho et al. (2013)
miRNAMap	Database of miRNAs and their targets	http://mirnamap.mbc.nctu.edu.tw/	Hsu et al. (2008)
miRDB	Webserver for animal miRNA target prediction and functional annotation	http://mirdb.org/miRDB/	Wang (2008)
RNA hybrid	Easy, fast, and flexible software for predicting miRNA targets	http://bibiserv.techfak.uni-bielefeld.de/rnahybrid/	Rehmsmeier et al. (2004)
miRU, psRNA target	Webserver: automated plant miRNA target prediction server	http://plantgrn.noble.org/psRNATarget/	Dai and Zhao (2011)
CleaveLand	Webserver for using degradome data to find cleaved small RNA targets	http://www.bio.psu.edu/people/faculty/Axtell/AxtellLab/Software.html	Addo-Quaye et al. (2009)

ginseng, 73 conserved mRNAs from 33 families and 28 nonconserved mRNAs belonging to 9 families were identified (Wu et al. 2012). Solexa sequencing of both an early flowering mutant of trifoliate orange (precocious trifoliate orange, *Poncirus trifoliata*) and its wild-type counterpart at different developmental stages identified 141 known miRNAs belonging to 99 families; 75 novel miRNAs and a total of 317 potential target genes were predicted based on the 51 novel miRNA families (Sun et al. 2012).

Genome-wide sequencing has also been used to identify 216 novel and 196 conserved miRNAs from five types of tissues in *Brassica rapa* that were predicted to target approximately 20% of the genome's protein-coding genes (Kim et al. 2012). In radish, sequence alignment and secondary structure prediction of the high-throughput Solexa sequencing data identified a total of 545 conserved miRNA families, 15 novel and 64 potentially novel miRNAs (Xu et al. 2013). Quantitative

RT-PCR analysis confirmed that both conserved and novel miRNAs were expressed in radish, and some of them were preferentially expressed in certain tissues. Furthermore, a total of 196 potential target genes were predicted for 42 novel radish miRNAs (Xu et al. 2013).

In total, 5940 plant miRNAs from 66 species have been identified and deposited in miRBase (miRBase Release 19.0, http://www.mirbase.org/). Next-generation-based sequencing is beginning to be used as a promising platform to comprehensively and accurately assess the entire miRNA repertoire, including isomiRs (Ahmed and Zhao 2011; Colaiacovo et al. 2012; Neilsen et al. 2012). IsomiRs with slight variation at the terminal sequence of the annotated matured miRNA arise due to variation in the site of cleavage by Dicer-Like enzyme. Interestingly, isomiRs can be expressed in a cell-specific manner, and different transcripts may be targeted by slight variations of miRNA sequence; however, their broader biological significance is yet to be fully resolved.

4.3.2 Degradome Sequencing

Degradome sequencing is also known as parallel analysis of RNA ends (PARE), a modified 5′-rapid amplification of cDNA ends (RACE) or genome-wide mapping of uncapped transcripts (GMUCT) (Addo-Quaye et al. 2008; German et al. 2008, 2009). Axtell et al. (2011) reviewed the target recognition process between plants and animals and near-perfect base complementation between the microRNA and its target, leading to the direct mRNA cleavage of a microRNA target in plants. This process is different in animals, in which protein repression is believed to occur by translational inhibition as well as mRNA degradation. In plants, miRNAs degrade their mRNA targets by precise cleavage between the 10th and 11th nucleotides from the 5′ end of the miRNA in the complementary region of the target transcript, resulting in a distinct peak of degradome sequence tag at the predicted cleavage site relative to other regions of the transcript (German et al. 2008, 2009). For construction of degradome libraries, polyA-RNA samples are ligated to a custom RNA adaptor containing a 3′*Mme* I site, and reverse transcribed. After second-strand synthesis and *Mme* I digestion, dsDNA adaptor is ligated, followed by gel purification and PCR amplication (German et al. 2008). A combined approach of high-throughput sequencing and degradome analysis globally identifies remnants of small RNA-directed target cleavage by sequencing the 5′ ends of uncapped RNAs in plants. Data analysis includes summarizing the mapped degradome reads into a degradome density file, generating miRNA prediction targets, and comparing the degradome density file with the target prediction.

One hundred sixty targets of 53 miRNA families (24 conserved and 29 specific) were identified in rice (Li et al. 2010). Degradome sequencing was used to identify 29 known miRNA families and 2 novel miRNA families containing a total of 64 miRNAs, showing differential expression in specific tissues of cucumber (Mao et al. 2012). These miRNAs targeted 21 mRNAs, associated with development, reactive oxygen species (ROS) scavenging, signal transduction, and transcriptional regulation (Mao et al. 2012). In a recent study by Zeng et al. (2012), microRNA and degradome sequencing were applied to investigate the expression of miRNAs and their targets responsive to aluminum stress in soybean. In total, 97 known miRNAs, 31 novel miRNAs, and 91 genes cleaved by miRNAs were detected through degradome sequencing, and 52 of these genes were transcription regulators. In *B. napus*, 41 conserved miRNAs (with 33 nonredundant mRNA targets) and 62 brassica-specific miRNAs (with 19 new nonredundant mRNA targets) were identified by genome-scale sequencing of the mRNA degradome (Xu et al. 2012). Using this approach, Yang et al. (2013) identified a total of 36 differentially expressed known miRNA families, including 25 novel miRNAs, with 476 gene targets, providing new information about the regulatory network of miRNAs during somatic embryogenesis in cotton. The differential expression of miRNAs and their targets using degradome sequencing in *B. juncea* suggested differences in biogenesis during reproductive development of cytoplasmic male sterility and its maintainer fertile lines (Yang et al. 2013).

All these large-scale miRNA profiling and miRNA–target pair validation, combined with transcriptome-wide sequencing, has greatly advanced our understanding of the regulatory roles of plant miRNAs in several biological processes. Furthermore, deep insights into miRNA precursor processing and miRNA-mediated self-regulation of their host precursors will also be possible by combining high-throughput degradome sequencing data with constantly evolving advanced bioinformatics tools.

4.4 ROLE OF miRNAs IN PLANT GROWTH AND DEVELOPMENT

miRNAs are responsible for very specific temporal and spatial control of their targets, and altered miRNA metabolism is known to cause a myriad of developmental defects (Palatnik et al. 2003). For example, miR156 targets Squamosa Promoter binding protein-like (SPL), controlling early branching events, especially in cereals (Xie et al. 2012). Overexpression of miR156 has also been shown to impart juvenile characteristics to maize, such as absence of epidermal hairs and deposition of epicuticular wax (Chuck et al. 2007). In *Arabidopsis*, the disruption of miR156 through a miRNA mimic resulted in the precocious initiation of adult leaves (Franco-Zorrilla et al. 2007). Similar studies by Smith et al. (2009) demonstrated the suppression of precocious phenotype through overexpression of miR156, as well as precocious initiation of adult traits in plants carrying miR156-insensitive versions of SPLs (Wang et al. 2008), substantiating its role in quantitatively repressing vegetative phase change. The overexpression of miR156 in transgenic *Arabidopsis* has been reported to increase the number of shoots through the activation of lateral meristems in the axils of rosette leaves (Schwarz et al. 2008), and the number of tillers with reduced spikelets and grains per panicle (Xie et al. 2006), clearly suggesting the participation of miR156 in both vegetative developments and in the reproductive phase transition through lateral meristem and organ identity manifestation.

miRNA159 targets a number of Myeloblastosis (MYB) transcription factor genes involved in flowering and male fertility. One known target of miR159 in *Arabidopsis* is *MYB33* (Palatnik et al. 2003), which belongs to a gibberellic acid (GA) MYB (*GAMYB*)-*like* family of transcription factors. miR159 has been shown to regulate the activity of *MYB33* and *MYB65* transcription factors, related to gibberellin-responsive elements (Gubler et al. 1999). miR159 levels were also shown to be decreased in GA-deficient mutants, and increased after exogenous GA application (Achard et al. 2004), indicating that this miRNA is gibberellin regulated. Overexpression of miR159 in transgenic plants delayed flowering under short-day conditions, possibly due to lowered levels of *MYB* and the floral meristem identity gene *LEAFY*, and exhibited reduced fertility due to defective anther development (Achard et al. 2004). Subsequent studies demonstrated the expression of *MYB33:GUS* in the anthers, indicating the importance of miR159 in restricting spatial expression of *MYB33* in anthers, thereby regulating plant growth and development (Millar and Gubler 2005). The loss-of-function strategy adopted by Allen et al. (2007) has also indicated the role of miR159a/miR159b in regulating *MYB33* and *MYB65*-induced altered growth habits, including curled leaves, small siliques, and small seeds.

miRNA160 targets and posttranscriptionally regulates the expression of auxin response transcription factors (ARFs) *ARF10*, *16*, and *17*, which are involved in auxin signal transduction during several stages of plant growth and development (Mallory et al. 2005). In *A. thaliana*, 23 ARF genes are thought to be posttranscriptionally regulated by miR160 (Liu et al. 2010). The modification of *ARF17* and *ARF10* to miR-resistant forms leads to embryonic vegetative and developmental defects in plants, including serrated leaves, curled stems, contorted flowers, and twisted siliques, indicating their role in plant development (Mallory et al. 2005). Expression of miR160-resistant silent mutation containing *ARF10* in *Arabidopsis* indicated the importance of auxin signaling in the early stages of plant growth, involving both *ARF10*-dependent auxin and abscisic acid (ABA) pathways (Liu et al. 2007).

miRNA164 cleaves NAC domains (from the first letters of three genes: *NAM* [no apical meristem], *ATAF* [*Arabidopsis* transcription activation factor], and *CUC* [cup-shaped cotyledon], which have conserved domains) such as *NAC1*, *CUC1*, and *CUC2* (Laufs et al. 2004; Mallory et al. 2004). These are plant-specific transcription factors and are involved in regulating embryonic, vegetative, and floral development. Plants overexpressing miR164 had reduced transcript levels of *CUC1* and *CUC2* (Laufs et al. 2004) and were phenotypically similar to *cuc1* and *cuc2* double mutants. miR164 has also been implicated in the control of boundary size in meristems and in the formation and control of vegetative and reproductive development, including the number of petals (Baker et al. 2005; Mallory et al. 2004). The balance between *CUC2* and miR164 regulates the serration of *Arabidopsis* leaves (Nikovics et al. 2006). In addition, miR164 regulates leaflet formation in a wide variety of dicotyledonous plants, including pea, tomato, and potato, through the modulation of miR164, thereby affecting *CUC* expression (Blein et al. 2008). miRNAs165 and 166 regulate the expression of developmental proteins such as HD-ZIP III (Homeodomain zinc finger protein III) and confer pleiotropic phenotypes affecting vascular patterning, the establishment/maintenance of abaxial–adaxial polarity, and the function of meristems (Jover-Gil et al. 2005).

miRNA172 is predicted to target members of the Apetala2 (*AP2*) gene family, such as the floral homeotic gene (*AP2*) in *Arabidopsis*, Intermediate spikelet and Glossy15 in maize, and TARGET OF EAT1 (*TOE1*), *TOE2*, and *TOE3* in *Arabidopsis* (Park et al. 2002). Interestingly, instead of the cleavage of the target transcripts, miR172 acts through translational repression, a mechanism that is common in most animals (Chen et al. 2004). Overexpression of miR172 caused translational repression of *AP2*, leading to the phenotypes of early flowering and floral abnormalities (Aukerman and Sakai 2003). One effect of miR172 has been reported to be an increase in seed weight in *Arabidopsis*, possibly through suppression of *AP2* and due to its effect on hexose and sucrose levels in maturing seeds (Ohto et al. 2005). On the other hand, rice seeds exhibited a decrease in weight when miR172 was overexpressed (Zhu et al. 2009), which was speculated to be due to the irregular development of floral organs, which, in turn, affected seed development.

Increased levels of miR319 cause prominent jagged and waxy leaves (jaw) in *Arabidopsis*, with different phenotypes for four independent alleles (*jaw*-D1–D4) (Weigel et al. 2000). The jaw-D phenotype is correlated with a reduction of transcription factors TCPs (from the first letters of three genes: Teosinte Branched 1 [*TB1*] from maize, the Antirrhinum gene Cycloidea [*CYC*], and PCNA promoter binding factors *PCF1* and *PCF2* from rice) (Cubas et al. 1999). TCPs are known to regulate leaf development by affecting cell division, expansion, and differentiation (Koyama et al. 2007; Efroni et al. 2008); branching (Aguilar-Martinez et al. 2007); reproductive development (Takeda et al. 2006); and modulation of the levels of jasmonic acid (JA) (Schommer et al. 2008) and auxin (Koyama et al. 2010) biosynthesis and signaling.

miRNA319-guided degradation of *TCP* mRNA in shoot apical meristem has been indicated through microarray analysis of wild-type and jaw-D mutants (Palatnik et al. 2003; Efroni et al. 2008). Besides the jaw-D phenotype, miR319 has also been shown to regulate the complexity of leaf shape in tomato plants by affecting the lanceolate (*LA*) gene (Ori et al. 2007; Shleizer-Burko et al. 2011). Thus, a large number of miRNA species affect a wide variety of vegetative and reproductive developmental processes in a number of plant species.

4.5 PLANT miRNAs AND ABIOTIC STRESS

The first indication for the role of plant miRNAs in adaptive responses to abiotic stresses came from bioinformatics, miRNA target predictions, and cloning of miRNA from stressed *A. thaliana* plants (Jones-Rhoades and Bartel 2004; Sunkar and Zhu 2004). Since then, miRNAs have been implicated in mediating responses to various abiotic stresses through regulation of their predicted targets, including transcription factors, affecting developmental processes, nutrient transport, and metabolism under stressed conditions (Table 4.3).

4.5.1 miRNAs in Response to Drought and Salt Stress

Genome-wide profiling and analysis of miRNAs in drought-challenged rice across a wide range of developmental stages revealed 11 miRNAs (miR170, miR172, miR397, miR408, miR529, miR896, miR1030, miR1035, miR1050, miR1088, and miR1126) that were significantly downregulated in

Table 4.3   Potential Targets and Annotated Biological Function of Abiotic Stress-Responsive miRNAs

miRNA	Target Gene/Protein	Annotated Biological Function	References
miR156	Squamosa promoter binding-like (SPL) genes	Plant growth and development and phosphorus stress	Zhang et al. (2012)
miR159	Members of the gibberellic acid MYB (GAMYB) family	Seed germination and anther formation, cell proliferation and programmed cell death	Alonso-Peral et al. (2012)
miR160	Auxin response factor (ARF)	Auxin signal transduction during several stages of plant growth and development	Liu et al. (2007)
miR164	NAC1 transcription factor	Auxin signals and lateral root emergence	Nikovics et al. (2006)
miR167	ARF6 and ARF8 genes	Auxin hormone homeostasis during normal and stress condition, modulate lateral root formation	Wu et al. (2006) and Lobbes et al. (2006)
miR169	CCAAT-binding transcription factors, nuclear transcription factor YA (NF-YA) genes	Cold, drought, and salinity	Zhao et al. (2009) and Zhang et al. (2011)
miR172	*APETALA2*, *TOE1* and *TOE2*	Salinity, drought, modulate lateral root development	Wu et al. (2009) and Zhu et al. (2009)
miRNA393	Auxin signaling F-BOX (AFB)	Altered auxin signaling: enlarged flag leaf inclination and altered primary and crown root growth	Arenas-Huertero et al. (2009) and Chen et al. (2012)
miR396	Growth regulating factor genes (GRF), GRF1 and GRF3	Response to abscisic acid stimulus, root development, seed dormancy process, seed germination	Rodriguez et al. (2010) and Hewezi and Baum (2012)
miRNA398	Copper/zinc superoxide dismutases (SOD)	Response to ABA and abiotic stresses	Bouche (2010) and Sunkar et al. (2006)
miRNA399	Phosphate (Pi) homeostasis, phosphate 2 gene (PHO2)	Pi transporter and ubiquitin conjugating enzyme (UBC24) regulating protein degradation pathway	Bari et al. (2006), Franco-Zorrilla et al. (2007) and Zhang et al. (2012)
miRNA408	Plastocyanin-like protein, copper transporter P-type ATPase (*PAA2*) of *Arabidopsis* and copper-binding proteins	Drought and metal stress	Lu et al. (2011) and Lobbes et al. (2006)
miR508	L-ascorbate oxidase (L-AO) and copper ion binding protein genes, Auxin binding proteins	Oxidative stress and hormone homeostasis	Lima et al. (2011)
miR528	L-AO and copper ion binding protein genes	Hormone homeostasis Drought stress	Lima et al. (2011)
miR824	Agamous like 16 gene (*AGL16*)	Posttranscriptional regulation of the stomata opening and closing	Kutter et al. (2007)

response to drought stress. Furthermore, eight novel drought-responsive miRNAs (miR395, miR474, miR845, miR851, miR854, miR901, miR903, and miR1125) were detected (Zhou et al. 2010). In drought-stressed *Medicago truncatula*, eight new miRNAs were identified by high-throughput sequencing, with four of these miRNAs showing up- and downregulation, respectively (Wang et al. 2011). An attempt was made to identify and compare the drought-associated miRNA profiles in two cowpea genotypes (CB46, drought-sensitive, and IT93K503-1, drought-tolerant). Deep sequencing small reads were generated from both genotypes and subsequently mapped to cowpea genomic sequences. Forty-four drought-associated miRNAs were identified, 30 of which were upregulated and 14 downregulated in drought conditions. In addition, 11 miRNAs were drought-regulated in only one genotype, but not the other (Barrera-Figuero et al. 2012).

High salt concentration has been implicated in adversely affecting plant developmental processes, including seed germination, seedling growth and vigor, vegetative growth, flowering, and fruit set, which, ultimately, negatively affect crop productivity and quality (Bartels and Sunkar 2005). These processes are modulated through a number of genes and pathways (Zhu 2002). In *Arabidopsis*, 10 miRNAs (miR396, miR168, miR167, miR165, miR319, miR159, miR394, miR156, miR393, miR171, miR158, and miR169) were upregulated in response to salt stress (Liu et al. 2008). In *Populus trichocarpa*, miR530a, miR1445, miR1446a-e, miR1447, and miR1711-n were downregulated, whereas miR482.2 and miR1450 were upregulated during salt stress (Lu et al. 2008). In *Phaseolus vulgaris*, Arenas-Huertero et al. (2009) observed increased accumulation of miRS1 and miR159.2 in response to NaCl. Comparative miRNA profiling of salt-tolerant (NC286) and salt-sensitive (Huangzao4) lines of *Zea mays* using microarrays identified a total of 98 miRNAs belonging to 27 plant miRNA families with significantly altered expression after salt treatment (Ding et al. 2009). Interestingly, 18 miRNAs were found that were only expressed in the salt-tolerant maize line, and 25 miRNAs that showed a deferred regulation pattern in the salt-sensitive line. Using psRNATarget (plant miRNA target prediction server; http://plantgrn.noble.org/psRNATarget/), 32 potential target genes likely to be involved in a broad spectrum of cellular, metabolic, physiological, and developmental processes, including energy metabolism, signal transduction, transcriptional regulation, and cell defense, were identified. Members of the miR396 family were found to be downregulated, whereas members of the miR395 and miR474 families were found to be upregulated in both maize lines upon exposure to salt. miRNA168 was also found to be upregulated under stress in *Arabidopsis* (Liu et al. 2008). However, the expression of zma-miR168 was repressed in the salt-sensitive line, and induced in the salt-tolerant line. Interestingly, the levels of *AGO1* mRNA levels decreased in the salt-tolerant and increased in the salt-sensitive line. Alteration of *AGO1* mRNA expression, a key regulator in the miRNA pathway in response to salt, may induce further changes of numerous miRNA activities (Ding et al. 2009). In *Arabidopsis*, Zhou et al. (2009) found miR169n and miR169g to be induced by salinity stress. The presence of a cis-acting ABA-responsive element (*ABRE*) in the upstream region of miR69n suggests that miR169 may be ABA regulated (Zhou et al. 2009). In a recent study, high-throughput sequencing was employed to identify miRNAs in control and salt-treated *P tomentosa* plantlets. Under salt stress, 19 miRNAs belonging to seven conserved miRNA families were significantly downregulated, and two miRNAs belonging to two conserved miRNA families were upregulated (Ren et al. 2013).

4.5.2 Involvement of miRNAs in Temperature Stress

An optimum temperature is required by most organisms for proper growth and development. Most plants can tolerate minor fluctuations in temperature, but fluctuations beyond a certain threshold lead to cold or heat stress (Sunkar 2010). Cold stress, including chilling and frost, cause tissue injury and delayed growth in plants, and affect the productivity of crops worldwide (Mahajan and Tuteja 2005). Response to cold temperature is a complex process, involving numerous genes and signaling pathways (Lv et al. 2010). The precise mechanism of cold regulation in plants is not fully understood.

A computational transcriptome-based approach was developed to annotate stress-inducible miRNAs in *Arabidopsis*. Using this approach, 19 miRNA genes of 11 microRNA families were found to be upregulated under cold stress. Northern blot analysis was carried out to further validate these results; of the eleven miRNAs, eight were differentially induced and three were expressed constitutively at low temperature (Zhou et al. 2008). In *Populus*, miR168a, b and miR477a, b were upregulated, while miR156g-j, miR475a, b and miR476a were downregulated under cold stress (Lu et al. 2008). As cold stress is one of the major abiotic stresses affecting rice yields, an attempt was made to understand the role of miRNAs in response to cold stress. Eighteen cold-responsive rice miRNAs were identified using microarrays, and most of them were found to be downregulated (Lv et al. 2010).

To identify genome-wide conserved and novel miRNAs that are responsive to heat stress in *B. rapa*, temperature thresholds of Chinese cabbage (*B. rapa* ssp. *chinensis*) were defined and small RNA libraries were constructed from seedlings exposed to high temperature (46°C) for 1 h. A series of conserved and novel miRNAs responding to heat stress were identified by deep sequencing and data analysis. It was found that Chinese cabbage shares at least 35 conserved miRNA families with *Arabidopsis thaliana*, and. among them, five miRNA families were responsive to heat stress (Yu et al. 2012). Solexa sequencing and analysis of two libraries from *Populus tomentosa* (with or without heat stress) identified 134 conserved miRNAs belonging to 30 miRNA families, and 16 novel miRNAs belonging to 14 families. Among these miRNAs, 52 miRNAs from 15 families were found to be responsive to heat stress. The expression of the conserved and novel miRNAs was confirmed by qRT-PCR analysis (Chen et al. 2012).

4.5.3 Role of miRNAs in Nutrient Stress

Plants require both macro- and micronutrients for their growth and development, acquired predominantly from the soil. As mineral nutrients are involved in basic cellular processes, a deficiency or excess of mineral nutrient elements or toxic ions, respectively, is deleterious for plant growth and development. The acquisition, assimilation, and distribution of these nutrients are controlled by a number of genes or gene products (Sunkar 2010), and understanding the regulatory mechanisms controlling the uptake and assimilation of key nutrient elements, such as phosphorus, sulfur, and copper, may offer avenues to improve plant nutrient efficiency (Chiou 2007).

Phosphorus as inorganic phosphate (Pi) is one of the essential macronutrients for plant growth, development, and reproduction. It is required for the synthesis of nucleic acids and membrane lipids, and is the least available essential nutrient in soil (Marschner 1995). In response to low Pi levels in soil, plants have adapted themselves in different ways, including altered root growth and exudation of organic acids to absorb and utilize Pi from the environment (Abel et al. 2002; Raghothama 1999). Recent findings have implicated the involvement of miR399 in Pi starvation responses (Bari et al. 2006). An inverse relation between the miRNA399 and mRNA levels of one of its targets, *UBC24* (ubiquitin-conjugating E2 enzyme), under normal and low-phosphorus conditions has also been documented in many studies (Chiou 2007; Fujii et al. 2005). The induction of miR399 expression in *Arabidopsis* in phosphorus-deficient conditions, with a corresponding decrease in *UBC24* (Fujii et al. 2005), results in expression of high-affinity Pi transporter such as *AtPR1* and alteration in root architecture. These adaptive changes are important for maintaining Pi levels in plants and indicate that miR399 plays a key role in phosphorus homeostasis.

Sulfur is the fourth important macronutrient for plants after N, P, and K, and is assimilated as the amino acid cysteine, which is important for several metabolic reactions, such as protein synthesis and formation of several low-molecular-weight compounds (Rausch and Wachter 2005). Sulfur is taken up as inorganic sulfate () by the plant roots. Under sulfate starvation conditions, several biological and physiological changes occur in order to sustain sulfate acquisition (Sunkar et al. 2006). The first indication of miR395 involvement in sulfate homeostasis came from its predicted targets (Jones-Rhoades et al. 2006). miR395 targets ATP sulfurylases (*APS1*, *APS3*, and *APS4*) and *AST68* (Lappartient et al. 1999). The ATP sulfurylase enzymes catalyze the first step of the sulfur assimilation pathway, while *AST68* is a low-affinity sulfate transporter implicated in the internal translocation of sulfate from roots to shoots (Allen et al. 2005; Takahashi et al. 1997). Under low sulfate conditions, it has been observed that the expression of miR395 is induced, while that of *APS1* is decreased. By contrast, under sulfur-sufficient conditions, miR395 expression is not detected, while *APS1* transcripts are abundantly expressed (Jones-Rhoades and Bartel 2004).

Copper (Cu) is an essential micronutrient for photosynthesis, oxidative responses, and other physiological processes (Marschner 1995). It is a cofactor for Cu/Zn superoxide dismutase (SOD), which is involved in plant defense systems against ROS (Chu et al. 2005). miR398 has been found to be a key regulator in copper homeostasis, and the conserved miR398 family consists of three genes: miR398a, miR398b, and miR398c (Burkhead et al. 2009). Under Cu-limiting conditions, the level of miR398 was found to be upregulated, with decreased levels of two Cu/Zn SOD mRNA targets (*CSD1* and *CSD2*) (Yamasaki et al. 2007), which resulted in reduced Cu allocation to *CSD*s, thus making it available for other essential processes (Abdel-Ghany and Pilon 2008). Other miRNAs, such as miR397, miR408, and miR587, were found to be upregulated under low-copper conditions (Yamasaki et al. 2007). The target genes for these miRNAs have been grouped into the Cu-miRNAs, which are upregulated by Cu-responsive transcription factor Squamosa promoter binding protein-like 7 (*SPL7*) under low-Cu conditions (Burkhead et al. 2009). Upregulation of *SPL7* activates expression of genes involved in Cu assimilation, demonstrating the role of miR398 in copper homeostasis (Burkhead et al. 2009).

4.5.4 miRNAs and Metal Stress

Metal accumulation in plants generally occurs by uptake, translocation, and sequestration, and can be considered as an integral system for the acquisition of metals by plants (Williams et al. 2000). Plants have acquired mechanisms to control the abundance of essential heavy metals (e.g., Zn and Cu) within the physiological range required for normal growth and development. Additionally, plants need to minimize the detrimental effects of nonessential heavy metals such as Cd and Pb, and an understanding of plant metal toxic/tolerant mechanisms is fundamentally important to agricultural practices (Huang et al. 2009).

Cadmium (Cd), a widespread heavy metal pollutant, is toxic to most plants at concentrations greater than 5–10 µg Cd g−1 leaf dry weight. Due to its high mobility (White and Brown 2010), it accumulates in all parts of the plant, resulting in toxicity symptoms, such as alterations in the chloroplast ultrastructure, inhibition of photosynthesis, inactivation of enzymes involved in CO_2 fixation, induction of lipid peroxidation, and cell death (Mishra et al. 2006; Mobin and Khan 2007; Rodriguez-Serrano et al. 2009). In a study to search for potential miRNAs in *B. napus* in response to cadmium (Xie et al. 2007), previously known miRNAs from *Arabidopsis*, rice, and other plant species were used against both *Brassica* genomic survey sequences (GSS) and EST databases, and five miRNAs were identified in response to cadmium stress (Xie et al. 2007). The regulation of miRNAs in *M. truncatula* under Cd, mercury (Hg), and aluminum (Al) stress was studied. Four miRNAs—miR171, miR319, miR393, and miR529—were found to be upregulated and miR166 and miR398 were observed to be downregulated by Hg, Cd, and Al exposure (Zhou et al. 2008). Huang et al. (2009) isolated 19 potential novel miRNAs from a library of small RNAs constructed from rice seedlings exposed to toxic levels of Cd. Following Cd exposure, miR601, miR602, and miR603 were upregulated in roots while miR602 and miR606 in leaves and miR604 in roots were downregulated (Huang et al. 2009). In order to investigate the responsive functions of miRNAs under Cd stress, Cd-stressed rice was profiled using a microarray assay (Ding et al. 2011).

4.6 miRNAs AND BIOTIC STRESS

Plants activate or evolve defense responses to protect themselves from viral, fungal, and bacterial pathogens. *Arabidopsis* miR393 was the first miRNA identified, which was induced by the flagellin-derived pathogen-associated molecular pattern (PAMP) peptide. miR393 negatively regulates the post-transcriptional expression of F-box auxin receptor transport inhibitor response 1 (*TIR1*) (Navarro et al. 2006). miRNA expression profiling of *Arabidopsis* inoculated with *Pseudomonas syringae* pv. tomato showed induced expression of miR160, 167, and 393 (Fahlgren et al. 2007). The role of these miRNAs was further supported by a mutagenesis study in which the constitutive overexpression of miR393 resistance *TIR1* paralog in *tir1-1* mutant background enhanced susceptibility to *P. syringae* in *Arabidopsis*; however, reduction of bacterial growth was observed in miR393-overexpressing lines (Navarro et al. 2006). The expression of miR156 and 164 was induced in transgenic *Arabidopsis*, infected with virus TYMV p69 (Kasschau et al. 2003). In soybean, Kulcheski et al. (2011) identified novel stress-responsive miRNAs in response to Asian soybean rust (*Phakopsora pachyrhizi*), and observed that *miR166a-5p*, *miR166f*, *miR169-3p*, *miR397ab*, and *miR-Seq13* were downregulated in the susceptible genotype during pathogen invasion and equally expressed in the resistant plants. The induction or expression of miRNA was shown to be pathogen and species specific. For instance, in *Brassica*, bra-miR158 and bra-miR1885 were induced only by infection with turnip mosaic virus, and not by cucumber mosaic virus or tobacco mosaic virus (He et al. 2008). In response to powdery mildew infection in wheat (*Triticum aestivum* L.), miR2001, miR827, miR2008, miR2012, and miR2013 showed increased expression, while miR156, miR159, miR164, miR171, and miR396 showed decreased expression after powdery mildew infection (Xin et al. 2010). In another study, miR156 and miR160 were repressed in pine stem when infected with the rust fungus (*Cronartium quercuum* f. sp. *fusiforme*) (Kasschau et al. 2003; Lu et al. 2007). All these studies signify the role of miRNAs in pathogen-based immune responses; however, further understanding of miRNA–mRNA interaction, and their regulation during pathogen attack, will increase our understanding of plant tolerance to biotic stress.

4.7 CONCLUSION AND FUTURE PERSPECTIVES

Abiotic and biotic stresses, as well as nutrient deficiency, are the major factors affecting crop productivity worldwide. Various molecular processes are involved in stress perception, signal transduction, and the expression of specific stress-induced genes in plants resulting in adaptation to environmental stresses. Recently, miRNA and short-interfering RNA (siRNA) have gained considerable attention because they act as important regulators of gene expression in animals and plants during development or in response to environmental challenges. miRNAs function through targeting genes involved in development, transcription factors, antioxidant enzymes, transporter genes, genes involved in nutrient transport, and metabolic production, or through changing the expression of other targets that are part of regulatory networks, remaining to be fully elucidated. Studies of in-depth annotation of small RNAs obtained from high-throughput sequencing data, using rapidly evolving bioinformatics approaches, will identify novel functions of currently known miRNAs as well as leading to the discovery of more classes of novel miRNAs. It is crucial to identify and validate a full set of miRNA targets in order to completely understand their involvement in different cellular functions. miRNAs may modulate these targets not only at the mRNA level but also at the level of protein expression, which could be investigated by employing quantitative proteomic approaches. It would also be interesting to identify the role in plants of long noncoding RNAs, which may play an important role in epigenetic gene regulation. Furthermore, miRNA studies will help in developing molecular markers, which will probably lead to new ways to enhance crop tolerance to different biotic and abiotic problems in the agricultural sector.

REFERENCES

Abdel-Ghany, S. E. and M. Pilon. 2008. MicroRNA-mediated systemic down-regulation of copper protein expression in response to low copper availability in *Arabidopsis. J Biol Chem* 283:15932–15945.

Abel, S., C. A. Ticconi, and C. A. Delatorre. 2002. Phosphate sensing in higher plants. *Physiol Plant* 115:1–8.

Achard, P., A. Herr, D. C. Baulcombe, and N. P. Harrberd. 2004. Modulation of floral development by a gibberellin-regulated microRNA. *Development* 131:3357–3365.

Adai, A., C. Johnson, S. Mlotshwa et al. 2005. Computational prediction of miRNAs in *Arabidopsis thaliana. Genome Res* 15:78–91.

Addo-Quaye, C., T. W. Eshoo, D. P. Bartel, and M. J. Axtell. 2008. Endogenous siRNA and miRNA targets identified by sequencing of the *Arabidopsis* degradome. *Curr Biol* 18:758–762.

Addo-Quaye, C., W. Miller, and M. J. Axtell. 2009. CleaveLand: A pipeline for using degradome data to find cleaved small RNA targets. *Bioinformatics* 25:130–131.

Aguilar-Martinez, J. A., C. Poza-Carrion, and P. Cubas. 2007. *Arabidopsis* BRANCHED1 acts as an integrator of branching signals within axillary buds. *Plant Cell* 19:458–472.

Ahmed, F. and P. X. Zhao. 2011. A comprehensive analysis of isomers and their targets using high-throughput sequencing data for *Arabidopsis thaliana. J Nat Sci Biol Med* 2:32–33.

Aldridge, S. and J. Hadfield. 2012. Introduction to miRNA profiling technologies and cross-platform comparison. *Methods Mol Biol* 822:19–31.

Alexiou, P., T. Vergoulis, M. Gleditzsch et al. 2010. miRGen 2.0: A database of microRNA genomic information and regulation. *Nucleic Acids Res* 38:D137–D141.

Allen, E., Z. Xie, A. M. Gustafson et al. 2004. Evolution of microRNA genes by inverted duplication of target gene sequences in *Arabidopsis thaliana.* Nat Genet 36:1282–1290.

Allen, E., Z. Xie, A. M. Gustafson, and J. C. Carrington. 2005. MicroRNA-directed phasing during trans-acting siRNA biogenesis in plants. *Cell* 121:207–221.

Allen, R. S., X. Li, M. I. Stahle et al. 2007. Genetic analysis reveals functional redundancy and the major target genes of the *Arabidopsis* miR159 family. *Proc Natl Acad Sci USA* 104:16371–16376.

Alonso-Peral, M. M., C. Sun, and A. A. Millar. 2012. MicroRNA159 can act as a switch tuning microRNA independently of its abundance in *Arabidopsis. PLoS ONE* 7:e34751.

Arenas-Huertero, C., B. Perez, F. Rabanal et al. 2009. Conserved and novel miRNAs in the legume *Phaseolus vulgaris* in response to stress. *Plant Mol Biol* 70:385–401.

Aukerman, M. J. and H. Sakai. 2003. Regulation of flowering time and floral organ identity by a microRNA and its APETALA2-Like target genes. *Plant Cell* 15:2730–2741.

Axtell, M. J., J. O. Westholm, and E. C. Lai. 2011. Vive la difference: Biogenesis and evolution of microRNAs in plants and animals. *Genome Biol* 12:221–234.

Baker, C. C., P. Sieber, F. Wellmer et al. 2005. The early extra petals1 mutant uncovers a role for microRNA miR164c in regulating petal number in *Arabidopsis. Curr Biol* 15:303–315.

Baker, M. 2010. MicroRNA profiling: Separating signal from noise. *Nat Methods* 7:687–691.

Balcells, I., S. Cirera, and P. K. Busk. 2011. Specific and sensitive quantitative RT-PCR of miRNAs with DNA primers. *BMC Biotechnol* 11:70–81.

Bari, R., B. P. Datt, M. Stitt et al. 2006. PHO2, microRNA399, and PHR1 define a phosphate-signaling pathway in plants. *Plant Physiol* 141:988–999.

Barrera-Figueroa, B. E., L. Gao, N. N. Diop et al. 2012. Identification and comparative analysis of drought-associated microRNAs in two cowpea genotypes. *BMC Plant Biol* 11:127–137.

Bartels, D. and R. Sunkar. 2005. Drought and salt tolerance in plants. *Crit Rev Plant Sci* 24:23–58.

Betel, D., M. Wilson, and A. Gabow. 2008. The microRNA.org resource: Targets and expression. *Nucleic Acids Res* 36:D149–D153.

Blein, T., A. Pulido, A. Vialette-Guiraud et al. 2008. A conserved molecular framework for compound leaf development. *Science* 322:1835–1839.

Bonnet, E., J. Wuyts, P. Rouze, and Y. Van de Peer. 2004. Detection of 91 potential conserved plant microRNAs in *Arabidopsis thaliana* and *Oryza sativa* identifies important target genes. *Proc Natl Acad Sci USA* 101:11511–11516.

Boualem, A., P. Laporte, M. Jovanovic et al. 2008. MicroRNA166 controls root and nodule development in *Medicago truncatula. Plant J* 54:876–887.

Bouche, N. 2010. New insights into miR398 functions in *Arabidopsis*. *Plant Signal Behav* 5:684–686.

Boutet, S., F. Vazquez, J. Liu et al. 2003. *Arabidopsis* HEN1: A genetic link between endogenous miRNA controlling development and siRNA controlling transgene silencing and virus resistance. *Curr Biol* 13:843–848.

Breakfield, N. W., D. L. Corcoran, J. J. Petricka et al. 2012. High-resolution experimental and computational profiling of tissue-specific known and novel miRNAs in *Arabidopsis*. *Genome Res* 22:163–176.

Burkhead, J. L., K. A. G. Reynolds, S. E. Abdel-Ghany et al. 2009. Copper homeostasis. *New Phytol* 182:799–816.

Chambers, C. and B. Shuai. 2009. Profiling microRNA expression in *Arabidopsis* pollen using microRNA array and real-time PCR. *BMC Plant Biol* 9:87.

Chen, C., D. A. Ridzon, A. J. Broomer et al. 2005. Real-time quantification of microRNAs by stem-loop RT-PCR. *Nucleic Acids Res* 33:e179.

Chen, C. Z., L. Li, H. F. Lodish, and D. P. Bartel. 2004. MicroRNAs modulate hematopoietic lineage differentiation. *Science* 303:83–86.

Chen, K. and N. Rajewsky. 2007. The evolution of gene regulation by transcription factors and microRNAs. *Nat Rev Genet* 8:93–103.

Chen, L., Y. Ren, Y. Zhang et al. 2012. Genome-wide identification and expression analysis of heat-responsive and novel microRNAs in *Populus tomentosa*. *Gene* 504:160–165.

Chiou, T. J. 2007. The role of microRNAs in sensing nutrient stress. *Plant Cell Environ* 30:323–332.

Cho, S., I. Jang, S. Yoon et al. 2013. MiRGator v3.0: A microRNA portal for deep sequencing, expression profiling and mRNA targeting. *Nucleic Acids Res* 41:D252–D257.

Chu, C. C., W. C. Lee, W. Y. Guo et al. 2005. A copper chaperone for superoxide dismutase that confers three types of copper/zinc superoxide dismutase activity in *Arabidopsis*. *Plant Physiol* 139:425–436.

Chuck, G., A. M. Cigan, K. Saeteurn et al. 2007. The heterochronic maize mutant Corngrass1 results from overexpression of a tandem microRNA. *Nat Genet* 39:544–549.

Colaiacovo, M., I. Bernardo, I. Certomani et al. 2012. Survey of microRNA length variants contributing to miRNome complexity in peach (*Prunus Persica* L.). *Front Plant Sci* 3:165–182.

Cubas, P., N. Lauter, J. Doebley, and E. Coen. 1999. The TCP domain: A motif found in proteins regulating plant growth and development. *Plant J* 18:215–222.

Dai, X. and P. X. Zhao. 2011. psRNATarget: A plant small RNA target analysis server. *Nucleic Acids Res* 39:W155–W159.

Ding, D., Z. Lifang, H. Wang et al. 2009. Differential expression of miRNAs in response to salt stress in maize roots. *Annals Bot* 103:29–38.

Ding, Y., Z. Chen, and C. Zhu. 2011. Microarray-based analysis of cadmium-responsive microRNAs in rice (*Oryza sativa*). *J Exp Bot* 62:3563–3573.

Dweep, H., C. Sticht, P. Pandey, and N. Gretz. 2011. miRWalk-database prediction of possible miRNA binding sites by walking the genes of 3 genomes. *J Biomed Inform* 44:839–847.

Efroni, I., E. Blum, A. Goldshmidt, and Y. Eshed. 2008. A protracted and dynamic maturation schedule underlies *Arabidopsis* leaf development. *Plant Cell* 20:2293–2306.

Elefant, N., A. Berger, H. Shein et al. 2011. RepTar: A database of predicted cellular targets of host and viral miRNAs. *Nucleic Acids Res* 39:D188–D194.

Fahlgren, N., M. D. Howell, K. D. Kasschau et al. 2007. High-throughput sequencing of *Arabidopsis* microRNAs: Evidence for frequent birth and death of MIRNA genes. *PLoS ONE* 2:e219.

Fang, Y. and D. L. Spector. 2007. Identification of nuclear dicing bodies containing proteins for microRNA biogenesis in living *Arabidopsis* plants. *Curr Biol* 17:818–823.

Franco-Zorrilla, J. M., P. A. Reis, M. T. Reis et al. 2007. Target mimicry provides a new mechanism for regulation of microRNA activity. *Nat Genet* 39:1033–1037.

Fujii, H., T. J. Chiou, S. I. Lin et al. 2005. A miRNA involved in phosphate starvation response in *Arabidopsis*. *Curr Biol* 15:2038–2043.

Gao, J. P., D. Y. Chao, and H. X. Lin. 2007. Understanding abiotic stress tolerance mechanisms: Recent studies on stress response in rice. *J Integr Biol* 49:742–750.

German, M. A., S. Luo, G. Schroth et al. 2009. Construction of parallel analysis of RNA ends (PARE) libraries for the study of cleaved miRNA targets and the RNA degradome. *Nat Protoc* 4:356–362.

German, M. A., M. Pillay, D. H. Jeong et al. 2008. Global identification of microRNA-target RNA pairs by parallel analysis of RNA ends. *Nat Biotechnol* 26:941–946.

Git, A., H. Dvinge, M. Salmon-Divon et al. 2010. Systematic comparison of microarray profiling, real-time PCR, and next-generation sequencing technologies for measuring differential microRNA expression. *RNA* 16:991–1006.

Griffiths-Jones, S., H. K. Saini, S. van Dongen et al. 2006. miRBase: Tools for microRNA genomics. *Nucleic Acids Res* 39:D154–D158.

Grigg, S. P., C. Canales, A. Hay et al. 2005. SERRATE co-ordinates shoot meristem function and leaf axial patterning in *Arabidopsis*. *Nature* 437:1022–1026.

Gubler, F., D. Raventos, M. Keys et al. 1999. Target genes and regulatory domains of the GAMYB transcriptional activator in cereal aleurone. *Plant J* 17:1–9.

Han, M. H., S. Goud, L. Song et al. 2004. The *Arabidopsis* double-stranded RNA-binding protein HYL1 plays a role in microRNA-mediated gene regulation. *Proc Natl Acad Sci USA* 101:1093–1098.

He, X. F., Y. Y. Fang, L. Feng et al. 2008. Characterization of conserved and novel microRNAs and their targets, including a TuMV-induced TIR-NBS-LRR class R gene-derived novel miRNA in Brassica. *FEBS Lett* 582:2445–2452.

Hewezi, T. and T. J. Baum. 2012. Complex feedback regulations govern the expression of miRNA396 and its GRF target genes. *Plant Signal Behav* 7:749–751.

Hsu, S. D., C. H. Chu, A. P. Tsou et al. 2008. miRNAMap 2.0: Genomic maps of microRNAs in metazoan genomes. *Nucleic Acids Res* 36:D165–D169.

Hsu, S. D., F. M. Lin, W. Y. Wu et al. 2011. miRTarBase: A database curates experimentally validated microRNA-target interactions. *Nucleic Acids Res* 39:D163–D169.

Huang, S. Q., J. Peng, C. X. Qiu et al. 2009. Heavy metal-regulated new microRNAs from rice. *J Inorg Biochem* 103:282–287.

Ivashuta, S., I. R. Banks, B. E. Wiggins et al. 2011. Regulation of gene expression in plants through miRNA inactivation. *PLoS ONE* 6:e21330.

Jagadeeswaran, G., Y. Zheng, Y. F. Li et al. 2009. Cloning and characterization of small RNAs from *Medicago truncatula* reveals four novel legume-specific microRNA families. *New Phytol* 184:85–98.

Jones-Rhoades, M. W. and D. P. Bartel. 2004. Computational identification of plant microRNAs and their targets, including a stress-induced miRNA. *Mol Cell* 14:787–799.

Jones-Rhoades, M. W., D. P. Bartel, and B. Bartel. 2006. MicroRNAs and their regulatory roles in plants. *Ann Rev Plant Biol* 57:19–53.

Jover-Gil, S., H. Candela, and M. R. Ponce. 2005. Plant microRNAs and development. *Int J Dev Biol* 49:733–744.

Kasschau, K. D., Z. Xie, E. Allen et al. 2003. P1/HC-Pro, a viral suppressor of RNA silencing, interferes with *Arabidopsis* development and miRNA function. *Dev Cell* 4:205–217.

Kertesz, M., N. Iovino, U. Unnerstall et al. 2007. The role of site accessibility in microRNA target recognition. *Nat Genet* 39:1278–1284.

Kidner, C. A. and R. A. Martienssen. 2005. The developmental role of microRNA in plants. *Curr Opin Plant Biol* 8:38–44.

Kim, B., H. J. Yu, S. G. Park et al. 2012. Identification and profiling of novel miRNAs in the *Brassica rapa* genome based on small RNA deep sequencing. *BMC Plant Biol* 12:218–231.

Koyama, T., M. Furutani, M. Tasaka et al. 2007. TCP transcription factors control the morphology of shoot lateral organs via negative regulation of the expression of boundary-specific genes in *Arabidopsis*. *Plant Cell* 19:473–484.

Koyama, T., N. Mitsuda, M. Seki, K. Shinozaki, and M. Ohme-Takagi. 2010. TCP transcription factors control the morphology of shoot lateral organs via negative regulation of the expression of boundary specific genes in *Arabidopsis*. *Plant Cell* 19:473–484.

Krek, A., D. Grun, M. N. Poy et al. 2005. Combinatorial microRNA target predictions. *Nat Genet* 37:495–500.

Kulcheski, F. R., L. F. deOliveira, L. G. Molina et al. 2011. Identification of novel soybean microRNAs involved in abiotic and biotic stresses. *BMC Genomics* 12:307.

Kutter, C., H. Schöb, M. Stadler, F. Meins Jr, and A. Si-Ammour. 2007. MicroRNA-mediated regulation of stomatal development in *Arabidopsis*. *Plant Cell* 19:2417–2429.

Lappartient, A. G., J. J. Vidmar, T. Leustek et al. 1999. Inter-organ signaling in plants: Regulation of ATP sulfurylase and sulfate transporter genes expression in roots mediated by phloem-translocated compound. *Plant J* 18:89–95.

Laufs, P., A. Peaucelle, H. Morin, and J. Traas. 2004. MicroRNA regulation of the CUC genes is required for boundary size control in *Arabidopsis* meristems. *Development* 131:4311–4322.

Lee, R. C., R. L. Feinbaum, and V. Ambros. 1993. The C. elegans heterochronic gene lin-4 encodes small RNAs with antisense complementarity to lin-14. *Cell* 75:843–854.

Lelandais-Briere, C., L. Naya, E. Sallet et al. 2009. Genome-wide *Medicago truncatula* small RNA analysis revealed novel microRNAs and isoforms differentially regulated in roots and nodules. *Plant Cell* 21:2780–2796.

Lewis, B. P., C. B. Burge, and D. P. Bartel. 2005. Conserved seed pairing, often flanked by adenosines, indicates that thousands of human genes are microRNA targets. *Cell* 120:15–20.

Li, Y. F., Y. Zheng, C. Addo-Quaye et al. 2010. Transcriptome-wide identification of microRNA targets in rice. *Plant J* 62:742–759.

Lima, J. C., R. A. Arenhart, M. Margis Pinheiro, and R. Margis. 2011. Aluminium triggers broad changes in microRNA expression in rice roots. *Genet Mol Res* 10:2817–2832.

Liu, H. H., X. Tian, Y. J. Li et al. 2008. Microarray-based analysis of stress regulated microRNAs in *Arabidopsis thaliana*. *RNA* 14:836–843.

Liu, P. P., T. A. Montgomery, N. Fahlgren et al. 2007. Repression of AUXIN RESPONSE FACTOR10 by microRNA160 is critical for seed germination and post-germination stages. *Plant J* 52:133–146.

Liu, X., J. Huang, Y. Wang et al. 2010. The role of floral organs in carpels, an *Arabidopsis* loss-of-function mutation in MicroRNA160a, in organogenesis and the mechanism regulating its expression. *Plant J* 62:416–428.

Lobbes, D., G. Rallapalli, and D. D. Schmidt. 2006. SERRATE: A new player on the plant microRNA scene. *EMBO Rep* 7:1052–1058.

Lu, S., Y. H. Sun, H. Amerson, and V. L. Chiang. 2007. MicroRNAs in loblolly pine (*Pinus taeda* L.) and their association with fusiform rust gall development. *Plant J* 51:1077–1098.

Lu, S., Y. H. Sun, and V. L. Chiang. 2008. Stress-responsive microRNAs in *Populus*. *Plant J* 55:131–151.

Lu, S., Y.-H. Sun, R. Shi et al. 2005. Novel and mechanical stress-responsive microRNAs in *Populus trichocarpa* that are absent from *Arabidopsis*. *Plant Cell* 17:2186–2203.

Lu, Y., Z. Hao, C. Xie et al. 2011. Large scale screening for maize drought resistance using multiple selection criteria evaluated under water stressed and well-watered environment. *Field Crops Res* 124:37–45.

Lv, D. K., X. Bai, Y. Li et al. 2010. Profiling of cold-stress-responsive miRNAs in rice by microarrays. *Gene* 459:39–47.

Mahajan, S. and N. Tuteja. 2005. Cold, salinity and drought stresses: An overview. *Arch Biochem Biophys* 444:139–158.

Mallory, A. C., D. P. Bartel, and B. Bartel. 2005. MicroRNA-directed regulation of *Arabidopsis* AUXIN RESPONSE FACTOR17 is essential for proper development and modulates expression of early auxin response genes. *Plant Cell* 17:1360–1375.

Mallory, A. C., D. V. Dugas, D. P. Bartel, and B. Bartel. 2004. MicroRNA regulation of NAC-domain targets is required for proper formation and separation of adjacent embryonic, vegetative and floral organs. *Curr Biol* 14:1035–1046.

Mao, W., Z. Li, X. Xia et al. 2012. A combined approach of high-throughput sequencing and degradome analysis reveals tissue specific expression of microRNAs and their targets in cucumber. *PloS ONE* 7:e33040.

Maragkakis, M., M. Reczko, V. A. Simossis et al. 2009. DIANA-microT web server: Elucidating microRNA functions through target prediction. *Nucleic Acids Res* 37:W273–W276.

Marschner, H. 1995. *Mineral Nutrition of Higher Plants*, 2nd edn. Academic Press, Germany.

Merchan, F., B. Adnane, C. Martin, and F. Florian. 2009. Plant polycistronic precursors containing non-homologous microRNAs target transcripts encoding functionally related to proteins. *Gen Biol* 10:R136.

Mestdagh, P., T. Feys, N. Bernard et al. 2008. High-throughput stem-loop RT-qPCR miRNA expression profiling using minute amounts of input RNA. *Nucleic Acids Res* 36:e143.

Metzker, M. L. 2010. Sequencing technologies-the next generation. *Nat Rev Genet* 11:31–46.

Millar, A. A. and F. Gubler. 2005. The *Arabidopsis* GAMYB-like genes, MYB33 and MYB65, are microRNA-regulated genes that redundantly facilitate anther development. *Plant Cell* 17:705–721.

Miranda, K. C., T. Huynh, Y. Tay et al. 2006. A pattern-based method for the identification of MicroRNA binding sites and their corresponding heteroduplexes. *Cell* 126:1203–1217.

Mishra, S., S. Srivastava, R. D. Tripathi et al. 2006. Phytochelatin synthesis and response of antioxidants during cadmium stress in *Bacopa monnieri* L. *Plant Physiol Biochem* 44:25–37.

Mobin, M. and N. A. Khan. 2007. Photosynthetic activity, pigment composition and antioxidative response of two mustard (*Brassica juncea*) cultivars differing in photosynthetic capacity subjected to cadmium stress. *J Plant Physiol* 164:601–610.

Navarro, L., P. Dunoyer, F. Jay et al. 2006. A plant miRNA contributes to antibacterial resistance by repressing auxin signaling. *Science* 312:436–439.

Neilsen, C. T., G. J. Goodall, and C. P. Bracken. 2012. IsomiRs: The overlooked repertoire in the dynamic microRNAome. *Trends Genet* 28:544–549.

Newman, M. A., V. Mani, and S. M. Hammond. 2011. Deep sequencing of microRNA precursors reveals extensive 3′ end modification. *RNA* 17:1795–1803.

Nikovics, K., T. Blein, A. Peaucelle et al. 2006. The balance between the *MIR164A* and *CUC2* genes controls leaf margin serration in *Arabidopsis*. *Plant Cell* 18:2929–2945.

Ohto, M. A., R. L. Fischer, R. B. Goldberg et al. 2005. Control of seed mass by APETALA2. *Proc Natl Acad Sci USA* 102:3123–3128.

Ori, N., A. R. Cohen, A. Etzioni et al. 2007. Regulation of LANCEOLATE by miR319 is required for compound-leaf development in tomato. *Nat Genet* 39:787–789.

Palatnik, J. F., E. Allen, X. Wu et al. 2003. Control of leaf morphogenesis by microRNAs. *Nature* 425:257–263.

Park, M. Y., G. Wu, A. Gonzalez-Sulser et al. 2005. Nuclear processing and export of microRNAs in *Arabidopsis*. *Proc Natl Acad Sci USA* 102:3691–3696.

Park, W., J. Li, R. Song et al. 2002. CARPEL FACTORY, a Dicer homolog, and HEN1, a novel protein, act in microRNA metabolism in *Arabidopsis thaliana*. *Curr Biol* 12:1484–1495.

Pritchard, C. C., H. H. Cheng, and M. Tewari. 2012. MicroRNA profiling: Approaches and considerations. *Nat Rev Genet* 13:358–369.

Raghothama, K. G. 1999. Phosphate acquisition. *Annu Rev Plant Physiol Plant Mol Biol* 50:665–693.

Rajagopalan, R., H. Vaucheret, J. Trejo et al. 2006. A diverse and evolutionarily fluid set of microRNAs in *Arabidopsis thaliana*. *Genes Dev* 20:3407–3425.

Rausch, T. and A. Wachter. 2005. Sulfur metabolism: A versatile platform for launching defence operations. *Trends Plant Sci* 10:503–509.

Rehmsmeier, M., S. Peter, H. Matthias et al. 2004. Fast and effective prediction of microRNA/target duplexes. *RNA* 10:1507–1517.

Reinhart, B. J., E. G. Weinstein, M. W. Rhoades, B. Bartel, and D. P. Bartel. 2002. MicroRNAs in plants. *Genes Dev* 16:1616–1626.

Ren, Y., L. Chen, Y. Zhang, X. Kang, Z. Zhang, and Y. Wang. 2013. Identification and characterization of salt-responsive microRNAs in *Populus tomentosa* by high-throughput sequencing. *Biochimie* 4:743–750.

Rodriguez, R. E., M. A. Mecchia, J. M. Debernardi, C. Schommer, D. Weigel, and J. F. Palatnik. 2010. Control of cell proliferation in *Arabidopsis thaliana* by microRNA miR396. *Development* 137:103–112.

Rodriguez-Serrano, M., M. C. Romero, D. M. Pazmino et al. 2009. Cellular response of pea plants to cadmium toxicity: Cross talk between reactive oxygen species, nitric oxide, and calcium. *Plant Physiol* 150:229–243.

Sablok, G., A. L. Pérez-quintero, M. Hassan, T. V. Tatarinova, and C. López. 2011. Artificial microRNAs (amiRNAs) engineering: On how microRNA-based silencing methods have affected current plant silencing research. *Biochem Biophys Res Commun* 406:315–319.

Schommer, C., J. F. Palatnik, P. Aggarwal et al. 2008. Control of jasmonate biosynthesis and senescence by miR319 targets. *PLoS Biol* 6:e230.

Schwarz, S., A. V. Grande, N. Bujdoso, H. Saedler, and P. Huijser. 2008. The microRNA regulated SBP-box gene SPL9 and SPL15 control shoot maturation in *Arabidopsis*. *Plant Mol Biol* 67:183–195.

Sethupathy, P., B. Corda, and A. G. Hatzigeorgiou. 2006. TarBase: A comprehensive database of experimentally supported animal microRNA targets. *RNA* 12:192–197.

Shleizer-Burko, S., Y. Burko, O. Ben-Herzel, and N. Ori. 2011. Dynamic growth program regulated by LANCEOLATE enables flexible leaf patterning. *Development* 138:695–704.

Smith, M. R., M. R. Willmann, G. Wu et al. 2009. Cyclophilin 40 is required for microRNA activity in *Arabidopsis*. *Proc Natl Acad Sci USA* 106:5424–5429.

Song, L., M. H. Han, J. Lesicka, and N. Fedoroff. 2007. *Arabidopsis* primary microRNA processing proteins HYL1 and DCL1 define a nuclear body distinct from the Cajal body. *Proc Natl Acad Sci USA* 104:5437–5442.

Subramanian, S., Y. Fu, R. Sunkar, W. B. Barbazuk, J. K. Zhu, and O. Yu. 2008. Novel and nodulation-regulated microRNAs in soybean roots. *BMC Genomics* 9:160.

Sun, L.-M., X.-Y. Ai, W.-Y. Li et al. 2012. Identification and comparative profiling of miRNAs in an early flowering mutant of trifoliate orange and its wild type by genome-wide deep sequencing. *PLoS ONE* 7:e43760.

Sunkar, R. 2010. MicroRNAs with macro-effects on plant stress responses. *Semin Cell Dev Biol* 21:805–811.

Sunkar, R., A. Kapoor, and J. K. Zhu. 2006. Post-transcriptional induction of two Cu/Zn superoxide dismutase genes in *Arabidopsis* is mediated by down regulation of miR398 and important for oxidative stress tolerance. *Plant Cell* 18:2051–2065.

Sunkar, R., X. F. Zhou, Y. Zheng, W. X. Zhang, and J. K. Zhu. 2008. Identification of novel and candidate miRNAs in rice by high throughput sequencing. *BMC Plant Biol* 8:25.

Sunkar, R. and J. K. Zhu. 2004. Novel and stress-regulated microRNAs and other small RNAs from *Arabidopsis*. *Plant Cell* 16:2001–2019.

Takahashi, H., M. Yamazaki, N. Sasakura et al. 1997. Regulation of sulfur assimilation in higher plants: A sulfate transporter induced in sulfate-starved roots plays a central role in *Arabidopsis thaliana*. *Proc Natl Acad Sci USA* 94:11102–11107.

Takeda, T., K. Amano, M. A. Ohto et al. 2006. RNA interference of the *Arabidopsis* putative transcription factor TCP16 gene results in abortion of early pollen development. *Plant Mol Biol* 61:165–177.

Varkonyi-Gasic, E. and R. P. Hellens. 2011. Quantitative stem-loop RT-PCR for detection of microRNAs. RNAi and plant gene function analysis. *Methods Mol Biol* 744:145–157.

Vazquez, F., V. Gasciolli, and P. Cre. 2004. The nuclear dsRNA binding protein HYL1 is required for microRNA accumulation and plant development, but not posttranscriptional transgene silencing. *Curr Biol* 14:346–351.

Wang, F., L. Li, and L. Liu. 2012a. High-throughput sequencing discovery of conserved and novel MicroRNAs in Chinese cabbage (*Brassica rapa* L. ssp. pekinensis). *Mol Genet Genomics* 287:555–563.

Wang, J. W., R. Schwab, B. Czech, E. Mica, and D. Weigel. 2008. Dual effects of miR156-targeted SPL genes and CYP78A5/KLUH on plastochron length and organ size in *Arabidopsis thaliana*. *Plant Cell* 20:1231–1243.

Wang, T., L. Chen, M. Zhao, Q. Tian, and W. Zhang. 2011. Identification of drought-responsive microRNAs in *Medicago truncatula* by genome-wide high-throughput sequencing. *BMC Genomics* 12:367.

Wang, X. 2008. miRDB: A microRNA target prediction and functional annotation database with a wiki interface. *RNA* 14:1012–1017.

Wang, Y., D. Deng, Y. Shi, N. Miao, Y. Bian, and Z. Yin. 2012b. Diversification, phylogeny and evolution of auxin response factor (ARF) family: Insights gained from analyzing maize ARF genes. *Mol Biol Rep* 39:2401–2415.

Weigel, D., J. H. Ahn, M. A. Blazquez et al. 2000. Activation tagging in *Arabidopsis*. *Plant Physiol* 122:1003–1013.

White, P. J. and P. H. Brown. 2010. Plant nutrition for sustainable development and global health. *Ann Bot* 105:1073–1080.

Wightman, B., I. Ha, and G. Ruvkun. 1993. Posttranscriptional regulation of the heterochronic gene lin-14 by lin-4 mediates temporal pattern formation in *C. elegans*. *Cell* 75:855–862.

Williams, L. E., J. K. Pittman, and J. L. Hall. 2000. Emerging mechanisms for heavy metal transport in plants. *Biochim Biophys Acta* 1465:104–126.

Wu, B., M. Wang, Y. Ma, L. Yuan, and S. Lu. 2012. High-throughput sequencing and characterization of the small RNA transcriptome reveal features of novel and conserved microRNAs in *Panax ginseng*. *PLoS ONE* 7:e44385.

Wu, G., Y. P. Mee, R. C. Susan et al. 2009. The sequential action of miR156 and miR172 regulates developmental timing in *Arabidopsis*. *Cell* 138:750–759.

Wu, G., Q. Tian, and J. W. Reed. 2006. *Arabidopsis* microRNA167 controls patterns of ARF6 and ARF8 expression, and regulates both female and male reproduction. *Development* 133:4211–4218.

Xiao, F., Z. Zuo, G. Cai, S. Kang, X. Gao, and T. Li. 2009. miRecords: An integrated resource for microRNA-target interactions. *Nucleic Acids Res* 37:D105–D110.

Xie, F. L., S. Q. Huang, K. Guo, Y. Y. Zhu, L. Nie, and Z. M. Yang. 2007. Computational identification of novel microRNAs and targets in *Brassica napus*. *FEBS Lett* 581:1464–1473.

Xie, K., J. Shen, X. Hou et al. 2012. Gradual increase of miR156 regulates temporal expression changes of numerous genes during leaf development in rice. *Plant Physiol* 158:1382–1394.

Xie, K., C. Wu, and L. Xiong. 2006. Genomic organization, differential expression, and interaction of SQUAMOSA promoter binding-like transcription factors and microRNA156 in rice. *Plant Physiol* 142:280–293.

Xie, Z., E. Allen, N. Fahlgren, A. Calamar, S. A. Givan, and J. C. Carrington. 2005. Expression of *Arabidopsis* miRNA genes. *Plant Physiol* 138:2145–2154.

Xie, Z., K. Khanna, and S. Ruan. 2010. Expression of microRNAs and its regulation in plants. *Semin Cell Dev Biol* 21:790–797.

Xin, M., Y. Wang, Y. Yao et al. 2010. Diverse set of microRNAs are responsive to powdery mildew infection and heat stress in wheat (*Triticum aestivum* L.). *BMC Plant Biol* 10:123.

Xu, L., Y. Wang, and Y. Xu. 2013. Identification and characterization of novel and conserved microRNAs in radish (*Raphanus sativus* L.) using high-throughput sequencing. *Plant Sci* 64:4271–4287.

Xu, M. Y., Y. Dong, Q. X. Zhang et al. 2012. Identification of miRNAs and their targets from *Brassica napus* by high-throughput sequencing and degradome analysis. *BMC Genomics* 13:421.

Xu, Q., Y. Liu, A. Zhu et al. 2010. Discovery and comparative profiling of microRNAs in a sweet orange red-flesh mutant and its wild type. *BMC Genomics* 11:246.

Yamasaki, H., S. E. Abdel-Ghany, C. M. Cohu, Y. Kobayashi, T. Shikanai, and M. Pilon. 2007. Regulation of copper homeostasis by micro-RNA in *Arabidopsis*. *J Biol Chem* 282:16369–16378.

Yang, J. H., J. H. Li, P. Shao et al. 2011. starBase: A database for exploring microRNA-mRNA interaction maps from Argonaute CLIP-Seq and Degradome-Seq data. *Nucleic Acids Res* 39:D202–D209.

Yang, J. H., P. Shao, H. Zhou et al. 2010. deepBase: A database for deeply annotating and mining deep sequencing data. *Nucleic Acids Res* 38:D123–D130.

Yang, X., L. Wang, D. Yuan, K. Lindsey, and X. Zhang. 2013. Small RNA and degradome sequencing reveal complex miRNA regulation during cotton somatic embryogenesis. *J Exp Bot* 64:1521–1536.

Yu, B., Z. Yang, J. Li et al. 2005. Methylation as a crucial step in plant microRNA biogenesis. *Science* 307:932–935.

Yu, X., W. Han, L. Yizhen et al. 2012. Identification of conserved and novel microRNAs that are responsive to heat stress in *Brassica rapa*. *J Exp Bot* 63:1025–1038.

Zeng, Q. Y., C. Y. Yang, Q. B. Ma, X. P. Li, W. W. Dong, and H. Nian. 2012. Identification of wild soybean miRNAs and their target genes responsive to aluminum stress. *BMC Plant Biol* 12:182.

Zhang, B. H., X. P. Pan, Q. L. Wang, G. P. Cobb, and T. A. Anderson. 2005. Identification and characterization of new plant microRNAs using EST analysis. *Cell Res* 15:336–360.

Zhang, X., Z. Zou, P. Gong et al. 2011. Over-expression of microRNA169 confers enhanced drought tolerance to tomato. *Biotechnol Lett* 33:403–409.

Zhang, Z., Y. Jingyin, L. Daofeng et al. 2010. PMRD: Plant microRNA database. *Nucleic Acids Res* 38:D806–D813.

Zhao, B., G. Liangfa, L. Ruqiang et al. 2009. Members of miR-169 family are induced by high salinity and transiently inhibit the NF-YA transcription factor. *BMC Mol Biol* 10:29.

Zhou, L. G., Y. H. Liu, Z. C. Liu, D. Y. Kong, M. Duan, and L. J. Luo. 2010. Genome-wide identification and analysis of drought-responsive microRNAs in *Oryza sativa*. *J Exp Bot* 61:4157–4168.

Zhou, X., J. Ruan, G. Wang, and W. Zhang. 2007. Characterization and identification of microRNA core promoters in four model species. *PLoS Comput Biol* 3:e37.

Zhou, X., R. Sunkar, H. Jin, J. K. Zhu, and W. Zhang. 2009. Genome-wide identification and analysis of small RNAs originated from natural antisense transcripts in *Oryza sativa*. *Genome Res* 19:70–78.

Zhou, X., G. Wang, K. Sutoh, J. K. Zhu, and W. Zhang. 2008. Identification of cold-inducible microRNAs in plants by transcriptome analysis. *Biochim Biophys Acta* 1779:780–788.

Zhu, J. K. 2002. Salt and drought stress signal transduction in plants. *Annu Rev Plant Biol* 53:247–273.

Zhu, Q. H., N. M. Upadhyaya, F. Gubler, and C. A. Helliwell. 2009. Over-expression of miR172 causes loss of spikelet determinacy and floral organ abnormalities in rice (*Oryza sativa*). *BMC Plant Biol* 9:149.

CHAPTER **5**

Genome-Wide View of the Expression Profiles of NAC-Domain Genes in Response to Infection by Rice Viruses

Shoshi Kikuchi

CONTENTS

ABBREVIATIONS

ABA   Abscisic acid, a phytohormone extensively involved in responses to
           abiotic stresses
ATAF   *Arabidopsis* transcription activation factor
bHLH   Basic helix-loop-helix
BMV   Brome mosaic virus
bZIP   Basic region/leucine zipper
CGSNL   The Committee on Gene Symbolization, Nomenclature and Linkage
CUC   Cup-shaped cotyledon, an NAC gene involved in organ boundary
           specification

DEG	Differentially expressed gene
GLH	Green leafhopper
HR	Hypersensitive response
JA	Jasmonic acid
NAC	NAM, ATAF1,2, CUC2, a multifunctional plant TF family
NAM	No apical meristem (NAM), an NAC gene involved in the development of the shoot apical meristem
PR protein	Pathogenesis-related protein
RDV	Rice dwarf virus
REn	Replication Enhancer
RGSV	Rice grassy stunt virus
RSV	Rice stripe virus
RTBV	Rice tungro bacilliform virus
RTD	Rice tungro disease
RTSV	Rice tungro spherical virus
SNAC	Stress-responsive NAC
TCV	Turnip crinkle virus
TF	Transcription factor
TLCV	Tomato leaf curl virus
TMV	Tobacco mosaic virus
WDV	Wheat dwarf geminivirus

5.1 NAC TRANSCRIPTION FACTORS

Transcription factors (TFs) and the corresponding *cis*-elements in the promoter region of the transcription unit are important switches that regulate temporal and spatial gene expression. Genome sequence analysis of many species has identified many of the genes encoding transcription factors. The transcription factor NAC (no apical meristem [NAM], *Arabidopsis* transcription activation factor [ATAF], and cup-shaped cotyledon [CUC]) (Aida et al. 1997; Souer et al. 1996) is a superfamily of transcription factors that belongs to the helix-turn-helix protein family, which contains five groups of proteins that have basic (bHLH and bZIP), zinc coordinating factor, helix-turn-helix, beta-scaffold (beta-ribbon), and other domain structures. Genes encoding NAC proteins are not found in bacteria, yeast, fungi, algae, or animal genomes, but are found in the genome of spike moss (*Selaginella moellendorffii*) and those of higher plants. Extensive investigation, aided by the availability of several complete plant genome sequences, has identified 117 NAC genes in *Arabidopsis*, 151 in rice, 79 in grape, 26 in citrus, 163 in poplar, and 152 each in soybean and tobacco (Hu et al. 2010; Le et al. 2011; Nuruzzaman et al. 2010, 2012b; Rushton et al. 2008).

5.1.1 Structure and Phylogenic Classification

NAC proteins commonly possess a conserved NAC domain at the N-terminus that consists of approximately 150–160 amino acids and is divided into five subdomains (Figure 5.1A–E). The function of the NAC domain has been associated with nuclear localization, DNA binding, and the formation of homo- and heterodimers with other NAC-domain-containing proteins (Olsen et al. 2005). The highly conserved, positively charged subdomains C and D bind to DNA, while subdomain A may be involved in the formation of a functional dimer, and the divergent subdomains B and E may be responsible for the functional diversity of NAC genes (Chen et al. 2011; Ernst et al. 2004; Jensen et al. 2010; Ooka et al. 2003).

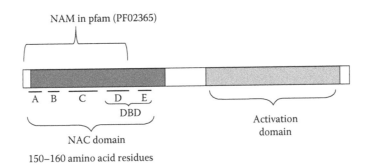

Figure 5.1 Representative molecular structure of NAC protein. Five subdomains (A–E) are present in the N-terminal 150–160 amino acids. Subdomains A–D are registered as pfam domain NAM (PF02365); the C-terminal region is recognized as an activation domain. (This figure is modified from figure 1 of Ooka, H., K. Satoh, K. Doi et al. 2003. Comprehensive analysis of NAC family genes in *Oryza sativa* and *Arabidopsis thaliana. DNA Res.* 10: 239–47, by permission of Oxford University Press.)

The structure of the DNA-binding domain of *Arabidopsis ANAC019* has been solved by x-ray crystallography (Ernst et al. 2004), and the functional dimer formed by the NAC domain was identified in the structural analysis. The NAC domain structure of a rice stress-responsive NAC protein (SNAC1; STRESS-RESPONSIVENAC1) was also reported (Chen et al. 2011) and shares structural similarity with the NAC domain from *Arabidopsis ANAC019*. In contrast, the C-terminal regions of NAC proteins are highly divergent (Ooka et al. 2003) and are responsible for the observed regulatory differences in the transcriptional activation activity of NAC proteins (Jensen et al. 2010; Xie et al. 2000; Yamaguchi et al. 2008). The divergent C-terminal region of these proteins generally operates as a functional domain, acting as a transcriptional activator or repressor (Hu et al. 2006; Kim et al. 2007; Tran et al. 2004). The C-terminal region is large and possesses protein-binding activity.

The NAC TF family has experienced extensive expansion through gene duplication events. Although NAC structural diversity has been constrained within the 60-amino acid conserved domain, which comprises a unique DNA-interacting β-sheet structure, structural conservation outside this conserved domain is extremely limited. As shown in Figure 5.2 (Nuruzzaman et al. 2010), clustering analysis of rice and *Arabidopsis* NAC proteins revealed 18 independent phylogenetic subgroups. Additional highly conserved motifs can be identified only within specific groups (e.g., SNAC, TIP, and SND), and most members in the same group share one or more motifs outside the NAC domain (Nuruzzaman et al. 2012b). A phylogeny of the SNAC group, which includes the *ANAC019* and *OsNAC6* genes, indicates the existence of multiple orthologs in dicots and monocots (Figure 5.2).

Indeed, the SNAC group has some highly conserved motifs (Figure 5.3) in regions outside the conserved domain. A 28-amino acid (WVLCR) motif (RSARKKNSLRLDDWVLCRIYNKKGGLEK in OsNAC) is found amino-terminal to the conserved DNA-binding domain in monocots and in dicots. We first identified putative conserved motifs outside the NAC domain in rice and compared them with those of *Arabidopsis* and citrus. Outside the NAC domain, rice-specific conserved motifs were detected (Nuruzzaman et al. 2012b). These conserved motifs are likely to be involved in the recruitment of proteins that are involved in activating gene expression or perhaps in the control of protein stability. It is notable that only some of these motifs are conserved in both dicots and monocots, suggesting that protein function has both diverged and been conserved even within this evolutionarily conserved NAC family. Further analysis of motif function via protein interaction analyses of TF complexes is needed.

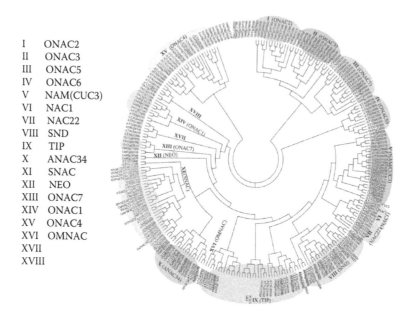

Figure 5.2 (See color insert) An unrooted phylogenetic tree of the NAC transcription factors of rice and *Arabidopsis* classified into 18 groups. The amino acid sequences of the NAC domain of 135 rice NAC-family proteins and 117 *Arabidopsis* NAC proteins were aligned by Clustal W, and a phylogenetic tree was constructed using MEGA4.0 and the NJ method. Bootstrap values from 1000 replicates were used to assess the robustness of the trees. (The classification by Nuruzzaman et al. (2010) is indicated in parentheses.) (Reprinted from *Gene* 465, Nuruzzaman, M., R. Manimekalai, and A. M. Sharoni et al. . Genome-wide analysis of NAC transcription factor family in rice. 30–44, Copyright 2010, with permission from Elsevier.)

	I	ONAC2
	II	ONAC3
	III	ONAC5
	IV	ONAC6
	V	NAM(CUC3)
	VI	NAC1
	VII	NAC22
	VIII	SND
	IX	TIP
	X	ANAC34
	XI	SNAC
	XII	NEO
	XIII	ONAC7
	XIV	ONAC1
	XV	ONAC4
	XVI	OMNAC
	XVII	
	XVIII	

```
AAM65392    156  1.10e-28  NLSTAAKPPD  LTTTRKNSLRLDDWVLCR IYKKNSSQRP  TMERVLLRED
AAB71483    153  1.10e-28  NLSTAAKPPD  LTTTRKNSLRLDDWVLCR IYKKNSSQRP  TMERVLLRED
Os01g66120  138  2.10e-27  MHEYRLADVD  RSARKKNSLRLDDWVLCR IYNKKGGLEK  PPAAAVAAAG
AT1G01720   136  3.73e-27  HEYRLADVDR  SVRKKKNSLRLDDWVLCR IYNKKGAERR  GPPPPVVYGD
AT5G63790   179  6.51e-27  EYRLANVDRS  ASTNKKNNLRLDDWVLCR IYNKKGTMEK  YLPAAAEKPT
Os05g34830  147  7.81e-27  MHEYRLADVD  RSARKKNTLRLDDWVLCR IYNKKGGVEK  PSGGGGGERS
AT3G15510   155  1.89e-26  ENKPNNRPPG  CDFGNKNSLRLDDWVLCR IYKKNNASRH  VDNDKDHDMI
Os03g60080  149  7.19e-26  EYRLADAGRA  AAGAKKGSLRLDDWVLCR IYLKKNEWEK  MQQGKEVKEE
Os07g48450  156  4.56e-25  AAANTYKPSS  SSRFRNVSMRLDDWVLCR IYKKSGQASP  MMPPLAADYD
AT5G08790   137  9.38e-25  EYRLANVDRS  ASVNKKNNLRLDDWVLCR IYNKKGMEKY  FPADEKPRTT
AAM63330    140  1.24e-24  HEYRLHDSRK  ASTKRSGSMRLDEWVLCR IYKKRGASKL  LNEQEGFMDE
AT1G61110   154  3.25e-24  GNLSTAAKPP  DLTTTRKSLRLDDWVLCR IYKKNSSQRP  TMERVLLRED
AT1G52880   154  4.24e-24  RLTDNKPTHI  CDFGNKKSLRLDDWVLCR IYKKNNSTAS  RHHHHLHHIH
Os11g08210  145  1.05e-23  DVDRSAAARK  LSKSSHNALRLDDWVLCR IYNKKGVIER  YDTVDAGEDV
Os12g03040  151  4.63e-23  TSANNTTTTK  QRRASSMTMRLDDWVLCR IHKKSNDFNS  SDQHDQEPEG
Os11g03300  151  4.63e-23  TSANSTTTTK  QRRASSMTMRLDDWVLCR IHKKSNDFNS  SDQHDQEPEE
Os01g01430  177  4.63e-23  SNMKQLASSS  SSSSSASMRLDEWVLCR IYKKKEANQQ  LQHYIDMMMD
AT1G69490   139  9.42e-23  HEYRLHDSRK  ASTKRNGSMRLDEWVLCR IYKKGASKLL  NEQEGFMDEV
AT3G04070   163  1.19e-22  SGGSEVNNFG  DRNSKEYSMRLDDWVLCR IYKKSHASLS  SPDVALVTSN
Os07g12340  149  1.19e-22  EYRLAKKGGA  AAAAGAGALRLDDWVLCR LYNKKNEWEK  MQSRKEEEEA
Os03g21060  152  2.10e-22  ADAHAANTYR  PMKFRNTSMRLDDWVLCR IYKKSSHASP  LAVPPLSDHE
Os01g60020  154  9.84e-22  HEYRLADADR  APGGKKGSQKLDDWVLCR LYNKKNNWEK  VKLEQQDVAS
AT1G52890   140  1.86e-21  NWIMHEYRLI  EPSRRNGSTKLDDWVLCR IYKQSSAQKQ  VYDNGIANAR
AT4G27410   140  2.54e-21  NWIMHEYRLI  EHSRSHGSSKLDDWVLCR IYKTSGSQRQ  AVTPVQACRE
Os07g37920  179  1.57e-20  TTRRPPPPIT  GGSKGAVSLRLDDWVLCR IYKKTNKAGA  GQRSMECEDS
Os05g34310  152  7.16e-17  GGSTASHPSL  SSSTAHPSVKLDEWVLCR IFNKSPEPDN  TAPPSNVVSR
AT2G33480   138  5.05e-14  NWVLHEYRLV  DSQQDSLYGQNMNWVLCR VFLKKSNSNS  KRKEDEKEEV
                                                     WVLCR
```

Figure 5.3 Conserved motifs outside the NAC domain of the SNAC/(IX) group in rice and *Arabidopsis*.

5.2 ROLES PLAYED BY NAC TRANSCRIPTION FACTORS

5.2.1 Interaction between Host NAC and Viral Proteins

Several NAC proteins can either enhance or inhibit viral multiplication by directly interacting with virally encoded proteins (Figure 5.4a–e; Jeong et al. 2008; Ren et al. 2000, 2005; Selth et al. 2005; Yoshii et al. 2009, 2010; Wang et al. 2009; Xie et al. 1999), and increases in the expression level of *NAC* genes have been observed in response to attack by viruses. Such dual modulation of plant defense implies an association between NAC proteins and distinct regulatory complexes. Transgenic rice plants overexpressing *OsNAC6* exhibited tolerance to blast disease (Nakashima et al. 2007). *ATAF2* overexpression resulted in increased susceptibility to the necrotrophic fungus *Fusarium oxysporum* under sterile conditions due to the repression of pathogenesis-related (*PR*) genes (Delessert et al. 2005), but in a nonsterile environment it induced *PR* genes, reducing tobacco mosaic virus accumulation (Wang et al. 2009).

5.3 RICE NAC GENE RESPONSE TO RICE VIRAL INFECTION

For several years, we have analyzed the gene expression profiles induced by the infection of rice viruses using the Agilent rice 44 k microarray system in collaboration with Dr. Omura's group at the National Agriculture Research Center and Dr. Choi's group at the International Rice Research Institute. We have published reports on rice dwarf virus (RDV) (Shimizu et al. 2007; Satoh et al. 2011), rice stripe virus (RSV) (Satoh et al. 2010), rice tungro spherical virus (RTSV) (Satoh et al. 2013a), and rice glassy stunt virus (RGSV) (Satoh et al. 2013b). In our microarray system (NCBI-GEO platform no. GPL7252), the expression of 130 rice NAC genes is detectable (Table 5.1).

By comparing the expression of NAC genes in virus-infected plants and mock-infected plants, this review analyzes how the 130 NAC genes respond to the infection and the commonality of changes with infection by different viruses.

5.3.1 Rice Dwarf Virus

Rice dwarf virus (RDV) is one of the viruses that cause the most economic damage to crop plants in northern Asian countries. RDV cannot be transmitted mechanically and is transmitted to rice plants exclusively by insect vectors, leafhoppers in particular (*Nephotettix* spp.), after multiplication of the virus in the insect. Infection of rice plants by RDV leads to the appearance of white chlorotic specks on leaves and stunting of plant growth, which result in considerable decreases in grain yield. RDV, which belongs to the genus Phytoreovirus in the family Reoviridae, is an icosahedral double-shelled particle of approximately 70 nm in diameter (Omura and Yan 1999). Twelve segments of double-stranded RNA (namely, the genome) and three proteins are encapsidated with a thin layer of the core capsid protein P3. The core is further encapsidated by the outer capsid proteins P2, P8, and P9. The roles of the nonstructural proteins after infection of insect vector cells as well as infected plant cells have been well characterized (Wei et al. 2006a,b,c). Some of these structural and nonstructural proteins are thought to be recognized by host rice plants and elicit responses including the induction of the expression of factors that mediate viral infection, multiplication, and movement.

Using the 22 k microarray system, RDV-infected rice plants showed significant decreases in the expression levels of genes involved in the formation of cell walls, explaining the stunted growth of diseased plants. The expression of plastid-related genes was also suppressed, as expected based on the white chlorotic appearance of infected leaves. In contrast, the expression of defense- and stress-related genes was enhanced after viral infection (Shimizu et al. 2007). Three strains of RDV,

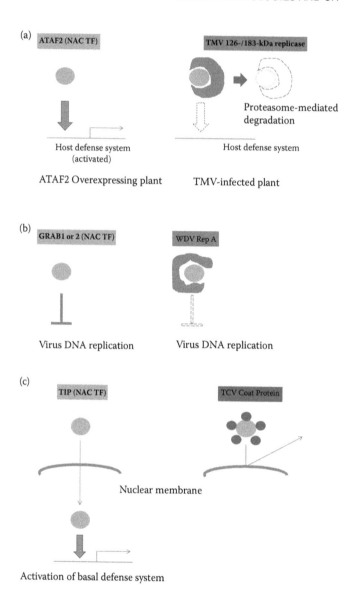

Figure 5.4 Interactions between NAC proteins and virally encoded gene products. (a) Overexpression of ATAF2 (NAC TF) is known to induce host defense systems, while in tobacco mosaic virus (TMV)-infected plant host cells ATAF2 is bound to 126/183 kDa replicase gene product of TMV and targeted to the proteasome-mediated degradation system. (From Wang, X., S. P. Goregaoker, and J. N. Culver. 2009. Interaction of the Tobacco mosaic virus replicase protein with a NAC domain transcription factor is associated with the suppression of systemic host defenses. *J. Virol* 83: 9720–30.) (b) GRAB1 or GRAB2 (NAC TF) is known to suppress DNA replication of the wheat dwarf geminivirus (WDV). The binding of the RepA protein with GRAB 1 or GRAB2 was shown to inhibit the ability of GRAB1 or GRAB 2 to suppress viral DNA replication by Xie et al. (1999). (From Xie, Q., A. P. Sanz-Burgos, H. Guo, J. A. García, C. Gutiérrez. GRAB proteins, novel members of the NAC domain family, isolated by their interaction with a geminivirus protein. *Plant Mol Biol* 39:647–56, 1999.) (c) TIP (NAC TF) is known to activate the basal defense system by passing through the nuclear membrane, but in turnip crinkle virus (TCV)-infected cells binding of the coat proteins of TCV inhibits the movement of TIP through the nuclear membrane and the host basal defense system is not activated, as shown by Ren et al. (2000). (From Ren, T., F. Qu, and T. J. Morris. HRT gene function requires interaction between a NAC protein and viral capsid protein to confer resistance to turnip crinkle virus. *Plant Cell* 12: 1917–26, 2000.)

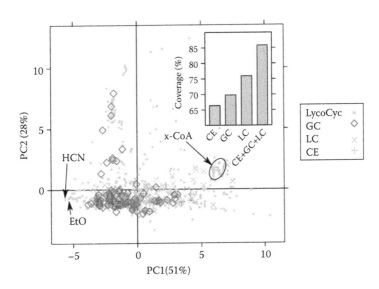

Figure 2.4 Evaluation of the achieved coverage. PCA was performed on the predicted physicochemical properties of the detected metabolites and the metabolites in the LycoCyc database. The score plots show that the distribution of the detected metabolites occupies a similar space as the reference metabolites. (Reprinted from Kusano, M., H. Redestig, T. Hirai et al. 2011. Covering chemical diversity of genetically-modified tomatoes using metabolomics for objective substantial equivalence assessment. *PLoS ONE* 16: e16989.)

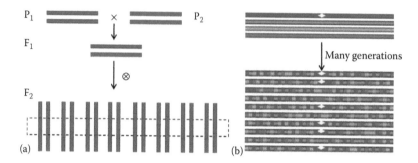

(a)   (b)

Figure 3.1 A schematic representation of traditional QTL mapping (a) and association mapping (b). QTL mapping can suffer from low resolution as single markers are typically in LD with large genomic regions. In association mapping, higher resolution is obtained because markers are in LD with much smaller genomic regions, primarily due to many generations of recombination and high genetic diversity in large, unstructured populations. A mutated allele (gold diamond) will be in LD only with closely located markers. (Reprinted with permission from Zhu, C., Gore, M., Buckler, E.S., and Yu, J., *Plant Genome*, 1, 5–20, 2008.)

(a)   (b)   (c)

Figure 3.6 (a) Chestnut trees are cork-borer inoculated with three different strains of chestnut blight: a mildly virulent strain at the top, a moderately virulent strain in the middle, and a highly virulent strain at the bottom. (b) Blight infections on a resistant tree are characterized by fungal cankers having defined margins of swollen calluses and living tissue. (c) Blight-susceptible trees show sunken cankers with necrotic tissue at the margins, which quickly girdle and kill the infected stem. (Photo by J. Hill Craddock.)

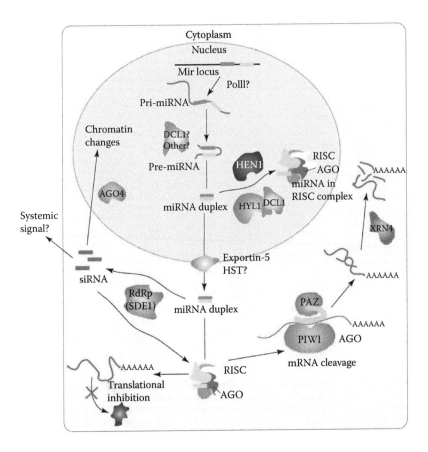

Figure 4.2 Model for miRNA biogenesis and activity in plants. (Reprinted from *Curr Opin Plant Bio*, 8, Kidner, C. A. and R. A. Martienssen. The developmental role of microRNA in plants, 38–44. Copyright 2005, with permission from Elsevier.)

I	ONAC2
II	ONAC3
III	ONAC5
IV	ONAC6
V	NAM(CUC3)
VI	NAC1
VII	NAC22
VIII	SND
IX	TIP
X	ANAC34
XI	SNAC
XII	NEO
XIII	ONAC7
XIV	ONAC1
XV	ONAC4
XVI	OMNAC
XVII	
XVIII	

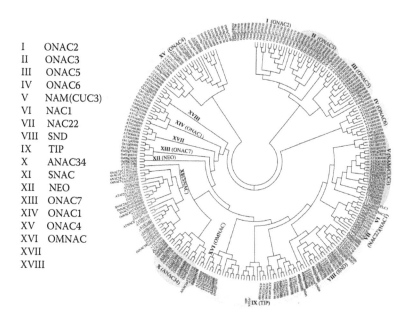

Figure 5.2 An unrooted phylogenetic tree of the NAC transcription factors of rice and *Arabidopsis* classified into 18 groups. The amino acid sequences of the NAC domain of 135 rice NAC-family proteins and 117 *Arabidopsis* NAC proteins were aligned by Clustal W, and a phylogenetic tree was constructed using MEGA4.0 and the NJ method. Bootstrap values from 1000 replicates were used to assess the robustness of the trees. (The classification by Nuruzzaman et al. (2010) is indicated in parentheses.) (Reprinted from *Gene* 465, Nuruzzaman, M., R. Manimekalai, and A. M. Sharoni et al. Genome-wide analysis of NAC transcription factor family in rice. 30–44, Copyright 2010, with permission from Elsevier.)

Figure 8.2 *Vaccinium vitis-idaea* L. plantlets at the end of their growth cycle in a RITA temporary immersion system.

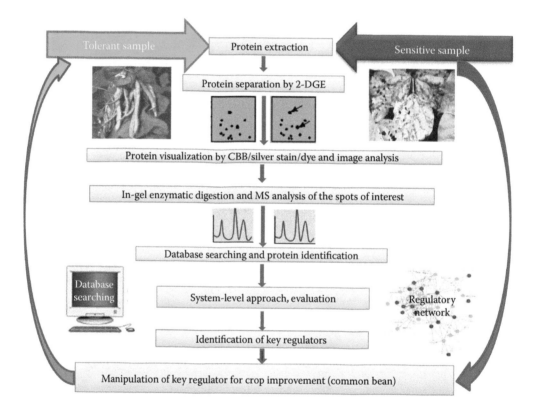

Figure 12.3 Proteomics approach for identification of novel proteins in the common bean. A gel-based proteomics approach can be used to identify the differentially expressed proteins among tolerant and susceptible cultivars. Identification of these novel proteins will help in revealing the metabolic pathways involved in inducing tolerance.

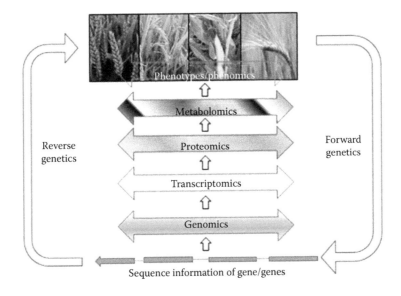

Figure 13.3 Flowchart of omic tools and strategies in forward and reverse genetic approaches.

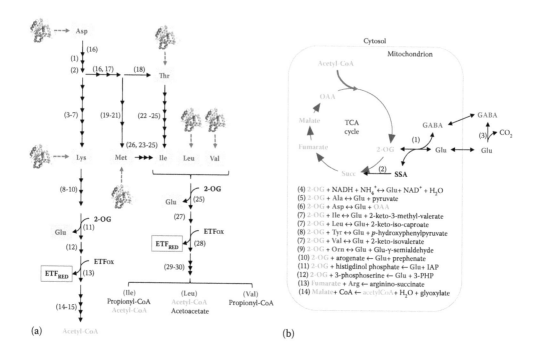

Figure 15.1 (a) Catabolic pathways of aspartate family and branched-chain amino acids. Black arrows represent single biochemical reactions. Red dotted arrows represent protein degradation. Involved enzymes: (1) monofunctional Asp kinase AK1; (2) Asp-semialdehyde dehydrogenase ASD; (3) dihydrodipicolinate synthase DHDPS1; (4) dihydrodipicolinate reductase DHDPR; (5) L,L-diaminopimelate aminotransferase AGD2; (6) diaminopimelate epimerase DAPE; (7) diaminopimelate decarboxylase DAPD; (8) Lys-ketoglutarate reductase/saccharopine dehydrogenase LKR/SDH; (9) saccharopine dehydrogenase SDH; (10) 3-chloroallyl aldehyde dehydrogenase ALDH; (11) NA; (12) NA; (13) NA; (14) enoyl-CoA hydratase MFP2; (15) acetoacetyl-CoA thiolase AACT1; (16) Asp kinase/homo-Ser dehydrogenase AK/HSDH; (17) homo-Ser kinase HSK; (18) Thr synthase TS; (19) cystathionine g-synthase CGS; (20) cystathionine b-lyase CBL; (21) Met synthase MS; (22) Thr deaminase TD; (23) acetolactate synthase AHASS; (24) ketol acid reductoisomerase KARI; (25) branched-chain amino acid aminotransferase BCAT; (26) Met g-lyase MGL; (27) branched-chain keto-acid dehydrogenase BCE2/LPD2; (28) NA; (29) enoyl-CoA hydratase MFP2; (30) acetoacetyl-CoA thiolase AACT1. (b) TCA cycle scheme, GABA shunt, and basic reactions of TCA intermediates involving amino acids and AA-derived acetyl-CoA. Enzymes involved: (1) GABA transaminase; (2) succinic semialdehyde dehydrogenase; (3) glutamate decarboxylase; (4) glutamate dehydrogenase; (5) alanine transaminase; (6) aspartate transaminase; (7) branched-chain amino acid transaminase; (8) aromatic amino acid transaminase; (9) ornithine transaminase; (10) glutamate-prephenate aminotransferase; (11) histidinol-phosphate transaminase; and (12) phosphoserine aminotransferase; (13) arginosuccinate lyase; (14) malate synthase. Metabolite abbreviations: IAP, imidazole acetol-phosphate; 3-PHP, 3-phosphohydroxypyruvate. The list of the enzymatic reactions indicated in the bottom of panel b was principally based on a previous review. (Sweetlove, L. J., K. F. Beard, A. Nunes-Nesi, A. R. Fernie, and R. G. Ratcliffe. 2010. Not just a circle: flux modes in the plant TCA cycle. *Trends Plant Sci* 15: 462–70.)

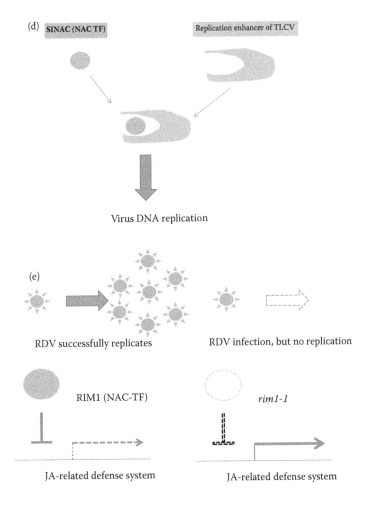

Figure 5.4 **(Continued)** (d) SINAC (NAC TF) is known to bind the replication enhancer (REn) encoded by tomato leaf curl virus (TLCV), which activates viral DNA replication. (From Selth, L. A., S. C. Dogra, M. S. Rasheed et al. A NAC domain protein interacts with tomato leaf curl virus replication accessory protein and enhances viral replication. *Plant Cell* 17:311–25.) (e) RIM1 (NAC TF) is known to act as a negative effector of the JA-related defense system, but in *rim1-1*, the loss-of-function mutant, if the JA-related defense system is mobilized quickly after RDV infection, RDV cannot replicate. (From Yoshii, M., T. Shimizu, M. Yamazaki et al. 2009. Disruption of a novel gene for a NAC-domain protein in rice confers resistance to Rice dwarf virus. *Plant J* 57:615–25. Yoshii, M., M. Yamazaki, R. Rakwal et al. 2010. The NAC transcription factor RIM1 of rice is a new regulator of jasmonate signaling. *Plant J* 61:804–15.)

differentiated by the severity of the symptoms they cause, have been reported (Kimura et al. 1987). Rice plants infected with the severe strain of RDV (RDV-S) were significantly more stunted than those infected with the ordinary strain of RDV (RDV-D84). Another strain, RDV-O, originating from RDV-D84, causes weaker symptoms than those caused by RDV-D84. Global gene expression analysis with a 44 k microarray system revealed that the gene responses to RDV infection were regulated differently depending on the gene group and RDV strain, and that symptom severity is associated with the degree of gene response in defense-, development-, and morphogenesis-related processes (Satoh et al. 2011).

Table 5.1   List of Rice NAC Genes

RAP-ID	MSU-ID	44 k Array	Subgroup
Os01g0104200	LOC_Os01g01430	On array	SNAC
Os01g0104500	LOC_Os01g01470	On array	NAM/CUC3
Os01g0191300	LOC_Os01g09550	On array	ONAC4
Os01g0261200	LOC_Os01g15640	On array	TIP
Os01g0378400	LOC_Os01g28050	On array	ONAC2
Os01g0393100	LOC_Os01g29840	On array	NAM/CUC3
Os01g0672100	LOC_Os01g48130	On array	ONAC4
Os01g0675800	LOC_Os01g48460		ONAC3
Os01g0811500	LOC_Os01g59640	On array	ONAC3
Os01g0816100	LOC_Os01g60020	On array	SNAC
Os01g0862800	LOC_Os01g64310	On array	ONAC7
Os01g0884300	LOC_Os01g66120	On array	SNAC
Os01g0888300	LOC_Os01g66490	On array	ANAC34
Os01g0925400	LOC_Os01g70110	On array	ONAC1
Os01g0946200	LOC_Os01g71790	On array	ONAC2
Os02g0165400	LOC_Os02g06950	On array	NAC1
Os02g0214500	LOC_Os02g12310	On array	SNAC
Os02g0252200	LOC_Os02g15340	On array	SND
Os02g0285900	LOC_Os02g18460	On array	ONAC2
Os02g0286000	LOC_Os02g18470	On array	ONAC2
Os02g0555300	LOC_Os02g34970	On array	ONAC6
Os02g0579000	LOC_Os02g36880	On array	NAM/CUC3
Os02g0594800	LOC_Os02g38130	On array	ONAC4
Os02g0623300	LOC_Os02g41450	On array	NAC22
Os02g0643600	LOC_Os02g42970	On array	SND
Os02g0745200	LOC_Os02g51120		ANAC34
Os02g0810900	LOC_Os02g56600	On array	NAC22
Os02g0822400	LOC_Os02g57650	On array	TIP
Os03g0109000	LOC_Os03g01870	On array	NAC22
Os03g0119966	LOC_Os03g02800	On array	TIP
None	LOC_Os03g03540	On array	
Os03g0133000	LOC_Os03g04070	On array	ANAC34
Os03g0221300	LOC_Os03g12120	On array	ONAC4
Os03g0327100	LOC_Os03g21030	On array	NAM/CUC3
Os03g0327800	LOC_Os03g21060	On array	SNAC
Os03g0587700	LOC_Os03g39050	On array	ONAC2
Os03g0588000	LOC_Os03g39100	On array	ONAC2
Os03g0624600	LOC_Os03g42630	On array	NAM/CUC3
Os03g0777000	LOC_Os03g56580	On array	ANAC34
Os03g0811850	LOC_Os03g59730	On array	ONAC3
Os03g0815100	LOC_Os03g60080	On array	SNAC
None	LOC_Os03g61249		
None	LOC_Os03g61319	On array	
None	LOC_Os03g62470	On array	
Os04g0437000	LOC_Os04g35660	On array	ONAC6
Os04g0460600	LOC_Os04g38720	On array	NAM/CUC3
Os04g0475400	LOC_Os04g39960	On array	ONAC3
Os04g0477300	LOC_Os04g40140	On array	ONAC4

Table 5.1 (Continued) List of Rice NAC Genes

RAP-ID	MSU-ID	44 k Array	Subgroup
Os04g0515900	LOC_Os04g43560	On array	NAC22
Os04g0619000	LOC_Os04g52810	On array	NAC1
Os04g0691300	LOC_Os04g59470	On array	SND
Os05g0194500	LOC_Os05g10620	On array	ONAC4
Os05g0325300	LOC_Os05g25960	On array	ONAC2
Os05g0415400	LOC_Os05g34310	On array	SNAC
Os05g0418800	LOC_Os05g34600	On array	ANAC34
Os05g0421600	LOC_Os05g34830	On array	SNAC
Os05g0426200	LOC_Os05g35170	On array	TIP
Os05g0442700	LOC_Os05g37080	On array	ONAC7
Os05g0515800	LOC_Os05g43960	On array	ONAC3
Os05g0563000	LOC_Os05g48850	On array	ONAC4
Os06g0101800	LOC_Os06g01230	On array	TIP
Os06g0104200	LOC_Os06g01480	On array	SND
Os06g0131700	LOC_Os06g04090	On array	SND
None	LOC_Os06g15690	On array	
Os06g0344900	LOC_Os06g23650	On array	NAM/CUC3
Os06g0530400	LOC_Os06g33940	On array	SND
Os06g0560300	LOC_Os06g36480	On array	ONAC4
Os06g0675600	LOC_Os06g46270	On array	NAC1
Os06g0726300	LOC_Os06g51070	On array	ANAC34
Os07g0138200	LOC_Os07g04560	On array	ANAC34
Os07g0195600	LOC_Os07g09740	On array	ONAC2
Os07g0196500	LOC_Os07g09830	On array	ONAC2
Os07g0196800	LOC_Os07g09860	On array	ONAC2
Os07g0225300	LOC_Os07g12340	On array	SNAC
Os07g0242800	LOC_Os07g13920	On array	ONAC2
Os07g0272700	LOC_Os07g17180	On array	ONAC2
Os07g0456900	LOC_Os07g27330	On array	ONAC2
Os07g0457100	LOC_Os07g27340	On array	ONAC2
None	LOC_Os07g31410	On array	
Os07g0566500	LOC_Os07g37920	On array	SNAC
Os07g0683200	LOC_Os07g48450	On array	SNAC
Os07g0684800	LOC_Os07g48550	On array	NAM/CUC3
Os08g0103900	LOC_Os08g01330	On array	SND
None	LOC_Os08g02160	On array	
Os08g0115800	LOC_Os08g02300	On array	SND
Os08g0157900	LOC_Os08g06140	On array	TIP
Os08g0200600	LOC_Os08g10080	On array	NAC1
Os08g0327800	LOC_Os08g23880	On array	ONAC2
Os08g0433500	LOC_Os08g33670	On array	NEO
Os08g0436700	LOC_Os08g33910	On array	ANAC34
Os08g0511200	LOC_Os08g40030	On array	NAM/CUC3
Os08g0535800	LOC_Os08g42400	On array	ONAC1
Os08g0562200	LOC_Os08g44820	On array	TIP
Os09g0295000	LOC_Os09g12380	On array	ONAC3
Os09g0411900	LOC_Os09g24560	On array	NEO

(continued)

Table 5.1   (Continued) List of Rice NAC Genes

RAP-ID	MSU-ID	44 k Array	Subgroup
Os09g0493700	LOC_Os09g32040	On array	OMNAC
Os09g0497900	LOC_Os09g32260	On array	NAM/CUC3
Os09g0509100	LOC_Os09g33490	On array	ONAC1
Os09g0552800	LOC_Os09g38000	On array	TIP
Os09g0552900	LOC_Os09g38010	On array	TIP
Os10g0177000	LOC_Os10g09820	On array	ONAC5
Os10g0359500	LOC_Os10g21560	On array	ONAC5
Os10g0395650	LOC_Os10g25620	On array	ONAC4
Os10g0396050	LOC_Os10g25640	On array	ONAC4
Os10g0401550	LOC_Os10g26240	On array	ONAC3
Os10g0402150	LOC_Os10g26270	On array	ONAC4
Os10g0413700	LOC_Os10g27360	On array	ONAC4
Os10g0414000	LOC_Os10g27390	On array	ONAC4
Os10g0477600	LOC_Os10g33760	On array	NAC22
Os10g0532000	LOC_Os03g03540	On array	SND
Os10g0571600	LOC_Os10g42130	On array	TIP
Os11g0126900	LOC_Os11g03300	On array	SNAC
Os11g0127000	LOC_Os11g03310	On array	NAM/CUC3
Os11g0127600	LOC_Os11g03370	On array	NAM/CUC3
Os11g0140500	LOC_Os11g04470		ONAC3
Os11g0146900	LOC_Os11g04960	On array	ONAC5
Os11g0154500	LOC_Os11g05614	On array	ONAC7
Os11g0179300	LOC_Os11g07700	On array	ONAC2
Os11g0184900	LOC_Os11g08210	On array	SNAC
Os11g0512000	LOC_Os11g31330	On array	ONAC1
Os11g0512100	LOC_Os11g31340	On array	ONAC1
Os11g0512200	LOC_Os11g31360	On array	ONAC1
Os11g0512600	LOC_Os11g31380	On array	ONAC1
Os11g0686700	LOC_Os11g45950	On array	ONAC7
Os12g0123700	LOC_Os12g03040	On array	SNAC
Os12g0123800	LOC_Os12g03050	On array	NAM/CUC3
Os12g0137000	LOC_Os12g04230		ONAC3
Os12g0156100	LOC_Os12g05990	On array	ONAC7
Os12g0177550	LOC_Os12g07790	On array	ONAC3
Os12g0415000	LOC_Os12g22630	On array	ONAC3
Os12g0418300	LOC_Os12g22940	On array	ONAC3
Os12g0419600	LOC_Os12g23090	On array	ONAC3
Os12g0477400	LOC_Os12g29330	On array	ONAC1
Os12g0610600	LOC_Os12g41680	On array	NAC1
Os12g0630800	LOC_Os12g43530	On array	ANAC34

To identify host factors involved in the multiplication of RDV, Tos17 insertion mutant lines of rice for mutants with reduced susceptibility to RDV were screened. One mutant, designated *rim1-1*, did not show typical disease symptoms upon infection with RDV. The accumulation of RDV capsid proteins was also drastically reduced in inoculated *rim1-1* mutant plants. Cosegregation and complementation analyses revealed that the *rim1-1* mutation had been caused by the insertion of Tos17 in an intron of a novel NAC gene (Yoshii et al. 2009). RIM1 functions as a transcriptional regulator of JA signaling and is degraded in response to JA treatment via a 26S proteasome-dependent pathway. Plants with *rim1* mutations show a phenotype of root growth inhibition. The expression profiles of the mutants were significantly correlated with those of JA-treated wild-type plants without the accumulation of endogenous JA, indicating that RIM1 functions as a component of JA signaling. The expression of genes encoding JA biosynthetic enzymes (lipoxygenase [LOX], allene oxide synthase 2 [AOS2], and OPDA reductase 7 [OPR7]) was upregulated in *rim1* mutants under normal conditions, and a rapid and massive accumulation of endogenous JA was detected in the mutants after wounding. These results suggest that RIM1 may be a new molecular link in jasmonate signaling and may therefore provide new insights into the well-established coronatine-insensitive 1 (COI1)-Jasmonate ZIM-domain (JAZ) JA-signaling pathway (Yoshii et al. 2010). Considering that RIM1 functions in the JA signaling pathway as a negative regulator, and our microarray-based gene expression profile after RDV infection, the reduced susceptibility to RDV in the *rim1-1* mutant may be caused by defense actions prepared before infection with RDV (Satoh et al. 2011).

In our 44 k array data, 16 NAC genes show a significant change of expression in infected plants compared with the control plants; 11 are upregulated and 5 are downregulated, including RIM1 (Os03g0119966) (Table 5.2).

5.3.2 Rice Stripe Virus

Rice stripe disease is the most severe viral disease in rice in eastern Asia. Typical symptoms are chlorosis and weakness in newly emerged leaves. The plant becomes considerably stunted when affected in the early growth stages (Ou 1972). Rice stripe disease also causes necrosis of newly emerged leaves (Takahashi et al. 1991). The causal virus is rice stripe virus (RSV), which belongs to the genus Tenuivirus (Falk and Tsai 1998). RSV is transmitted by the small brown planthopper (SBPH; *Laodelphax striatellus*), *Terthron albovittatum*, *Unkanodes sapporonus*, and *Unkanodes albifascia* (Falk and Tsai 1998; Ou 1972). RSV has a thin, filamentous shape and no envelope. The genome consists of four single-stranded RNA segments; RNA1 is negative-sense and RNAs 2–4 are ambisense. Viral mRNAs transcribed from viral RNA or viral cRNA by RNA-dependent RNA polymerase are released into the cytoplasm. Subsequently, a 59-nucleotide capped short ribonucleotide leader cleaved from the host mRNA is added to the viral mRNAs by cap-snatching. The 59-nucleotide capped RSV RNA is efficiently transcribed in host cells (Falk and Tsai 1998; Shimizu et al. 1996). Genes encoding a gene-silencing suppressor and movement proteins were also identified in the RSV genome (Lu et al. 2009; Xiong et al. 2008, 2009). To characterize host responses to RSV infection at the gene expression level, the time course of transcriptome changes in RSV-infected rice plants was examined using a 44 k microarray. Time-course gene expression analysis in rice infected with RSV and concurrent observations of viral concentration and symptom development indicate that timely modifications of the expression of genes involved in selective cellular processes may be associated with RSV propagation and RSV-induced symptoms (Satoh et al. 2010).

Forty-three NAC genes showed a change in expression in plants infected with RSV compared with mock-infected plants. Twenty-five genes were upregulated, and 18 genes were downregulated (Table 5.3).

5.3.3 Rice Tungro Spherical Virus

Rice tungro disease (RTD) is a major constraint in the production of rice (*Oryza sativa*) in southern and southeastern Asia. Rice plants affected by RTD show symptoms such as stunting and yellow-orange discoloration of leaves (Hibino 1983). RTD is caused by two viruses: rice tungro spherical virus (RTSV) and rice tungro bacilliform virus (RTBV). Both viruses are transmitted by green leafhoppers (GLH) such as *Nephotettix virescens*, but RTBV can be transmitted by GLH only in the presence of RTSV (Hibino 1983). RTBV is mainly responsible for causing the disease symptoms, while RTSV plays the role of a helper virus for insect transmission of RTBV and

Table 5.2 List of NAC Genes That Show a Change in Expression in RDV-Infected Rice

RAP-ID	MSU-ID	Subgroup	RDV Line and dpi	Up or Down	RDV Line and dpi	Up or Down	RDV Line and dpi	Up or Down	Evaluation
Os01g0191300	LOC_Os01g09550	ONAC4			D21	Down	S21	Down	D
Os01g0261200	LOC_Os01g15640	TIP			D21	Up	S21	Up	U
Os01g0862800	LOC_Os01g64310	ONAC7	O21	Up	D21	Up	S21	Up	U
Os02g0594800	LOC_Os02g38130	ONAC4	O21	Up	D21	Up	S21	Up	U
Os03g0119966	LOC_Os03g02800	TIP	O21	Down	D21	Down	S21	Down	D
Os03g0327100	LOC_Os03g21030	NAM/CUC3			D21	Up			U
Os04g0619000	LOC_Os04g52810	NAC1	O21	Down					D
Os05g0421600	LOC_Os05g34830	SNAC			D21	Up	S21	Up	U
Os07g0566500	LOC_Os07g37920	SNAC	O21	Up	D21	Up	S21	Up	U
Os08g0200600	LOC_Os08g10080	NAC1					S21	Down	D
Os10g0359500	LOC_Os10g21560	ONAC5			D21	Up	S21	Up	U
Os11g0127000	LOC_Os11g03310	NAM/CUC3	O21	Up	D21	Up	S21	Up	U
Os11g0127600	LOC_Os11g03370	NAM/CUC3	O21	Up	D21	Up	S21	Up	U
Os11g0154500	LOC_Os11g05614	ONAC7	O21	Up	D21	Up	S21	Up	U
Os12g0123800	LOC_Os12g03050	NAM/CUC3	O21	Up	D21	Up	S21	Up	U
Os12g0610600	LOC_Os12g41680	NAC1					S21	Up	U

Table 5.3 List of NAC Genes That Show a Change in Expression in RSV-Infected Rice

RAP-ID	MSU-ID	Subgroup	3 DAI	6 DAI	9 DAI	12 DAI	Evaluation
Os01g0191300	LOC_Os01g09550	ONAC4	ns	ns	−0.62	−0.73	D
Os01g0261200	LOC_Os01g15640	TIP	ns	ns	ns	0.91	U
Os01g0672100	LOC_Os01g48130	ONAC4	ns	ns	ns	−1.81	D
Os01g0862800	LOC_Os01g64310	ONAC7	ns	1.75	ns	ns	U
Os01g0888300	LOC_Os01g66490	ANAC34	ns	ns	ns	−0.75	D
Os01g0946200	LOC_Os01g71790	ONAC2	ns	ns	ns	1.09	U
Os02g0555300	LOC_Os02g34970	ONAC6	ns	ns	ns	1.48	U
Os02g0594800	LOC_Os02g38130	ONAC4	ns	ns	0.84	1.04	U
Os02g0623300	LOC_Os02g41450	NAC22	ns	ns	−0.71	−1.75	D
Os02g0810900	LOC_Os02g56600	NAC22	ns	ns	−0.68	−1.43	D
Os03g0133000	LOC_Os03g04070	ANAC34	ns	ns	ns	1.38	U
Os03g0624600	LOC_Os03g42630	NAM/CUC3	ns	ns	ns	−1.16	D
Os03g0777000	LOC_Os03g56580	ANAC34	ns	ns	ns	−0.6	D
Os03g0815100	LOC_Os03g60080	SNAC	ns	ns	ns	−1.02	D
Os04g0477300	LOC_Os04g40140	ONAC4	ns	−0.62	0.87	2.66	U
Os04g0515900	LOC_Os04g43560	NAC22	ns	ns	ns	−1.13	D
Os04g0619000	LOC_Os04g52810	NAC1	ns	ns	ns	−0.69	D
Os05g0194500	LOC_Os05g10620	ONAC4	ns	ns	0.87	2.46	U
Os05g0325300	LOC_Os05g25960	ONAC2	ns	ns	ns	−0.7	D
Os05g0421600	LOC_Os05g34830	SNAC	ns	ns	−0.65	ns	D
Os05g0426200	LOC_Os05g35170	TIP	ns	ns	ns	1.28	U
Os05g0563000	LOC_Os05g48850	ONAC4	ns	ns	ns	−1.65	D
Os06g0131700	LOC_Os06g04090	SND	ns	ns	ns	−1.27	D
Os06g0726300	LOC_Os06g51070	ANAC34	ns	ns	ns	0.59	U
Os07g0138200	LOC_Os07g04560	ANAC34	ns	ns	ns	0.73	U
Os08g0115800	LOC_Os08g02300	SND	ns	ns	ns	−0.77	D
Os08g0433500	LOC_Os08g33670	NEO	ns	ns	−0.65	−1.33	D
Os08g0562200	LOC_Os08g44820	TIP	ns	ns	ns	0.85	U
Os09g0493700	LOC_Os09g32040	OMNAC	ns	ns	ns	0.81	U
Os09g0552900	LOC_Os09g38010	TIP	ns	ns	ns	−1	D
Os10g0359500	LOC_Os10g21560	ONAC5	−0.71	ns	ns	ns	D
Os10g0413700	LOC_Os10g27360	ONAC4	ns	ns	ns	1.27	U
Os10g0571600	LOC_Os10g42130	TIP	ns	ns	ns	1.3	U
Os11g0126900	LOC_Os11g03300	SNAC	ns	ns	ns	0.75	U
Os11g0127000	LOC_Os11g03310	NAM/CUC3	ns	ns	ns	1.95	U
Os11g0127600	LOC_Os11g03370	NAM/CUC3	ns	ns	ns	1.32	U
Os11g0154500	LOC_Os11g05614	ONAC7	ns	ns	0.83	2.93	U
Os12g0123700	LOC_Os12g03040	SNAC	−0.59	ns	ns	0.76	U
Os12g0123800	LOC_Os12g03050	NAM/CUC3	ns	ns	ns	2.7	U
Os12g0156100	LOC_Os12g05990	ONAC7	ns	ns	0.86	3.31	U
Os12g0177550	LOC_Os12g07790	ONAC3	ns	ns	0.74	1.75	U
Os12g0418300	LOC_Os12g22940	ONAC3	ns	ns	ns	1.14	U
Os12g0610600	LOC_Os12g41680	NAC1	ns	ns	ns	0.72	U

enhances the disease symptoms caused by RTBV (Hibino 1983). The Indonesian rice cultivar Utri Merah is highly resistant to RTSV and RTBV (Encabo et al. 2009). RTSV resistance in Utri Merah is controlled by a single recessive locus (*tsv1*) mapped to approximately 22.1 Mb of chromosome 7 (Lee et al. 2010). A survey of various rice cultivars for the allele types of the gene for translation initiation factor 4G (*eIF4G^{tsv1}*) located within *tsv1* indicated an association between RTSV resistance and single-nucleotide polymorphisms (SNPs) in the *eIF4G^{tsv1}* gene (Lee et al. 2010). RTSV synergistically enhances symptoms caused by RTBV in rice, although RTSV itself rarely causes any visible symptoms (Encabo et al. 2009). The rice cultivar Taichung Native 1 (TN1) is susceptible to RTSV. TW16 is a backcross line developed from TN1 and the RTSV-resistant cultivar Utri Merah. RTSV accumulation in TW16 was significantly lower than in TN1, although both TN1 and TW16 remained asymptomatic. Microarray-based gene expression analysis revealed that approximately 11% and 12% of the genome was differentially expressed after infection with RTSV in TN1 and TW16, respectively. Approximately 30% of the differentially expressed genes (DEGs) were detected in both TN1 and TW16. DEGs related to development and stress-response processes were significantly overrepresented in both TN1 and TW16. Differences in gene expression between TN1 and TW16 instigated by RTSV infection included the following: (1) suppression of more genes for development-related transcription factors in TW16; (2) activation of more genes for the development-related peptide hormone Rapid Alkalinization Factors (RALF) in TN1; (3) TN1- and TW16-specific regulation of genes for jasmonate synthesis and the pathway and genes for stress-related transcription factors such as WRKY, SNAC, and AP2-EREBP; (4) activation of more genes for glutathione S-transferase in TW16; (5) activation of more heat shock protein genes in TN1; and (6) suppression of more genes for Golden2-like transcription factors involved in plastid development in TN1. A total of 27 NAC genes showed changes in gene expression in TW16 or TN1 infected with RTSV compared with mock-infected plants.

Among these NAC genes, four were upregulated in both TW16 and TN1, six were downregulated in both TW16 and TN1, five were upregulated only in TW16, five were only upregulated in TN1, one was upregulated in TW16 and downregulated in TN1, two were only downregulated in TW16, and four were only downregulated in TN1 (Table 5.4).

5.3.4 Rice Grassy Stunt Virus

The grassy stunt disease in rice caused by rice grassy stunt virus (RGSV) is a severe viral disease in several southeast Asian countries (Shikata et al. 1980; Ramirez 2008). RGSV is a member of the genus Tenui and is transmitted by the brown plant hopper (BPH, *Nilaparvata lugens*) and two other *Nilaparvata* spp. (Hibino 1996). RGSV has a thin filamentous-shaped virion, and the genome is composed of six ambisense single-stranded RNA segments (RNA1–6) containing 12 open reading frames (Ramirez 2008). RGSV RNA1, 2, 5, and 6 correspond to the four RNA segments of the RSV type of Tenuiviruses. RGSV RNA 3 and 4 are unique in this genus. The phylogenic relationship among Tenuiviruses, including RGSV and RSV, indicates that RGSV is distinct from the other Tenuiviruses (Ramirez 2008). Typical symptoms induced by RGSV infection are leaf yellowing (chlorosis), stunting, and excess tillering (branching) (Shikata et al. 1980). Chlorosis and stunting are also observed in plants infected with other Tenuiviruses, whereas excess tillering is a symptom that is specific to RGSV infection. The gene expression profile of plants infected with RGSV suggests that symptoms such as stunting and leaf chlorosis are associated with the suppression of genes related to cell wall, hormone, and chlorophyll synthesis, while the excess tillering symptom specific to RGSV infection is associated with the suppression of strigolactone signaling and gibberellic acid (GA) metabolism (Satoh et al. 2013b).

Thirty-eight NAC genes are upregulated, including RIM1 (Os03g0119966, LOC_Os03g02800), and 13 NAC genes are downregulated in RGSV-infected plants compared with mock-infected plants (Table 5.5).

Table 5.4 List of NAC Genes That Show a Change in Expression in RTSV-Infected Rice

RAP-ID	MSU-ID	Subgroup	6 DAI on TW16	9 DAI on TW16	15 DAI on TW16	6 DAI on TN1	9 DAI on TN1	15 DAI on TN1	Evaluation
Os01g0191300	LOC_Os01g09550	ONAC4	-0.800666667	-0.901	-1.117666667	-0.189666667	-0.441	-0.464666667	D
Os01g0261200	LOC_Os01g15640	TIP	0.631666667	0.347666667	0.576	ns	ns	ns	U (TW16)
Os01g0862800	LOC_Os01g64310	ONAC7	1.305666667	1.013666667	1.240333333	0.754333333	0.563666667	0.48	U
Os02g0594800	LOC_Os02g38130	ONAC4	ns	ns	ns	0.098333333	0.358333333	0.706	U (TN1)
Os02g0810900	LOC_Os02g56600	NAC22	0.787333333	ns	ns	-0.503666667	-1.072333333	-1.171	U (TW16) D (TN1)
Os03g0119966	LOC_Os03g02800	TIP	ns	ns	ns	-0.493333333	-0.243	-0.665333333	D (TN1)
Os03g0327100	LOC_Os03g21030	NAM/CUC3	0.64	0.425333333	0.075	ns	ns	0.696666667	U (TW16)
Os04g0460600	LOC_Os04g38720	NAM/CUC3	ns	ns	ns	0.491	0.281	1.356666667	U (TN1)
Os04g0619000	LOC_Os04g52810	NAC1	0.943666667	0.346666667	0.502333333	0.707333333	-0.018333333	-0.189	U
Os05g0325300	LOC_Os05g25960	ONAC2	-0.745666667	-0.585666667	-0.548333333	-0.357333333	-0.098666667	-0.437	D
Os05g0418800	LOC_Os05g34600	ANAC34	-0.497333333	-0.732	-0.713	-0.231333333	-0.579333333	ns	D
Os05g0421600	LOC_Os05g34830	SNAC	ns	ns	ns	-0.714666667	ns	ns	D (TN1)
Os05g0442700	LOC_Os05g37080	ONAC7	ns	ns	ns	1.101	0.397	0.827333333	U (TN1)
Os07g0566500	LOC_Os07g37920	SNAC	0.634333333	0.522333333	0.324666667	ns	ns	0.519333333	U
None	LOC_Os08g02160	ANAC34	-0.906	-0.740333333	-0.900333333	ns	ns	ns	D (TW16)
Os08g0200600	LOC_Os08g10080	NAC1	-0.949333333	-1.050666667	-0.114333333	-1.021333333	-1.166	-0.899	D
Os08g0535800	LOC_Os08g42400	ONAC1	-0.614	-0.950333333	-0.733333333	ns	ns	ns	D (TW16)
Os09g0509100	LOC_Os09g33490	ONAC1	-0.752	-1.003666667	-0.873	ns	-0.766	ns	D
Os10g0359500	LOC_Os10g21560	ONAC5	0.382333333	0.288666667	0.622666667	ns	ns	ns	U
Os11g0126900	LOC_Os11g03300	SNAC	ns	ns	ns	-1.378666667	-0.063333333	-0.592	D (TN1)
Os11g0127000	LOC_Os11g03310	NAM/CUC3	0.749333333	0.016	0.429333333	ns	ns	ns	U (TW16)
Os11g0127600	LOC_Os11g03370	NAM/CUC3	ns	ns	ns	ns	ns	1.115666667	U (TN1)
Os11g0154500	LOC_Os11g05614	ONAC7	ns	ns	ns	0.715666667	0.056333333	1.475	U (TN1)
Os11g0184900	LOC_Os11g08210	SNAC	1.186666667	0.658	0.263	ns	ns	ns	U (TW16)
Os12g0123700	LOC_Os12g03040	SNAC	ns	ns	ns	-1.363333333	-0.042266667	-0.513	D (TN1)
Os12g0123800	LOC_Os12g03050	NAM/CUC3	0.898	0.135666667	0.541333333	ns	ns	ns	U (TW16)
Os12g0610600	LOC_Os12g41680	NAC1	-0.317	-0.845333333	-0.586333333	-0.274	-0.921333333	-0.347	D

Table 5.5   List of NAC Genes That Show a Change in Expression in RGSV-Infected Rice

RAP-ID	MSU-ID	Subgroup	Value	Evaluation
Os04g0477300	LOC_Os04g40140	ONAC4	1.53	U
Os04g0619000	LOC_Os04g52810	NAC1	2.29	U
Os05g0194500	LOC_Os05g10620	ONAC4	1.78	U
Os05g0325300	LOC_Os05g25960	ONAC2	−0.79	D
Os05g0421600	LOC_Os05g34830	SNAC	0.74	U
Os05g0426200	LOC_Os05g35170	TIP	0.89	U
Os05g0442700	LOC_Os05g37080	ONAC7	2.92	U
Os05g0563000	LOC_Os05g48850	ONAC4	−1.72	D
Os06g0675600	LOC_Os06g46270	NAC1	1.35	U
Os06g0726300	LOC_Os06g51070	ANAC34	1.81	U
Os07g0138200	LOC_Os07g04560	ANAC34	3.63	U
Os07g0225300	LOC_Os07g12340	SNAC	2.99	U
Os07g0566500	LOC_Os07g37920	SNAC	0.84	U
Os07g0684800	LOC_Os07g48550	NAM/CUC3	1.94	U
Os08g0103900	LOC_Os08g01330	SND	1.07	U
Os08g0200600	LOC_Os08g10080	NAC1	−2.73	D
Os08g0436700	LOC_Os08g33910	ANAC34	1.39	U
Os08g0535800	LOC_Os08g42400	ONAC1	0.72	U
Os08g0562200	LOC_Os08g44820	TIP	1.16	U
Os09g0552800	LOC_Os09g38000	TIP	−0.59	D
Os09g0552900	LOC_Os09g38010	TIP	1.49	U
Os10g0395650	LOC_Os10g25620	ONAC4	−0.65	D
Os10g0532000	LOC_Os03g03540	SND	2.33	U
Os10g0571600	LOC_Os10g42130	TIP	5.46	U
Os11g0126900	LOC_Os11g03300	SNAC	1.41	U
Os11g0127000	LOC_Os11g03310	NAM/CUC3	1.79	U
Os11g0127600	LOC_Os11g03370	NAM/CUC3	1.5	U
Os11g0154500	LOC_Os11g05614	ONAC7	4.05	U
Os11g0184900	LOC_Os11g08210	SNAC	0.67	U
Os12g0123700	LOC_Os12g03040	SNAC	1.05	U
Os12g0123800	LOC_Os12g03050	NAM/CUC3	2.27	U
Os12g0610600	LOC_Os12g41680	NAC1	1.66	U
None	LOC_Os07g31410	ONAC3	−0.92	D
Os01g0104200	LOC_Os01g01430	SNAC	−1.02	D
Os01g0104500	LOC_Os01g01470	NAM/CUC3	−1.25	D
Os01g0261200	LOC_Os01g15640	TIP	0.86	U
Os01g0672100	LOC_Os01g48130	ONAC4	−0.78	D
Os01g0816100	LOC_Os01g60020	SNAC	2.36	U
Os01g0888300	LOC_Os01g66490	ANAC34	−0.68	D
Os02g0555300	LOC_Os02g34970	ONAC6	3.21	U
Os02g0579000	LOC_Os02g36880	NAM/CUC3	1.97	U
Os02g0623300	LOC_Os02g41450	NAC22	−2.75	D
Os03g0109000	LOC_Os03g01870	NAC22	3.33	U
Os03g0119966	LOC_Os03g02800	TIP	0.85	U
Os03g0133000	LOC_Os03g04070	ANAC34	1.12	U
Os03g0327800	LOC_Os03g21060	SNAC	2.69	U

Table 5.5 (Continued) List of NAC Genes That Show a Change in Expression in RGSV-Infected Rice

RAP-ID	MSU-ID	Subgroup	Value	Evaluation
Os03g0587700	LOC_Os03g39050	ONAC2	−0.84	D
Os03g0588000	LOC_Os03g39100	ONAC2	−0.66	D
Os03g0624600	LOC_Os03g42630	NAM/CUC3	0.67	U
Os03g0815100	LOC_Os03g60080	SNAC	1.51	U

5.4 COMPARISON OF EXPRESSION OF NAC GENES IN RESPONSE TO FOUR VIRAL INFECTIONS

As described above, infection by each virus induces similar symptoms, such as stunting and chlorosis, and the expression of the genes related to defense systems and cell wall formation is commonly changed. The commonality of the changes in NAC gene expression in plants infected with each of the four viruses is also surveyed. The list consists of a total of 70 NAC genes reported in Tables 5.2 through 5.5. Genes were classified into four groups: genes that show a change in expression after infection by (A) all four viruses, (B) three viruses, (C) two viruses, and (D) only one virus. It is very interesting that group B consists of only two groups of viruses: RSV, RTSV, and RGSV (B1) and RDV, RTSV, and RGSV (B2). Group C also consists of two types, those changed by infection with RSV and RGSV (C1) and RTSV and RGSV (C2). NAC genes whose expression was changed by infection with only one virus exist only for RGSV (D1) or RSV (D2) (Table 5.6). This nonneutral distribution of NAC genes suggests that there might be a common mechanism among viruses to induce a change in the expression of NAC genes.

Group A consists of 12 NAC genes: Os01g0191300, Os01g0261200, Os01g0862800, Os02g0594800, Os04g0619000, Os05g0421600, Os10g0359500, Os11g0127000, Os11g0127600, Os11g0154500, Os12g0123800, and Os12g0610600. Among these genes, six genes (Os01g0261200, Os02g0594800, Os11g0127000, Os11g0127600, Os11g0154500, and Os12g0123800) are upregulated by all viral infections, while Os01g0862800 is upregulated in infections with RDV, RSV, and RTSV but downregulated by RGSV. Os04g0619000 is upregulated by RTSV and RGSV but downregulated by RDV and RSV. Os05g0421600 is upregulated by RDV and RGSV but downregulated by RSV and RTSV. Os12g0610600 is upregulated by RDV, RSV, and RSGV but downregulated by RTSV. Os01g0191300 is downregulated by RDV, RSV, and RTSV but upregulated by RGSV. These NAC genes might contribute to the common symptoms induced by the infection of these viruses or the fundamental defense response against viral infection.

Cold, drought, submergence, and laid-down submergence treatments were reported to induce changes in the gene expression of Os05g34830 and Os12g41680 based on microarray-based gene expression analysis (Nuruzzaman et al. 2010, 2012a). Os11g0127600 (ONAC045) was induced by drought, high-salt, and low-temperature stresses and abscisic acid (ABA) treatment in leaves and roots. Transgenic rice plants overexpressing ONAC045 showed enhanced tolerance to drought and salt treatments (Zheng et al. 2009). These data suggest that some NAC genes have a common role among many kinds of stress responses including abiotic and biotic stress responses.

Group B1 consists of four NAC genes: Os02g0810900, Os05g0325300, Os11g0126900, and Os12g0123700. Os05g0325300 is downregulated by RSV, RTSV, and RGSV, while Os02g0810900 is downregulated by infection with RSV, RTSV, and RGSV in TN1 but upregulated in RTSV in TW16. Os11g0126900 and Os12g0123700 are upregulated by infection by RSV and RGSV but downregulated by RTSV. Among the four viruses, only RDV is known to construct its viroplasm so as to protect the double-stranded viral RNA from the host gene-silencing system and to be the facility for the production of progeny viruses. The other three viruses are thought to utilize the host's enzymes and proteins. Therefore, the NAC genes in this group might be related to the utilization of the host system for the production of progeny viruses.

Table 5.6 Classification of NAC Genes by the Gene Expression Profile after Infection by Four Different Viruses

Class	RAP-ID	MSU-ID	Subgroup	RDV	RSV	RTSV(TW16)	RTSV(TN1)	RGSV	CGSNL
A	Os01g0191300	LOC_Os01g09550	ONAC4	D	D	D	D	s U	ANAC075
A	Os01g0261200	LOC_Os01g15640	TIP	U	U	U	–	U	
A	Os01g0862800	LOC_Os01g64310	ONAC7	U	U	U	U	D	NAC59
A	Os02g0594800	LOC_Os02g38130	ONAC4	U	U	–	U	s U	
A	Os04g0619000	LOC_Os04g52810	NAC1	D	D	U	U	U	
A	Os05g0421600	LOC_Os05g34830	SNAC	U	D	–	D	U	
A	Os10g0359500	LOC_Os10g21560	ONAC5	U	D	U	U	U	NAC61
A	Os11g0127000	LOC_Os11g03310	NAM/CUC3	U	U	U	–	U	
A	Os11g0127600	LOC_Os11g03370	NAM/CUC3	U	U	–	U	U	NAC45
A	Os11g0154500	LOC_Os11g05614	ONAC7	U	U	–	U	U	NAC17
A	Os12g0123800	LOC_Os12g03050	NAM/CUC3	U	U	U	U	U	ONAC300
A	Os12g0610600	LOC_Os12g41680	NAC1	U	U	U		U	NAC60
B1	Os02g0810900	LOC_Os02g56600	NAC22	–	D	D	D	D	NAC32
B1	Os05g0325300	LOC_Os05g25960	ONAC2	–	D	D	D	D	
B1	Os11g0126900	LOC_Os11g03300	SNAC	–	U	D	D	U	ONAC122/OsNAC10
B1	Os12g0123700	LOC_Os12g03040	SNAC	–	U		D	U	ONAC131
B2	Os03g0119966	LOC_Os03g02800	TIP	D	–	–	D	U	RIM1
B2	Os03g0327100	LOC_Os03g21030	NAM/CUC3	U	–	U	–	U	NAC39
B2	Os07g0566500	LOC_Os07g37920	SNAC	U	–	U	U	U	WheatGPC
B2	Os08g0200600	LOC_Os08g10080	NAC1	D	–	D	D	D	
C1	Os01g0672100	LOC_Os01g48130	ONAC4	–	D	–	–	D	
C1	Os01g0888300	LOC_Os01g66490	ANAC34	–	D	–	–	D	
C1	Os01g0946200	LOC_Os01g71790	ONAC2	–	U	–	–	D	
C1	Os02g0555300	LOC_Os02g34970	ONAC6	–	U	U	–	D	
C1	Os02g0623300	LOC_Os02g41450	NAC22	–	D	–	–	D	
C1	Os03g0133000	LOC_Os03g04070	ANAC34	–	U	–	–	U	
C1	Os03g0624600	LOC_Os03g42630	NAM/CUC3	–	D	–	–	U	

Group	Gene	Locus	Name						Alias
C1	Os03g0777000	LOC_Os03g56580	ANAC34	–	D	–	–	U	
C1	Os03g0815100	LOC_Os03g60080	SNAC	–	D	–	–	U	OsNAC19
C1	Os04g0477300	LOC_Os04g40140	ONAC4	–	U	–	–	U	BET1
C1	Os04g0515900	LOC_Os04g43560	NAC22	–	D	–	–	U	
C1	Os05g0194500	LOC_Os05g10620	ONAC4	–	U	–	–	U	
C1	Os05g0426200	LOC_Os05g35170	TIP	–	U	–	–	U	
C1	Os05g0563000	LOC_Os05g48850	ONAC4	–	D	–	–	D	
C1	Os06g0131700	LOC_Os06g04090	SND	–	D	–	–	D	
C1	Os06g0726300	LOC_Os06g51070	ANAC34	–	U	–	–	U	
C1	Os07g0138200	LOC_Os07g04560	ANAC34	–	U	–	–	U	
C1	Os08g0115800	LOC_Os08g02300	SND	–	U	–	–	U	
C1	Os08g0562200	LOC_Os08g44820	TIP	–	D	–	–	D	
C1	Os09g0493700	LOC_Os09g32040	OMNAC	–	U	–	–	U	
C1	Os09g0552900	LOC_Os09g38010	TIP	–	U	–	–	U	
C1	Os10g0413700	LOC_Os10g27360	ONAC4	–	U	–	–	U	
C1	Os10g0571600	LOC_Os10g42130	TIP	–	U	–	–	U	
C1	Os12g0156100	LOC_Os12g05990	ONAC7	–	U	–	–	U	
C1	Os12g0177550	LOC_Os12g07790	ONAC3	–	U	–	–	U	
C1	Os12g0418300	LOC_Os12g22940	ONAC3	–	U	–	–	D	
C2	Os04g0460600	LOC_Os04g38720	NAM/CUC3	–	–	–	U	D	Ostil1
C2	Os05g0418800	LOC_Os05g34600	ANAC34	–	–	D	D	U	
C2	Os05g0442700	LOC_Os05g37080	ONAC7	–	–	–	U	D	
C2	None	LOC_Os08g02160	ANAC34	–	–	D	–	U	
C2	Os08g0535800	LOC_Os08g42400	ONAC1	–	–	D	D	D	
C2	Os09g0509100	LOC_Os09g33490	ONAC1	–	–	D	–	sU	
D1	Os01g0104200	LOC_Os01g01430	SNAC	–	–	–	–	D	
D1	Os01g0104500	LOC_Os01g01470	NAM/CUC3	–	–	–	–	D	
D1	Os01g0816100	LOC_Os01g60020	SNAC	–	–	–	–	U	
D1	Os02g0579000	LOC_Os02g36880	NAM/CUC3	–	–	–	–	U	

(continued)

Table 5.6 (Continued) Classification of NAC Genes by the Gene Expression Profile after Infection by Four Different Viruses

Class	RAP-ID	MSU-ID	Subgroup	RDV	RSV	RTSV(TW16)	RTSV(TN1)	RGSV	CGSNL
D1	Os03g0109000	LOC_Os03g01870	NAC22	–	–	–	–	U	
D1	Os10g0532000	LOC_Os03g03540	SND	–	–	–	–	U	
D1	Os03g0327800	LOC_Os03g21060	SNAC	–	–	–	–	U	
D1	Os03g0587700	LOC_Os03g39050	ONAC2	–	–	–	–	D	
D1	Os03g0588000	LOC_Os03g39100	ONAC2	–	–	–	–	D	
D1	Os06g0675600	LOC_Os06g46270	NAC1	–	–	–	–	U	
D1	Os07g0225300	LOC_Os07g12340	SNAC	–	–	–	–	U	
D1	None	LOC_Os07g31410	ONAC3	–	–	–	–	D	
D1	Os07g0684800	LOC_Os07g48550	NAM/CUC3	–	–	–	–	U	
D1	Os08g0103900	LOC_Os08g01330	SND	–	–	–	–	U	
D1	Os08g0436700	LOC_Os08g33910	ANAC34	–	–	–	–	U	
D1	Os09g0552800	LOC_Os09g38000	TIP	–	–	–	–	D	
D1	Os10g0395650	LOC_Os10g25620	ONAC4	–	–	–	–	D	
D2	Os08g0433500	LOC_Os08g33670	NEO	–	D	–	–		

Os11g0126900 (ONAC122) and Os12g0123700 (ONAC131) have been reported to be induced by cold, drought, submergence, and laid-down submergence treatments (Nuruzzaman et al. 2010). ONAC122 and ONAC131 expression was induced after infection by *Magnaporthe grisea*, the causal agent of rice blast disease, and *M. grisea*-induced expression of both genes was faster and higher in the incompatible interaction compared with the compatible interaction during the early stages of infection. ONAC122 and ONAC131 were also induced by treatment with salicylic acid, methyl jasmonate, and 1-aminocyclopropane-1-carboxylic acid (a precursor of ethylene). Silencing ONAC122 or ONAC131 expression using a modified Brome mosaic virus (BMV)-based silencing vector resulted in enhanced susceptibility to *M. grisea*. Furthermore, the expression levels of several other defense- and signaling-related genes (OsLOX, OsPR1a, OsWRKY45, and OsNH1) were downregulated in plants with silenced ONAC122 or ONAC131 expression via the BMV-based silencing system (Sun et al. 2013).

Os11g0126900 (also OsNAC10) is predominantly expressed in roots and panicles and is induced by drought, high salinity, and abscisic acid. Overexpression of OsNAC10 in rice under the control of the constitutive promoter GOS2 and the root-specific promoter RCc3 increased the plant tolerance to drought, high salinity, and low temperature at the vegetative stage. More importantly, the RCc3:OsNAC10 plants showed significantly enhanced drought tolerance at the reproductive stage, increasing grain yield by 25%–42% and by 5%–14% over controls in the field under drought and normal conditions, respectively. The grain yield of GOS2: OsNAC10 plants in the field, in contrast, remained similar to that of controls under both normal and drought conditions (Jeong et al. 2010). This evidence suggests that some NAC genes are common to the abiotic and biotic stress responses except for the RDV-induced responses.

Group B2 consists of four NAC genes: Os03g0119966, Os03g0327100, Os07g0566500, and Os08g0200600. Two genes (Os03g0327100 and Os07g0566500) are upregulated by three viruses (RDV, RTSV, and RGSV), Os03g0119966 (RIM1) is upregulated by RGSV but downregulated by RDV and RTSV, and Os08g0200600 is downregulated by all three viruses.

As described before, the RIM1 gene was characterized as a transcriptional regulator of JA signaling and is degraded in response to JA treatment via a 26S proteasome-dependent pathway (Yoshii et al. 2010). Infection with RDV and RTSV in TN1 downregulates the expression of RIM1 to activate the JA-related defense system, but, in the case of RGSV, the defense system may be inactivated. Os07g0566500 is the closest ortholog of the wheat GPC-B1 gene, which is an early regulator of senescence and affects the remobilization of protein and minerals to the grain. RNA interference transgenic rice lines (Os07g37920-RNAi) and 10 overexpressing transgenic lines (Os07g37920-OE) were constructed, but none of them showed differences in senescence. Transgenic Os07g37920-RNAi rice plants had reduced proportions of viable pollen grains and were male-sterile, but were able to produce seeds by cross-pollination. Analysis of the flower morphology of the transgenic rice plants showed that the anthers failed to dehisce. Transgenic Os07g37920-OE lines showed no sterility or anther dehiscence problems. The Os07g37920 transcript levels were higher in stamens compared with leaves and significantly reduced in transgenic Os07g37920-RNAi plants. Wheat GPC genes showed the opposite transcription profile (higher transcript levels in leaves than in flowers), and plants carrying knockout mutations of all GPC-1 and GPC-2 genes exhibited delayed senescence but normal anther dehiscence and fertility. Functional divergence of homologous NAC genes in wheat and rice was observed, and the necessity for separate studies of the function and targets of these transcription factors in wheat and rice has been suggested (Distelfeld et al. 2012).

Group C1 consists of 26 NAC genes, as shown in Table 5.5. They are divided into four groups by the pattern of the expression changes. The groups are genes that are upregulated by both RSV and RGSV (twelve genes), downregulated by both RSV and RGSV (five genes), upregulated by RSV but downregulated by RGSV (three genes), and downregulated by RSV but upregulated by RGSV (six genes). As described before, RSV and RGSV are members of the same genus as the Tenuivirus. Therefore, a common gene-silencing system or similar proteins involved in the replication of the single-stranded viral RNA genome may induce a change in NAC gene expression.

So far, Os02g0555300 has been reported to be induced by drought, submergence, and laid-down submergence (Nuruzzaman et al. 2010), and Os07g0138200 has been reported to be induced by severe drought stress in root (Nuruzzaman et al. 2012a). Os03g0815100 (OsNAC19, SNAC1) shows high expression in rice seedling roots, culms, and blade sheaths, but its expression in rice leaves is low. The expression of OsNAC19 in rice leaves could be induced by the infection of blast fungus and by application of exogenous methyl jasmonate (MeJA), ABA, and ethylene, but ethylene had a relatively weak induction effect (Lin et al. 2007). Overexpression of this gene significantly enhances drought resistance in transgenic rice (22%–34% higher seed setting than control) in the field under severe drought-stress conditions at the reproductive stage while showing no phenotypic changes or yield penalty. The transgenic rice also shows significantly improved drought resistance and salt tolerance at the vegetative stage. Compared with wild type, the transgenic rice is more sensitive to abscisic acid and loses water more slowly because it closes more stomatal pores, yet it displays no significant difference in the rate of photosynthesis (Hu et al. 2006). A rice homolog of SRO (similar to RCD), called OsSRO1c, was identified as a direct target gene of stress-responsive NAC 1 (SNAC1) and is involved in the regulation of stomatal aperture and the oxidative response. SNAC1 could bind to the promoter of OsSRO1c and activate the expression of OsSRO1c. OsSRO1c was induced in guard cells by drought stress (You et al. 2013). Recently, Os04g0477300 (BET1: BORON EXCESS TOLERANT1) was found to act to increase the sensitivity of rice plants to excess boron. In B-toxicity-sensitive rice, excess B might trigger a fatal cascade. However, a loss-of-function mutation in BET1 interrupts this cascade in B-toxicity-tolerant cultivars, allowing the tolerant cultivars to continue growing. Continuous growth also confers the benefit of diluting the internal B concentrations. BET1 might make the rice hypersensitive to excess B (Ochiai et al. 2011).

Group C2 consists of six genes, as shown in Table 5.5. Os04g0460600 and Os05g0442700 are upregulated by RTSV and RGSV, while Os05g0418800 and LOC_Os08g02160 (MSU-ID) are downregulated. Os08g0535800 and Os09g0509100 are upregulated by RGSV but downregulated by RTSV.

An activation-tagging mutant *Ostil1* (*Oryza sativa* tillering1) was characterized as having increased tillers, an enlarged tiller angle, and a semidwarf phenotype. The flanking sequence was obtained by plasmid rescue. Transgenic RNA interference (RNAi) and overexpression rice plants were produced using *Agrobacterium*-mediated transformation. The mutant phenotype cosegregated with the reallocation of the Ds element, and the flanking region of the reallocated Ds element was identified as part of the OsNAC2 gene (Os04g0460600). Northern blot analysis showed that the expression of OsNAC2 was greatly induced in the mutant plants. Transgenic rice overexpressing OsNAC2 resulted in recapture of the mutant phenotype, while downregulation of OsNAC2 in the Ostil1 mutant through RNAi complemented the mutant phenotype, confirming that the Ostil1 was caused by overexpression of OsNAC2. Overexpression of OsNAC2 regulates shoot branching in rice. Overexpression of OsNAC2 contributes to tiller bud outgrowth but does not affect tiller bud initiation (Mao et al. 2007). One of the particular symptoms induced by RGSV infection is excess tillering and upregulation of the *Ostil1* gene (Os04g0460600). The fact that the Ostil1 gene is an NAC gene is a very interesting correlation.

Infection by RGSV specifically induces changes in 17 NAC genes, as shown in Table 5.5 (Group D1). Ten genes are upregulated and seven are downregulated. In the D1 group, Os01g0104200 is reported to be induced by drought, submergence, and laid-down submergence (Nuruzzaman et al. 2010). Os07g0684800 is reported to be induced by cold, drought, submergence, and laid-down submergence (Nuruzzaman et al. 2010). Os10g0532000 is reported to be induced by severe drought stress in roots (Nuruzzaman et al. 2012a). Overexpression of Os01g0816100 (OsNAC4) leads to hypersensitive response (HR) cell death accompanied by loss of plasma membrane integrity, nuclear DNA fragmentation, and typical morphological changes. In OsNAC4 knockdown lines, HR cell death is markedly decreased in response to avirulent bacterial strains. After induction by an avirulent pathogen-recognition signal, OsNAC4 is translocated into the nucleus in a phosphorylation-dependent manner (Kaneda et al. 2009). Among the transgenic *Arabidopsis* population expressing full-length rice cDNAs, a thermotolerant line, R08946, was identified. The rice cDNA inserted in R08946 encoded an NAC transcription factor, Os08g0436700 (ONAC063). This protein was localized in the nucleus and showed transactivation activity at the C-terminus. ONAC063 expression was not induced by high temperatures but was highly induced by high salinity in rice roots. High osmotic pressure and levels of reactive oxygen species also induced ONAC063 expression. The seeds of ONAC063-expressing transgenic *Arabidopsis* showed enhanced tolerance to high salinity and osmotic pressure. Microarray and real-time reverse transcription-polymerase chain reaction analyses showed upregulated expression of some salinity-inducible genes, including the amylase gene AMY1, in ONAC063-expressing transgenic *Arabidopsis* (Yokotani et al. 2009).

Infection with RSV specifically downregulated the expression of the Os08g0433500 gene, and the change in expression of this gene was also observed in the panicle during severe drought-stress treatment (Nuruzzaman et al. 2012a).

5.5 CONCLUSION

Temporal and spatial regulation of gene expression is controlled by transcription factors. As an example, about 3000 genes encoding transcription factors are known to exist in the rice genome. They are classified into five major groups by the structure of the DNA-binding domain and into 15 superfamilies. Among the genes encoding transcription factor proteins, the NAC genes were, interestingly, thought to play roles in developmental processes and environmental stress responses. Genome analysis of many higher plants revealed that the NAC genes commonly exist in the genomes of higher plant species, and, through gene duplication, more than 100 NAC genes exist. Currently, researchers are interested in how each NAC gene is related to various biological processes, why so many members exist, and how they cooperate in biological functions. We recently published a review of the NAC transcription factors (Nuruzzaman et al. 2013). Overexpression or knockdown transgenic analyses have generated data on each NAC gene, but the data are still fragmentary. To clearly understand why there are so many NAC genes in the genome and how they are organized, it is important to perform a genome-wide, gene family-based functional search.

In this review, the NAC genes in the rice genome and how they respond to infection with rice viruses were summarized. Infection by four different rice viruses, RDV, RSV, RTSV, and RGSV, induces different symptoms in the host plants. NAC gene expression in host plants exhibits a great deal of variation. By clustering NAC genes with similar gene expression profiles and referring to data from the literature, it is possible to classify the NAC genes into virus response-specific groups that commonly respond to viral infection and other biotic stress treatments. There is still much to be done to elucidate the biological contribution of each NAC gene to a given signal transduction pathway, but even slow accumulation of new data is very important in the understanding of the biological function of the NAC genes.

ACKNOWLEDGMENTS

I am grateful to Dr. Omura and his laboratory members and Dr. Choi and his laboratory members for their collaboration, and to my former laboratory members Dr. Satoh, for the microarray data analysis, Dr. Nuruzzaman, for the NAC and WRKY gene family analysis, Mr. Kondou and Ms. Hosaka, for data analysis and laboratory work, and Ms. Kimura and Ms. Satoh, for their technical assistance. This work was financially supported by a grant from PROBRAIN.

REFERENCES

Aida, M., T. Ishida, H. Fukaki, H. Fujisawa, and M. Tasaka. 1997. Genes involved in organ separation in *Arabidopsis*: An analysis of the cup-shaped cotyledon mutant. *Plant Cell* 9:841–857.

Chen, Q., Q. Wang, L. Xiong, and Z. Lou. 2011. A structural view of the conserved domain of rice stress-responsive NAC1. *Protein Cell* 2:55–63.

Delessert, C., K. Kazan, I. W. Wilson et al. 2005. The transcription factor ATAF2 represses the expression of pathogenesis-related genes in *Arabidopsis*. *Plant J* 43:745–757.

Distelfeld, A., S. P. Pearce, R. Avni et al. 2012. Divergent functions of orthologous NAC transcription factors in wheat and rice. *Plant Mol Biol* 78:515–524.

Encabo, J. R., P. Q. Cabauatan, R. C. Cabunagan et al. 2009. Suppression of two tungro viruses in rice by separable traits originating from cultivar Utri Merah. *Mol Plant Microbe Interact* 22:1268–1281.

Ernst, H. A., A. N. Olsen, S. Larsen et al. 2004. Structure of the conserved domain of ANAC, a member of the NAC family of transcription factors. *EMBO Rep* 5:297–303.

Falk, B. W. and J. H. Tsai. 1998. Biology and molecular biology of viruses in the genus Tenuivirus. *Annu Rev Phytopathol* 36:139–163.

Hibino, H. 1983. Relations of rice tungro bacilliform and spherical viruses with their vector *Nephotettix virescens*. *Ann Phytopathol Soc Jpn* 49:545–553.

Hibino, H. 1996. Biology and epidemiology of rice viruses. *Annu Rev Phytopathol* 34:249–274.

Hu, H., M. Dai, J. Yao et al. 2006. Overexpressing a NAM, ATAF, and CUC (NAC) transcription factor enhances drought resistance and salt tolerance in rice. *Proc Natl Acad Sci USA* 103:12987–12992.

Hu, R., G. Qi, Y. Kong et al. 2010. Comprehensive analysis of NAC domain transcription factor gene family in *Populus trichocarpa*. *BMC Plant Biol* 10:145.

Jensen, M. K., T. Kjaersgaard, M. M. Nielsen et al. 2010. The *Arabidopsis thaliana* NAC transcription factor family: Structure-function relationships and determinants of ANAC019 stress signaling. *Biochem J* 426:83–96.

Jeong, J. S., Y. S. Kim, K. H. Baek et al. 2010. Root-specific expression of OsNAC10 improves drought tolerance and grain yield in rice under field drought conditions. *Plant Physiol* 153:185–197.

Jeong, R. D., A. C. Chandra-Shekara, A. Kachroo, D. F. Klessig, and P. Kachroo. 2008. HRT-mediated hypersensitive response and resistance to Turnip crinkle virus in *Arabidopsis* does not require the function of TIP, the presumed guardee protein. *Mol Plant Microbe Interact* 21:1316–1324.

Kaneda, T., Y. Taga, R. Takai et al. 2009. The transcription factor OsNAC4 is a key positive regulator of plant hypersensitive cell death. *EMBO J* 28:926–936.

Kim, S. Y., S. G. Kim, Y. S. Kim et al. 2007. Exploring membrane-associated NAC transcription factors in *Arabidopsis*: Implications for membrane biology in genome regulation. *Nucleic Acids Res* 35:203–213.

Kimura, I., Y. Minobe, and T. Omura. 1987. Changes in a nucleic acid and a protein component of rice dwarf virus particles associated with an increase in symptom severity. *J Gen Virol* 68:3211–3215.

Le, D. T., R. Nishiyama, Y. Watanabe et al. 2011. Genome-wide survey and expression analysis of the plant-specific NAC transcription factor family in soybean during development and dehydration stress. *DNA Res* 18:263–276.

Lee, J. H., M. Muhsin, G. A. Atienza et al. 2010. Single nucleotide polymorphisms in a gene for translation initiation factor (eIF4G) of rice (*Oryza sativa*) associated with resistance to rice tungro spherical virus. *Mol Plant Microbe Interact* 23:29–38.

Lin, R., W. Zhao, X. Meng, M. Wang, and Y. Peng. 2007. Rice gene OsNAC19 encodes a novel NAC-domain transcription factor and responds to infection by *Magnaporthe grisea*. *Plant Sci* 172:120–130.

Lu, L., Z. Du, M. Qin et al. 2009. Pc4, a putative movement protein of rice stripe virus, interacts with a type I DnaJ protein and a small Hsp of rice. *Virus Genes* 38:320–327.

Mao, C., W. Ding, Y. Wu et al. 2007. Overexpression of a NAC-domain protein promotes shoot branching in rice. *New Phytol* 176:288–298.

Nakashima, K., L. S. Tran, D. Van Nguyen et al. 2007. Functional analysis of a NAC-type transcription factor OsNAC6 involved in abiotic and biotic stress-responsive gene expression in rice. *Plant J* 51:617–630.

Nuruzzaman, M., R. Manimekalai, A. M. Sharoni et al. 2010. Genome-wide analysis of NAC transcription factor family in rice. *Gene* 465:30–44.

Nuruzzaman, M., A. M. Sharoni, and S. Kikuchi. 2013. Roles of NAC transcription factors in the regulation of biotic and abiotic stress responses in plants. *Front Microbiol* 4:248.

Nuruzzaman, M., A. M. Sharoni, K. Satoh et al. 2012a. Comprehensive gene expression analysis of the NAC gene family under normal growth conditions, hormone treatment, and drought stress conditions in rice using near-isogenic lines (NILs) generated from crossing Aday Selection (drought tolerant) and IR64. *Mol Genet Genomics* 287:389–410.

Nuruzzaman, M., A. M. Sharoni, K. Satoh, H. Kondoh, A. Hosaka, and S. Kikuchi. 2012b. A genome-wide survey of the NAC transcription factor family in monocots and eudicots. In J. Wan (ed.), *Introduction to Genetics. DNA Methylation, Histone Modification and Gene Regulation*, pp. 3–23. Hong Kong: iConcept Press.

Ochiai, K., A. Shimizu, Y. Okumoto, T. Fujiwara, and T. Matoh. 2011. Suppression of a NAC-like transcription factor gene improves boron-toxicity tolerance in rice. *Plant Physiol* 156:1457–1463.

Olsen, A. N., H. A. Ernst, L. L. Leggio, and K. Skriver. 2005. NAC transcription factors: Structurally distinct, functionally diverse. *Trends Plant Sci* 10:79–87.

Omura, T. and J. Yan. 1999. Role of outer capsid proteins in transmission of Phytoreovirus by insect vectors. *Adv Virus Res* 54:15–43.

Ooka, H., K. Satoh, K. Doi et al. 2003. Comprehensive analysis of NAC family genes in *Oryza sativa* and *Arabidopsis thaliana*. *DNA Res* 10:239–247.

Ou, S. H. 1972. *Rice Diseases*. Slough: Commonwealth Mycological Institute.

Ramirez, B. C. 2008. Tenuivirus. In B. W. J. M. Mahyand and H. V. van Regenmortel (eds), *Encyclopedia of Virology*, pp. 24–27. Paris: CNRS.

Ren, T., F. Qu, and T. J. Morris. 2000. HRT gene function requires interaction between a NAC protein and viral capsid protein to confer resistance to turnip crinkle virus. *Plant Cell* 12:1917–1926.

Rushton, P. J., M. T. Bokowiec, S. Han et al. 2008. Tobacco transcription factors: Novel insights into transcriptional regulation in the Solanaceae. *Plant Physiol* 147:280–295.

Satoh, K., H. Kondoh, T. B. De Leon et al. 2013a. Gene expression responses to rice tungro spherical virus in susceptible and resistant near-isogenic rice plants. *Virus Res* 171:111–120.

Satoh, K., H. Kondoh, T. Sasaya et al. 2010. Selective modification of rice (*Oryza sativa*) gene expression by rice stripe virus infection. *J Gen Virol* 91:294–305.

Satoh, K., T. Shimizu, H. Kondoh et al. 2011. Relationship between symptoms and gene expression induced by the infection of three strains of rice dwarf virus. *PLoS One* 6:e18094.

Satoh, K., K. Yoneyama, H. Kondoh et al. 2013b. Relationship between gene responses and symptoms induced by Rice grassy stunt virus. *Front Microbiol* 4:313.

Selth, L. A., S. C. Dogra, M. S. Rasheed et al. 2005. A NAC domain protein interacts with tomato leaf curl virus replication accessory protein and enhances viral replication. *Plant Cell* 17:311–325.

Shikata, E., T. Senboku, and T. Ishimizu. 1980. The causal agent of rice grassy stunt disease. *Proc Jpn Acad Series B* 56:89–94.

Shimizu, T., K. Satoh, S. Kikuchi, and T. Omura. 2007. The repression of cell wall- and plastid-related genes and the induction of defense-related genes in rice plants infected with rice dwarf virus. *Mol Plant Microbe Interact* 20:247–254.

Shimizu, T., S. Toriyama, M. Takahashi, K. Akutsu, and K. Yoneyama. 1996. Non-viral sequences at the 5′ termini of mRNAs derived from virus-sense and virus-complementary sequences of the ambisense RNA segments of rice stripe tenuivirus. *J Gen Virol* 77:541–546.

Souer, E., A. van Houwelingen, D. Kloos, J. Mol, and R. Koes. 1996. The no apical meristem gene of Petunia is required for pattern formation in embryos and flowers and is expressed at meristem and primordia boundaries. *Cell* 85:159–170.

Sun, L., H. Zhang, D. Li et al. 2013. Functions of rice NAC transcriptional factors, ONAC122 and ONAC131, in defense responses against *Magnaporthe grisea*. *Plant Mol Biol* 81:41–56.

Takahashi, Y., T. Omura, K. Shohara, and T. Tsuchizaki. 1991. Comparison of four serological methods for practical detection of ten viruses of rice in plants and insects. *Plant Dis* 75:458–461.

Tran, L. S., K. Nakashima, Y. Sakuma et al. 2004. Isolation and functional analysis of *Arabidopsis* stress-inducible NAC transcription factors that bind to a drought-responsive cis-element in the early responsive to dehydration stress 1 promoter. *Plant Cell* 16:2481–2498.

Wang, X., S. P. Goregaoker, and J. N. Culver. 2009. Interaction of the tobacco mosaic virus replicase protein with a NAC domain transcription factor is associated with the suppression of systemic host defenses. *J Virol* 83:9720–9730.

Wei, T., A. Kikuchi, Y. Moriyasu et al. 2006a. The spread of rice dwarf virus among cells of its insect vector exploits virus-induced tubular structures. *J Virol* 80:8593–8602.

Wei, T., A. Kikuchi, N. Suzuki et al. 2006b. Pns4 of rice dwarf virus is a phosphoprotein, is localized around the viroplasm matrix, and forms minitubules. *Arch Virol* 151:1701–1712.

Wei, T., T. Shimizu, K. Hagiwara et al. 2006c. Pns12 protein of rice dwarf virus is essential for formation of viroplasms and nucleation of viral-assembly complexes. *J Gen Virol* 87:429–438.

Xie, Q., G. Frugis, D. Colgan, and N. H. Chua. 2000. *Arabidopsis* NAC1 transduces auxin signal downstream of TIR1 to promote lateral root development. *Genes Dev* 14:3024–3036.

Xie, Q., A. P. Sanz-Burgos, H. Guo, J. A. García, and C. Gutiérrez. 1999. GRAB proteins, novel members of the NAC domain family, isolated by their interaction with a geminivirus protein. *Plant Mol Biol* 39:647–656.

Xiong, R., J. Wu, Y. Zhou, and X. Zhou. 2008. Identification of a movement protein of the tenuivirus rice stripe virus. *J Virol* 82:12304–12311.

Xiong, R., J. Wu, Y. Zhou, and X. Zhou. 2009. Characterization and subcellular localization of an RNA silencing suppressor encoded by rice stripe tenuivirus. *Virology* 387:29–40.

Yamaguchi, M., M. Kubo, H. Fukuda, and T. Demura. 2008. Vascular-related NAC-DOMAIN7 is involved in the differentiation of all types of xylem vessels in *Arabidopsis* roots and shoots. *Plant J* 55:652–664.

Yokotani, N., T. Ichikawa, Y. Kondou et al. 2009. Tolerance to various environmental stresses conferred by the salt-responsive rice gene ONAC063 in transgenic *Arabidopsis*. *Planta* 229:1065–1075.

Yoshii, M., T. Shimizu, M. Yamazaki et al. 2009. Disruption of a novel gene for a NAC-domain protein in rice confers resistance to rice dwarf virus. *Plant J* 57:615–625.

Yoshii, M., M. Yamazaki, R. Rakwal et al. 2010. The NAC transcription factor RIM1 of rice is a new regulator of jasmonate signaling. *Plant J* 61:804–815.

You, J., W. Zong, X. Li et al. 2013. The SNAC1-targeted gene OsSRO1c modulates stomatal closure and oxidative stress tolerance by regulating hydrogen peroxide in rice. *J Exp Bot* 64:569–583.

Zheng, X., B. Chen, G. Lu, and B. Han. 2009. Overexpression of a NAC transcription factor enhances rice drought and salt tolerance. *Biochem Biophys Res Commun* 379:985–989.

CHAPTER 6

Plant Molecular Breeding
Perspectives from Plant Biotechnology and Marker-Assisted Selection

Ashwani Kumar, Manorma Sharma, Saikat Kumar Basu,
Muhammad Asif, Xian Ping Li, and Xiuhua Chen

CONTENTS

ABBREVIATIONS

MAS	Marker-assisted selection
ODM	Oligonucleotide-directed mutagenesis
QTL	Quantitative trait loci
RAPD	Randomly amplified polymorphic DNA
RFLP	Restriction fragment length polymorphism
SDS-PAGE	Sodium dodecyl sulfate polyacrylamide gel electrophoresis
SRFA	Selective restriction fragment amplification

SSRs	Simple sequence repeats
STRs	Short tandem repeats
ZFN	Zinc finger nuclease

6.1 INTRODUCTION

Food scarcity caused by growing populations is a compelling and increasingly important issue, and plant biotechnology could offer a quick and reliable method of plant breeding
(Bhowmik and Basu 2008; Kumar and Roy 2006). Genetic transformation has represented a
more precise and predictable method for producing plants with new and desirable traits (Basu
et al. 2010; Vasil 2007). Plant tissue culture has invariably been used to obtain genetically
transformed plants by direct gene transfer (Kumar and Roy 2006, 2011; Kumar and Sopory
2010; Stasolla and Thorpe 2009). One of the most important emergent research platforms in the
realm of plant biotechnology is "molecular farming," which has reportedly synthesized several
value-added products, such as nutraceuticals and pharmaceuticals (Acharya et al. 2010; Kumar
and Sopory 2008).

The development of transgenic plants resistant to abiotic and biotic stress has been rendered
possible through improvement and optimization of tissue culture conditions and by the refinement
of transformation techniques (Sharma et al. 2011). A single gene or a gene cassette can be transferred into cells which become stably integrated into the nuclear genome of the host cells, and this
ensures the transmission of the inserted gene(s) to the progeny through two distinct methods: vector-
independent and vector-dependent (Bhowmik and Basu 2008). Success of these transgenic plants
is dependent upon molecular markers, which helps plant breeders to develop modified crops. The
present review will provide an insight into breeding through tissue culture and the role of molecular
markers in establishing clonal fidelity.

6.2 PLANT BREEDING

Plant breeding is a science for changing the genetics of plants in order to produce desired
characters such as high resistance, high yield, quality improvement, and stress tolerance (Nandy
et al. 2008). The success of plant breeding is dependent upon genetic variability of traits of interest and efficient markers for selection of the traits (Abdullah 2001). Traditional methods of plant
breeding include testing, selection, backcross, linkage analysis, emasculation, hybridization, trip-
loids, wide crossing, selection, chromosome doubling, male sterility, chromosome counting, and
statistical tools (Acharya et al. 2010; Badea and Basu 2009, 2010). All the traditional methods of
plant breeding required several years to obtain plants with the required characteristics, which is
a costly and laborious process (Basu et al. 2011). The development of plant biotechnology techniques has overcome the drawbacks associated with traditional plant breeding (Basu et al. 2010).
Modern plant breeding is an amalgamation of genetics, statistics, molecular biology, botany,
forestry, bioinformatics, pathology and entomology, agronomy, and plant physiology (Basu et al.
2011).

Advanced techniques of plant breeding include DNA markers, tissue culture, mutagenesis, marker-assisted selection, DNA sequencing, microarray analysis, plant genomic analysis,
primer design, plant transformation, and in situ hybridization (Bhowmik et al. 2009; Bussell
et al. 2005; Goyal et al. 2009; Venugopal et al. 1992; Wolff et al. 1993). Some more advanced
plant breeding techniques include oligonucleotide-directed mutagenesis (ODM), cisgenesis
and intragenesis, zinc finger nuclease (ZFN) technology (such as ZFN-1, ZFN-2, and ZFN-3),

RNA-dependent DNA methylation (RdDM), reverse breeding, agro-infiltration, and synthetic genomics (Lusser et al. 2011). In plant breeding, selection of plants with desired characteristics is an important step. The following types of selection are available:

1. Phenotypic selection (PS)
2. Marker-based selection (MBS)
3. Marker-assisted selection (MAS)
4. Marker-assisted backcrossing (MABC)
5. Marker-assisted recurrent selection (MARS)
6. Genomic selection or genome-wide selection (GS or GWS)

6.3 PLANT TISSUE CULTURE

Totipotency, dedifferentiation, and competency are the three main features that gave rise to the concept of plant tissue culture. In 1902, G. Haberlandt first reported success in culturing isolated palisade cells from leaves in sucrose-enriched Knop's salt solution. Most current plant tissue culture media compositions have been derived from the extensive research of Skoog et al. conducted in the 1950s–1960s (George et al. 2008). In 1965, French botanist George Morel introduced micropropagation to obtain virus-free orchid plants. Since 1965, many techniques and protocols have been proposed to obtain monoclonal plants with desired features. In plant breeding, tissue culture is used as an essential tool for crop improvement; later techniques have been largely reliant on this foundation. There are many benefits associated with plant breeding through plant tissue culture: increasing and stabilizing yields, developing stress-tolerant varieties, and enhancing the nutritional content of foods. Barley, wheat, potato, blackberry, flax, celery, tomato, and rice are the main commercial crops that have been improved through tissue culture. In plant tissue culture, whole plants or parts of plants can be regenerated. Under *in vitro* conditions, culture of cells is affected by various factors: growth media, environmental factors, explant source, and genetics (Bhowmik and Basu 2008; Bhowmik et al. 2009).

During the development of a desired plant by tissue culture, the plants are extensively analyzed at various stages to reject those plants that possess undesired characteristics. Obviously all these processes require a great deal of time and money, but once an effective protocol has been established it can be utilized for commercial conventional breeding operations. The main applications of tissue culture in plant breeding are:

1. Micropropagation
2. *In vitro* mutagenesis
3. Embryo culture
4. Somaclonal and gametoclonal variation
5. Germplasm preservation
6. Haploid production
7. Doubled haploid production
8. Somatic embryogenesis
9. Micrografting
10. Somatic hybridization

6.4 MARKER-ASSISTED SELECTION (MAS)

To increase the speed and efficiency of selection, markers are used in breeding program to predict whether an individual plant carries a particular gene or particular traits (Nandy et al. 2007).

Marker-assisted approaches can be used in many steps in breeding, such as selection, management of germplasm identity, and quality control. The different types of markers are:

1. Morphological markers
2. Biochemical markers (protein diversity)
3. Cytological markers
4. Molecular markers (DNA sequence diversity)

6.4.1 Morphological Markers

Morphological markers are the traditional markers, generally representing genetic polymorphisms that are visible as differences in appearance. In this technique, desired morphological characteristics are generated in plants using tissue culture and mutation breeding. After many generations through successive breeding these mutant characteristics are selected using the physically identified mutant for the traits (Acharya et al. 2010; Akhter et al. 2010). Selected genetic stable mutant traits can be used as morphological markers. Markers of this type are not successful because of the following limitations:

1. These characteristics are limited in number.
2. They are highly dependent on environmental factors.
3. Some morphological characteristics appear late in the development of the plant, which makes early scoring impossible.
4. It is a time-consuming, labor-intensive process requiring a large field for planting.
5. Due to pleiotropic gene action, one molecular marker can affect others.

6.4.2 Biochemical Markers

Isozymes are commonly used protein markers in plant breeding (Nandy et al. 2007). Most isozyme markers have been derived from enzymes of intermediary metabolism; however, conceivably, any enzyme could be used as an isozyme genetic marker (White et al. 2007). It is a very simple technique, in which a small amount of tissue is required for preparing a crude extract to produce isozymes (Nandy et al. 2008). The technique is based on the principle that allelic forms of enzyme variation exist among several proteins. These variations can be distinguished by electrophoretic separation based on the different net charges and masses of the molecules (Bhowmik and Basu 2008; Bhowmik et al. 2009). The basic steps of biochemical markers are shown in Figure 6.1.

Differences in the amino acid sequences are utilized as a polymorphic biochemical marker; for example, between the two species *Oryza minuta* J. S. Presl ex. C. B. Presl and *O. sativa* L., six isozymes (*Amp-4, Got-1, Got-3, Pgd-1, Pgd-3*, and *Sdh-1*) were reported to be polymorphic (Abdullah 2001).

Glucosinolates are secondary metabolites of the order Brassicales and genus *Drypetes* (Abdel-Farid et al. 2009; Rodman et al. 1996). Myrosinase enzyme hydrolysis cleaves off the glucose group from a glucosinolate, synthesizing nitriles and thio- and iso-thiocyanates. These active defense molecules are toxic to humans and animals (Johnson 2002). Hence, selective breeding was used to produce crops with low concentrations of glucosinolates. Now, however, glucosinolates are used as biochemical markers for selection. In cauliflower they are used as resistance markers against downy mildew disease outbreaks caused by the fungus *Peronospora parasitica* (Ménard et al. 1999). In plants, nitro blue tetrazolium reduction is used as a biochemical marker for rapid identification of aluminum-tolerant plants within a segregating population (Maltais and Houde 2002).

Likewise, pathogen resistance biochemical markers have been utilized to select salinity stress-tolerant plants (Badea and Basu 2010). In stress-tolerant plants, the levels of osmolytes, soluble sugars and proteins, polyamines, amides, quaternary ammonium compounds, amino acids, polyols, antioxidants, and ATPase are utilized as biochemical markers (Ashraf and Harris 2004). An important

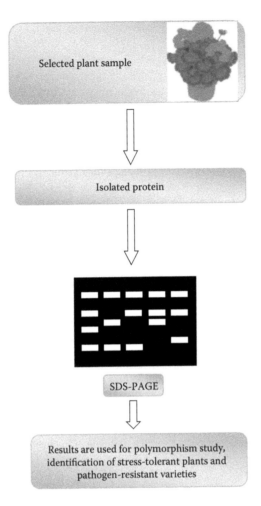

Figure 6.1 Isozyme analysis of plant samples (isolated proteins of examined plants separated by SDS-PAGE for various purposes).

protein, allene oxide cyclase, has been reported to be related to salt stress tolerance in the mangrove *Bruguiera sexangula* (Yamada et al. 2002). The expression of the same protein in commercial yeast and tobacco cell lines has been reported to increase tolerance to higher concentrations of salt (Ashraf and Harris 2004). In salt-stressed plants, proline has been used as a biochemical marker that accumulates in higher concentrations (Abraham et al. 2003). In the case of lentil (*Lens culinaris* Medik.), a drought tolerance biochemical marker was identified through seed proteins using sodium dodecyl sulfate–polyacrylamide gel electrophoresis (SDS-PAGE) (Bhowmik et al. 2009; El-nahas et al. 2011). Wheat seed storage proteins (gliadins and glutenins) demonstrated higher polymorphism, which can be used in investigating evolution, breeding, and certification of cultivated wheat (Xynias et al. 2007).

Different stages of regeneration under *in vitro* conditions can be identified using isozyme analyses (Badea and Basu 2009). In maize, glutamate dehydrogenase, peroxidase, and acid phosphatase were used as biochemical markers to distinguish nonembryogenic and embryogenic calli (Fransz et al. 1989). In horseradish (*Armoracial pathifolia* Gilib.) tissue culture, biochemical markers were used for selection over a long period. Hyperhydrated teratoma and tumor tissues showed changes in peroxidase activities and ascorbate peroxidase isoform pattern, which were used to confirm the tissue culture stability (Balen et al. 2009). These types of marker were also found to be useful in tissue culture of *Euphorbia millii* Des Moul, *Mammillaria gracilis* Salm-Dyck, *Crotalaria juncea* L., and

Craterostigma plantagineum Hochst tissue lines (Balen et al. 2003; Dewir et al. 2006; Ohara et al. 2000; Toldi et al. 2002).

In plant breeding, selection of pathogen-free crops is of great importance. In plants, attack by pathogens is a stress stimulus that stimulates the plant's defense system, leading to accumulation of specific proteins that can be used as biochemical markers. The western conifer seed bug (*Leptoglossus occidentalis* Heidemann) causes damage to the seeds of Douglas fir (*Pseudotsuga menziesii* [Mirbel] Franco). Through antibody-based assay, crystalloid storage reserve proteins, specific antibodies, and salivary protein constituents represent potential markers for spotting damage caused by insect pests (Lait et al. 2001).

In addition to identification of diseases, it is also important to find specific disease-causing pathogens. Protozoan species such as *Phytomonas* have been reported in over a dozen different plant families. Such protozoan pathogens are disseminated through plant-specific Hemipteran insect vectors representing family cohorts such as Coreidae, Lygaeidae, Pyrrhocoridae, and Pentatomidae (Uttaro et al. 1997). This specific plant pathogen was classified through isoenzyme polymorphisms using as a biochemical marker the enzyme isopropanol dehydrogenase in conjunction with a hydroxyacid dehydrogenase and malate dehydrogenase (Uttaro and Opperdoes 1997; Uttaro et al. 1997). Other fruit pathogens, Trypanosomatids (kinetoplastid protozoa), were classified using arginase, citrulline hydrolase, arginine deiminase, and ornithine carbamoyltransferase as biochemical markers (Catarino et al. 2001).

6.4.3 Molecular Markers

Genetic markers or molecular markers are "gene tags" that are situated in close proximity to the gene of interest. A DNA sequence located close to or within the gene of interest is known as a marker (Figure 6.2). A marker located within the gene of interest is known as a direct marker. These are uncommon and difficult to find. The recombination rate between the marker and the target gene is zero, and the probability of the marker in each backcross generation with respect to that of the target gene is 50%, while a linked or indirect marker is located close to the gene of interest (Nandy et al. 2008). Reliability of selection is increased if we use two markers flanking quantitative trait loci (QTL) (Figure 6.3).

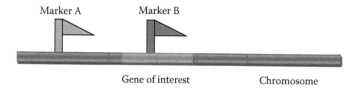

Figure 6.2   Direct marker B located within gene of interest; indirect marker A located close to gene of interest.

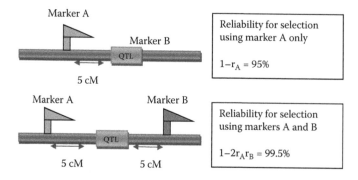

Figure 6.3   Reliability for selection increased when two markers, A and B, are used instead of one marker, A.

To avoid the risk of losing the target gene, the marker must be as close as 1 cM to the target gene. Characteristics of the perfect marker are:

1. Randomly and frequently distributed throughout the genome
2. Highly polymorphic
3. Highly reproducible
4. Detectable by an easy, rapid, and economical assay
5. Not affected by environmental conditions

In practice, it is impossible to find a marker that possesses all the characteristics described above. Depending on requirements, some of the characteristics are considered for selection of markers. Based on hybridization, molecular markers are of two types:

1. Non-PCR based: RFLP
2. PCR based: RAPDs, SSRs, AFLPs

6.4.3.1 Restriction Fragment Length Polymorphism (RFLP)

The first molecular marker technique to be developed and used in MAS was RFLP. It is the most widely used hybridization-based molecular marker (Semagn et al. 2006). In 1980, White described the first polymorphic RFLP marker; later, various other markers were discovered year by year (Table 6.1). The main steps involved in RFLP are as follows:

1. Isolation of DNA
2. Restriction digestion and gel electrophoresis
3. DNA transfer by Southern blotting
4. DNA hybridization

A significant amount of clean and high-molecular-weight DNA is extracted from plant cells. We can also use chloroplast and mitochondrial DNA (de Vicente and Fulton 2003). Isolated DNA is restricted using molecular scissors: restriction endonuclease enzymes (Bhowmik et al. 2008). These enzymes are isolated from several microbes and are able to cut DNA at specific sites known as "recognition sites," which are randomly distributed throughout the genome. Recognition sites are usually four-, six-, or eight-base-pair (bp) sequences in the DNA. The restriction endonuclease cleaves double-stranded DNA when these specific sequences are identified in the molecular sequence (Semagn et al. 2006). Various sizes of restricted DNA fragments are separated through electrophoresis (Bhowmik et al. 2008).

The whole process of RFLP is shown in Figure 6.4. Agarose or polyacrylamide gel can be used for electrophoresis, depending upon the restriction enzyme used. If we are using a high-resolution four-base-pair cutter, we have to use polyacrylamide gel, whereas agarose gel is preferred for large restriction fragments. The velocity of migration in the gel is dependent on the net molecular charge, the particular frictional coefficient within the gel matrix, and the strength of the electric field. Restricted fragments are denatured to single strands and then transferred onto nitrocellulose membrane, a process known as Southern hybridization after its inventor E.M. Southern (1975). After transfer, the nitrocellulose membrane is incubated with the DNA probe. DNA probes are available from three sources: genomic, cDNA, and cytoplasmic DNA (mitochondrial and chloroplast DNA) libraries (Bhowmik et al. 2009).

Through hybridization, labeled DNA probe binds to complementary DNA on the filter. Labeling can be achieved using radioactive nucleotides or by nonradioactive means. The digoxigenin system is commonly used for nonradioactive labeling (Holtke et al. 2009). Radioactive labeling can be visualized by autoradiography. The RFLP markers are codominantly inherited and are reported to be highly polymorphic, reproducible, heritable, and locus specific. By using different RFLP probes, the

Table 6.1   Time Line of Development of Molecular Markers

Acronym	Nomenclature	Year	References
RFLP	Restriction fragment length polymorphism	1974	Grodzicker et al. (1974)
VNTR	Variable number tandem repeats	1985	Jeffrey (1979)
ASO	Allele-specific oligonucleotides	1986	Saiki et al. (1986)
AS-PCR	Allele-specific polymerase chain reaction	1988	Landegren et al. (1988)
OP	Oligonucleotide polymorphism	1988	Beckmann (1988)
SSCP	Single-stranded conformational polymorphism	1989	Orita et al. (1989)
STS	Sequence-tagged site	1989	Olsen et al. (1989)
STMS	Sequence-tagged microsatellite site	1990	Beckmann and Soller (1990)
AP-PCR	Arbitrarily primed polymerase chain reaction	1990	Welsh and McClelland (1990)
RAPD	Randomly amplified polymorphic DNA	1990	Williams et al. (1990)
RLGS	Restriction landmark genome scanning	1991	Hatada et al. (1991)
CAPS	Cleaved amplified polymorphic sequence	1992	Akopyanz et al. (1992)
SSR	Simple sequence repeats	1992	Akkaya et al. (1992)
DOP-PCR	Degenerate oligonucleotide primer-PCR	1992	Telenius et al. (1992)
SCAR	Sequence-characterized amplified region	1993	Paran and Michelmore (1993)
MAAP	Multiple arbitrary amplicon profiling	1993	Caetano-Anollés and Gresshoff (1996)
SNP	Single-nucleotide polymorphisms	1994	Jordan and Humphries (1994)
ISSR	Inter simple sequence repeats	1994	Zietkiewicz et al. (1994)
SAMPL	Selective amplification of microsatellite polymorphic loci	1994	Morgante and Vogel (1994)
ASAP	Allele-specific associated primers	1995	Gu et al. (1995)
AFLP (SRFA)	Amplified fragment length polymorphism (selective restriction fragment amplification)	1995	Vos et al. (1995)
ISTR	Inverse sequence-tagged repeats	1996	Rohde (1996)
CFLP	Cleavage fragment length polymorphism	1996	Brow et al. (1996)
S-SAP	Sequence-specific amplified polymorphism	1997	Waugh et al. (1997)
DAMD-PCR	Directed amplification of mini satellite DNA-PCR	1997	Bebeli et al. (1997)
RBIP	Retrotransposon-based insertional polymorphism	1998	Flavell et al. (1998)
MSAP	Methylation-sensitive amplification polymorphism	1999	Xiong et al. (1999)
IRAP	Inter-retrotransposon amplified polymorphism	1999	Kalendar et al. (1999)
REMAP	Retrotransposon-microsatellite amplified polymorphism	1999	Kalendar et al. (1999)
TE-AFLP	Three endonuclease AFLP	2000	van der Wurff et al. (2000)
MITE	Miniature inverted-repeat transposable element	2000	Casa et al. (2000)
SRAP	Sequence-related amplified polymorphism	2001	Li and Quiros (2001)
IMP	Inter-MITE polymorphisms	2001	Chang et al. (2001)
EST-SSRs	Expressed sequence tags, simple sequence repeats	2002	Eujayl et al. (2002)

DNA blots can be analyzed quite easily by reprobing and stripping (Agarwal et al. 2008). Besides the advantages, there are some limitations of this technique:

1. The level of polymorphism in RFLP is low.
2. It is a time-consuming, laborious, and expensive technique, which requires high quantity and quality of DNA.
3. Specific probe libraries for the species are necessary.

These limitations paved the way for the development of PCR-based techniques.

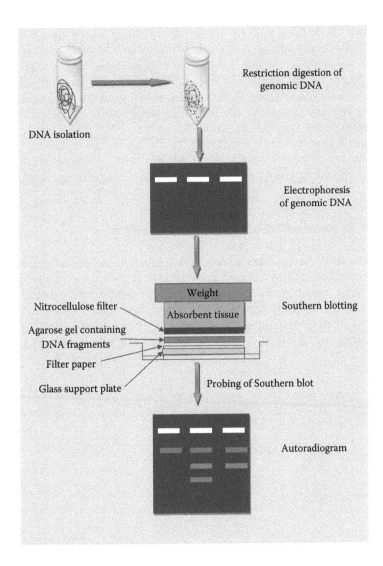

Figure 6.4 Steps involved in RFLP analysis with PCR-based markers. PCR-based markers are collectively known as arbitrarily amplified DNA (AAD) markers. (Adapted from Bussell, J. D., M. Waycott, and J. A. Chappill. 2005. Arbitrarily amplified DNA markers as characters for phylogenetic inference. *Perspect. Plant Ecol* 7: 3–26.)

6.4.3.2 Randomly Amplified Polymorphic DNA (RAPD)

RAPD markers are polymorphic DNA molecules separated by gel electrophoresis following PCR using short, random, and oligonucleotide primers (Williams et al. 1990). This method was developed in 1990 (Williams et al. 1990). RAPD analysis is popular due to its various applications: the identification and characterization of genotypes (Arif et al. 2010), plant breeding and the study of genetic relationships (Bhowmik et al. 2009; Ranade et al. 2001), evaluation of intraline uniformity of plants (Staniaszek and Habdas 2006), determination of genetic fidelity (Kawiak and Ojkowska 2004), and development of genetic markers linked to the trait in question (Bardakci 2001).

The primers can be designed without prior genetic information on the plant; thus, the method is suited to those plants on which no previous genetics research has been conducted (Nybom and Bartish 2000; Weder and Jargen 2002). Primers are complementary to the sequence of genomic DNA. About 3–20 loci can be amplified through RAPD. For amplification, opposite-oriented loci need to be close together (<3000 bp). PCR products can be visualized using simple agarose gel electrophoresis or ethidium bromide or silver-stained PAGE (Weising et al. 1995), radioactivity, or fluorescently labeled primers/nucleotides (Corley-Smith et al. 1997). The basic steps of RAPD are described in Figure 6.5.

A major limitation of RAPD markers is the fact that they are dominant alleles that cannot differentiate between homozygous and heterozygous individuals carrying a specific allele (Liu et al. 1999).

6.4.3.3 Simple Sequence Repeats (SSR)/Microsatellites

SSRs, also known as short tandem repeats (STRs) or microsatellites, are repeating sequences of 2–6 base pairs of DNA tandem and randomly spread in genome. In microsatellite terminology, "micro" means short sequences while "satellite" means repetitive sequences that can be separated from the whole genome by centrifugation. Based on the length of the sequence, alternative terminologies are used: "satellites," "minisatellites," or "microsatellites." These markers are codominant in nature and are usually visualized on agarose or polyacrylamide gels (Goyal et al.

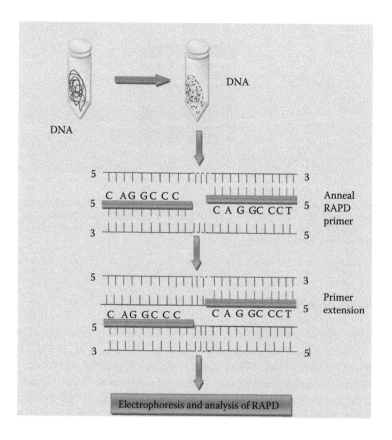

Figure 6.5 Steps involved in RAPD marker system.

2009; Yazdani et al. 2003). The sequences flanking these microsatellites can be used to design primers. SSRs are highly polymorphic in nature, showing high levels of instability, incorporations, or omissions of repeated units, resulting in variations in the copy number of repeated sequences (Viguera et al. 2001). These variations were therefore named "dynamic mutations" (Richards and Sutherland 1992). A dynamic mutation of repetitive sequences was proposed by a replication slippage model or copy-choice recombination (Petruska et al. 1998). According to this model, slippage strand mispairing is a specific error that occurs during replication of short repetitive sequences, such as microsatellites.

During slippage, the new strand mispairs with the template strand. The DNA polymerase enzyme might either incorporate too many bases by inserting a complete repeat (insertion) or miss a repeat (deletion). Thus, SSR allelic variations are results of difference in the numbers of repeating units due to insertion/deletion within the microsatellite architecture (Bhowmik et al. 2009; Semagn et al. 2006). This polymorphism of SSR can be identified by primers with sequences flanking the repeat unit following PCR amplification and subsequent visualization in a suitable gel matrix (Alzohairy 2005).

6.4.3.4 Amplified Fragment Length Polymorphism (AFLP)

AFLP is the combination of PCR and RFLP. It was originally described by Zabeau and Vos (1993) as selective restriction fragment amplification (SRFA). Mueller and Wolfenbarger (1999) provide an overview of AFLP and its application to a broad array of research topics. In this technique, restricted fragments are selectively amplified by PCR. As with RAPD, no prior sequence knowledge is required in AFLP. "Genome representation" is the key feature of AFLP; it can simultaneously screen representative DNA regions distributed randomly throughout the genome (Bhowmik et al. 2008; Semagn et al. 2006). The AFLP procedure involves the following steps (Figure 6.6):

1. Restriction digestion of genomic DNA with specific restriction enzymes such as the rare cutters (*EcoRI* or *PstI*) or frequent cutters (*MseI* or *TaqI*).
2. Fusion of double-stranded oligonucleotide adapters with known sequence for PCR amplifications to either end of the fragments.
3. Selective amplification of restriction fragments using primers that can specifically identify the adapter and restriction site sequences.
4. The AFLP generated can be easily detected by labeled primers and can be subsequently separated by electrophoresis in denaturing polyacrylamide gels.

In a single PCR reaction, 50–100 fragments are amplified with high specificity and reproducibility in the absence of any genome sequence data. This technique is useful in addressing genetic variation within a species or among closely related species. AFLP can also be used to study organisms whose genome has not been sequenced.

6.5 FUTURE INSIGHT

It is evident that various plant breeding techniques have been discovered from the earliest human civilization until today. Earlier techniques were based on Mendelian principles, while present techniques are based on genetics. The new plant breeding techniques ZFN-3 technology, cisgenesis/intragenesis, and floral dip, a variant of agro-infiltration, are used for stable insertion of a new gene. However, despite all these developments in plant breeding, it is still a major challenge to satisfy the appetite of a large population without disturbing the ecosystem.

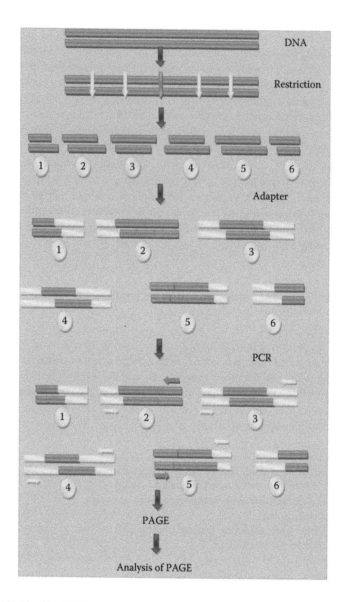

Figure 6.6 Steps involved in AFLP.

6.6 SUMMARY

Plant tissue culture is an important method of obtaining valuable single plants or hybrids produced through protoplast culture, anther culture, or genetic recombination. This can provide an easy method for clonal propagation of selected plants. However, the chance occurrence of "ploidy chimeras," as the late Professor Neumann called them (Neumann et al. 2009), may be a problem. Recently, molecular characterization of plants produced by tissue culture has been attempted in order to check and monitor clonal fidelity (Kumar 2010; Kumar and Shekhawat 2009). This is of great importance in the case of woody plants and fruit trees with immense economic potential, as well as horticultural crops and medicinal plants. Different molecular markers have been employed to confirm the efficacy of the technique and genetic fidelity of the regenerants in order to make them commercially viable.

REFERENCES

Abdel-Farid, I. B., M. Jahangir, C. Van Den Hondel, H. K. Kim, Y. H. Choi, and R. Verpoorte. 2009. Fungal infection-induced metabolites in *Brassica rapa*. *Plant Sci* 176:608–615.

Abdullah, B. 2001. The use of isozymes as biochemical markers in rice research. *Bull Agro Bio* 4:39–44.

Abraham, E., G. Rigo, G. Szekely, R. Nagy, C. Koncz, and L. Szabados. 2003. Light-dependent induction of proline biosynthesis by abscisic acid and salt stress is inhibited by brassinosteroid in *Arabidopsis*. *Plant Mol Biol* 51:363–372.

Acharya, S. N., S. K. Basu, S. DattaBanik, and R. Prasad. 2010. Genotype X environment interactions and its impact on use of medicinal plants. *Open Nutraceuticals J* 3:47–54.

Agarwal, M., N. Shrivastava, and H. Padh. 2008. Advances in molecular marker techniques and their applications in plant sciences. *Plant Cell Rep* 27:617–631.

Akhter, J., P. Monneveux, S. A. Sabir, M. Y. Ashraf, Z. Lateef, and R. Serraj. 2010. Selection of drought tolerant and high water use efficient rice cultivars through 13c isotope discrimination technique. *Pakistan J Bot* 42:3887–3897.

Akkaya, M., A. Bhagwat, and P. Cregan. 1992. Length polymorphisms of simple sequence repeat DNA in soybean. *Genet Mol Res* 132:1131–1139.

Akopyanz, N., N. Bukanov, T. Westblom, and D. Berg. 1992. PCR-based RFLP analysis of DNA sequence diversity in the gastric pathogen *Helicobacter pylori*. *Nucleic Acid Res* 20:6221–6225.

Alzohairy, A. M. M. M. 2005. *PCR Techniques, in Genetics Department, Faculty of Agriculture*. Zagazig University, Egypt.

Arif, I. A., M. A. Bakir, H. A. Khan et al. 2010. Application of RAPD for molecular characterization of plant species of medicinal value from an arid environment. *Genet Mol Res* 9:2191–2198.

Ashraf, M. and P. J. C. Harris. 2004. Potential biochemical indicators of salinity tolerance in plants. *Plant Sci* 166:3–16.

Badea, C. and S. K. Basu. 2009. The effect of low temperature on metabolism of membrane lipids in plants and associated gene expression. *Plant Omics J* 2:78–84.

Badea, C. and S. K. Basu. 2010. Impact of drought on plant proteome and metabolome. In *Proceedings of the UGC State Level Seminar (India)*. Section II, pp. 104–120.

Balen, B., D. Pavokovic, P. Peharec, and M. Krsnik-Rasol. 2009. Biochemical markers of morphogenesis in long term horseradish (*Armoraciala pathifolia* Gilib.) tissue culture. *Sci Hortic* 119:88–97.

Balen, B., M. Tkalec, and M. Krsnik-Rasol. 2003. Peroxidase activity and isoenzymes in relation to morphogenesis in *Mammillaria gracilis* tissue culture. *Period Biol* 105:281–286.

Bardakci, F. 2001. Random amplified polymorphic DNA (RAPD) markers. *Turkish J Biol* 25(1):2185–2196.

Basu, S. K., M. Dutta, A. Goyal et al. 2010. Is genetically modified crop the answer for the next green revolution? *GM Crops* 1:1–12.

Basu, S. K., M. Dutta, M. Sharma, and A. Kumar. 2011. Haploid production technology in wheat and some selected higher plants. *Aust J Crop Sci* 5:1087–1093.

Bebeli, P. J., Z. Zhou, D. J. Somers, and J. P. Gustafson. 1997. PCR primed with minisatellite core sequences yields DNA fingerprinting probes in wheat. *Theor Appl Genet* 95:276–283.

Beckmann, J. S. 1988. Oligonucleotide polymorphisms: A new tool for genomic genetics. *Nat Biotechnol* 6:1061–1064.

Beckmann, J. and M. Soller. 1990. Toward a unified approach to genetic mapping of eukaryotes based on sequence tagged micro satellite sites. *Biotechnology* 8:930–932.

Bhowmik, P. K. and S. K. Basu. 2008. Current developments, progress, issues and concerns in producing transgenic peas (*Pisum sativum* L.). *Transgenic Plant J* 2:138–150.

Bhowmik, P. K., S. K. Basu, and A. K. M. S. Alam. 2008. Changes in enzyme activities during postharvest deterioration of green asparagus spears. *Open Hort J* 1:1–6.

Bhowmik, P. K., S. K. Basu, and A. Goyal. 2009. *Advances in Biotechnology*. Bentham Science Publishers, Oak Park.

Brow, M. A., M. C. Oldenburg, V. Lyamichev et al. 1996. Differentiation of bacterial 16S rRNA genes and intergenic regions and *Mycobacterium tuberculosis* katG genes by structure-specific endonuclease cleavage. *J Clin Microbiol* 34:3129–3137.

Bussell, J. D., M. Waycott, and J. A. Chappill. 2005. Arbitrarily amplified DNA markers as characters for phylogenetic inference. *Perspect Plant Ecol* 7:3–26.

Caetano-Anollés, G. and P. M. Gresshoff. 1996. Generation of sequence signatures from DNA amplification fingerprints with mini-hairpin and microsatellite primers. *Biotechniques* 20:1044–1057.

Casa, A. M., C. Brouwer, A. Nagel et al. 2000. The MITE family Heartbreaker (Hbr): Molecular markers in maize. *Proc Natl Acad Sci USA* 97:10083–10089.

Catarino, L. M., M. G. Serrano Jr., M. Cavazzana et al. 2001. Classification of trypanosomatids from fruits and seeds using morphological, biochemical and molecular markers revealed several genera among fruit isolates. *FEMS Microbiol Lett* 201:65–72.

Chang, R. Y., L. S. O' donoughue, and T. E. Bureau. 2001. Inter-MITE polymorphisms (IMP): A high throughput transposon-based genome mapping and fingerprinting approach. *Theor Appl Genet* 102:773–781.

Corley-Smith, G. E., C. J. Lim, G. B. Kalmar, and B. P. Brandhorst. 1997. Efficient detection of DNA polymorphisms by fluorescent RAPD analysis. *Biotechniques* 22:690.

de Vicente, M. C. and T. Fulton. 2003. *Using Molecular Marker Technology in Studies on Plant Genetic Diversity*. Illus. Nelly Giraldo. IPGRI, Rome, Italy.

Dewir, Y. H., D. Chakrabarty, M. B. Ali, E. J. Hahn, and K. Y. Paek. 2006. Lipid peroxidation and antioxidant enzyme acitivities of *Euphorbia milli* hyperhydric shoots. *Environ Exp Bot* 58:93–99.

El-Nahas, A. I., H. H. El-Shazly, S. M. Ahmed, and A. A. A. Omran. 2011. Molecular and biochemical markers in some lentil (*Lens culinaris* Medik.) genotypes. *Ann Agric Sci* 56:105–112.

Eujayl, I., M. E. Sorrells, M. Baum, P. Wolters, and W. Powell. 2002. Isolation of EST-derived microsatellite markers for genotyping the A and B genomes of wheat. *Theoret Appl Genet* 104:399–407.

Flavell, A., M. Knox, S. Pearce, and T. Ellis. 1998. Retro transposon-based insertion polymorphisms (RBIP) for high throughput marker analysis. *Plant J* 16:643–650.

Fransz, P. F., N. C. A. Ruijter, and J. H. N. Schel. 1989. Isozymes as biochemical and cytochemical markers in embryogenic callus cultures of maize (*Zea mays* L.). *Plant Cell Rep* 8:67–70.

George, E. F., M. A. Hall, and G. J. Klerk. 2008. *Plant Propagation by Tissue Culture. Volume 1: The Background*. Springer-Verlag, Berlin.

Goyal, A., P. K. Bhowmik, and S. K. Basu. 2009. Minichromosomes: The second generation genetic engineering tool. *Plant Omics J* 2:1–8.

Grodzicker, T., J. Williams, P. Sharp, and J. Sambrook. 1974. Physical mapping of temperature sensitive mutations. *Cold Spring Harbor Symp Quart Biol* 39:439–446.

Gu, W., N. Weeden, J. Yu, and D. Wallace. 1995. Large-scale, cost-effective screening of PCR products in marker-assisted selection applications. *Theor Appl Genet* 91:465–470.

Hatada, I., Y. Hayashizaki, S. Hirotsune, H. Komatsubara, and T. Mukai. 1991. A genome scanning method for higher organisms using restriction sites as landmarks. *Proc Natl Acad Sci USA* 88:397–400.

Holtke, H. J., R. Seibl, J. Burg, K. Muhlegger, and C. Kessler. 2009. Non-radioactive labeling and detection of nucleic acids. II. Optimization of the digoxigenin system. *Biol Chem Hoppe-Seyler* 371:929–938.

Jeffrey, A. 1979. DNA sequence variants in the G gamma A gamma-, delta- and beta-globin genes of man. *Cell* 18:1–10.

Johnson, I. T. 2002. Glucosinolates: Bioavailability and importance to health. *Int J Vitam Nutr Res* 72:26–31.

Jordan, S. A. and P. Humphries. 1994. Single nucleotide polymorphism in exon 2 of the BCP gene on 7q31-q35. *Human Mol Genet* 3:1915.

Kalendar, R., T. Grob, M. Regina, A. Suoniemi, and A. Schulman. 1999. IRAP and REMAP: Two new retrotransposon-based DNA fingerprinting techniques. *Theoret Appl Genet* 98:704–711.

Kawiak, A. and E. Łojkowska. 2004. Application of RAPD in the determination of genetic fidelity in micropropagated *Drosera* plantlets. *In Vitro Cell Dev Biol-Plant* 40:592–595.

Kumar, A. 2010. *Plant Genetic Transformation and Molecular Markers*. Pointer Publishers, Jaipur.

Kumar, A. and S. Roy. 2006. *Plant Biotechnology and Its Applications in Tissue Culture*. I.K. International, New Delhi.

Kumar, A. and S. Roy. 2011. *Plant Tissue Culture and Applied Plant Biotechnology*. Avishkar Publishers, Jaipur.

Kumar, A. and N. S. Shekhawat. 2009. *Plant Tissue Culture and Molecular Markers: Their Role in Improving Crop Productivity*. I.K. International, New Delhi.

Kumar, A. and S. K. Sopory. 2008. *Recent Advances in Plant Biotechnology and Its Applications*. I.K. International, New Delhi.

Kumar, A. and S. K. Sopory. 2010. *Applications of Plant Biotechnology: In Vitro Propagation, Plant Transformation and Secondary Metabolite Production.* I.K. International, New Delhi.

Lait, C. G., S. L. Bates, A. R. Kermode, K. K. Morrissette, and J. H. Borden. 2001. Specific biochemical marker-based techniques for the identification of damage to Douglas-fir seed resulting from feeding by the western conifer seed bug, *Leptoglossus occidentalis* Heidemann (*Hemiptera: Coreidae*). *Insect Biochem Molec* 31:739–746.

Landegren, U., R. Kaiser, J. Sanders, and L. Hood. 1988. DNA diagnostics. Molecular techniques and automation. *Science* 241:1077–1080.

Li, G. and C. F. Quiros. 2001. Sequence-related amplified polymorphism (SRAP), a new marker system based on a simple PCR reaction: Its application to mapping and gene tagging in Brassica. *Theoret Appl Genet* 103:455–461.

Liu, Z. J., P. Li, B. J. Argue, and R. A. Dunham. 1999. Random amplified polymorphic DNA markers: Usefulness for gene mapping and analysis of genetic variation of catfish. *Aquaculture* 174:59–68.

Lusser, M., C. Parisi, D. Plan et al. 2011. *New Plant Breeding Techniques. State-of-the-Art and Prospects for Commercial Development.* EU Publication, Brussels.

Maltais, K. and M. Houde. 2002. A new biochemical marker for aluminum tolerance in plants. *Physiol Plantarum* 115:81–86.

Ménard, R., J. P. Larue, D. Silua, and D. Thouvenot. 1999. Glucosinolates in cauliflower as biochemical markers for resistance against downy mildew. *Phytochemistry* 52:29–35.

Morgante, M. and J. Vogel. 1994. Compound micro satellite primers for the detection of genetic polymorphisms. U.S. Patent Application No. 08/326-456.

Mueller, U. G. and L. L. Wolfenbarger. 1999. AFLP genotyping and fingerprinting. *Trends Ecol Evol* 14(10):389–394.

Nandy, S., P. K. Bhowmik, S. K. Basu, P. K. Sarkar, and A. K. M. S. Alam. 2008. Current role and future implications of genomics and proteomics in jute and allied fiber (JAF) improvement. *Bang J Prog Sci Tech* 6:293–296.

Nandy, S., N. Mandal, S. Mitra, and S. K. Basu. 2007. Current status of biotechnological research on rice (*Oryza sativa* L.). *Ind J Biol Sci* 13:14–26.

Neumann, K. H., A. Kumar, and J. Imani. 2009. *Plant Cell and Tissue Culture—A Tool in Biotechnology: Basics and Application.* Springer Verlag, Berlin.

Nybom, H. and I. V. Bartish. 2000. Effects of life history traits and sampling strategies on genetic diversity estimates obtained with RAPD markers in plants. *Perspect Plant Ecol* 3:93–114.

Ohara, A., Y. Akasaka, H. Daimon, and M. Mii. 2000. Plant regeneration from hairy roots induced by infection with *Agrobacterium rhizogenes* in *Crotalaria juncea* L. *Plant Cell Rep* 19:563–568.

Olsen, M., L. Hood, C. Cantor, and D. Botstein. 1989. A common language for physical mapping of the human genome. *Science* 245:1434–1435.

Orita, M., Y. Suzuki, T. Sekiya, and K. Hayashi. 1989. Rapid and sensitive detection of point mutations and DNA polymorphisms using polymerase chain reaction. *Genomics* 5:874–879.

Paran, I. and R. Michelmore. 1993. Development of reliable PCR-based markers linked to downy mildew resistance genes in lettuce. *Theor Appl Genet* 85:985–993.

Petruska, J., M. J. Hartenstine, and M. F. Goodman. 1998. Analysis of strand slippage in DNA polymerase expansions of CAG/CTG triplet repeats associated with neurodegenerative disease. *J Biol Chem* 273:5204–5210.

Ranade, S. A., N. Farooqui, E. Bhattacharya, and A. Verma. 2001. Gene tagging with random amplified polymorphic DNA (RAPD) markers for molecular breeding in plants. *Crit Rev Plant Sci* 20:251–275.

Richards, R. I. and G. R. Sutherland. 1992. Dynamic mutations: A new class of mutations causing human disease. *Cell* 70:709.

Rodman, J. E., K. G. Karol, R. A. Price, and K. J. Sytsma. 1996. Molecules, morphology, and Dahlgren's expanded order Capparales. *Syst Bot* 21(3): 289–307.

Rohde, W. 1996. Inverse sequence-tagged repeat (ISTR) analysis, a novel and universal PCR-based technique for genome analysis in the plant and animal kingdom. *J Genet Breed* 50:249–262.

Saiki, R., T. Bugawan, G. Horn, K. Mullis, and H. Erlich. 1986. Analysis of enzymatically amplified beta-globin and HLA-DQ alpha DNA with allele-specific oligonucleotide probes. *Nature* 324:163–166.

Semagn, K., A. Bjrnstad, and M. N. Ndjiondjop. 2006. An overview of molecular marker methods for plants. *Afr J Biotechnol* 5:2540–2568.

Sharma, M., A. Sharma, A. Kumar, and S. K. Basu. 2011. Enhancement of secondary metabolites in cultured plant cells through stress stimulus. *Am J Plant Physiol* 6:50–71.

Staniaszek, M. and H. Habdas. 2006. RAPD technique application for intraline evaluation of androgenic carrot plants. *Folia Hort* 18:87–97.

Stasolla, C. and T. A. Thorpe. 2009. Tissue culture: Historical perspectives and applications. In K.-H. Neumann, A. Kumar, and J. Imani (eds), *Principles and Practice: Plant Cell and Tissue Culture—A Tool in Biotechnology Basics and Applications*, pp. 229–255. Springer, Berlin.

Telenius, H., N. Carter, C. Bebb, M. Nordenskjold, B. Ponder, and A. Tunnacliffe. 1992. Degenerate oligonucleotide-primed PCR: General amplification of target DNA by a single degenerate primer. *Genomics* 13:718–725.

Toldi, O., S. Toth, T. Ponyi, and P. Scott. 2002. An effective and reproducible transformation protocol for the model resurrection plant *Craterostigma plantagineum* Hochst. *Plant Cell Rep* 21:63–69.

Uttaro, A. D. and F. R. Opperdoes. 1997. Characterisation of the two malate dehydrogenases from *Phytomonas* sp.: Purification of the glycosomal isoenzyme. *Mol Biochem Parasitol* 89:51–59.

Uttaro, A. D., M. Sanchez-Moreno, and F. R. Opperdoes. 1997. Genus-specific biochemical markers for *Phytomonas* spp. *Mol Biochem Parasitol* 90:337–342.

van der Wurff, A. W. G., Y. L. Chan, N. M. Van Straalen, and J. Schouten. 2000. TE-AFLP: Combining rapidity and robustness in DNA fingerprinting. *Nucleic Acids Res* 28:e105.

Vasil, I. K. 2007. Molecular genetic improvement of cereals: Transgenic wheat (*Triticum aestivum* L.). *Plant Cell Rep* 26:1133–1154.

Venugopal, K., A. Buckley, H. W. Reid, and E. A. Gould. 1992. Nucleotide sequence of the envelope glycoprotein of negishi virus shows very close homology to louping III virus. *Virology* 190:515–521.

Viguera, E., D. Canceill, and S. D. Ehrlich. 2001. Replication slippage involves DNA polymerase pausing and dissociation. *EMBO J* 20:2587–2595.

Vos, P., R. Hogers, M. Bleeker et al. 1995. AFLP: A new technique for DNA fingerprinting. *Nucleic Acids Res* 23:4407–4414.

Waugh, R., K. McLean, A. J. Flavell et al. 1997. Genetic distribution of Bare-1like retrotransposable elements in the barley genome revealed by sequence-specific amplification polymorphisms (S-SAP). *Mol Gen Genet* 253:687–694.

Weder, P. and K. Jargen. 2002. Identification of food and feed legumes by RAPD-PCR. *Lebensm Wiss Technol* 35:504–511.

Weising, K., H. Nybom, K. Wolff, and W. Meyer. 1995. *DNA Fingerprinting in Plants and Fungi*. CRC Press, Boca Raton.

Welsh, J. and M. McClelland. 1990. Fingerprinting genomes using PCR with arbitrary primers. *Nucleic Acids Res* 18:7213–7218.

White, T. L., W. T. Adams, and D. B. Neale. 2007. Genetic markers-morphological, biochemical and molecular markers. In T. L. White, W. T. Adams, and D. B. Neale (eds), *Forest Genetics*, pp. 53–74. CABI Publisher, Wallingford.

Williams, J. G. K., A. R. Kubelik, K. J. Livak, J. A. Rafalski, and S. V. Tingey. 1990. DNA polymorphisms amplified by arbitrary primers are useful as genetic markers. *Nucleic Acids Res* 18:6531–6535.

Wolff, K., E. D. Schoen, and J. P. V. Rijn. 1993. Optimizing the generation of random amplified polymorphic DNAs in chrysanthemum. *Theoret Appl Genet* 86:1033–1037.

Xiong, L. Z., C. G. Xu, M. A. Saghai-Maroof, and Q. Zhang. 1999. Patterns of cytosine methylation in an elite rice hybrid and its parental lines, detected by a methylation-sensitive amplification polymorphism technique. *Mol Gen Genet* 261:439–446.

Xynias, I. N., N. O. Kozub, I. A. Sozinov, and A. A. Sozinov. 2007. Biochemical markers in wheat breeding. *Int J Plant Breed* 1:1–9.

Yamada, A., T. Saitoh, T. Mimura, and Y. Ozeki. 2002. Expression of mangrove allene oxide cyclase enhances salt tolerance in *Escherichia coli*, yeast, and tobacco cells. *Plant Cell Physiol* 43:903–910.

Yazdani, R., I. Scotti, G. Jansson, C. Plomion, and G. Mathur. 2003. Inheritance and diversity of simple sequence repeat (SSR) microsatellite markers in various families of *Picea abies*. *Hereditas* 138:219–227.

Zabeau, M. and Vos, P. 1993. Selective restriction fragment amplification: A general method for DNA fingerprinting. European Patent Application, Publication 0534858-A1, Office Europeen Des Brevets, Paris.

Zietkiewicz, E., A. Rafalski, and D. Labuda. 1994. Genome fingerprinting by simple sequence repeat (SSR)-anchored polymerase chain reaction amplification. *Genomics* 20:176–183.

CHAPTER 7

A Comprehensive Forage Development Model for Advancing the Agricultural and Rural Economy of Pakistan through Integration of Agronomic and Omics Approaches

Mukhtar Ahmed, Muhammad Asif, Muhammad Kausar Nawaz Shah, Arvind H. Hirani, Muhammad Sajad, Fayyaz-ul-Hassan, and Saikat Kumar Basu

CONTENTS

ABBREVIATIONS

ARI	Agricultural Research Institute
FAO	Food and Agriculture Organization
FAOSTAT	Food and Agriculture Organization Corporate Statistical Database
FRI	Fodder Research Institute
GDP	Gross domestic product
IPCC	Intergovernmental Panel on Climate Change
MIG	Management-intensive grazing
NARC	National Agriculture Research Centre
NCFRP	National Coordinated Fodder Research Program
PARC	Pakistan Agricultural Research Council
SCARP	Salinity Control and Reclamation Project

7.1 INTRODUCTION

Agriculture is one of the main contributors to gross domestic product (GDP) in South Asia, and it has demonstrated impressive growth trends in past decades due to the success of the Green Revolution. However, the region is neglected in terms of infrastructure development and lack of growth in industry and agriculture as well as economic and social development. The majority of the local population live in rural areas. South Asia is the poorest region in the world after sub-Saharan Africa, and most of the countries in the region are among the least economically developed. The World Bank has indicated that 40% of the population in South Asia live below the poverty line. Pakistan is the fifth largest economy in the region in terms of GDP (FAO 2012).

The economy of Pakistan is dependent on agriculture, which constitutes the largest sector, providing employment to about 45% of the population. It is also the main source of input for different types of agro-based industries. The demand for industrial products has a strong link with agriculture, and additional supply of different types of products depends on agricultural production. Similarly, in order to deal with the issue of food security, it is important to pay more attention to the economic progress of rural communities. Climatic extremes (increasing global temperatures, enhanced incidences of drought and floods, and rapidly expanding deserts) will possibly impact agriculture, especially in the developing world (IPCC 2007). Pakistan, a developing country, is no exception to such challenges. Food security will be the foremost global issue, due mainly to increased population and climate change, and there is a dire need to mitigate the adverse effects of climate change by breeding drought-tolerant and low input-responsive crops for various agroecological regions of Pakistan.

The plants and crops that are used to feed animals (mostly cattle) are termed *forages*. They may be conserved by production of hay after drying of plants. Silage may be produced by fermentation, and cubes, pellets, and firmed hay can also be produced by compressing dried matter. The animals may graze on a meadow freely, or the crop may be cut and given to them in the form of fermented silage or dry feed (Yamada et al. 2010). These crops ensure a sufficient diet for the animals, even in warm and dry conditions, when there are few food sources. They may be used to make the animals healthier before slaughter or to ensure a sufficient diet to enable working animals to work in the field for longer periods of time (Yamada et al. 2010). The forage crops are able to cope with the challenges of abiotic stresses such as drought and salinity. Hence, considering future vulnerability from climatic fluctuations, forage crops could be an ideal solution for low-input agriculture systems (Dost 1989). In addition, due to the absence of proper grazing lands, domestic animals and livestock populations are grazed in forested areas, resulting in serious environmental disturbances in fragile ecological areas. Cultivation of forages will not only support agricultural growth but could also reduce stress on natural forest resources.

Rangelands could also be used to raise people's living standards by providing sites for growing different annual and perennial forages. In Pakistan, forages are grown in two seasons: Kharif (summer) and Rabi (winter). Among traditional forages, the Kharif fodders include sorghum (*Sorghum bicolor* L.), bajra (*Pennisetum americanum* [L.] Leeke), maize (*Zea mays* L.), sadabahar (*Sorghum* × Sudan grass), mazenta (maize × teosinte), cowpea (*Vigna ungiculata* [L.] Walp.), guar (*Cyamopsis tetragonoloba* [L.] Taub), Sudan grass (*Sorghum sudanenses* L.), and mottgrass (*Pennisetum purpureum* L.). The Rabi fodders are mainly berseem (*Trifolium alexandrinum* L.), clover or shaftal (*Trifolium resupinatum* L.), fenugreek (*Trigonella foenum-graecum* L.), lucerne (*Medicago sativa* L.), and oats (*Avena sativa* L.). However, there are some nonconventional forage crops, including Napier grass (*Pennisetum purpureum* L.), teosinte (*Zea mays* L. subsp. *mexicana*), rice bean (*Vigna umbellate* [Thunb]), sesbania (*Sesbania aegyptiaca* [Poir.] Pers.), and moth bean (*Vigna aconitifolia* [Jacq]), that are recommended for the summer season in Pakistan. Similarly, for the winter season (Rabi), barley (*Hordeum vulgare* L.), mattri (*Lathyrus sativus* L.), common vetch (*Vicia sativa* L.), rye (*Secale cereale* L.), senji or Indian clover (*Melilotus indica* L.), and triticale (*Triticosecale hexa/octo*) could be used as forage crops. Perennial grasses such as rye grass (*Lolium perenne* L.), salt grass (*Leptochloa fusca* [L.] Kunth), Rhodes grass (*Chloris gayana* Kunth), Johnson grass (*Sorghum halepense* [L.] Pers), Bahia grass (*Paspalum notatum* Flügge), and Dallis grass (*Paspalum dilatatum* Poir.) could also serve as potential sources of fodder.

7.2 CLIMATE

Pakistan (latitude 31.34 N and longitude 74.22 E) is located on the Tropic of Cancer. Therefore, it has a continental climate type, characterized by extreme variations in seasonal as well as diurnal temperature ranges. The climatic variability between the north and the south of Pakistan is very high. The climate is very cold at higher altitudes, while on the coastal strip the climate is moderate. Other parts of the country experience very high temperatures during summer, but winters in these areas are very cold. The seasons can be classified into hot, warm, mild, cool, and cold regions based upon temperature. Most of the country receives summer rainfall, called monsoon, the intensity and frequency of which are highly variable. In general, four rainy seasons are reported: winter, premonsoon, monsoon, and postmonsoon rainfall. Based upon rainfall, the region may be divided into arid, semiarid, subhumid, and humid regions (Khan et al. 2010).

7.3 SOILS

The mainly arid climate of Pakistan has a significant effect on soil characteristics. These soils are deficient in water and vegetation, resulting in low organic matter. There are three main types of parent materials from which soils are derived: alluvium, loess, and calcareous rocks. The distinct features of the soil are dependent upon the nature of the parent material and the mode of formation. The soil of Pakistan is classified as pedocal (dry soil with high $CaCO_3$ and low organic matter), which is the characteristic soil of most of the arid and semiarid regions of the world.

7.4 LAND DISTRIBUTION AND AGRICULTURE

The principal natural resources of Pakistan are arable land and water. The greatest portion of the land area is used for agriculture, forestry, and livestock production. The production of these crops depends upon availability of irrigation water and rainfall. The seasonal changes in temperature and great variability in rainfall are the main limiting factors for agricultural production. Pakistan has extensive irrigation systems, and more than 25% of the country's total area is under cultivation. According to the Food and Agriculture Organization Corporate Statistical Database (FAOSTAT 2012), the main use of land is as arable land (26.50%), followed by permanent meadows and pastures (6.49%), forested areas (2.24%), and permanent crops (1.10%). However, 63.66% of the total area is used for other purposes. About 60% of the total area of Pakistan consists of rangeland, which provides a mainstay for livestock. The productivity of rangeland is very low and unsustainable due to poor rainfall, less productive soil, and mismanagement such as overgrazing. Pakistan is one of the largest agricultural producers and suppliers of apricots (fourth in world), cotton (fourth), sugarcane (fifth), oranges (sixth), wheat (eighth), rice (eleventh) and chickpeas (second in world) (FAO 2012; FAOSTAT 2012). Agriculture has provided a significant boost to the local economy in the form of both the green and the gene revolution (Basu et al. 2010). The agricultural sector contributes 21% to the nation's GDP and generates 45% of employment; the livelihood of 60% of the rural population is dependent upon it. The main Kharif crops include rice, sugarcane, cotton, maize, mung, millet, and sorghum, while Rabi crops include wheat, gram, lentil, tobacco, rapeseed, mustard, and barley.

7.5 FORAGES AND LIVESTOCK INTERRELATIONSHIPS

Livestock, which is the largest source of foreign earnings, occupies a prime position in economic development. The livestock sector provides milk, meat, and other by-products, and its growth rate during 2011–2012 was 4% (Government of Pakistan 2012). This sector is largely managed by small-holders and can be considered as an important source of earnings for both poor and small-scale farmers. The livestock sector is also a source of income when crop failure occurs due to extreme climatic events. Livestock production practices in Pakistan are of three types: rural households integrated with the rural economy, livestock rearing on grown fodder, and large herds on commercial farms and rangelands. The contribution of livestock to agricultural value addition is 55.1%, while its contribution to GDP is 11.6% (Government of Pakistan 2012). Since Pakistan is a major producer of milk, it is important to maximize the output by promoting the forage sector. Options available to farmers for feeding their livestock are forages, crop residues, concentrates, and nonconventional feeds. Forage production could be increased by avoiding competition between forage and field crops (Sarwar et al. 2002). The limited availability of land results in lower production of forages and more cash crops, leading to negative impacts on the livestock sector. Therefore, it is necessary and mandatory to introduce new high-yielding forage crops that can be intercropped with the main cash crops so that the quality and quantity of the livestock sector can be improved. Multiple forage options are available, from diversified agriculture to commercial production (Yamada et al. 2010). Since green forages are a main source of milk production, fodders such as maize, berseem, lucerne, and sesbania are important. Similarly, the use of legumes as forages could be beneficial, as they are a good source of protein compared with other forages (Dost et al. 1993, 1994).

Livestock feed can be divided into roughages and concentrates, roughages being feedstuffs with high fiber content. These include green forages and dry roughages. Green forages in Pakistan are produced during both Kharif and Rabi seasons. The Kharif season forages include Sudan grass, sorghum, maize, millet, guar, cowpea, sadabhar, janter (*Sesbania aculeate*), and moth, while Rabi season forages are berseem, lucerne, oat, mustard, rapeseed, and turnip (Mohr et al. 1999). The dry roughages used by livestock are hay, straw, stover, hulls, and silages. Fodder shortage is the major limiting factor for production of livestock, especially during extreme seasons throughout the country. Therefore, growing new high-yielding forage crops will be necessary to obtain high livestock production (Sarwar et al. 2002). Most livestock nutrition comes from fodder, while some comes from grazing and different by-products. The production of green fodder in Pakistan is 53 Mt, from a 2.7 Mha area. The average yield of forage per hectare is 19.4 t, but the country is not yet self-sufficient in the production of livestock and its by-products. Since the nutritional requirements of animals in Pakistan are not properly met, due to deficiency in forage supply, compared with advanced countries, where animal requirements are met by grains and highly nutritive forages (Bulla et al. 1977), forage production could be considered to be an important new aspect for agricultural development in this country.

Fodder production is the major limiting factor for livestock production in Pakistan. Therefore, livestock production is directly related to the supply of good-quality forages. The coordinated effort to boost fodder production in Pakistan, the National Coordinated Fodder Research Program (NCFRP), works under the umbrella of the Pakistan Agricultural Research Council (PARC). It includes coordinated units from Punjab (University of Agriculture, Faisalabad and Fodder Research Institute/ FRI Sargodha), Khyber Pakhtunkwa (Agricultural Research Institute/ARI, D.I. Khan and NWFP Agricultural University, Peshawar), Sindh (ARI, Tandojam), and Baluchistan (ARI, Quetta).

7.6 IMPORTANCE OF ROTATION RESEARCH FOR RAINFED FORAGE CROPS

Crop rotation is the best way to maintain sustainability of the ecosystem, since it provides the soil with nutrients and maintains its physiochemical properties. It is an important tool used for healthy management of agricultural systems and biodiversity, sustainability of ecosystems, and efficient control of several field crop diseases (Howard 1996; Leake 2003). The different crop rotations, having different morphological features (e.g., different rooting patterns), might help to improve the physical properties of soil, helping with water infiltration to the deeper zone and increased microbial diversity. In addition to installing an environmentally friendly system to control weeds, forage rotations could be used, which can reduce the cost of herbicides by 10%–20% (Labrada 2003). The rotation system, having a competitive advantage over weeds, might minimize weed density and avoid weed buildup (Tabachnik and Fidell 1996). Similarly, perennial weeds could be controlled by using a rotation system (Tabachnik and Fidell 1996), since it reduces weed growth and reproduction (Liebman and Robichaux 1990). Legumes and forages have played a very significant role in weed control and nutrient management in earlier research. Forages have the potential to suppress weeds by competing for light and nutrients (Blackshaw et al. 1994). The rotation of alfalfa with wheat might control wild oats like a herbicide. However, the impact of forages in controlling weed populations varies. Earlier research reported that forages greatly reduce the population of wild oats and foxtail, but broadleaf weeds are less affected (Mohr 1996).

The role of leguminous plants as food and fodder has been recognized since ancient times. They are also given preference because of their potential to maintain soil fertility. Legumes have special features: first, they perform symbiotic nitrogen fixation, and, second, they are rich sources of protein. Therefore, they are responsible for maintaining sustainability of different cropping systems, particularly cereal-based rotations (Yamada et al. 2010). The importance of legumes in crop rotation was depicted by Stinner et al. (1992), who concluded that forage legumes are suitable for a large part of the world, which could maintain agricultural and cultural sustainability. Peters et al. (2001) further elaborated the importance of fodder crops in poverty reduction, and concluded that fodder legumes integrated with other crops, trees, or livestock might generate a synergistic effect which could minimize cost of inputs and increase land and resource use efficiency.

7.7 LEGUMES AS AN INTEGRATION MODEL TO MAINTAIN AGRICULTURAL SUSTAINABILITY

In the current scenario of changing climate, in which more emphasis is being placed on a bio-based economy (Fairley 2011), the importance of biomass production has increased manyfold. Therefore, the use of biorefineries to fulfill world demand has been a challenging task. The Leguminosae family members, like chickpea (*Cicer arietinum* L.), soybean (*Glycine max* [L.] Merr.), field pea (*Pisum sativum* L.), and faba bean (*Vicia faba* [L.]); clovers (*Trifolium* spp.) and alfalfa/lucerne (*Medicago sativa L.*); and trees and shrubs belonging to species representing genera such as *Gliricidia, Leucaena, Callinadra, Sesbania,* and *Acacia* might be good sources of biomass (forage) production in a wide range of agroecosystems. These crops could help in the improvement of soil fertility (Crews and Peoples 2004), soil structure improvement (McCallum et al. 2004), impact on edaphic fauna and flora (Osborne et al. 2010), reduction in groundwater pollution due to nitrates, and reduction in salinity levels (Lefroy et al. 2001). Legume-intense pastures and crop farming are common in the rainfed agriculture system of Australia (Kirkegaard et al. 2011), and this needs to be considered as a role model for other countries. Since legumes are good for the ecosystem and society, the integration of legumes with other crops might help to combat problems as well as feeding animals and human beings (Mohr et al. 1999).

Livestock is an essential component of agriculture in Pakistan, since 30–35 million pastoral people are involved in raising livestock (World Bank 2006). In Pakistan, generally two practices are adopted for the production of livestock: domesticating animals in rural areas using crop residues and growing fodder to feed livestock. The main source of economy generation for rural people is either rearing livestock in small farm units or maintaining giant herds in the rangelands. Fodders and crop residues constitute 50% of the animal feed. The remaining need for animal feed is met by unauthorized grazing in forested areas, canal banks, roadsides, rangeland, parks, and wasteland, and from the by-products of other crops (Ali 2012). This contributes about 11.5% of national GDP, and the contribution toward agricultural GDP is about 55%. The population of livestock in Pakistan is around 163.0 million head (Government of Pakistan 2012). The livestock population is growing at the rate of 4.2% annually, with a corresponding increase in feed requirement. The total area of Pakistan under cultivation is 22.54 MHa, and fodder is cultivated on 2.35 Mha only. This is very low in light of the trend of increasing demand for high-quality fodder (Waheed et al. 2007).

The performance of the animals can be maximized by nourishing them with fodder having high nutritional value for both pasture-based grazing systems and confinement feeding. This can be attained by growing genetically recent forage varieties and by improving harvest and storage techniques (Sarwar et al. 2002). The significance of forage management is increasing daily. Many farmhouses have adopted management-intensive grazing (MIG), which facilitates the reduction of cost for the farms as well as increasing production. Progress has been made by using current forage varieties. Due to lack of appropriate natural grazing lands, most of our livestock, particularly buffalo, are fed with cultivated forages (Sarwar et al. 2002). Many dairy animals are kept in confined areas in cities, and the forages need to be brought there. Due to lack of natural pastures, animals are kept in farm buildings or homesteads and fed on concentrates, cultivated fodder, and crop remnants. The practice of rearing dairy animals has been adopted in regions of intensive farming as well as in irrigated areas, with the exception of some arid areas such as Rawalpindi and Islamabad.

Forage quality may be defined as the ability of the forage to generate the preferred animal response. Usually the characteristics of forage can be summarized as:

7.7.1 Bulk

This means the unit mass of forage particles, and is affected by their length and concentration. If the forage is very bulky, its intake rate may be decreased.

7.7.2 High Fiber

The nutritive value of forage is also affected by the fiber content. Generally, forage contains 30% fiber. An increase in fiber will result in lower energy content of the forage.

7.7.3 Proteins

Protein content in the forages varies from variety to variety. At maturity, legume forages may contain 15%–23% crude protein, while in grasses crude protein content depends upon fertilization, but is generally about 8%–18%. Crop remnants, including straw, contain very low amounts (3%–4%) of crude proteins.

7.7.4 Vitamins

Forages contains higher concentrations of fat-soluble vitamins as compared with concentrates. Legumes are one good source of vitamin B.

7.7.5 Minerals

Usually, forages contain higher amounts of minerals, such as calcium, potassium, and trace minerals, as compared with the majority of concentrates that are used as feed for animals. Forages contain lower phosphorus levels than needed for the animals' dietary requirements (Waheed et al. 2007).

7.8 POTENTIAL FORAGE CROPS SPECIFIC FOR THE REGIONAL ECONOMY

7.8.1 Legume Forages

7.8.1.1 *Lucerne/Alfalfa* (Medicago sativa L.)

Alfalfa/lucerne (*Medicago sativa* L.) is called the *Queen of Forages* and is widely cultivated as a perennial legume in temperate zones (Erice et al. 2010). It is considered a good source for intercropping with cereal crops (Skelton and Barrett 2005). Multicropping techniques, such as intercropping of legumes with cereals, are among the most practical ways to increase crop and land-use efficiency (Gao et al. 2010; Li et al. 2006). Previous research related to legume–cereal intercropping demonstrated a higher yield compared with monocultures (Misra et al. 2006). Similarly, economic returns of this intercropping are higher than with monocultures, as reported by Smith and Carter (1998). The production stability of intercropping agroecosystems is higher than with conventional monocultures (Skelton and Barrett 2005). Alfalfa can withstand high as well as low temperatures, but the degree of adaptability varies across different germplasms. It can be grown on a variety of soils, but yields optimal results on well-drained loamy soils (Misra et al. 2006). It is more tolerant than wheat to frost and salt, but its yield decreases with increasing salinity. It has high digestible energy and protein, which makes it an important forage crop. The use of alfalfa in livestock feed makes it a protein-rich diet, eliminating the need for protein supplements to livestock. Similarly, it has high levels of Ca, P, and Mg, which could minimize mineral supplementation costs. Alfalfa can be used as hay, silage, or green chop, showing its versatility and compatibility under different conditions. Being a legume, it can contribute to sustainable crop and livestock production by fixing atmospheric N (Crews and Peoples 2004).

The life cycle of alfalfa is normally up to 4–8 years, depending upon climate and variety, but it can exist for more than 20 years. The height of the plant is up to 1 m, and roots extend more than 15 m. Therefore, alfalfa is highly resilient to multiple stresses, particularly due to drought (Gault et al. 1995). The chemical composition of alfalfa, as reported by Duke (1983), consists of 90.76% dry matter in pellet form, while green forage contains 80% moisture, 5.2 g protein, 0.9 g fat, 3.5 g fiber, and 2.4 g ash per 100 g. Similarly, the whole meal and leaf meal contain 66 and 77 calories, 7.5% and 8% moisture, 16 and 20.4 g protein, 2.5 and 2.6 g fat, 27.3 and 17.1 g fiber, and 9.1 and 11.5 g ash, respectively, per 100 g. The crop is also a valuable source of vitamins, such as A and E. The forage yield of alfalfa is 5–75 metric t/ha/year (8–12 cuttings/year) with annual seed yield of 186–280 kg/ha. Furthermore, it can fix 83–694 kg/ha of N in soil (Duke 1983). The sowing of lucerne requires good seedbed preparation with proper field levelling so that all weeds are destroyed, resulting in a good crop stand. It is important to apply fertilizers such as P, S, and K after sowing. The sowing time needs to be decided based upon local climatic conditions and past experience. In areas where rainfall is high, sowing time needs to be adjusted according to the available rainfall so that irrigation costs can be minimized. A higher seed rate (25–30 kg/ha) was recommended under flood irrigation using the broadcast sowing method. However, for soils low in water, a lower seed rate (15–20 kg/ha) is recommended to give good economic returns. The seed rate used for sowing lucerne in Pakistan is 10–12 kg/ha, with sowing time from mid-October to mid-November and a line spacing of 45 cm to give a maximum yield of 110–125 t/ha. Fertilizer is applied at a ratio of 22:115:00 N:P:K kg/ha. The farmer can cut six or seven times, and lucerne forage remains available throughout the year (Dost 1996).

The forage yield of alfalfa could be improved by using genetic and molecular techniques. Efforts are ongoing to increase disease resistance and cold-hardiness in alfalfa. This includes identification of genes that control important agronomic traits that translate into enhanced forage quality and greater yield. A project like *Medicago truncatula* provides a large amount of sequence information, but limited information about gene regulation and protein products of these genes *in planta*. Similarly, uncertainty exists over the transfer of these genes to alfalfa to give the desired phenotype. Therefore, in future it will be necessary to identify and transfer genes that can improve the agronomic performance of alfalfa. Functional genomics has the potential to increase stress and drought tolerance in alfalfa. However, there is a great need for identification of genes that can perform under a range of environmental conditions (Volenec et al. 2002).

7.8.1.2 Berseem/Egyptian Clover (Trifolium alexandrinum L.)

This is one of the most important fodder crops of the winter season in India, Pakistan, and countries of the Mediterranean region. It is aptly called a "milk multiplier," as it appears to enhance the milk production of most lactating buffaloes and cows (Pathak et al. 1992). It is an annual forage crop that can be cut four to eight times, and its life cycle is from November to April (Malaviya et al. 2012). It can be grown successfully on a fertile, medium- to heavy-textured soil with slightly acidic to neutral pH. It can be cultivated in rainfed areas and requires at least 550 mm average rainfall annually for maximum production (Pathak et al. 1992). Leaflets are 2–3 cm in width and 4–5 cm in length. Application of P fertilizer at the rate of 10 kg/ha is recommended at the time of sowing (Malaviya et al. 2012). Loam to clay–loam soils are preferred for this crop. It is considered a good-quality crop from the nutritional point of view, as it contains 20% crude protein. It also improves the fertility of the soil due to its nitrogen-fixing ability. It is adapted to a wide range of environmental conditions, has the ability to tolerate salinity, and can play a vital role in soil bioremediation (Malaviya et al. 2012).

7.8.1.3 *Persian Clover or Shaftal* (Trifolium resupinatum *L.*)

This is an annual forage crop that gives two to three cuts. It recovers well from mowing. It has three subvarieties and is extensively cultivated in Iran, Afghanistan, and Pakistan. In cold winter regions of Asian countries it serves as a major hay crop for livestock. In irrigated areas it is grown as a forage crop. It is frost tolerant, but higher temperatures increase its growth rate. It prefers neutral to alkaline soils and can withstand high pH (Suttie 1999).

7.8.1.4 *Mungbean* (Vigna radiata *[L.] Wilczek*)

This is considered the major pulse crop of the Kharif season. In rainfed areas millet and sorghum are the crops that compete with the mungbean (Government of Pakistan 2002). The area under cultivation with mungbean has been slowly increasing. Large areas are being currently targeted for cultivation of mungbean as it is very economical. The production can be increased by adopting intercropping practices and by cultivation of mungbean on the area fallowed after the harvesting of wheat. Mungbean is valuable not only to human beings but also to livestock, because it provides 21%–28% protein. However, it is not economical to feed mungbean as a primary source fodder, as it is more expensive than the other available fodders (Khattak et al. 2006).

7.8.1.5 *Fenugreek/Methi* (Trigonella foenum-graecum *L.*)

This is an annual forage legume crop, traditionally grown in the subcontinent predominantly as a spice and aromatic crop (Basu et al. 2008). Fenugreek is a bloat-free forage crop and has been reported to increase milk and meat production in animals (Basu et al. 2009). The crop is also known as an important medicinal herb, due to the presence of steroidal sapogenins and galactomannan, and has been reported to be antidiabetic and hypocholesterolaemic (Acharya et al. 2008). The crop is well suited to semiarid regions and does well under both rainfed and irrigated conditions (Basu et al. 2009). It has traditionally been grown in dry, semiarid regions of the subcontinent and in the Mediterranean region and Australia (Acharya et al. 2008).

7.8.2 Graminaceous Forages

7.8.2.1 *Oat* (Avena sativa *L.*)

This was relatively unimportant in the past few decades as it was grown only in very few farm stations. However, due to the emergence of new and improved varieties, it has become one of the major winter season forage crops, and it is grown in both rainfed and irrigated areas of Pakistan. The highest fodder yield potential recorded in the northern areas of Pakistan was 80–120 t/ha. Subtropical and cool regions favor the production of oat. It is also a multicut crop. If a scarcity of fodder occurs, the oat can act as an immediate source of fodder for livestock within 60–70 days, but after 90–100 days of germination it produces excessive drymatter (Pathak et al. 1992). Oat straw is given more importance than wheat and barley straw as it contains higher amounts of digestible nutrients, easily digestible crude proteins, vitamin B1, fats, and minerals such as iron and phosphorus. The National Agriculture Research Centre (NARC) Pakistan assessed 400 cultivars throughout the country and selected the top 20 for further assessment on the basis of fodder production, dry matter yield potential, and maturity during the Rabi season under different climatic conditions. They have established the most appropriate varieties for the different agroecological zones of Pakistan (Dost 2002).

7.8.2.2 *Sorghum* (Sorghum bicolor *L.*)

This is grown on a large scale in irrigated as well as rainfed areas, and requires higher seed rates if it is sown using a broadcast method. Fresh fodder of sorghum can also be used, but fresh feeding may also cause toxicity. Over the past 20 years, the reputation of hybrid sweet *Sorghum* × Sudan grass has steadily increased, as multiple cuttings can be achieved during the summer season if the recommended dose of fertilizers is applied. It is a harmless feed, with a lower risk of HCN toxicity than conventional sorghum (Dost 2002). In Pakistan, sorghum is a main drought-tolerant forage crop, and meets more than half of the fodder requirements of arid regions of the country. If the sorghum fodder is preserved appropriately, with some enhancement of its protein value, it will maintain livestock in better condition during the whole winter season. Fresh fodder of sorghum contains minerals and crude fat as well as 12% protein, and a very high percentage of carbohydrates (70%). Maximum fodder production can be obtained if it is sown till the end of April. The proper application of fertilizers is essential to ensure maximum yield. The nutritive value and production potential of sorghum fodder can be improved by mixed cropping with crops of the Leguminosae family (Ali 2012).

7.8.2.3 *Forbs*

Forages other than Gramineae (Poaceae) family members are called forbs. The term forb is also used to represent herbaceous broadleaf plants on rangelands. Brassica crops are members of the mustard family and they come into the category of forbs. Forage brassicas, including turnips, swedes, forage rape, kale, and brassica hybrids, have been used for more than 600 years. There are many varieties of brassicas that are good sources of home feed for dairy and beef-producing animals. Forage brassicas are grown during the cool season, producing quality fodder from June to February or March. Forbs are a good source of hay as well as straw, but cows in milk should not be fed more than 30% per day as dry matter (British Seed Houses 2012). These are annual, highly productive crops that are generally ready for grazing 80–150 days after sowing, depending on the varieties. Their succulence, digestibility, and high nutritional values make brassicas a favorite forage. The moisture content may vary from 80% to 90% in roots and shoots of fresh forage, while crude protein ranges from 15% to 25% depending upon the available nitrogen supply. Antiquality constituents such as glucosinolates are disadvantages of these forages, but this can be managed by using genotypes that have a low glucosinolate content.

7.8.3 Types of Brassicas

7.8.3.1 *Kale* (Brassica oleracea *L.*)

The stems and leaves of kale are of prime importance as fodder. Kale can tolerate extremely cold seasons and is highly digestible with high protein content. It is an ideal feed for livestock, particularly as a late-season fodder. There are two kinds of varieties: narrow-stem and stemless varieties. Narrow-stem varieties can potentially grow up to 60 inches in height, with a 2-inch stem. The highest production of kale can be obtained in 150–180 days. Stemless varieties are shorter and mature earlier than the narrow-stem varieties. These varieties attain a maximum height of 25 inches, mature in 90 days, and contain 15%–17% protein (Lemus 2009).

7.8.3.2 Rape (Brassica napus L.)

Forage rape is often sown in late July–August as a fodder crop to meet the nutritional require-ments of sheep. Rape is also sown in late April–May by some farmers for grazing by lambs. In these conditions it can either be grown as a pure crop or be sown along with grass seed. One of the drawbacks of rape is its stiff, woody stem, which discourages lambs from grazing it (Gottstein 2008). The palatability of forage rape will be reduced if it is grazed too early, because at that stage it contains high levels of moisture and nitrate. The excessive grazing of forage rape results in illness or may even lead to the death of cows due to nitrate poisoning. If the animals keep grazing for a long time, they will remove the active growing parts, so the regrowth of the forage will be slower (Fulkerson 2008).

7.8.3.3 Turnip (Brassica rapa L.)

Turnip belongs to the well-known Cruciferae family. It is grown twice a year, with tender, curved or compacted roots for storage during the first year of growth. Turnip rape was grown as a vegetable in past years, but is now mostly grown for fodder purposes. It benefits from speedy growth, even in cold autumns, and it has the ability to resist low temperatures. Generally its growing period lasts from 60 to 80 days. Its cultivation begins from the first week of July to early August. It can be grazed 70–90 days after planting, and delay in grazing may result in attacks by several fungi, which ultimately reduce the yield (Undersander et al. 2012).

7.8.4 Other Forage Crops

7.8.4.1 Safflower (Carthamus tinctorius L.)

This annual, broadleaf crop from the Compositae family is primarily adapted to rainfed and irrigated systems (Rohini and Sankara 2000). The crop has been reported as drought resistant and suitable for dry or semiarid regions (Weiss 1983). However, it is sensitive to excess moisture stress and is also susceptible to a number of fungal diseases. The lack of high-yielding varieties, the spiny nature of cultivars interfering with manual harvesting and threshing, competition from major winter crops, long delay until crop maturity, and poor germination of seeds have been important factors behind the low success of the crop in rural areas (IDRC 1998). Farmers in rainfed areas may grow crops such as safflower rather than other fodder sources such as oat or rapeseed/mustard for their animals. Safflower can be stored as hay or silage. Hay from safflower is best suited to feed for sheep and goats. Feed value and yields of safflower forage are similar to those of oat or alfalfa (Mundel et al. 2004).

7.9 FORAGE PRODUCTION IN SALINITY-IMPACTED AREAS

The main factors responsible for decreasing soil fertility around the world are waterlogging and salinity, which have impacted more than 3% of total agricultural land (FAO 2006). Agriculture in Pakistan is heavily dependent on irrigation, since most of Pakistan is in the arid or semiarid zone. Irrigation water and the Indus Basin irrigation system are the main causes of soil saliniza-tion in Pakistan. Due to poor soil drainage, poor quality of water, and destruction of the dynamic equilibrium between groundwater discharge and recharge, the level of salinity is increasing every year. The Salinity Control and Reclamation Project (SCARP) began work in the 1960s to control salinity by installing tube wells (Aslam and Prathapar 2006). These also have adverse socioeconomic impacts, especially in an agricultural country like Pakistan.

The introduction of forage crops in saline and waterlogged areas might improve the socioeconomic status of the community living there. The fodder crops that could be used in the saline areas of Pakistan as potential sources of reclamation are salt bushes like river saltbush (*Atriplex amnicola* Paul G. Wilson), qualibrush (*A. lentiformis* [Torrey] S. Watson), four-wing saltbush (*A. camescens* [Torr. & Frém.] S. Wats.), gray saltbush (*A. cinerea* Poir.), Mediterranean saltbush (*A. halimus* L.), old man saltbush (*A. nummularia* Lindl.), *A. stocksii* Boiss, wavy leaf saltbush (*A. undulate* [Moq.] D. Dietr.), bladder saltbush (*A. vesicaria* [Heward ex Benth.]), karnal kallar grass (*Leptochloa fusca* [L.] Kunth), saltwater couch (*Paspalum vaginatum* w.), saltgrass (*Puccinellia ciliate* Bor.), and salt grass (*Salicornia bigelovii* Torr., *Distichlis spicata* [L.] Greene).

7.10 MOLECULAR APPROACHES FOR FORAGE IMPROVEMENT

Forage production is equally important as production of other grain and oilseed crops to maintain sustainable agricultural production and productivity. Numerous molecular approaches have been recently deployed for production and productivity improvement in various forage crops that can be utilized in second-generation biofuel production. Recent advances in genome sequencing technologies, molecular marker development, and genome-wide single-nucleotide polymorphism (SNP) scanning techniques have produced vast amounts of insight into genomic information that can be utilized in forage crop improvements. Availability and utilization of genome and expressed sequence tags (EST) sequence information from the public domain for several forage and turf grass species, such as ryegrass (*Lolium* spp.), red clover, and tall fescue (*Festuca* spp.) (Basu and Prasad 2011; Isobe et al. 2013), have opened avenues for new breeding strategies with the additional support of molecular tools. The following molecular approaches have so far been employed in forage crop genetic improvement.

7.10.1 QTL Mapping in Forage Crop Species

The majority of economically important traits in crop plants are controlled by multiple genes or loci that are commonly known as quantitative trait loci (QTL). It is therefore important to identify or map QTL within genomes to understand their effects on particular traits. Several researchers have conducted QTL mapping studies in forage crops for yield, quality, agronomic, and biotic and abiotic stress-related traits. Genomic regions involved in barley forage quality have been mapped through a QTL mapping study in a doubled haploid population derived from the Steptoe × Morex cross. Quality traits such as dry matter digestibility, acid detergent fiber, neutral detergent fiber, crude fiber, crude protein, water-soluble carbohydrates, and ash content were used for the QTL mapping study (Taleei et al. 2009). Similarly, four genomic regions on chromosomes 1, 3, 7, and 8 were identified for forage quality-related traits, protein content, and energy value in the RIL population of *M. truncatula*. These QTL and linked markers can be deployed in marker-assisted selection (MAS) for improvement of protein content (Lagunes and Julier 2013). A QTL mapping study has also been reported for forage quality traits in annual (*Lolium multiflorum* Lam.) and perennial (*Lolium perenne* L.) ryegrass species. In this study, a total of 63 QTL were identified for four forage quality traits in a population developed from interspecific crosses (Xiong et al. 2006). A QTL mapping study was carried out for biomass production in *L. perenne* F_2 population, and three QTL were reported on linkage groups 2, 3, and 7 for total biomass (Anhalt et al. 2009) (Table 7.1).

Table 7.1   QTL Mapping Studies for Forage Quality-Related Traits in Important Forage Crops

Forage Crop	Population Type (Parents)	Traits	Number of QTL Identified	References
Barley	Doubled haploid population (Steptoe × Morex)	Dry matter digestibility, acid detergent fiber, neutral detergent fiber, crude fiber, crude protein	3 major and several minor QTL	Taleei et al. (2009)
Alfalfa	Four RIL populations	Forage quality-related traits, protein content, and energy value	4	Lagunes and Julier (2013)
Annual and perennial ryegrass	F_2 population	Fiber components and crude proteins	63	Xiong et al. (2006)
Perennial ryegrass	F_2 population	Biomass production	3	Anhalt et al. (2009)
Tall fescue	Pseudo-testcross	Morphological, reproductive, and quality traits		Saha et al. (2009)
Red clover	Two full-sib populations (272 × WF1680 and HR × R130)	Plant persistency	10 and 23	Klimenko et al. (2010)
Italian ryegrass	Two-way pseudo-testcross F_1 population	Heading date, plant height, culm weight, culm diameter, culm strength, tiller number, and culm pushing resistance and evaluated lodging scores	33	Inoue et al. (2004)

7.10.2 Microarray and EST Gene Expression Analysis in Forage Crops

The DNA or cDNA microarray approach is useful for the functional characterization of genes and detection of expression patterns of large numbers of genes within a genome. Both oligonucleotide and cDNA microarrays have recently become available for gene expression studies in forages and turf. Suzuki et al. (2002) reported a study on the gene expression patterns of saponin biosynthesis in alfalfa. In addition, DNA microarrays developed for *Festuca* and *Lolium* consisted of 7680 probes. This DNA array offers high-thoughput and low-cost screening of a large number of genetic loci. It can be useful in biparental mapping, association mapping, and physical mapping studies in related forage crop species (Kopecký et al. 2013). The joint Pasture Plant Genomics Program generated high-density spotted cDNA microarrays (5000 spots/array), and is now being established as the principal screening tool for novel ryegrass and clover sequences with unknown functions (Spangernberg et al. 2003). Through the application of EST-based plant microarrays, global gene expression patterns for functional genomics can now be achieved, and further application of functional genes can be comprehensively studied in the case of several forage crops.

A large number of ESTs were generated for perennial ryegrass through single-pass DNA sequencing of randomly selected clones from as many as 29 libraries of cDNA (Sawbridge et al. 2003). Large sets of genomics and transcriptomics sequences would allow the generation of functional molecular markers that can be effectively employed for marker-assisted selection of perennial forage crop improvement.

7.10.3 Genetic Engineering for Forage Crop Improvement

Narrow genetic diversity, self-incompatibility, and the perennial nature of some forage species place restrictions on conventional and molecular breeding approaches that can only be overcome by genetic engineering of crops for specific traits. It is feasible to introduce a gene from any living organism into the plant by transformation. Genetic engineering of plants can be achieved by two approaches: (i) *Agrobacterium*-mediated transformation and (ii) the biolistic method. In forage crops, several studies have been reported for the improvement of cellulose, carbohydrate, and fatty acid composition, which could be beneficial to second-generation biofuel production. A transgenic approach was employed for *in vitro* dry matter digestibility in alfalfa, tall fescue, and perennial ryegrass (Guo et al. 2001; Chen et al. 2003; Tu et al. 2010). Similarly, a transgene technique was used for biotic and abiotic stress tolerance in alfalfa, white clover, perennial ryegrass, and tall fescue (Zhang et al. 2005; Jiang et al. 2009; Xiong et al. 2010). Numerous successful attempts have been reported using transgenic techniques in *in vitro* conditions; however, their application for crop improvement under field conditions is highly restricted by governmental regulation and negative public perception regarding genetically modified crops. Applications of transgenic approaches for forage and turf improvement have mainly concentrated on key target traits such as retained forage quality, disease and pest resistance, nutrient acquisition efficiency, and biomass production. In the near future, increasing application of transgenesis may address possible solutions to scenarios associated with animal health, such as foliar accumulation of proanthocyanidins in transgenic alfalfa, and white clover with "bloat safety" and foliar expression of foot and mouth disease antigens.

Forage production is as important as production of other grain and oilseed crops in maintaining sustainable agricultural production and productivity. Numerous molecular approaches have been recently deployed for production and productivity improvement in various forage crops that can be utilized in second-generation biofuel production. Recent advances in genome sequencing technologies, molecular marker development, and genome-wide SNP scanning techniques have produced vast insight into genomic information that can be utilized in forage crop improvements. The availability and utilization of genome and EST sequence information from the public domain for several forage and turf grass species such as ryegrass (*Lolium* spp.), red clover, and tall fescue (*Festuca* spp.) (Basu and Prasad 2011; Isobe et al. 2013) have opened avenues for new breeding strategies with the additional support of molecular tools.

7.11 CONCLUSION

We conclude that it is important to understand that there are great potential opportunities available for increasing forage production in Pakistan. The lack of availability of quality forage and fodder products has hindered a rapid increase in animal productivity in terms of average annual milk and meat production in the country. With the increase of acreage under forage production, there are definite indicators strongly suggesting an association with the social and economic advancement of the population in rural Pakistan, who are heavily dependent on agriculture. The targeted increase in forage production has twofold benefits: increased agronomic and livestock productivity as well as long-term environmental sustainability. Forage production will certainly reduce the pressure on forested land areas and could introduce minimal-input agricultural activities in fringe areas such as forest fringes, water and salinity-impacted areas, low fertility-level soil areas, and areas with climatic extremes by selecting and adapting the right forage species suitable for specific localities.

The forage production can be increased by adopting measures like (i) awareness among rural communities dependent on agriculture for their livelihood; (ii) scientific management; (iii) proper land evaluation; (iv) development of suitable locally adapted cultivars of target forage species for different agroclimatic zones of Pakistan; (v) training of local farmers; (vi) availability of cheaper bank loans; (vii) access to necessary agrochemicals; (viii) promotion of sustainable agropractices such as low-input agriculture/organic farming, proper storage facilities, and disease control programs; and (ix) implementation of a countrywide forage development program.

It will be important to marry traditional breeding approaches with the new omics-age technologies discussed above to bring in suitable developments for testing new germplasms and adapt them to local climatic and soil variabilities in order to optimize production. We have made a humble attempt to suggest a model for integrated forage development for agriculture in Pakistan utilizing both conventional agricultural practices and new molecular approaches for rapid crop development, germplasm screening, and establishing and releasing new varieties to supplement Pakistani agriculture using a low-input agriculture system to help millions of rural poor escape poverty, help increase income, and promote social development. We will be happy if, in the near future, forage production benefits from the projected drive in local agriculture and helps local rural communities in this region to explore a better quality of life. It will be interesting to observe how enhanced forage production helps in reducing dependence on available forest resources and protects the local environment through reduced grazing by farm animals. We hope that the proposed model for integrated forage development will help grow the rural economy in the near future.

7.12 SUMMARY

Forage development is a neglected area in Pakistan's agriculture, although there is substantial demand for forages within the country as well as in international markets. There is enormous potential for growing forage crops in abandoned and overgrazed marginal land and forest fringe areas in several parts of the country. A model using traditional breeding and agronomic approaches integrated with the tremendous development in the arenas of plant and crop omics could possibly help in developing and promoting the forage crop development platform in Pakistan. This will not only help address the challenges of restricting livestock grazing in prime forested areas of the country, but may also help reduce the economic hardship of many small farmers in resource-poor marginal lands. The model has the potential to develop into an effective tool for broadening the agricultural base and promoting economic opportunities for small farmers in the region while helping to build environmentally friendly and sustainable agricultural practices in Pakistan.

REFERENCES

Acharya, S. N., J. E. Thomas, and S. K. Basu. 2008. Fenugreek (*Trigonella foenum-graecum* L.) an alternative crop for semiarid regions of North America. *Crop Sci* 48:841–853.

Ali, M. A. 2012. Sorghum: A valuable forage crop. *Agri Hunt*. Available at: http://www.agrihunt.com/major-crops/41-sorghum-a-valuable-forage-crop.html (accessed September 17, 2012).

Anhalt, U. C. M., L. S. Heslop-Harrison, H. P. Piepho, S. Byrne, and S. Barth. 2009. Quantitative trait loci mapping for biomass yield traits in a *Lolium* inbred line derived F_2 population. *Euphytica* 170:99–107.

Aslam, M. and S. A. Prathapar. 2006. Strategies to mitigate secondary salinization in the Indus basin of Pakistan: A selective review, Research Report, International Water Management Institute.

Basu, S. K., S. N. Acharya, M. S. Bandara, D. Friebel, and J. E. Thomas. 2009. Effects of genotype and environment on seed and forage yield in fenugreek (*Trigonella foenum-graecum* L.) grown in western Canada. *Aus J Crop Sci* 3:305–314.

Basu, S. K., S. N. Acharya, and J. E. Thomas. 2008. Application of phosphate fertilizer and harvest management for improving fenugreek (*Trigonella foenum-graecum* L.) seed and forage yield in a dark brown soil zone of Canada. *KMITL Sci Technol J* 8:1–7.

Basu, S. K., M. Dutta, A. Goyal et al. 2010. Is genetically modified crop the answer for the next green revolution? *GM Crops* 1:1–12.

Basu, S. K. and R. Prasad. 2011. Trends in new technological approaches for forage improvement. *Aust J Agric Eng* 2:176–185.

Blackshaw, R. E., F. O. Larney, C. W. Lindwall, and G. C. Kozub. 1994. Crop rotation and tillage effects on weed populations on the semi-arid Canadian prairies. *Weed Technol* 8:231–237.

British Seed Houses. 2012. Forage brassicas: Year round cost savings and feeding solutions. British Seed Houses. Available at: www.britishseedhouses.com/files/Fodder_Brassicas_24pp_A5.pdf (accessed November 16, 2012).

Bulla, R. J., V. L. Lichtenberg, and D. A. Holt. 1977. *Potential of the World's Forages for Ruminant Animal Production*. Winrock International Livestock Research and Training Center, Petit Jean Mountain, Morrilton, Arkansas.

Chen, L., C. Auh, P. Dowling et al. 2003. Improved forage digestibility of tall fescue (*Festuca arundinacea*) by transgenic down-regulation of cinnamyl alcohol dehydrogenase. *Plant Biotech J* 1:437–449.

Crews, T. E. and M. B. Peoples. 2004. Legume versus fertilizer sources of nitrogen: Ecological tradeoffs and human needs. *Agric Ecosyst Environ* 102:279–297.

Dost, M. 1989. Country pasture/forage resource profiles. FAO. Available at: http://www.fao.org/ag/AGP/AGPC/doc/Counprof/Pakistan.htm (accessed September 10, 2012).

Dost, M. 1996. Improving fodder in smallholder livestock production in northern Pakistan. World Animal Review 1996/2 Rome. Available at: http://www.fao.org/docrep/W2650T/w2650t08.htm (accessed December 20, 2012).

Dost, M. 2002. Fodder production for peri-urban dairies in Pakistan. FAO. Available at: http://www.fao.org/ag/AGP/AGPC/doc/pasture/dost/fodderdost.htm (accessed September 10, 2012).

Dost, M., A. Hussain, S. Khan, and M. B. Bhatti. 1993. Genotype X Environment interaction in oats and their implications on forage oat breeding programmes in Pakistan. *Pak J Sci Ind Res* 36:22–24.

Dost, M., A. Hussain, S. Khan, and M. B. Bhatti. 1994. Green forage yield, dry matter yield, and chemical composition of oat with advances in maturity. *Pak J Sci Ind Res* 37:198–200.

Duke, J. A. 1983. *Medicago sativa* [L.]. in *Handbook of Energy Crops*. Purdue University. Available at: http://www.hort.purdue.edu/newcrop/duke_energy/Medicago_sativa.html (accessed December 20, 2012).

Erice, G., S. Louahlia, J. J. Irigoyen, M. Sanchez-Diaz, and J. C. Avice. 2010. Biomass partitioning, morphology and water status of four alfalfa genotypes submitted to progressive drought and subsequent recovery. *J Plant Physiol* 167:114–120.

Fairley, P. 2011. Next generation biofuels. *Nature* 474:S2–S5.

FAO (Food and Agricultural Organization). 2006. TerraSTAT database. FAO. Available at: http://www.fao.org/ag/agl/agll/terrastat/ (accessed August 30, 2012).

FAO (Food and Agricultural Organization). 2012. *Summary of the World Food and Agricultural Statistics*. FAO, Rome. Available at: http://faostat.fao.org (accessed August 30, 2012).

FAOSTAT. 2012. FAO database. Available at: http://faostat.fao.org/site/339/default.aspx (accessed September 22, 2012).

Fulkerson WJ. 2008. Growing forage rape in autumn. Future Dairy Tech Note, pp. 1–6. Available at http://www.futuredairy.com.au/documents/TechNoteBrassica2008.pdf (accessed June 7, 2014).

Gao, Y., A. Duan, X. Qiu, Z. Liu, J. Sun, and J. Zhang, and H. Wang. 2010. Distribution of roots and root length density in a maize/soybean strip intercropping system. *Agric Water Manage* 98:199–212.

Gault, P. R., M. B. Peoples, G. L. Turner, D. M. Lilley, J. Brockwell, and F. J. Bergenson. 1995. Nitrogen fixation by irrigated lucerne during the first, third year after establishment. *Aust J Agri Res* 46:1401–1425.

Gottstein, M. 2008. Alternative forage crops: An option to replace meal? *Independent.ie* November 26. Available at: http://www.independent.ie/farming/sheep/alternative-forage-crops-an-option-to-replace-meal-1361718.html (accessed June 30, 2012).

Government of Pakistan. 2002. *Agricultural Statistics of Pakistan 2000–2001*. Ministry of Food, Agriculture and Livestock, Islamabad, Pakistan.

Government of Pakistan. 2012. *Economic Survey of Pakistan 2011–2012, Chapter 2: Agriculture*. Ministry of Finance, Islamabad, Pakistan.

Guo, D. J., F. Chen, J. Wheeler et al. 2001. Improvement of in-rumen digestibility of alfalfa forage by genetic manipulation of lignin O-methyltransferases. *Transgenic Res* 10:457–464.

Howard, R. J. 1996. Cultural control of plant diseases: A historical perspective. *Can J Plant Path* 18:145–150.

IDRC. 1998. *Oilseed Processing—Oilseed Crops of Pakistan*. IDRC, Ottawa, Canada.

Inoue, M., Z. Gao, and H. Cai. 2004. QTL analysis of lodging resistance and related traits in Italian ryegrass (*Lolium multiflorum* Lam.). *Theor Appl Genet* 109:1576–1585.

Intergovernmental Panel on Climate Change (IPCC). 2007. IPCC fourth assessment report: Climate change 2007. IPCC, Geneva.

Isobe, S., B. Boller, I. Klimenko, S. Kolliker, J. C. Rana, and T. T. Sharma. 2013. Genome-wide SNP marker development and QTL identification for genomic selection in red clover. In S. Barth and D. Milbourne (eds), *Breeding Strategies for Sustainable Forage and Turf Grass Improvement*, pp. 29–36. Springer, New York.

Jiang, Q. J., Y. Zhang, X. Guo, M. Monteros, and Z. Y. Wang. 2009. Physiological characterization of transgenic alfalfa (*Medicago sativa*) plants for improved drought tolerance. *Int J Plant Sci* 170:969–978.

Khan, S. U., M. U. Hasan, F. K. Khan et al. 2010. *Climate Classification of Pakistan*. BALWOIS, Ohrid.

Khattak, G. S. S., R. Zamir, and M. S. Khan. 2006. Area planted to MYMV resistant mungbean and future needs in Pakistan. In S. Shanmugasundaram (ed.), *Proceedings of Final Workshop and Planning Meeting of Improving Income and Nutrition by Incorporating Mungbean in Cereal Fallows in the Indo-Gangetic Plains of South Asia DFID Mungbean Project for 2002-2004*, pp. 283–295. Punjab Agricultural University, Ludhiana, Punjab, India, 27-31 May 2004. AVRDC, The World Vegetable Center. AVRDC Publication No. 06–682.

Kirkegaard, J. A., M. B. Peoples, J. E. Angus, and M. J. Unkovich. 2011. Diversity and evolution of rainfed farming systems in Southern Australia. In P. Tow, I. Cooper, I. Partridge, and C. Birch (eds), *Rainfed Farming Systems*, pp. 715–754. Springer, Dordrecht.

Klimenko, I., N. Razgulayeva, M. Gau et al. 2010. Mapping candidate QTLs related to plant persistency in red clover. *Theor Appl Genet* 120:1253–1263.

Kopecký, D., J. Bartos, A. J. Lukaszewski et al. 2013. DArTFest DNA Array—Applications and perspectives for grass genetics, genomics and breeding. In S. Barth and D. Milbourne (eds), *Breeding Strategies for Sustainable Forage and Turf Grass Improvement*, pp. 115–120. Springer, New York.

Labrada, R. 2003. *Weed Management for Developing Countries*. FAO, Rome.

Lagunes, E. L. and B. Julier. 2013. QTL detection for forage quality and stem histology in four connected mapping populations of the model legume *Medicago truncatula*. *Theor Appl Genet* 126:497–509.

Leake, A. R. 2003. Integrated pest management for conservation agriculture. In L. Garcia-Torres, J. Benites, V. A. Martinez, and C. A. Holgado (eds), *Conservation Agriculture: Environment, Farmers Experiences, Innovations, Socio-Economy, Policy*, pp. 271–279. Kluwer Academic, Dordrecht.

Lefroy, E. C., R. J. Stirzaker, and J. S. Pate. 2001. The influence of tagasaste (*Chamaecytisus profliferus*) trees on the water balance of an alley cropping system on deep sand in south-western Australia. *Aust J Agric Res* 52:235–246.

Lemus, R. 2009. Forage brassicas for winter grazing systems. *Forage News*. Mississippi State University. Available at: http://msucares.com/crops/forages/newsletters/09/9.pdf (accessed November 15, 2012).

Li, L., J. Sun, F. Zhang, T. Guo, X. Bao, and F. A. Smith, and S. E. Smith. 2006. Root distribution and interactions between intercropped species. *Oecologia* 147:280–290.

Liebman, M. and R. H. Robichaux. 1990. Competition by barley and pea against mustard: Effects on resource acquisition, photosynthesis and yield. *Agric Ecosyst Environ* 31:155–172.

Malaviya, D. R., A. K. Roy, P. Kaushal, S. N. Tripathi, and S. Natrajan. 2012. Berseem. Indian Grassland and Fodder Research Institute. Available at: http://www.igfri.res.in/ (accessed May 5, 2012).

McCallum, M. H., J. A. Kirkegaard, T. Green, H. P. Cresswell, S. L. Davies, J. F. Angus, and M. B. Peoples. 2004. Improved subsoil macro porosity following perennial pastures. *Aust J Exp Agric* 44:299–307.

Misra, A. K., C. L. Acharya, and A. S. Rao. 2006. Inter specific interaction and nutrient use in soybean/sorghum intercropping system. *Agron J* 98:1097–1108.

Mohr, R. 1996. Entrepreneurship in intra-organizational networks: An entrepreneurial perspective on the management of multi-unit organizational systems. PhD Thesis, University of Manitoba.

Mohr, R. M., M. H. Entz, H. H. Janzen, and W. J. Bullied. 1999. Plant-available nitrogen supply as affected by method and timing of alfalfa termination. *Agron J* 91:622–630.

Mundel, H. H., R. E. Blackshaw, J. R. Byers et al. 2004. *Safflower Production on the Canadian Prairies: Revisited in 2004*. Agriculture and Agri Food, Canada.

Osborne, C. S., M. B. Peoples, and P. H. Janssen. 2010. Exposure of soil to a low concentration of hydrogen elicits a reproducible, single member shift in the bacterial community. *Appl Environ Microbiol* 76:1471–1479.

Pathak, N. N., N. Kewalramani, and D. N. Kamra. 1992. Intake and digestibility of oats (*Avena sativa*) and berseem (*Trifolium alexandrinum*) in adult blackbuck (*Antilope cervicapra*). *Small Ruminant Res* 8:265–268.

Peters, M., P. Horne, A. Schmidt et al. 2001. The role of forages in reducing poverty and degradation of natural resources in tropical production systems. Paper # 117. Agricultural Research and Extension Network, London.

Rohini, V. K. and K. R. Sankara. 2000. Embryo transformation, a practical approach for realizing transgenic plants of safflower (*Carthamus tinctorius* L.). *Ann Bot* 86:1043–1049.

Saha, M., F. Kirigwi, K. Chekhovskiy, J. Black, and A. Hopkins. 2009. Molecular mapping of QTLs associated with important forage traits in tall fescue. In T. Yamada and G. Spangenberg (eds), *Molecular Breeding of Forage and Turf*, pp. 251–258. Springer, New York.

Sarwar, M., M. K. Khan, and Z. Iqbal. 2002. Feed resources for livestock in Pakistan. *Int J Agric Biol* 4:186–192.

Sawbridge, T., E. K. Ong, C. Binnion et al. 2003. Generation and analysis of expressed sequence tags in white clover (*Trifolium repens* L.). *Plant Sci* 165:1077–1087.

Skelton, L. E. and G. W. Barrett. 2005. A comparison of conventional and alternative agro-ecosystems using alfalfa (*Medicago sativa*) and winter wheat (*Triticum aestivum*). *Renew Agr Food Syst* 20:38–47.

Smith, M. A. and P. R. Carter. 1998. Strip intercropping corn and alfalfa. *J Prod Agric* 11:345–352.

Spangernberg, R., M. Emmerling, U. John et al. 2003. Transgenesis and genomics in molecular breeding of temperate pasture grasses and legumes. In I. K. Vasil (ed.), *Plant Biotechnology 2002 and Beyond. Proceedings of the 10th IAPTC7B Congress*, pp. 497–502. Kluwer Academic, Dordrecht.

Stinner, D. H., I. Glick, and B. R. Stinner. 1992. Forage legumes and cultural sustainability: Lessons from history. *Agric Ecosyst Environ* 40:233–248.

Suttie, M. 1999. *Trifolium resupinatum* L. FAO. Available at: http://www.fao.org/ag/AGP/AGPC/doc/Gbase/data/pf000415.htm (accesssed September 10, 2012).

Suzuki, H., L. Achnine, R. Xu, S. P. Matsuda, and R. A. Dixon. 2002. A genomics approach to the early stages of triterpene saponin biosynthesis in *Medicago truncatula*. *Plant J* 32:1033–1048.

Tabachnik, B. G. and L. S. Fidell. 1996. *Using Multivariate Statistics*, 3rd edn. Harper Collins College, New York.

Taleei, A., B. A. Siahsar, and S. A. Peighambari. 2009. QTL mapping for forage quality-related traits in barley. *Biosci Biotech Comm Comp Inform Sci* 57:53–62.

Tu, Y., S. Rochfort, Z. Liu et al. 2010. Functional analyses of caffeic acid O-methyltransferase and cinnamoyl-CoA-reductase genes from perennial ryegrass (*Lolium perenne*). *Plant Cell* 22:3357–3373.

Undersander, D. J., A. R. Kaminski, E. A. Oelke et al. 2012. Turnip. Alternative Field Crops Manual. Available at: http://corn.agronomy.wisc.edu/Crops/Turnip.aspx (accessed November 15, 2012).

Volenec, J. J., S. M. Cunningham, D. M. Haagenson, W. K. Berg, B. C. Joern, and D. W. Wiersma. 2002. Physiological genetics of alfalfa improvement: Past failures, future prospects. *Field Crops Res* 75:97–110.

Waheed, A., A. Hannan, and A. M. Ranjha. 2007. Forage crops for cattle. Available at: www.pakissan.com/english/news/newsDetail.php?newsid=13927 (accessed September 10, 2012).

Weiss, E. A. 1983. *Oil Seed Crops*. Longman, London.

World Bank. 2006. *Pakistan Strategic Country Environmental Assessment*, vol. II. World Bank, Washington, DC.

Xiong, X., V. James, H. Zhang, and F. Altpeter. 2010. Constitutive expression of the barley *HvWRKY38* transcription factor enhances drought tolerance in turf and forage grass (*Paspalum notatum* Flugge). *Mol Breed* 25:419–432.

Xiong, Y., S. Z. Fei, E. C. Brummer, K. J. Moore, R. E. Barker, and G. Jung. 2006. QTL analyses of fiber components and crude protein in an annual perennial ryegrass interspecific hybrid population. *Mol Breed* 18:327–340.

Yamada, T., K. Tamura, X. Wang, and Y. Aoyagi. 2010. Transgenesis and genomics in forage crops. In M. S. Jain and D. S. Brar (eds), *Molecular Techniques in Crop Improvement*, pp. 719–744. Springer, New York.

Zhang, J. Y., C. D. Broeckling, E. B. Blancaflor, M. Sledge, L. W. Sumner, and Z. Y. Wang. 2005. Overexpression of *WXP1*, a putative *Medicago truncatula* AP2 domain-containing transcription factor gene, increases cuticular wax accumulation and enhances drought tolerance in transgenic alfalfa (*Medicago sativa*). *Plant J* 42:689–707.

CHAPTER **8**

New Approaches for Detection of Unique Qualities of Small Fruits

Teodora Dzhambazova, Ilian Badjakov, Ivayla Dincheva, Maria Georgieva, Ivan Tsvetkov, Atanas Pavlov, Andrey Marchev, Kiril Mihalev, Galin Ivanov, Violeta Kondakova, Rossitza Batchvarova, and Atanas Atanassov

CONTENTS

8.1 INTRODUCTION

The omics disciplines applied in the context of nutrition and health have the potential to deliver biomarkers for health and comfort, reveal early indicators of disposition to disease, and discover bioactive, beneficial food components. These technologies are aimed at unraveling the overall expression of genes, proteins, and metabolites in a functionally relevant context, and provide insights into the molecular basis of various fundamental processes involved in growth and development of plants and their environment.

Genomics is an entry point for looking at the other omics sciences. The information in the genes of an organism, its genotype, is largely responsible for the final physical makeup of the organism, referred to as the *phenotype*. The main purpose of the application of genomics is to gain a better understanding of the whole genome of plants. Agronomically important genes may be identified and targeted to produce more nutritious and safe food while at the same time preserving the environment.

Proteomics is known as protein *expression profiling*, whereby the expression of proteins in an organism resulting from a stimulus is identified at a certain time. Proteomics can also be applied to map protein modification in order to determine the difference between a wild type and a genetically modified organism. It is also used to study protein–protein interactions involved in plant defense reactions.

Metabolomics approaches enable the parallel assessment of the levels of a broad range of metabolites, and have been documented to have great value in both phenotyping and diagnostic analyses in plants. These tools have recently been used for evaluation of the natural variance apparent in metabolite composition. Here, we describe exciting progress made in the identification of the genetic determinants of plant chemical composition, focusing on the application of metabolomics strategies and their integration with other high-throughput technologies.

Phytochemical studies of small fruits (raspberry, bilberry, lingonberry, strawberry, and grapevine) can be used for evaluating the level of beneficial polyphenolics in different fruit breeding populations and how the levels of these components are genetically controlled and also influenced by environmental conditions.

In this context, we will evaluate both the technologies for propagation of small fruits by bioreactors and opportunities for processing and storage of fruits to preserve valuable compounds and their quality.

8.2 APPLICATION OF OMICS TECHNOLOGY FOR DISCOVERY OF UNIQUE QUALITIES OF SMALL FRUITS

The rapid developments of omics technology in biological research have a relationship with the development of analytical methods and a new generation of scientific infrastructure. Accumulated knowledge in various fields will help consumers to assess the quality of individual foods and their impact on human health.

Omics technologies are relatively new biomarker discovery tools that can be applied to studying large sets of biological molecules. The number of these technologies is constantly expanding. The field of technology includes genomics, proteomics, and metabolomics. Usually research is divided into genomics (genotyping [focused on the genome sequence] and transcriptomics [focused on genomic expression]) and epigenomics (focused on epigenetic regulation of genome expression) (Vlaanderen et al. 2010).

At the beginning of the twenty-first century, modern technologies opened up new horizons for plant science. Since the beginning of the new millennium, *omics* sciences have been rapidly developed and optimized, generating a large amount of data that could be used for crop improvement. The main technologies—genomics, proteomics, and metabolomics—can give a close-up view of the pathways that link genotype to phenotype.

Grapevine (*Vitis vinifera* L.) is one of the earliest domesticated fruit crops. Due to its economic and historical importance, grape became a subject of the new biotechnology tools such as omics technologies (Di Gaspero and Cattonaro 2010). The main application of these new high-throughput technologies is to help the selection process to improve grape taste and disease resistance of cultivars by exploiting the high genetic diversity of grapevine (Topfer et al. 2009).

Due to its importance, the grape has become one of the model species for perennial fruit crops (Grimplet et al. 2012). Grapevines (*Vitis* spp.) are one of the most ancient plants that still inhabit the Earth. The excavation findings of leaf and seed prints show the existence of *Vitis* species since the Eocene (Kirchheimer 1939), which has enabled the genus to accumulate tremendous genetic diversity over several million years. Across the *Vitis* genus, Eurasian grapevine (*V. vinifera* L.) is the only species grown commercially for consumption as fresh fruit and wine. *Vinifera* species harbors high genetic polymorphism, with a large number of cultivars, estimated between 6,000 and 11,000 (Maul and Eibach 2003). The uncertainty about the exact number of cultivars is due to intense grape propagation and human migration over the millennia, which resulted in numerous synonymies and homonymies around the world. Despite the great diversity of extant cultivars, only a small number of commercial varieties, such as Cabernet Sauvignon, Merlot, Syrah, and Chardonnay, are grown for the wine industry worldwide (This et al. 2006), resulting in a reduction of the overall genetic diversity of cultivated grape. However, thousands of local cultivars still exist and are maintained in germplasm collections (Bacilieri et al. 2010), thus becoming a valuable source of genetic resources for grape breeding through genomic techniques.

Grape breeding is focused on developing new varieties, with an emphasis on combining the fruit quality with disease resistance and environmental tolerance. Concerning the breeding, grapevine is a difficult species because of its long generation time and high level of heterozygosity (Meredith 2001), and, therefore, the conventional breeding programs are complex and time consuming. The high heterozygosity of grape determinates the polygenic nature of many traits of interest (such as mildew resistance), which additionally hampers the breeding process. However, the modern genetic technologies have become very helpful in the determination and targeting of genes that code or regulate certain traits. The complex omics technologies will advance the understanding of gene organization and metabolite pathways of grape.

Other important perennial fruit species belong to the families *Fragaria* (strawberry), *Rubus* (raspberry, blackberry), and *Vaccinium* (blueberry, lingonberry), which are commercially important berry genera worldwide. Generally they are good sources of natural antioxidants, including vitamins, phenols, flavonoids, and endogenous metabolites.

The potential benefit of wild berries is their ability to be included in breeding programs as donors for economically important diseases and also for their use in future metabolic engineering.

8.2.1 Genomics

Basic experience in combination with technological improvements resulted in the development of assays that are able to assess variability in the DNA sequence of many thousands of genes in a single experiment. Knowing the exact sequence and location of all the genes of a given organism is only the first step toward understanding how all the parts of a biological system work together. In this respect, functional genomics is the key approach to transforming quantity into quality. Functional genomics is a general approach toward understanding how the genes of an organism work together by assigning new functions to unknown genes. The tremendous amount of biological diversity in plant systems will allow the identification of novel gene functions. Data from various approaches have to be interconnected and organized into central databases in order to allow easy extraction and comparison of data for meaningful analysis (Holtorf et al. 2002). Plant genomics research has entered the phase of fast functional characterization of all plant genes. For efficient gene function analysis, researchers can choose from a multitude of different methods from plant functional genomics. During recent years it has become increasingly clear, however, that each method has its inherent limitations, and none of them alone is sufficient to assign a function to a gene of interest. If one wants to take full advantage of the available genomic information on plant genes, only the multidisciplinary integrated approach will allow the functional characterization of plant genes.

One of the major steps in accelerating grape breeding involves current genomics technologies for better understanding of the genetic structure of quality and resistance traits in grape (Martínez-Zapater et al. 2010). In the past decades, molecular genetics techniques, especially DNA markers, have changed the face of grape genetics. The implementation of markers such as simple sequence repeats (SSRs) enabled the unambiguous identification of cultivars and their existing synonymies and homonymies (reviewed by Sefc et al. 2001) in different parts of the world, which shed light on the existing number of varieties worldwide. These markers proved to be helpful when elucidating the origin of some of the most economically important wine cultivars—for example, Cabernet Sauvignon was discovered to be a progeny of Cabernet Franc and Sauvignon Blanc (Bowers and Meredith 1997), while Chardonnay, Gamay Noir, and other important varieties occurred from a cross between Pinot Noir and Gouais Blanc (Bowers et al. 1999). The efforts of the international grape science community were focused on a construction of genetic linkage maps using a large collection of molecular markers—random amplified polymorphic DNA (RAPD) markers, SSRs, amplified fragment length polymorphisms (AFLPs), single-nucleotide polymorphisms (SNPs), and BAC end sequences (BES) (Adam-Blondon et al. 2004; Doligez et al. 2006; Grando et al. 2003; Lodhi et al. 1995; Riaz et al. 2004; Vezzulli et al. 2008a,b). At the same time, many of these molecular markers were assigned to several traits of interest in grape breeding, such as berry color (Doligez et al. 2002), muscat flavor (Doligez et al. 2006), seedlessness and berry weight (Cabezas et al. 2006; Doligez et al. 2002; Lahogue et al. 1998; Mejía et al. 2007), flesh development (Fernandez et al. 2006, 2007), fruit yield (Fanizza et al. 2005), fungal disease resistance (Dalbo et al. 2001; Di Gaspero et al. 2007; Donald et al. 2002; Fischer et al. 2004; Pauquet et al. 2001; Welter et al. 2007), and many others, thus associating genotype with phenotype and improving grape breeding by application of marker-assisted selection (MAS). In this way, MAS provides an easier method for screening the grape seedlings for desirable traits and assembling these traits in one genotype, thus accelerating breeding programs. So far, MAS has proven to be applicable for the identification of grapevine candidates that carry certain genes of interest (Adam-Blondon et al. 2001; Akkurt et al. 2012; Cabezas et al. 2006; Fatahi et al. 2003; Karaagac et al. 2012; Mejía and Hinrichsen 2002; Riaz et al. 2009).

Another highly exploited tool in the genomic era is expressed sequence tag (EST) analysis, which is utilized for gene discovery, organization, and genome characterization (Nagaraj et al. 2007). ESTs are considered to be the best method for exploration of the transcriptome, and the abundance of collected data is stored in few databases such as UniGene (Pontius et al. 2003) and DFCI Gene Indices (http://compbio.dfci.harvard.edu/tgi/tgipage.html). EST analysis was widely exploited for more than 20 years prior to the appearance of modern sequencing methods, which are faster and cheaper.

The development of sequencing technologies allowed the next step in modern grape breeding to be accomplished, that is, passing from MAS to genomics-assisted selection. The first approach was made a few years ago, when the genome sequences of two Pinot Noir accessions, the nearly homozygous line PN40024 and the heterozygous clone ENTAV115, were published (Jaillon et al. 2007; Velasco et al. 2007). Knowing the DNA sequence of a certain gene responsible for a desirable trait will allow development of DNA markers to precisely target this gene. The published reference genome sequences aim for better understanding of the molecular base of the grape phenotype. Velasco et al. (2007) found 29,585 genes in total in the grape genome, of which 16,859 genes are species specific. When the grape genome is compared with the sequenced genomes of other plant species (poplar, *Arabidopsis*, and rice), it is revealed that the genes responsible for some of the significant traits of wine, such as aroma and resveratrol content, are expanded by a higher number of gene copies (Jaillon et al. 2007). Because most of the cultivars of European grapevine are susceptible to a diversity of diseases (fungal diseases, Pierce's disease) and pests (*Phylloxera*), the finding that disease-resistance genes comprise a significant part of the grape genome (Velasco et al. 2007) is somewhat surprising. The authors explain this discrepancy by the presence of SNPs in the functional resistance domains, which leads to disease susceptibility in many grape varieties. Nevertheless, resistant varieties exist, and functioning genes have already been discovered. Access to the genome of resistant *vinifera* cultivars and other *Vitis* species, together with molecular markers, could provide a successful tool in the process of introduction of pathogen resistance traits in cultivars without altering grape taste and wine quality.

As we enter the postgenomic era, the need for genetic markers does not diminish, even in the species with fully sequenced genomes (raspberry and wild strawberry). Plant Markers is a genetic marker database that contains a comprehensive predicted molecular marker (Rudd et al. 2005). The availability of genetic markers is fundamental within plant biology and plant breeding. Some genes involved in the formation of volatile compounds in strawberry have been identified using DNA microarrays in combination with targeted analysis of volatile metabolites (Aharoni et al. 2000). Molecular markers are a useful tool for assaying genetic variation and have greatly enhanced the genetic analysis of a wide spectrum of small fruits. The choice of the most appropriate marker system needs to be made on a case-by-case basis, and will depend on many issues, including the availability of technology platforms, costs for marker development, species transferability, information content, and ease of documentation (Varshney et al. 2005).

Breeding in raspberry is a long process, due to a highly heterozygous perennial fruit crop with a relatively long period of juvenility (McNicol et al. 1992). The creation of genetic linkage maps can facilitate the development of diagnostic markers for polygenic traits and the identification of genes controlling complex phenotypes. Molecular marker applications have been reviewed in *Rubus* (Antonius-Klemola 1999) and in the small fruits (Hokanson 2001).

In the diploid strawberry (*Fragaria vesca*) and diploid blueberry (*Vaccinium* spp.), 445 and 950, and 1288 cM-long, respectively, linkage maps based on RAPD markers have been constructed (Davis and Yu 1997; Qu and Hancock 1997; Rowland and Levi 1994). *Rubus idaeus* has the potential to serve as a model species for the Rosaceae, since it is diploid ($2n = 2x = 140$) and has a very small genome (275 Mb).

A wide consortium selected *Fragaria vesca* ($2n = 2x = 14$) for sequencing as a genomic reference for the genus, as it has a small genome (240 Mb) (Shulaev et al. 2010). *Fragaria vesca* offers many other advantages as a reference genomic system for Rosaceae, including a short generation time for a perennial, ease of vegetative propagation, and small herbaceous stature compared with tree species such as apple. These properties render strawberry an attractive analog for testing gene function for all plants in the Rosaceae family.

The availability of a map would provide the basis to locate and hence manipulate quantitative traits in breeding programs. The availability of informative mapped microsatellite markers, including functional EST-derived SSRs, will also allow selective genotyping of the pedigrees within raspberry breeding programs. The dominant (AFLPs) and co-dominant (SSR) markers have been developed, providing a long-term resource for breeding. The utility of the map is demonstrated by easy mapping and labeling of morphological characteristics of commercial interest. Access to mapped markers will allow new approaches for breeding of complex traits that are difficult to manipulate in breeding programs (Graham et al. 2004).

8.2.2 Proteomics

The proteome consists of all proteins present in specific cell types or tissue. In contrast to the genome, the proteome is highly variable over time, between cell types, and in response to changes in its environment (Fliser et al. 2007). The function of cells can be described by the proteins that are present in the intra- and intercellular space and the abundance of these proteins (Sellers and Yates 2003). One focus of proteomics is quantification of the protein abundance. Protein expression levels represent the balance between translation and degradation of proteins in cells. A direct tool for assessment of protein function is studying the interaction partner of the protein of interest.

Studying grape and wine proteins using the methods of proteomics gives important knowledge about the biological transformations that affect plant and fruit development and winemaking. Moreover, the proteins play a crucial role in inducing resistance and tolerance to biotic and abiotic stress. The first approach to study the proteomic mechanisms of response to pathogens, specifically to Pierce's disease, was made by Basha et al. (2010). The proteomic viewpoint of the grape response to abiotic stress induced by high salinity, water deficit, or herbicide treatment is reported by a number of studies (Castro et al. 2005; Grimplet et al. 2009a; Jellouli et al. 2008; Vincent et al. 2007). Analyzing protein biosynthesis in grape could provide the necessary information to control the plant's response to different factors, which is marked by production of important metabolites that influence the wine composition. As an addition, some of the berry proteins do not ferment and tend to precipitate during winemaking, thus negatively affecting the quality of the wine (Bayly and Berg 1967; Lee 1985). However, wine proteins have been associated with positive features as well, especially in sparkling wines, where they promote foam formation (Brissonnet and Maujean 1993; Cilindre et al. 2007). Some of the peptides exhibit sensory properties and contribute to the organoleptic properties of wine, which is another reason to study grape and wine proteomics. Although this area of grape research is still developing, there is an increasing number of papers that discuss the peptide and protein content, analyzed with various methods, such as 2-D gel electrophoresis and immunoblotting, or mass spectrometry (MS) in combination with liquid chromatography (LC), electrospray ionization (ESI), or matrix-assisted laser desorption/ionization time-of-flight (MALDI-TOF) (Flamini and De Rosso 2006; Marangon et al. 2009; Moreno-Arribas et al. 2002; Nunes-Miranda et al. 2012; Wigand et al. 2009; and many others). A comprehensive review of studies on the proteomics of grape and wine in recent years was presented by Giribaldi and Giuffrida (2010).

The increasing knowledge about the grape genome, metabolome, and proteome due to the development of omics technologies has consolidated grapevine as a model plant. In their study, Grimplet et al. (2009a) made the first approach to integrate the grapevine omics data into a system of molecular networks, VitisNet, which will improve the understanding of structure and regulation molecular pathways. The authors have carried out exhaustive work to assemble 39,424 unique genomic sequences, most of which they have assigned to molecular networks: that is, 88 metabolic, 80 transcription factors, 21 transport, 15 genetic information-processing, 12 environmental information-processing, and three cellular process pathways. The integration of omics data into a coherent framework is necessary to gain a better understanding of grape biology and physiology and to incorporate this information into breeding programs to help improve grape and wine quality.

Fruit development and ripening are key processes in the production of the phytonutrients that are essential for a balanced diet and for disease prevention. The pathways involved in these processes are unique to plants and vary between species (Dincheva et al. 2013). The different species share common pathways; development programs, physiological, anatomical, and biochemical composition, and structural differences must contribute to the operation of unique pathways, genes, and proteins.

There are very few reports that describe protein and polypeptide composition in strawberry (Folta and Davis 2006). Proteomic investigation, together with future studies in the field of Rosaceae molecular biology, will certainly enable a deeper understanding of strawberry fruit metabolism and its regulation during ripening (Bianco et al. 2009). The roles of some of the identified proteins are discussed in relation to strawberry fruit ripening and to quality traits. This is one of the first characterizations of the strawberry fruit proteome and the time course of variation during maturation by using multiple approaches.

Small fruits are one of the most popular sources of vitamins and dietary beneficial compounds, and are appreciated worldwide for their unique flavor. Their favorable nutritional and taste features are closely related to the fruit ripening process. The comprehension of genetic regulatory elements is central to a full understanding of fruit ripening. Unfortunately, the small fruits have suffered from a dearth of molecular-genetic studies as compared with other fruit crops. The lack of species-specific sequences in the protein database has long discouraged proteomic studies.

8.2.3 Metabolomics

Metabolomics describes recent high-throughput approaches in the field of metabolic genomics that aim to identify gene function on the basis of analyzing the metabolome, the full complement of metabolites of an organism (Badjakov et al. 2011).

Metabolic phenotypes are the by-products that result from the interaction between genetic, environmental, lifestyle, and other factors (Holmes et al. 2008) Metabolomics, one of the most important parts of general omics technology, help us to assess better all the components that are present in our food, how they are synthesized, and the genetic and environmental factors that influence food composition and stability. Applications of metabolomic technologies in both applied and fundamental science strategies are therefore growing rapidly in popularity (Hall et al. 2008a). Natural metabolomic diversity and lack of unifying principles to help us detect and identify compounds will be major analytical challenges for many years to come. It is clear that metabolomics has quickly gained its place in modern plant science in both a fundamental and an applied context. Using a metabolomics approach to identify key loci will play a significant role in helping to deconstruct complex metabolic interactions and provide the knowledge to design better crops to feed the world (Baxter and Borevitz 2006).

The taste of grapes and wine is determined by their metabolite profile—the content of sugar, organic acids, anthocyanin pigments, and tannins, among others—which makes metabolite analysis an important part of wine production. Moreover, some of the secondary metabolites in plant are responsible for induction of defense pathways as a response to pathogens, while those in wine are thought to be beneficial for human health (Dzhambazova et al. 2011). Hence, the development of metabolomics is an important part of grape research.

Modern metabolite analyses are performed mainly using gas chromatography–mass spectrometry (GC–MS), high-performance liquid chromatography (HPLC), or nuclear magnetic resonance (NMR). Although research into the grape metabolome is still poor, it has rapidly evolved during the last few years, which is reflected in the increasing number of published studies. Most of the studies investigate the metabolites responsible for certain traits of interest. The study of Figueiredo et al. (2008) analyzed the possibility that metabolites in grape leaves might determine innate resistance against fungal pathogens. The leaf metabolite fingerprint of the resistant cv. Regent, obtained by NMR, revealed a significant content of caffeic acid and inositol, with the first compound reported to inhibit several pathogenic fungi (Harrison et al. 2003; Widmer and Laurent 2006) and the second to determine rapid response to fungal attacks (Figueiredo et al. 2008). Subsequent studies performed with NMR reported a few additional compounds, namely quercetin glucoside, *trans*-fertaric acid, *trans*-caftaric acid, alanine, and α-linolenic acid, that are responsible for the resistance of some grape cultivars (Ali et al. 2009, 2012; Lima et al. 2010). The reported results showed that fungal attacks activate different metabolite pathways in resistant and susceptible grape cultivars, and that metabolite analyses could be used to determinate the resistance ability of varieties. However, more studies in this area should be performed.

Another line of research into the grape metabolome follows the biochemical events that lead to ripening of grape berries. Most studies present a combined comprehensive analysis of grape metabolome and transcriptome, which gives a better understanding of biochemical pathways that lead to the phenotype differences between cultivars (Deluc et al. 2007; Tornielli et al. 2012). The published works present a good survey of carbohydrate, amino acid, and phenylpropanoid metabolisms during ripening. Especially for wine cultivars, it is very important to elucidate flavonoid and terpenoid metabolisms, which are responsible for the subsequent organoleptic characteristics of wine. This kind of study has been performed for Cabernet Sauvignon (Deluc et al. 2007, 2009; Grimplet et al. 2009b), Trincadeira—a Portuguese wine cultivar (Fortes et al. 2011), Chardonnay (Deluc et al. 2009), and Corvina—an Italian wine cultivar (Toffali et al. 2011).

The other important task is to study the metabolite changes that originate during winemaking. The obtained data will give information about which winemaking technique leads to better metabolite composition of the final product. For example, it is known that microoxygenation of wine increases volatile and nonvolatile compounds; it also influences the concentration of secondary metabolites (Arapitsas et al. 2012). Additionally, Arapitsas et al. (2012) reported that the concentration of not only secondary metabolites (tannins and pigments), but also primary metabolites (e.g., arginine, proline, tryptophan, and raffinose), is subject to dynamic change due to microoxygenation. The study of Son et al. (2009) revealed that the metabolite profile of wines produced from grapes with different geographical origin strongly depends on the clime-environmental conditions, especially on duration of sun exposure and precipitation. The metabolite analysis of high-quality white wines (Ali et al. 2011) revealed metabolites responsible for the superior taste of these wines, that is, a precise combination of 2,3-butanediol, malate, proline, gamma-aminobutyric acid (GABA) and wine phenolics, concentrations of which depend on the grape variety, terroir, vintage, and yeast strain and could be used for differentiation of wines. All these metabolomics studies elucidate the factors that influence wine quality and provide a better understanding of how to make wines of good quality.

The development of metabolomics technology provides a great opportunity for small fruits—strawberry, raspberry, and blueberry (red and black)—to be important components of a healthy diet because of their range of bioactive compounds (Garzón et al. 2009). Their high content of phenolic components contributes to protection against degenerative diseases, and their effects on health have been commonly attributed to their antioxidant properties (Seeram 2008). The content of phenolic compounds in berry fruits is determined by many factors: cultivars, climatic factors, ripening, harvesting time, and storage conditions (Castrejon et al. 2008). Obtaining basic data on their bioavailability is very important step in the effort to understand their possible mechanisms of action and their impact on human health (Szajdek and Borowska 2008). Many researchers have now proved the high content and extreme diversity of bioactive compounds in small berry fruits.

There is an interesting fact regarding small fruits that has fundamental importance for human health and for agronomic trials: berry fruits that grow in a cold climate without fertilizers and pesticides are characterized by a higher content of polyphenols than those growing in a milder climate (Szajdek and Borowska 2008). An excessive content of polyphenols, in particular tannins, may have adverse consequences because it inhibits the bioavailability of iron (House 1999) and thiamine (Veloz-García et al. 2010).

Anthocyanins are an important group of polyphenols in berry fruits. A high concentration of anthocyanins is reported in many small berries as representatives of family Rosaceae. Anthocyanins comprise aglycones (anthocyanidins) and their glycosides (anthocyanins) (Viskelis et al. 2009). They form a highly differentiated group of compounds. In berry fruits, anthocyanins are found in the form of mono-,di- or triglycosides, where glycoside residues are usually substituted at C_3 or, less frequently, at C_5 or C_7.

Ellagic acid is the predominant acid in strawberries, accounting for 51% of all acids found in this fruit (Aiyer et al. 2008). The total content of ellagic acid determined after acid hydrolysis ranged from 25.01 to 56.35 mg/100 g fresh weight (Gülçin et al. 2011). Ellagic acid is the predominant phenolic acid in raspberries, where it accounts for 88% of all phenolic acids. Significant amounts of p-coumaric acid and ferukic acid have been reported in blackcurrant (Häkkinen and Törrönen 2000).

Another important group in berries is tannins. They play an essential role in defining the sensory properties of fresh fruit and derived products. They are responsible for the tart taste and changes in the color of the fruit and fruit juice. Hydrolysable tannins are less frequently encountered and have been found in strawberries, raspberries, and blackberries (Holt et al. 2008).

Water deficit is one of the main factors affecting the content and stability of bioactive compounds. It exerts a significant effect on berry composition, promoting an improvement of quality traits such as color, flavor, and aroma.

Matthews et al. (1990) showed that the growth of berries and the concentration of flavonoids in fruit were inhibited when water deficits were imposed before version. Based on the observation of similar flavonoid content among small berries after different treatments, Kennedy et al. (2002) concluded that postvéraison water deficits only inhibited fruit growth. Other authors analyzed the effects of berry size on flavonoid concentration to show that there are effects on fruit composition that are independent of the inhibition of berry growth (Castellarin et al. 2007).

Different techniques can be used to characterize and quantify bioactive compounds in berries, although by far the most widely employed technique has been HPLC coupled with ultraviolet/visible (UV/Vis) and MS detection instruments (Escarpa and Gonzalez 2001; Vazquez-Cruz et al. 2012).

Several studies employ different methods of extraction, purification, and characterization of the phenolic compounds in berries (Naczk et al. 2004). Isolation of bioactive compounds from a sample matrix is generally a prerequisite for any comprehensive analytical scheme. Solid-phase extraction (SPE) techniques and fractionation based on acidity are commonly used to remove undesirable phenolic and nonphenolic substances or volatile organic compounds (VOCs).

The chemical profile of phenolic compounds from strawberries has been studied using liquid chromatography electrospray ionization mass spectrometry (LC-ESI-MS) methods. Since all phenolics contain at least one aromatic ring, they can absorb UV light effectively, and each class of phenolics has a distinctive UV or UV-Vis spectrum (Côté et al. 2010). The MS detection method is generally used to determine molecular masses and to establish the distribution of substituents on the phenolic rings (Harnly et al. 2006).

Strawberries contain a wide variety of phenolics, such as ellagic acid, ellagitannins, gallotannins, anthocyanins, and flavonols (Zhang et al. 2008). The correlation between qualitative and quantitative composition of some berry fruits has been well characterized and included among the different traits that describe the fruit quality (Zabetakis and Holden 1997). The different amounts of sugars and organic acids largely contribute to taste perception. Berry fruits generally accumulate sugars (glucose and fructose). Sugar level is a very important component for consumers of berries, and its level depends on various factors: maturity stage and genetic and environmental factors (Moises et al. 2012). Analyzing the relationship between sugar and acid levels of berries would help in developing optimum quality standards for harvesting dates.

As we mentioned above, the antioxidative and antimicrobial activities of small fruits depend on cultivar, growth conditions, and storage of raw material. The composition of pigments may also differ, depending on the same factors and the method of extraction of bioactive compounds. Some anthocyanins are more stable than others, depending on their molecular structure. The stability of anthocyanin pigments depends on many factors, such as structure, concentration, pH, temperature, light intensity, and presence of another pigment, along with metal ions, enzymes, oxygen, ascorbic acid, and sugars (Mazza and Miniati 1993).

During recent years, a great amount of knowledge has been accumulated describing the relationship between berry phytochemicals and human health (Beattie et al. 2005). Progress in research into the potential health benefits of berries continues in the postgenomic era; as a consequence, there will be increasing demand for observation and characterization of variations within biological systems. Research focusing on nutrigenomics (effects of nutrients on the genome, proteome, and metabolomics) and the effects of genetic variation on the interaction between diet and disease will be essential (Hall et al. 2008b). The biosynthetic capacity of plants in relation to their secondary metabolites should be evaluated in a more careful way, especially concerning their outstanding functional potential.

Plant metabolites represent tremendous chemical diversity. Each plant has its own specific metabolites. The advances in metabolomics and its integration into other omics—genomics, transcriptomics, and proteomics—have brought us closer to understanding the links between different levels of biological systems, leading to the realization of systems biology (Fernie et al. 2004). Today we need to understand the role of metabolites, many of which are involved in adaptations to specific ecological niches and some of which find beneficial use as pharmaceuticals. Another aspect will be to understand the relationship between different omics technologies.

8.2.4 Nutrigenomics

The amassed knowledge and methodology has been used for the creation of the new field of *nutrigenomics*. This is also a rapidly developing field, which has been designed to offer the prospect of a greater understanding of the relationship between human genome and diet (Davies 2007). Nutrigenomics is the future of nutritional science. It has significant impacts on society—from medicine, to agricultural and dietary practices, to social and public policies.

The omics technologies—genomics and metabolomics—are the basis for the rapid development of the new interdisciplinary field of nutrigenomics.

Anthocyanins are water-soluble plant pigments that belong to the large group of polyphenols and, more specifically, to the subclass of flavonoids. They are abundant in the human diet due to their wide occurrence in fruits, such as berries, and fruit-based beverages (Clifford 2000). Bilberry (*Vaccinium myrtillus* L.) is one of the richest sources of anthocyanins, containing 300–600 mg/100 g fresh weight (Lätti et al. 2007). Dietary intake of anthocyanin-rich foods has been associated with a reduced risk of coronary heart disease in the Iowa Women's Health Study. Increasing evidence supports an effect of berry anthocyanin on vascular protection through reduced lipid peroxidation. The overall results suggest that berry anthocyanins may also prevent atherosclerosis by reducing the release of proinflammatory mediators. The results allow new hypotheses to be formulated on the mechanisms of action of anthocyanins in the prevention of atherosclerosis. Further work is required to evaluate whether the observed changes in mRNA levels are translated into biochemical and physiological processes relevant for protection against atherosclerosis (Mauray et al. 2010).

Nutrigenomics is being applied to agriculture (plants and animal food sources) and to human health. The grand challenge is to develop an overarching and integrated framework for thinking about how gene–nutrient interactions influence metabolism. To address this challenge, nutrigenomics has to build a cohort of multidisciplinary scientists who can talk each other's language (Zeisel 2010).

Nutrigenomics is a very active field of research, and clinical studies are ongoing. Many scientists believe that nutrigenomics has tremendous potential for improving public health. The application of genomics technologies in food technology will reduce research and development times, thereby reducing costs and shortening time to market.

In the future, berry research should be focused on gene–nutrient interactions and health outcomes to achieve individual dietary intervention strategies directed at preventing human chronic diseases, improving life, and promoting healthy aging. The accumulation of knowledge, the carrying out of great numbers of research experiments, and practical methodology are the basis for development of different omics technologies.

The quality of small fruits depends on several influences, and one of the most essential production factors is planting material. For this purpose, the European Union has formulated a uniform Community certification scheme that aims to ensure varietal identity, purity, and health status. The basis of any certification scheme is the establishment of a foundation or prebasic block of berry varieties, clones, or accessions. One way to achieve this is through selection of true-to-type and disease-free planting material using different techniques for *in vitro* propagation as well as the development of suitable processing technologies for preserving the fruit quality.

In the long run, the wild biodiversity of small fruits is expected to diminish significantly. By imitating or mimicking the natural conditions, one could develop, preserve, and even produce the relevant natural substances that have an effect on human health. Such approaches could assist in preserving the natural biodiversity.

8.3 DIFFERENT APPROACHES FOR *IN VITRO* PROPAGATION OF SMALL FRUITS

Propagation of plants in *in vitro* conditions usually involves periodic transfers of plant material to fresh media due to exhaustion of the nutrients in the medium and also because of volume limitations of the culture devices used for the purpose. Traditional approaches to micropropagation, mainly based on cultivation on solid media, lead to high production costs, and therefore the commercial use of this technology is limited (Berthouly and Etienne 2005). Labor costs are generally between 40% and 60% of total production costs (Afreen 2006; Hanhineva and Karenlampi 2007). Other major limitations are acclimatization and stem and root hyperhydricity. It has been concluded for various species that extensive expansion of micropropagation would only take place if new technologies became available to automate procedures, and if acclimatization protocols were improved (Berthouly and Etienne 2005; Debnath 2009).

Using liquid media in micropropagation processes is considered to be the prospective solution for reducing plantlet production costs and for enabling automation. Liquid culture systems provide uniform culturing conditions; the media can easily be replaced without changing the cultivation apparatus (Berthouly and Etienne 2005).

Bioreactors developed in the past are not suitable for micropropagation; they are mainly designed for bacterial culture and do not take into account the specific requirements of plant cells. The advantages of *in vitro* culture in a liquid medium are therefore often counterbalanced by technical problems such as hyperhydricity, shear stress, and so on. To solve such problems, a number of temporary immersion techniques have been developed. It has been clearly outlined that plant tissues prefer temporary contact with liquid medium to permanent immersion (Afreen 2006; Berthouly and Etienne 2005).

Besides reduction of hyperhydricity, temporary immersion cultivation systems may also reduce shear stress and improve gas exchange, as well as overcoming the liquid–solid mass transfer limitations.

8.3.1 Case Study

Influence of immersion frequency on *in vitro* biomass multiplication and secondary metabolism of cowberry (*Vaccinium vitis-idaea* L.) cultivated in a temporary immersion system, RITA® (Figure 8.1).

In vitro explants of *Vaccinium vitis-idaea* L. were obtained two years ago by placing fresh leaf material on solid basal woody plant medium (WPM) with vitamin C (McCown and Lloyd 1981). The medium was supplemented with 3 mg/l zeatin, 2 mg/l 2ip, and 20 g/l sucrose, and pH was 4.2. The *in vitro* plantlets were subcultivated on the same medium three times, and the reported coefficient of multiplication was 7. This coefficient varies from 3.4 to 7 depending on the geographical location of lingonberry (unpublished results). Based on intensification of the propagation process experiment, the influence of immersion frequency on biomass multiplication and polyphenol biosynthesis during *V. vitis-idaea* L. cultivation in a temporary immersion RITA system was investigated. For this purpose, three regimes of immersion were tested: 15 immersion periods followed by 4, 8, and 12 h standby periods.

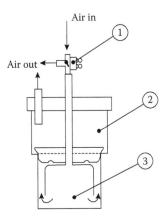

Figure 8.1 Schematic diagram of temporary immersion RITA system: (1) three-way solenoid valve; (2) culture chamber; (3) medium compartment.

The adaptation of lingonberry plantlets to temporary immersion conditions of cultivation was easy. They grew fast, with an intensive rate of proliferation and no morphological changes, and by the end of cultivation they filled the whole vessel (Figure 8.2).

The highest amounts of biomass (0.86 g/RITA) of lingonberry plantlets accumulated at the end of their cultivation in a temporary immersion RITA system at 15 min immersion periods followed by 4 h standby periods (Figure 8.3). The maximum growth index was also achieved in these conditions. During the cultivation of plantlets in the other tested regimes, observed accumulated dry biomass and growth indexes were significantly lower (0.5 and 1.55, and 0.5 and 0.95 for cultivations with 15 min immersion periods and 8 and 12 h standby periods, respectively). It should be emphasized that the plantlets formed during cultivation under temporary immersion conditions were longer that those obtained by cultivation on solid medium.

Because of the specific morphology of plantlets growing *in vitro*, nutrient availability during their cultivation in bioreactor systems is one of the most important factors that influence biomass accumulation and secondary metabolite production (Steingroewer et al. 2013). In contrast to cultivation on solid media or in conventional bioreactor systems, in the temporary immersion RITA systems the plantlets are in contact with the medium for only a short period of time; therefore, ensuring nutrient availability is a critical point for the success of the micropropagation process. From this perspective, we investigated the catabolism of the main nutrients of the medium during the cultivation of *Vaccinium vitis-idaea* L. plantlets in a temporary immersion RITA system. No significant differences were observed in the remaining amounts of the main macronutrients in the medium at the end of plantlet cultivation. Phosphate ions were almost completely exhausted; by the

Figure 8.2 (See color insert) *Vaccinium vitis-idaea* L. plantlets at the end of their growth cycle in a RITA temporary immersion system.

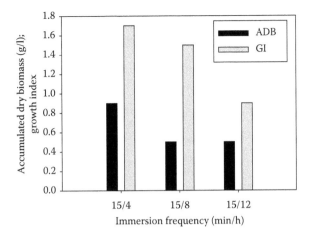

Figure 8.3 Maximal amounts of accumulated dry biomass (ADB) and growth indexes (GI) at the end of cultivation of *Vaccinium vitis-idaea* L. plantlets in a RITA temporary immersion system.

end of cultivation about 2% of phosphate ions remained unutilized, while the remaining amounts of ammonium and nitrate ions were about 25% and 50%, respectively, of initial. These results clearly showed that the temporary immersion RITA system, in the regimes investigated, fully provided the main inorganic components needed by *Vaccinium vitis-idaea* L. plantlets.

For correct and detailed evaluation of applicability of a certain cultivation system to a defined biological system, as well as evaluation of mass propagation and bioavailability of the main macro-nutrients, it is also necessary to evaluate the response of secondary metabolism of the investigated biological system to the new conditions of cultivation. For this purpose, the metabolism of some of the main polyphenols synthesized by *Vaccinium vitis-idaea* L. plantlets cultivated in a temporary immersion RITA system was investigated.

The content of total phenolic components biosynthesized by *Vaccinium vitis-idaea* L. plantlets varied slightly at different regimes of immersion—between 2.0 and 1.7 mg gallic acid equivalents (GAE)/g dry weight (dw) (Figure 8.4). The same insignificant influence of immersion frequency was

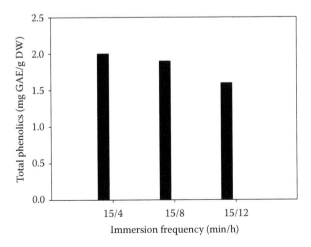

Figure 8.4 Total phenols synthesized by *Vaccinium vitis-idaea* L. cultivated in a RITA temporary cultivation system.

Table 8.1   HPLC Analysis of Polyphenolic Content in Methanol Extracts of *Vaccinium vitis-idaea* L. Cultivated in RITA Temporary Cultivation System

Compound (µg/g dw)		Immersion Frequency (Immersion Period [min])/ (Standby Period [h])		
		15/4	15/8	15/12
Flavonoids				
Flavonols	Myricetin	27.5	38.6	27.6
	Kaempferol	10.5	16.4	10.9
	Quercetin	30.4	37.6	30.0
Quercetin glycosides	Rutin	157.8	203.7	185.8
	Hyperoside	385.4	479.1	379.9
Flavanone glycoside	Hesperidin	103.1	78.5	79.7
Phenolic Acids				
Hydroxybenzoic Acids				
Protocatechuic acid		107.9	115.9	103.7
Salicylic acid		880.1	508.52	700.5
Chlorogenic acid		913.6	288.1	606.4
Vanillic acid		329.0	333.8	285.8
Syringic acid		468.4	456.8	375.6
Hydroxycinnamic Acids				
Caffeic acid		1274.0	1191.2	963.4
Cinnamic acid		17.0	16.3	17.2
p-Coumaric acid		172.3	162.6	133.8
Ferulic acid		259.4	269.4	282.4
Sinapic acid		483.7	490.4	398.6

observed concerning anabolism of the main phenolic acids and flavonoids (Table 8.1). These results clearly outlined that temporary immersion conditions of cultivation do not affect cell metabolism, which is especially important from the point of view of the quality of the obtained plant material.

However, it should be noted that *Vaccinium vitis-idaea* L. biosynthesized significant amounts of hyperoside (between 385 and 479 µg/g dw), rutin (between 158 and 204 µg/g dw), 2-OH benzoic acid (between 509 and 880 µg/g dw), chlorogenic acid (between 288 and 914 µg/g dw), caffeic acid (between 963 and 1274 µg/g dw), ferulic acid (between 259 µg/g dw and 282 dw), and sinapic acid (between 399 and 484 µg/g dw) during its cultivation in the RITA temporary immersion system (Table 8.1).

All these secondary metabolites are strong antioxidants, and consequently could be used as important protective agents for human health against oxidative stress disorders, and as protectors against oxidative damage in food systems as well. Hence, the RITA temporary immersion system is also appropriate for production of valuable bioactive secondary metabolites.

In conclusion, the data presented show that temporary immersion cultivation systems are appropriate for mass propagation of cowberry and ensure a good environment for plantlets during their cultivation. Beside mass propagation, this cultivation system and *Vaccinium vitis-idaea* L. plantlets formed a prospective biological system for the production of a complex of bioactive polyphenols with potential applications in the food industry and pharmacy. The results obtained are a good basis for the development of complex biotechnology, both for *in vitro* multiplication of this important plant and for a production process for an antioxidant mixture.

8.4 BERRY POLYPHENOLS: PROCESSING AND FOOD MATRIX EFFECTS

Berries, including grapes, generally have a short shelf life, and they are widely processed into juices, wines, liquors, jams, and fruit preparations for ice cream, yoghurt, and confectionery. Therefore, the effects of processing and subsequent storage on the bioactive compounds, particularly polyphenols, need to be considered when assessing the potential health benefits of berry-derived foodstuffs and beverages. On the other hand, efficient, inexpensive, and environmentally friendly utilization of the by-products of berry processing is becoming increasingly important.

Several studies have investigated the effects of processing on berry anthocyanins. Freezing and subsequent frozen storage have been shown to have minimal effects on red raspberry anthocyanins (de Ancos et al. 2000; Mullen et al. 2002). However, significant losses of anthocyanins have been observed in blueberry juices (Brownmiller et al. 2008; Lee et al. 2002; Rossi et al. 2003; Skrede et al. 2000), raspberry puree (Ochoa et al. 1999) and jams (García-Viguera et al. 1998), strawberry jams (García-Viguera et al. 1999; Ngo et al. 2007), canned fruits (Ngo et al. 2007), juice (Klopotek et al. 2005), and nectar (Klopotek et al. 2005). Despite heat treatment, these losses are most likely due to enzymatic polymerization and/or degradation of anthocyanins prior to pasteurization or polymerization reactions with anthocyanins and other phenolic compounds during storage of the thermally processed products. Interestingly, the polymeric compounds formed have been suggested to compensate for the loss of *in vitro* antioxidant capacity due to degradation of monomeric anthocyanins (Brownmiller et al. 2008). However, more studies are needed to identify the anthocyanin polymers and to evaluate their bioavailability and *in vivo* activity.

While almost equal distributions between the juice and press-cake were observed for the lingonberry, the bilberry juice possessed 1.5 and 3.2 times lower recovery rates for the polyphenolic and anthocyanin content, respectively (Dinkova et al. 2012). Correspondingly, 1.8–3.0 times higher values, on a fruit weight basis, were obtained for the antioxidant capacity of the bilberry press-cake. Therefore, additional processing steps are required to increase the recovery of anthocyanins and polyphenols during bilberry juice extraction. On the other hand, the wild berry processing waste represents a potential source for manufacturing extracts rich in anthocyanins and/or polyphenols.

Pectolytic enzymes are commonly used in industrial berry processing to facilitate juice extraction. These enzymes cause degradation of the cell wall matrix, enhancing the juice yield. Concomitantly, an increased extractability of phenolic compounds, particularly anthocyanins (Buchert et al. 2005) and flavonols (Koponen et al. 2008), has been observed in bilberry and blackcurrant enzyme-aided processing. However, careful selection of enzyme preparations is required, as some glycosidase activities present can hydrolyze anthocyanins to the corresponding aglycones, thus negatively affecting the color stability (Wrolstad et al. 1994).

Even though grape pomace has been recognized as a rich source of polyphenols, comprehensive data have only recently been presented on the contents of individual phenolic compounds in the press residues of different cultivars (Kammerer et al. 2004). Further, a novel process for enzyme-assisted extraction of grape pomace has been established (Kammerer et al. 2005), offering a valuable alternative to the recovery of polyphenols using sulfites. Thus, problems originating from potential pseudoallergic reactions due to the consumption of foods with added sulfites and labeling of the suspected allergen (Directive 2000/13/EC) could be avoided. Preextraction of the grape pomace followed by resuspension of the residue and enzymatic treatment with a combination of pectinases and cellulases results in notably higher recovery rates. Together with the use of absorber technology for the concentration of phenolic compounds, this two-step extraction process may be a very helpful tool for obtaining highly concentrated pigment extracts and functional food ingredients. The effectiveness of enzymatic treatment has also been demonstrated for producing polyphenol-rich extracts from blueberry (Lee and Wrolstad 2004) and blackcurrant (Landbo and Meyer 2001) processing waste.

Besides processing, the food matrix can also have a dramatic effect on berry anthocyanin stability. Skrede et al. (1992) compared the stability of blackcurrant and strawberry syrups, where strawberry syrups were fortified with purified strawberry anthocyanins and ascorbic acid to the same level as that present in blackcurrant syrup. While the strawberry syrups fortified with anthocyanins had the same stability as blackcurrant syrup, fortifying strawberry syrup with ascorbic acid greatly accelerated pigment degradation. Interestingly, higher anthocyanin stability has been observed in directly extracted strawberry juice compared with juice prepared from concentrate, which might be due to the retention of polymeric matrix compounds in the former (Sadilova et al. 2009). Recently, a stabilizing effect of rose petal polyphenols acting as copigments was demonstrated in heated model (Shikov et al. 2008) and real beverage (Mollov et al. 2007) systems of strawberry anthocyanins. Moreover, adding these natural copigments to the brine used in the fruit-firming process ensured reliable color retention of the texture-improved canned strawberries (Shikov et al. 2012). The possibilities for incorporation of blueberries or blueberry extracts into fermented milks have also been studied. Cinbas and Yazici (2008) suggested that blueberries could be added to the yoghurt formula at 25% level. The yoghurt matrix did not significantly affect bilberry anthocyanin stability (Ivanov et al. 2009). A slight decrease of polyphenols and anthocyanins during cold storage of fermented milks with added bilberry juice was found.

In the future, berry research should focus on gene–nutrient interactions and health outcomes to achieve individual dietary intervention strategies directed at preventing human chronic diseases, improving life, and promoting healthy aging. The accumulation of knowledge, the carrying out of great numbers of research experiments, and practical methodology are the basis of development of different omics technologies.

REFERENCES

Adam-Blondon, A.-F., C. Roux, D. Claux, G. Butterlin, D. Merdinoglu, and P. This. 2004. Mapping 245 SSR markers on the *Vitis vinifera* genome: A tool for grape genetics. *Theor Appl Genet* 109:1017–1027.

Adam-Blondon, A.-F., F. Lahogue-Esnault, A. Bouquet, J. M. Boursiquot, and P. This. 2001. Usefulness of two SCAR markers for marker-assisted selection of seedless grapevine cultivars. *Vitis* 40:147–155.

Afreen, F. 2006. Temporary immersion bioreactor. In S. D. Gupta and Y. Ibaraki (eds), *Plan Tissue Culture Engineering*, pp. 187–201. Amsterdam: Springer Verlag.

Aharoni, A., L. C. P. Keizer, H. J. Bouwmeester et al. 2000. Identification of the SAAT gene involved in strawberry flavor biogenesis by use of DNA microarrays. *Plant Cell Online* 12:647–661.

Aiyer, H. S., C. Srinivasan, and R. C. Gupta. 2008. Dietary berries and ellagic acid diminish estrogen-mediated mammary tumorigenesis in ACI rats. *Nutr Cancer* 60:227–234.

Akkurt, M., A. Çakır, M. Shidfar, B. P. Çelikkol, and G. Söylemezoğlu. 2012. Using SCC8, SCF27 and VMC7f2 markers in grapevine breeding for seedlessness via marker assisted selection. *Genet Mol Res* 11:2288–2294.

Ali, K., F. Maltese, A. Figueiredo et al. 2012. Alterations in grapevine leaf metabolism upon inoculation with *Plasmopara viticola* in different time-points. *Plant Sci* 191:100–107.

Ali, K., F. Maltese, R. Toepfer, Y. H. Choi, and R. Verpoorte. 2011. Metabolic characterization of Palatinate German white wines according to sensory attributes, varieties, and vintages using NMR spectroscopy and multivariate data analyses. *J Biomol NMR* 49:255–266.

Ali, K., F. Maltese, E. Zyprian, M. Rex, Y. H. Choi, and R. Verpoorte. 2009. NMR metabolic fingerprinting based identification of grapevine metabolites associated with downy mildew resistance. *J Agr Food Chem* 57:9599–9606.

Antonius-Klemola, K. 1999. Molecular markers in *Rubus* (Rosaceae) research and breeding. *J Hort Sci Biotech* 74:149–160.

Arapitsas, P., M. Scholz, U. Vrhovsek et al. 2012. A metabolomic approach to the study of wine microoxygenation. *PLoS ONE* 7:e37783.

Bacilieri, R., D. Maghradze, S. Grando et al. 2010. Conservation, characterisation and management of grapevine genetic resources: The European project grapegen06. In *10th International Conference on Grapevine Breeding and Genetics: Program and Abstracts*, pp. 141–142. August 1–5, 2010, Geneva, NY.

Badjakov, I., V. Kondakova, and A. Atanassov. 2011. Metabolomics: Current view on fruit quality relation with human health. In N. Benkeblia (ed.), *Sustainable Agriculture and New Biotechnologies*, pp. 303–319. Boca Raton: CRC Press.

Basha, S. M., H. Mazhar, and H. K. N. Vasanthaiah. 2010. Proteomics approach to identify unique xylem sap proteins in Pierce's disease-tolerant *Vitis* species. *Appl Biochem Biotechnol* 160:932–944.

Baxter, I. R. and J. O. Borevitz. 2006. Mapping a plant's chemical vocabulary. *Nat Genet* 38:737–738.

Bayly, F. C. and H. W. Berg. 1967. Grape and wine proteins of white wine varietals. *Am J Enol Vitic* 18:18–32.

Beattie, J., A. Crozier, and G. Duthie. 2005. Potential health benefits of berries. *Curr Nutr Food Sci* 1:71–86.

Berthouly, M. and H. Etienne. 2005. Temporary immersion system: A new concept for use liquid medium in mass propagation. In A. Hvoslef-Eide and W. Preil (eds), *Liquid Culture Systems for In Vitro Plant Propagation*, pp. 165–195. Amsterdam: Springer Verlag.

Bianco, L., L. Lopez, A. G. Scalone et al. 2009. Strawberry proteome characterization and its regulation during fruit ripening and in different genotypes. *J Proteomics* 72:586–607.

Bowers, J. E. and C. P. Meredith. 1997. The parentage of a classic wine grape, Cabernet Sauvignon. *Nat Genet* 16:84–87.

Bowers, J. E., G. S. Dangl, and C. P. Meredith. 1999. Development and characterization of additional microsatellite DNA markers for grape. *Am J Enol Vitic* 50:243–246.

Brissonnet, F. and A. Maujean. 1993. Characterization of foaming proteins in a Champagne base wine. *Am J Enol Vitic* 44:297–301.

Brownmiller, C., L. R. Howard, and R. L. Prior. 2008. Processing and storage effects on monomeric anthocyanins, percent polymeric color, and antioxidant capacity of processed blueberry products. *J Food Sci* 73:72–79.

Buchert, J., J. M. Koponen, M. Suutarinen et al. 2005. Effect of enzyme-aided pressing on anthocyanin yield and profiles in bilberry and blackcurrant juices. *J Sci Food Agric* 85:2548–2556.

Cabezas, J. A., M. T. Cervera, L. Ruiz-García, J. Carreño, and J. M. Martínez-Zapater. 2006. A genetic analysis of seed and berry weight in grapevine. *Genome* 49:1572–1585.

Castellarin, S. D., M. A. Matthews, G. Di Gaspero, and G. A. Gambetta. 2007. Water deficits accelerate ripening and induce changes in gene expression regulating flavonoid biosynthesis in grape berries. *Planta* 227:101–112.

Castrejon, A. D. R., I. Elchholz, S. Rohn, L. W. Kroh, and S. Huyskens-Keil. 2008. Phenolic profile and antioxidant activity of highbush blueberry (*Vaccinium corymbosum* L.) during fruit maturation and ripening. *Food Chem* 109:564–572.

Castro, A. J., C. Carapito, N. Zorn et al. 2005. Proteomic analysis of grapevine (*Vitis vinifera* L.) tissues subjected to herbicide stress. *J Exp Bot* 56:2783–2795.

Cilindre, C., A. J. Castro, C. Clement, P. Jeandet, and R. Marchal. 2007. Influence of *Botrytis cinerea* infection on Champagne wine proteins (characterized by two-dimensional electrophoresis/immunodetection) and wine foaming properties. *Food Chem* 103:139–149.

Cinbas, A. and F. Yazici. 2008. Effect of the addition of blueberries on selected physicochemical and sensory properties of yoghurts. *Food Technol Biotech* 46:434–441.

Clifford, M. N. 2000. Anthocyanins—Nature, occurrence and dietary burden. *J Sci Food Agr* 80:1063–1072.

Côté, J., S. Caillet, G. Doyon, J. F. Sylvain, and M. Lacroix. 2010. Analyzing cranberry bioactive compounds. *Crit Rev Food Sci Nutr* 50:872–888.

Dalbo, M. A., G. N. Ye, N. F. Weeden, W. F. Wilcox, and B. I. Reisch. 2001. Marker-assisted selection for powdery mildew resistance in grapes. *J Amer Soc Hort Sci* 126(1):83–89.

Davies, K. M. 2007. Genetic modification of plant metabolism for human health benefits. *Mutat Res* 622:122–137.

Davis, T. M. and H. Yu. 1997. A linkage map of the diploid strawberry, *Fragaria vesca*. *J Hered* 88:215–221.

de Ancos, B., E. Ibanez, G. Reglero, and M. P. Cano. 2000. Frozen storage effects on anthocyanins and volatile compounds of raspberry fruit. *J Agr Food Chem* 48:873–879.

Debnath, S. C. 2009. Characteristics of strawberry plants propagated by in vitro bioreactor culture and ex vitro propagation method. *Eng Life Sci* 9:239–246.

Deluc, L. G., J. Grimplet, M. D. Wheatley et al. 2007. Transcriptomic and metabolite analyses of Cabernet Sauvignon grape berry development. *BMC Genomics* 8:429.

Deluc, L. G., D. R. Quilici, A. Decendit et al. 2009. Water deficit alters differentially metabolic pathways affecting important flavor and quality traits in grape berries of Cabernet Sauvignon and Chardonnay. *BMC Genomics* 10:212.

Di Gaspero, G. and F. Cattonaro. 2010. Application of genomics to grapevine improvement. *Aust J Grape Wine R* 16:122–130.

Di Gaspero, G., G. Cipriani, A.-F. Adam-Blondon, and R. Testolin. 2007. Linkage maps of grapevine displaying the chromosomal locations of 420 microsatellite markers and 82 markers for R-gene candidates. *Theor Appl Genet* 114:1249–1263.

Dincheva, I., I. Badjakov, V. Kondakova, and R. Batchvarova. 2013. Metabolic profiling of raspberry (*Rubus idaeus*) during fruit development and ripening. *Int J Agr Sci Res* 3:71–78.

Dinkova, R., V. Shikov, K. Mihalev, Z. Velchev, H. Dinkov, and P. Mollov. 2012. Changes in the total anthocyanins and polyphenols during processing of wild berries into freshly pressed juices. *J EcoAgriTourism* 8:254–59.

Doligez, A., A.-F. Adam-Blondon, G. Cipriani et al. 2006. An integrated SSR map of grapevine based on five mapping populations. *Theor Appl Genet* 113:369–382.

Doligez, A., A. Bouquet, Y. Danglot et al. 2002. Genetic mapping of grapevine (*Vitis vinifera* L.) applied to the detection of QTLs for seedlessness and berry weight. *Theor Appl Genet* 105:780–795.

Donald, T. M., F. Pellerone, A.-F. Adam-Blondon, A. Bouquet, M. R. Thomas, and I. B. Dry. 2002. Identification of resistance gene analogs linked to a powdery mildew resistance locus in grapevine. *Theor Appl Genet* 104:610–618.

Dzhambazova, T., V. Kondakova, I. Tsvetkov, and R. Batchvarova. 2011. Advanced understanding of neurodegenerative diseases. In R. C.-C. Chang (ed.), *Grape Secondary Metabolites—Benefits for Human Health*, pp. 285–298. Rijeka: InTech.

Escarpa, A. and M. C. Gonzalez. 2001. An overview of analytical chemistry of phenolic compounds in foods. *Crit Rev Anal Chem* 31:57–139.

Fanizza, G., F. Lamaj, L. Costantini, R. Chaabane, and M. S. Grando. 2005. QTL analysis for fruit yield components in table grapes (*Vitis vinifera*). *Theor Appl Genet* 111:658–664.

Fatahi, R., Z. Zamani, A. Ebadi, and S. A. Mehlenbacher. 2003. The inheritance of seedless SCC8-SCAR and SSRS loci alleles in progeny of "Muscat Hamburg" × "Bidane Quermez" grapes. In *Proceedings of the 1st International Symposium on Grapevine Growing, Commerce and Research*, vol. 652, pp. 329–335. 30 June–2 July, Lisbon, Portugal.

Fernandez, L., A. Doligez, G. Lopez, M. R. Thomas, A. Bouquet, and L. Torregrosa. 2006. Somatic chimerism, genetic inheritance, and mapping of the fleshless berry (flb) mutation in grapevine (*Vitis vinifera* L.). *Genome* 49:721–728.

Fernandez, L., L. Torregrosa, N. Terrier et al. 2007. Identification of genes associated with flesh morphogenesis during grapevine fruit development. *Plant Mol Biol* 63:307–323.

Fernie, A. R., R. N. Trethewey, A. J. Krotzky, and L. Willmitzer. 2004. Metabolite profiling: From diagnostics to systems biology. *Nat Rev Mol Cell Biol* 5:763–769.

Figueiredo, A., A. M. Fortes, S. Ferreira et al. 2008. Transcriptional and metabolic profiling of grape (*Vitis vinifera* L.) leaves unravel possible innate resistance against pathogenic fungi. *J Exp Bot* 59:3371–3381.

Fischer, B. M., I. Salakhutdinov, M. Akkurt et al. 2004. Quantitative trait locus analysis of fungal disease resistance factors on a molecular map of grapevine. *Theor Appl Genet* 108:501–515.

Flamini, R. and M. De Rosso. 2006. Mass spectrometry in the analysis of grape and wine proteins. *Expert Rev Proteomic* 3:321–331.

Fliser, D., J. Novak, V. Thongboonkerd et al. 2007. Advances in urinary proteome analysis and biomarker discovery. *J Am Soc Nephrol* 18:1057–1071.

Folta, K. M. and T. M. Davis. 2006. Strawberry genes and genomics. *Crit Rev Plant Sci* 25:399–415.

Fortes, A., P. Agudelo-Romero, M. Silva et al. 2011. Transcript and metabolite analysis in Trincadeira cultivar reveals novel information regarding the dynamics of grape ripening. *BMC Plant Biol* 11:149.

García-Viguera, C., P. Zafrilla, F. Artés, F. Romero, P. Abellán, and F. A. Tomás-Barberán. 1998. Colour and anthocyanin stability of red raspberry jam. *J Sci Food Agr* 78:565–573.

García-Viguera, C., P. Zafrilla, F. Romero, P. Abellán, F. Artés, and F. A. Tomás-Barberán. 1999. Color stability of strawberry jam as affected by cultivar and storage temperature. *J Food Sci* 64:243–247.

Garzón, G. A., K. M. Riedl, and S. J. Schwartz. 2009. Determination of anthocyanins, total phenolic content, and antioxidant activity in Andes berry (*Rubus glaucus* Benth). *J Food Sci* 74:C227–C232.

Giribaldi, M. and M. G. Giuffrida. 2010. Heard it through the grapevine: Proteomic perspective on grape and wine. *J Proteomics* 73:1647–1655.

Graham, J., K. Smith, K. MacKenzie, L. Jorgenson, C. Hackett, and W. Powell. 2004. The construction of a genetic linkage map of red raspberry (*Rubus idaeus* subsp. *idaeus*) based on AFLPs, genomic-SSR and EST-SSR markers. *Theor Appl Genet* 109:740–749.

Grando, M. S., D. Bellin, K. J. Edwards, C. Pozzi, M. Stefanini, and R. Velasco. 2003. Molecular linkage maps of *Vitis vinifera* L. and *Vitis riparia* Mchx. *Theor Appl Genet* 106:1213–1224.

Grimplet, J., G. R. Cramer, J. A. Dickerson, K. Mathiason, J. Van Hemert, and A. Y. Fennell. 2009a. VitisNet:"Omics" integration through grapevine molecular networks. *PLoS ONE* 4:e8365.

Grimplet, J., J. Van Hemert, P. Carbonell-Bejerano et al. 2012. Comparative analysis of grapevine whole-genome gene predictions, functional annotation, categorization and integration of the predicted gene sequences. *BMC Res Notes* 5:213.

Grimplet, J., M. D. Wheatley, H. Ben Jouira, L. G. Deluc, G. R. Cramer, and J. C. Cushman. 2009b. Proteomic and selected metabolite analysis of grape berry tissues under well-watered and water-deficit stress conditions. *Proteomics* 9:2503–2528.

Gülçin, I. T., R. Çakmakç, M. Bilsel, A. C. Gören, and U. Erdogan. 2011. Pomological features, nutritional quality, polyphenol content analysis, and antioxidant properties of domesticated and 3 wild ecotype forms of raspberries (*Rubus idaeus* L.). *J Food Sci* 76:C585–C593.

Häkkinen, S. H. and A. R. Törrönen. 2000. Content of flavonols and selected phenolic acids in strawberries and *Vaccinium* species: Influence of cultivar, cultivation site and technique. *Food Res Int* 33:517–524.

Hall, E. K., C. Neuhauser, and J. B. Cotner. 2008a. Toward a mechanistic understanding of how natural bacterial communities respond to changes in temperature in aquatic ecosystems. *ISME J* 2:471–481.

Hall, R. D., I. D. Brouwer, and M. A. Fitzgerald. 2008b. Plant metabolomics and its potential application for human nutrition. *Physiol Plantarum* 132:162–175.

Hanhineva, K. J. and S. O. Karenlampi. 2007. Production of transgenic strawberries by temporary immersion bioreactor system and verification by TAIL-PCR. *BMC Biotechnol* 7:11.

Harnly, J. M., R. F. Doherty, G. R. Beecher et al. 2006. Flavonoid content of US fruits, vegetables, and nuts. *J Agr Food Chem* 54:9966–9977.

Harrison, H. F., J. K. Peterson, M. E. Snook, J. R. Bohac, and D. M. Jackson. 2003. Quantity and potential biological activity of caffeic acid in sweet potato (*Ipomoea batatas* [L.] Lam.) storage root periderm. *J Agr Food Chem* 51:2943–2948.

Hokanson, S. C. 2001. SNiPs, Chips, BACs, and YACs: Are small fruits part of the party mix? *HortScience* 36:859–871.

Holmes, E., R. L. Loo, J. Stamler et al. 2008. Human metabolic phenotype diversity and its association with diet and blood pressure. *Nature* 453:396–400.

Holt, H. E., I. L. Francis, J. Field, M. J. Herderich, and P. G. Iland. 2008. Relationships between berry size, berry phenolic composition and wine quality scores for Cabernet Sauvignon (*Vitis vinifera* L.) from different pruning treatments and different vintages. *Aust J Grape Wine Res* 14:191–202.

Holtorf, H., M.-C. Guitton, and R. Reski. 2002. Plant functional genomics. *Naturwissenschaften* 89:235–249.

House, W. A. 1999. Trace element bioavailability as exemplified by iron and zinc. *Field Crop Res* 60:115–141.

Ivanov, G., K. Mihalev, N. Yoncheva, K. Beeva, and P. Mollov. 2009. Changes in the content of total anthocyanins and polyphenols during cold storage of fruit milk beverages containing bilberry juice. In *Proceedings of the Scientific Conference of Union of Scientists in Bulgaria*, vol. 12, pp. 169–173. November 5–6, 2009, Plovdiv, Bulgaria.

Jaillon, O., J.-M. Aury, B. Noel et al. 2007. The grapevine genome sequence suggests ancestral hexaploidization in major angiosperm phyla. *Nature* 449:463–467.

Jellouli, N., H. Ben Jouira, H. Skouri, A. Ghorbel, A. Gourgouri, and A. Mliki. 2008. Proteomic analysis of Tunisian grapevine cultivar Razegui under salt stress. *J Plant Physiol* 65:471–481.

Kammerer, D., A. Claus, R. Carle, and A. Schieber. 2004. Polyphenol screening of pomace from red and white grape varieties (*Vitis vinifera* L.) by HPLC-DAD-MS/MS. *J Agr Food Chem* 52:4360–4367.

Kammerer, D., A. Claus, A. Schieber, and R. Carle. 2005. A novel process for the recovery of polyphenols from grape (*Vitis vinifera* L.) pomace. *J Food Sci* 70:C157–C163.

Karaagac, E., A. M. Vargas, M. T. de Andrés et al. 2012. Marker assisted selection for seedlessness in table grape breeding. *Tree Genet Genomes* 8:1003–1015.

Kennedy, J. A., M. A. Matthews, and A. L. Waterhouse. 2002. Effect of maturity and vine water status on grape skin and wine flavonoids. *Am J Enol Viticult* 53:268–274.

Kirchheimer, F. 1939. 24 Plantae Rhamnales I. In *Vitaceae*, pp. 85–112. Berlin, Germany: W. Junk.

Klopotek, Y., K. Otto, and V. Böhm. 2005. Processing strawberries to different products alters contents of vita-min C, total phenolics, total anthocyanins, and antioxidant capacity. *J Agr Food Chem* 53:5640–5646.

Koponen, J. M., A. M. Happonen, S. Auriola et al. 2008. Characterization and fate of black currant and bilberry flavonols in enzyme-aided processing. *J Agr Food Chem* 56:3136–3144.

Lahogue, F., P. This, and A. Bouquet. 1998. Identification of a codominant scar marker linked to the seedless-ness character in grapevine. *Theor Appl Genet* 97:950–959.

Landbo, A.-K. and A. S. Meyer. 2001. Enzyme-assisted extraction of antioxidative phenols from black currant juice press residues (*Ribes nigrum*). *J Agr Food Chem* 49:3169–3177.

Lätti, A. K., K. R. Riihinen, and P. S. Kainulainen. 2007. Analysis of anthocyanin variation in wild populations of bilberry (*Vaccinium myrtillus* L.) in Finland. *J Agr Food Chem* 56:190–196.

Lee, J., R. W. Durst, and R. E. Wrolstad. 2002. Impact of juice processing on blueberry anthocyanins and poly-phenolics: Comparison of two pretreatments. *J Food Sci* 67:1660–1667.

Lee, J. and R. E. Wrolstad. 2004. Extraction of anthocyanins and polyphenolics from blueberry processing waste. *J Food Sci* 69:564–573.

Lee, T. H. 1985. Protein instability: Nature, characterization and removal by bentonite. In T. H. Lee (ed.), *Physical Stability of Wine*, pp. 23–39.

Lima, M. R. M., M. L. Felgueiras, G. Graça et al. 2010. NMR metabolomics of esca disease-affected *Vitis vinifera* cv. Alvarinho leaves. *J Exp Bot* 61:4033–4042.

Lodhi, M. A., G.-N. Ye, N. F. Weeden, B. I. Reisch, and M. J. Daly. 1995. A molecular marker based linkage map of *Vitis. Genome* 38:786–794.

Marangon, M., S. C. Van Sluyter, P. A. Haynes, and E. J. Waters. 2009. Grape and wine proteins: Their fraction-ation by hydrophobic interaction chromatography and identification by chromatographic and proteomic analysis. *J Agr Food Chem* 57:4415–4425.

Martínez-Zapater, J. M., M. J. Carmona, J. Díaz-Riquelme, L. Fernandez, and D. Lijavetzky. 2010. Grapevine genetics after the genome sequence: Challenges and limitations. *Austr J Grape Wine R* 16:33–46.

Matthews, M. A., R. Ishii, M. M. Anderson, and M. O'Mahony. 1990. Dependence of wine sensory attributes on vine water status. *J Sci Food Agr* 51:321–335.

Maul, E. and R. Eibach. 2003. *Vitis* international variety catalogue. http://www.vivc.bafz.de/index.php.

Mauray, A., C. Felgines, C. Morand, A. Mazur, A. Scalbert, and D. Milenkovic. 2010. Nutrigenomic analysis of the protective effects of bilberry anthocyanin-rich extract in apo E-deficient mice. *Genes Nutr* 5:343–353.

Mazza, G. and E. Miniati. 1993. Anthocyanins in fruits, vegetables and grains. In H. Böhm (ed.), *Molecular Nutrition and Food Research*, vol. 3, pp. 343–343. London: CRC Press.

McCown, B. H. and G. Lloyd. 1981. Woody plant medium (WPM)-a mineral nutrient formulation for micro-culture of woody plant species. *HortScience* 16:453.

McNicol, R. J., J. Graham, F. A. Hammerschlag, and R. E. Litz. 1992. Temperate small fruits. In F. A. Hammerschlag and R. E. Litz (eds), *Biotechnology of Perennial Fruit Crops*, pp. 303–321. Oxford University Press.

Mejía, N., M. Gebauer, L. Muñoz, N. Hewstone, C. Muñoz, and P. Hinrichsen. 2007. Identification of QTLs for seedlessness, berry size, and ripening date in a seedless × seedless table grape progeny. *Am J Enol Viticult* 58:499–507.

Mejía, N. and P. Hinrichsen. 2002. A new, highly assertive SCAR marker potentially useful to assist selection for seedlessness in table grape breeding. In *Proceedings of the 8th International Conference on Grape Genetics and Breeding.* vol. 603, pp. 559–564. August 26–31, Kecskemét, Hungary.

Meredith, C. P. 2001. Grapevine genetics: Probing the past and facing the future. *Agric Conspec Sci (ACS)* 66:21–25.

Moises, A., M. A. Vazquez-Cruz, S. N. Jimenez-Garcia, I. Torres-Pacheco, S. H. Guzman-Maldonado, R. G. Guevara-Gonzalez, and R. Miranda-Lopez. 2012. Effect of maturity stage and storage on flavor com-pounds and sensory description of Berrycactus (*Myrtillocactus geometrizans*). *J Food Sci* 77:C366–C373.

Mollov, P., K. Mihalev, V. Shikov, N. Yoncheva, and V. Karagyozov. 2007. Colour stability improvement of strawberry beverage by fortification with polyphenolic copigments naturally occurring in rose petals. *Innov Food Sci Emerg Tech* 8:318–321.

Moreno-Arribas, M. V., E. Pueyo, and M. C. Polo. 2002. Analytical methods for the characterization of proteins and peptides in wines. *Anal Chim Acta* 458:63–75.

Mullen, W., A. J. Stewart, M. E. J. Lean, P. Gardner, G. G. Duthie, and A. Crozier. 2002. Effect of freezing and storage on the phenolics, ellagitannins, flavonoids, and antioxidant capacity of red raspberries. *J Agr Food Chem* 50:5197–5201.

Naczk, M. and F. Shahidi. 2004. Extraction and analysis of phenolics in food. *J Chromatogr A* 1054:95–111.

Nagaraj, S. H., R. B. Gasser, and S. Ranganathan. 2007. A hitchhiker's guide to expressed sequence tag (EST) analysis. *Brief Bioinform* 8:6–21.

Ngo, T., R. E. Wrolstad, and Y. Zhao. 2007. Color quality of Oregon strawberries-impact of genotype, composition, and processing. *J Food Sci* 72:C025–C032.

Nunes-Miranda, J. D., G. Igrejas, J. C. Mejuto, M. Reboiro-Jato, and J. L. Capelo. 2012. Mass spectrometry-based fingerprinting of proteins and peptides in wine quality control: A critical overview. *Crit Rev Food Sci* 57:751–759.

Ochoa, M. R., A. G. Kesseler, M. B. Vullioud, and J. E. Lozano. 1999. Physical and chemical characteristics of raspberry pulp: Storage effect on composition and color. *LWT-Food Sci Technol* 32:149–153.

Pauquet, J., A. Bouquet, P. This, and A.-F. Adam-Blondon. 2001. Establishment of a local map of AFLP markers around the powdery mildew resistance gene Run1 in grapevine and assessment of their usefulness for marker assisted selection. *Theor Appl Genet* 103:1201–1210.

Pontius, J. U., L. Wagner, and G. D. Schuler. 2003. 21. UniGene. In J. McEntyre and J. Ostell (eds), *A Unified View of the Transcriptome*, pp. 1–11. Bethesda: National Library of Medicine.

Qu, L. and J. F. Hancock. 1997. Randomly amplified polymorphic DNA-(RAPD-) based genetic linkage map of blueberry derived from an interspecific cross between diploid *Vaccinium darrowi* and tetraploid *V. corymbosum*. *J Am Soc Hort Sci* 122:69–73.

Riaz, S., G. S. Dangl, K. J. Edwards, and C. P. Meredith. 2004. A microsatellite marker based framework linkage map of *Vitis vinifera* L. *Theor Appl Genet* 108:864–872.

Riaz, S., A. C. Tenscher, R. Graziani, A. F. Krivanek, D. W. Ramming, and M. A. Walker. 2009. Using marker-assisted selection to breed Pierce's disease-resistant grapes. *Am J Enol Viticult* 60:199–207.

Rossi, M., E. Giussani, R. Morelli, R. Lo Scalzo, R. C. Nani, and D. Torreggiani. 2003. Effect of fruit blanching on phenolics and radical scavenging activity of highbush blueberry juice. *Food Res Int* 36:999–1005.

Rowland, L. J. and A. Levi. 1994. RAPD-based genetic linkage map of blueberry derived from a cross between diploid species (*Vaccinium darrowi* and *V. elliottii*). *Theor Appl Genet* 87:863–868.

Rudd, S., H. Schoof, and K. Mayer. 2005. PlantMarkers: A database of predicted molecular markers from plants. *Nucleic Acids Res* 33:D628–D632.

Sadilova, E., F. C. Stintzing, D. R. Kammerer, and R. Carle. 2009. Matrix dependent impact of sugar and ascorbic acid addition on color and anthocyanin stability of black carrot, elderberry and strawberry single strength and from concentrate juices upon thermal treatment. *Food Res Int* 42:1023–1033.

Seeram, N. P. 2008. Berry fruits: Compositional elements, biochemical activities, and the impact of their intake on human health, performance, and disease. *J Agr Food Chem* 56:627–629.

Sefc, K. M., F. Lefort, M. S. Grando, K. D. Scott, H. Steinkellner, and M. R. Thomas. 2001. Microsatellite markers for grapevine: A state of the art. In K. A. Roubelakis-Angelakis (ed.), *Molecular Biology and Biotechnology of Grapevine*, pp. 433–463. The Netherlands: Kluwer Academic Publishers.

Sellers, T. A. and J. R. Yates. 2003. Review of proteomics with applications to genetic epidemiology. *Genet Epidemiol* 24:83–98.

Shikov, V., D. R. Kammerer, K. Mihalev, P. Mollov, and R. Carle. 2008. Heat stability of strawberry anthocyanins in model solutions containing natural copigments extracted from rose (*Rosa damascena* Mill.) petals. *J Agr Food Chem* 56:8521–8526.

Shikov, V., D. R. Kammerer, K. Mihalev, P. Mollov, and R. Carle. 2012. Antioxidant capacity and colour stability of texture-improved canned strawberries as affected by the addition of rose (*Rosa damascena* Mill.) petal extracts. *Food Res Int* 46:552–556.

Shulaev, V., D. J. Sargent, R. N. Crowhurst et al. 2010. The genome of woodland strawberry (*Fragaria vesca*). *Nat Genet* 43:109–116.

Skrede, G., R. E. Wrolstad, and R. W. Durst. 2000. Changes in anthocyanins and polyphenolics during juice processing of highbush blueberries (*Vaccinium corymbosum* L.). *J Food Sci* 65:357–364.

Skrede, G., R. E. Wrolstad, P. Lea, and G. Enersen. 1992. Color stability of strawberry and blackcurrant syrups. *J Food Sci* 57:172–177.

Son, H.-S., G.-S. Hwang, K. M. Kim et al. 2009. Metabolomic studies on geographical grapes and their wines using 1H NMR analysis coupled with multivariate statistics. *J Agr Food Chem* 57:1481–1490.

Steingroewer, J., T. Bley, V. Georgiev et al. 2013. Bioprocessing of differentiated plant in vitro systems. *Eng Life Sci* 13:26–38.

Szajdek, A. and E. J. Borowska. 2008. Bioactive compounds and health-promoting properties of berry fruits: A review. *Plant Foods Hum Nutr* 63:147–156.

This, P., T. Lacombe, and M. R. Thomas. 2006. Historical origins and genetic diversity of wine grapes. *Trends Genet* 22:511–519.

Toffali, K., A. Zamboni, A. Anesi et al. 2011. Novel aspects of grape berry ripening and post-harvest withering revealed by untargeted LC-ESI-MS metabolomics analysis. *Metabolomics* 7:424–436.

Topfer, R., K. N. Sudharma, S. Kecke et al. 2009. The Vitis International Variety Catalogue (VIVC): New design and more information. *Bull de l'OIV* 82:45.

Tornielli, G. B., A. Zamboni, S. Zenoni, M. Delledonne, and M. Pezzotti. 2012. Transcriptomics and metabolomics for the analysis of grape berry development. In M. M. C. H. Gerós and S. Delrot (eds), *The Biochemistry of the Grape Berry*, Bentham Science Publishers.

Varshney, R. K., A. Graner, and M. E. Sorrells. 2005. Genic microsatellite markers in plants: Features and applications. *Trends Biotechnol* 23:48–55.

Vazquez-Cruz, M. A., S. N. Jimenez-Garcia, I. Torres-Pacheco, S. H. Guzman-Maldonado, R. G. Guevara-Gonzalez, and R. Miranda-Lopez. 2012. Effect of maturity stage and storage on flavor compounds and sensory description of berrycactus (*Myrtillocactus geometrizans*). *J Food Sci* 77:C366–C373.

Velasco, R., A. Zharkikh, M. Troggio et al. 2007. A high quality draft consensus sequence of the genome of a heterozygous grapevine variety. *PLoS ONE* 2:e1326.

Veloz-García, R., R. Marín-Martínez, R. Veloz-Rodríguez et al. 2010. Antimicrobial activities of cascalote (*Caesalpinia cacalaco*) phenolics-containing extract against fungus *Colletotrichum lindemuthianum*. *Indl Crop Prod* 31:134–138.

Vezzulli, S., D. Micheletti, S. Riaz et al. 2008a. A SNP transferability survey within the genus *Vitis*. *BMC Plant Biol* 8:128.

Vezzulli, S., M. Troggio, G. Coppola et al. 2008b. A reference integrated map for cultivated grapevine (*Vitis vinifera* L.) from three crosses, based on 283 SSR and 501 SNP-based markers. *Theor Appl Genet* 117:499–511.

Vincent, D., A. Ergül, M. C. Bohlman et al. 2007. Proteomic analysis reveals differences between *Vitis vinifera* L. cv. Chardonnay and cv. Cabernet Sauvignon and their responses to water deficit and salinity. *J Exp Bot* 58:1873–1892.

Viskelis, P., M. Rubinskiené, I. Jasutiené, A. Šarkinas, R. Daubaras, and L. Česoniené. 2009. Anthocyanins, antioxidative, and antimicrobial properties of American cranberry (*Vaccinium macrocarpon* Ait.) and their press cakes. *J Food Sci* 74:C157–C161.

Vlaanderen, J., L. E. Moore, M. T. Smith et al. 2010. Application of OMICS technologies in occupational and environmental health research; current status and projections. *Occup Environ Med* 67:136–143.

Welter, L. J., N. Göktürk-Baydar, M. Akkurt et al. 2007. Genetic mapping and localization of quantitative trait loci affecting fungal disease resistance and leaf morphology in grapevine (*Vitis vinifera* L). *Mol Breed* 20:359–374.

Widmer, T. L. and N. Laurent. 2006. Plant extracts containing caffeic acid and rosmarinic acid inhibit zoospore germination of *Phytophthora* spp. pathogenic to *Theobroma cacao*. *Eur J Plant Pathol* 115:377–388.

Wigand, P., S. Tenzer, H. Schild, and H. Decker. 2009. Analysis of protein composition of red wine in comparison with rosé and white wines by electrophoresis and high-pressure liquid chromatography-mass spectrometry (HPLC-MS). *J Agr Food Chem* 57:4328–4333.

Wrolstad, R. E., J. D. Wightman, R. W. Durst. 1994. Glycosidase activity of enzyme preparations used in fruit juice processing. *Food Technol* 48:90, 92–94, 96, 98.

Zabetakis, I. and M. A. Holden. 1997. Strawberry flavour: Analysis and biosynthesis. *J Sci Food Agr* 74:421–434.

Zeisel, S. H. 2010. A grand challenge for nutrigenomics. *Front Genet* 1:1–3.

Zhang, Y., N. P. Seeram, R. Lee, L. Feng, and D. Heber. 2008. Isolation and identification of strawberry phenolics with antioxidant and human cancer cell antiproliferative properties. *J Agr Food Chem* 56:670–675.

CHAPTER **9**

Marker-Assisted Selection in Coffee

Sarada Krishnan

CONTENTS

9.1 INTRODUCTION

Coffee is an important agricultural commodity contributing significantly to the economies of many developing countries. *Coffea* L. consists of at least 125 species distributed throughout Africa, Madagascar, the Comoros Islands, the Mascarene Islands (La Réunion and Mauritius), tropical Asia, and Australia (Davis 2010, 2011; Davis et al. 2006, 2011). The two main commercial species used in the production of the beverage coffee are *Coffea arabica* L. (arabica coffee) and *C. canephora* A. Froehner (robusta coffee), with arabica coffee accounting for about 70% of total coffee production (International Coffee Organization 2013a).

The low genetic diversity of *C. arabica* is attributed to its origin, reproductive biology, and evolution (Lashermes et al. 1999), which limit the potential for germplasm improvement (Pearl et al. 2004). This species is an allotetraploid ($2n = 4x = 44$), originating from two different diploid wild ancestors ($2n = 2x = 22$), *C. canephora* and *C. eugenioides* S. Moore, or ecotypes related to these species (Lashermes et al. 1999). It is self-compatible and mostly reproduces by self-fertilization, which occurs in about 90% of the flowers (Fazuoli et al. 2000), one of the reasons for its low genetic diversity. Additionally, the spread of coffee to other countries originated from very few plants and from self-fertilized seeds, leading to drastic reduction in genetic diversity of cultivated varieties (Anthony et al. 2002). Enlarging the genetic base and improvement of arabica cultivars, characterized by homogeneous agronomic behavior with high susceptibility to pests and diseases, have become high priorities for researchers (Lashermes et al. 2000b). Though *C. canephora* has been utilized in breeding programs, other taxa such as *C. eugenioides*, *C. congensis* A. Froehner, or

Coffea sp. Moloundou have been neglected as desirable gene sources (Lashermes et al. 1999). Many of the resistance traits to diseases and pests, such as coffee leaf rust (*Hemileia vastatrix* Berkeley and Broome), coffee berry disease caused by *Colletotrichum kahawae* Bridge and Waller, and root-knot nematode (*Meloidogyne* sp.), not found in *C. arabica* have been found in *C. canephora* (Lashermes et al. 2000b). *Coffea racemosa* Lour. also constitutes a promising source of coffee leaf miner (*Perileucoptera coffeella* Guerin-Meneville) resistance (Guerreiro Filho et al. 1999).

Conventional breeding of coffee suffers limitations due to the long regeneration time of the coffee tree (approximately 5 years), the high cost of field trials (Lashermes et al. 2000b), and dependence on environmental conditions. In cases where backcrossing is done over five generations, a minimum of 25 years after initial hybridization will be required to ensure that the desirable trait for improved quality or disease or pest resistance has been assimilated in the progeny (Lashermes et al. 2000b). The development of marker-assisted selection (MAS) provides an alternative to overcome the limitations of conventional coffee breeding (Lashermes et al. 2000b). The general principle of MAS is the use and selection of an identified molecular marker linked to a gene for a specific trait rather than selection for the trait itself (Lashermes et al. 2000b). In coffee, one of the main breeding objectives has been the transfer of desirable characters from diploid wild relatives into *C. arabica* cultivars without affecting quality traits (Herrera et al. 2002). The identification of markers linked to specific traits represents an important starting point for early selection of seedlings with these specific traits through enhanced backcross programs (Noir et al. 2003).

Marker maps provide the framework needed for eventual applications of MAS (Ruane and Sonnino 2007). Using marker maps, marker variants associated with traits of interest can be detected, allowing MAS through the selection of identifiable marker variants in order to select for nonidentifiable favorable variants of genes of interest (Ruane and Sonnino 2007). In a breeding context, MAS involves identifying the desirable phenotype and scoring indirectly for the presence or absence of this phenotype based on the sequences or banding patterns of molecular markers located in or near the genes controlling the phenotype (Edwards and McCouch 2007). The presence or absence of the specific gene or chromosomal segment that is known to carry the desirable allele is indicated by the sequence polymorphism or banding pattern of the molecular marker (Edwards and McCouch 2007). Edwards and McCouch (2007) outline a number of ways in which DNA markers can increase screening efficiency in breeding programs, such as

- The ability to screen for traits expressed late in the life of the organism in the juvenile stage (e.g., fruit quality, photoperiod sensitivity)
- The ability to screen for traits that are extremely difficult, expensive or time-consuming to score phenotypically (e.g., resistance to pests and diseases, tolerance to abiotic stresses such as drought, salt, mineral deficiencies)
- The ability to distinguish the homozygous from the heterozygous condition in a single generation using codominant molecular markers
- The ability to perform simultaneous MAS for several characters

9.2 TRAITS TARGETED FOR MAS IN COFFEE

Intensive research programs gained importance in the early 1950s with increased commercial significance of coffee (Pinto-Maglio 2006). Most breeding programs between 1950 and 1978 involved intraspecific and interspecific hybridization between the two cultivated species, *C. arabica* and *C. canephora*, with the use of some wild species of coffee (Pinto-Maglio 2006). Spontaneous interspecific hybrids that occur occasionally have also been widely used for improving disease and pest resistance in arabica coffee (Pearl et al. 2004). An example is the Timor Hybrid, a spontaneous interspecific cross between *C. arabica* and *C. canephora*, discovered on the island of Timor, which has been used intensively in coffee breeding programs for resistance to diseases such as coffee leaf rust (Lashermes et al. 2000b; Pearl et al. 2004). Introgressed lines of *C. liberica* from a spontaneous interspecific hybrid between *C. arabica* and *C. liberica* have been used as a main source of rust resistance in breeding programs in India (Prakash et al. 2002).

Molecular markers have been used in coffee for introgression assessment, determination of mode of inheritance of disease and pest resistance, assessment of beverage quality, and analysis of quantitative trait loci (QTL), all of which have great implications for future breeding. Efficient use of the genetic variation in wild species involves the genetic determination of the desirable trait and the ability to introgress the desirable DNA segments from wild species to the genome of the cultivated species (Prakash et al. 2004) without affecting quality traits (Herrera et al. 2002). Using amplified fragment length polymorphism (AFLP) markers, introgressed genotypes derived from the Timor Hybrid were evaluated and compared with parental genotypes of *C. arabica* and *C. canephora* to estimate the amount of introgression present in order to gain insights into the mechanism of introgression in *C. arabica* (Lashermes et al. 2000a). These researchers concluded that AFLP is an extremely efficient technique for DNA marker generation in coffee and offers an efficient way of distinguishing and fingerprinting coffee germplasm collections. Markers linked to specific traits such as resistance to leaf rust (de Brito et al. 2010; Diola et al. 2011; Herrera et al. 2009; Mahe et al. 2008; Prakash et al. 2004, 2011), berry disease (Agwanda et al. 1997; Gichuru et al. 2008), and root rot nematode (Noir et al. 2003) have been identified. Additionally, considerable research has been conducted to identify markers associated with quantitative traits such as fructification time (Akaffou et al. 2003), cherry and green bean characters (Priyono and Sumirat 2012), inheritance of chlorogenic acid (Ky et al. 1999), and other key agronomic and quality traits (Akaffou et al. 2012; Leroy et al. 2011). Advances in coffee marker technology and MAS are reviewed below.

9.2.1 Coffee Leaf Rust (*Hemileia vastatrix*)

Coffee leaf rust caused by the obligate parasitic fungus *Hemileia vastatrix* causes considerable economic losses to coffee producers (Diola et al. 2011), especially in *C. arabica*. First observed in 1861 near Lake Victoria, the fungus has now spread throughout coffee-growing countries and led to significant economic impact in Sri Lanka in 1868 (Silva et al. 2006). In India, crop losses due to coffee rust amount to about 70% in susceptible *C. arabica* cultivars (Prakash et al. 2004). A recent outbreak of coffee rust in Central America has resulted in more than 60% of the trees exhibiting 80% defoliation in Mexico (Cressey 2013). Crop devastation in Nicaragua, El Salvador (Cressey 2013), Guatemala, and Honduras has also been reported. According to the International Coffee Organization, this 2012/2013 outbreak of coffee rust in Central America is expected to cause crop losses of $500 million and cost 374,000 jobs (International Coffee Organization 2013b). Taking economics and minimization of chemical input for disease management into consideration, the most viable and effective option is the development and cultivation of tolerant varieties (Prakash et al. 2004). Hence, breeding for varieties resistant to coffee leaf rust has been one of the highest priorities in many countries (Prakash et al. 2004).

Nine resistance genes designated as S_H1–S_H9, either singly or in combination, confer leaf rust resistance on coffee plants, with the corresponding virulence genes designated as V1–V9 (Bettencourt and Rodrigues 1988 as cited in Prakash et al. 2004). Of these resistance factors, S_H1, S_H2, S_H4, and S_H5 have been found in *C. arabica*, S_H6, S_H7, S_H8, and S_H9 have been introgressed from the diploid species *C. canephora*, and S_H3 possibly originates from *C. liberica* Hiern (as cited in Prakash et al. 2004). In early breeding programs in India, S.26, a putative natural hybrid between *C. arabica* and a diploid species, has been used as a main source of rust resistance (Prakash et al. 2002). Using AFLP markers, Prakash et al. (2002) deduced that the polymorphism identified in this natural hybrid and its derivatives was a consequence of introgressive hybridizations involving *C. liberica*. Using the selfed offspring of S.26, the S.288 line carrying the S_H3 gene, Prakash et al. (2004) identified three distinct introgressed fragments of 52.8 cM length. A total of 21 AFLP markers tightly linked to the S_H3 rust resistance gene were identified, of which marker M8 was found to cosegregate perfectly with S_H3, making it highly desirable for MAS (Prakash et al. 2004).

In another study by Mahe et al. (2008) focusing on the leaf rust-resistant S_H3 gene, they identified a total of ten sequence-characterized genetic markers associated with this resistance gene, which included simple sequence repeats (SSR) markers as well as sequence-characterized amplified regions (SCAR) markers. Of these, two markers appeared to be very closely associated with the S_H3 leaf rust resistance gene and cosegregated perfectly with this gene in two populations analyzed, making them useful for MAS, for fine-mapping studies, and ultimately for cloning the S_H3 gene (Mahe et al. 2008).

Using a segregating F_2 population derived from a cross between the susceptible C. arabica cv. Caturra and a C. canephora-introgressed arabica line exhibiting high partial leaf rust resistance, Herrera et al. (2009) identified five AFLP and two SSR markers exhibiting significant association with partial resistance. de Brito et al. (2010) amplified 176 AFLP primer combinations using bulked segregant analysis (BSA) in the Timor Hybrid (Hibrido de Timor) and its derivatives and identified three markers linked to a coffee leaf rust resistance gene, of which two were distributed on either side flanking the resistance gene. In another study by Diola et al. (2011), the researchers used an F_2 population derived from a cross between Hibrido de Timor UFV 427-15 (resistant) and Catuai Amarelo IAC 30 (susceptible), and their genetic analysis showed that one dominant gene conferred resistance to race II of H. vastatrix on coffee. Development of a high-density genetic map showed six SCAR markers delimiting a chromosomal region of 9.45 cM and flanking the dominant gene at 0.7 and 0.9 cM (Diola et al. 2011).

More recently, Prakash et al. (2011) have successfully applied MAS in achieving durable rust resistance. Using two SCAR markers closely linked to the S_H3 gene (Sat244 and BA-124-12K-f), they were able to distinguish the presence or absence of the S_H3 gene using the arabica cultivar S.795, a cultivar derived from S.26, a spontaneous hybrid of C. arabica and C. liberica. The marker Sat244 was more efficient in distinguishing the homozygous and heterozygous status of the S_H3 gene (Prakash et al. 2011). This study is the first successful report of the use of MAS for maintenance breeding and gene pyramiding of coffee leaf rust resistance.

9.2.2 Coffee Berry Disease (*Colletotrichum kahawae*)

Coffee berry disease (CBD) caused by the fungus *Colletotrichum kahawae* was first identified in Kenya in 1922 (as cited in Silva et al. 2006). It infects all stages of the crop, from flowers to ripe fruits and occasionally leaves, and may cause up to 70%–80% crop losses if no control measures are adopted, with maximum crop losses occurring following infection of green berries, leading to formation of dark sunken lesions and premature dropping and mummification of the fruits (as cited in Silva et al. 2006). Annual economic impact on arabica coffee production in Africa is estimated to be US$300–500 million due to crop losses by CBD and cost of chemical control (van der Vossen and Walyaro 2009). Although CBD is currently restricted to Africa, precautions to prevent introduction of this disease should be taken in other coffee-producing countries (Silva et al. 2006).

It is reported that CBD resistance appears to be complete in C. canephora and partial in C. arabica (Silva et al. 2006). Breeding for CBD resistance in C. arabica was initiated in response to severe disease epidemics about 35–40 years ago in Kenya, Ethiopia, and Tanzania, with release of resistant cultivars to coffee growers since 1985 (van der Vossen and Walyaro 2009). Under field and laboratory conditions, differences in resistance of coffee trees to CBD have been observed, with higher resistance in Geisha 10, Blue Mountain, K7, Rume Sudan, and progenies of Hibrido de Timor than in Harar and Bourbon in Kenya (Silva et al. 2006). Resistance to CBD appears to be controlled by major genes at three different loci, with the highly resistant variety Rume Sudan carrying the dominant *R-* and the recessive *K-*genes (van der Vossen and Walyaro 1980). The variety Pretoria also has the *K-*gene, and the moderately resistant variety K7 carries only the recessive *K*-gene (van der Vossen and Walyaro 1980). One gene for CBD resistance on the *T*-locus with intermediate gene action is reported to be carried by Hybrido de Timor (van der Vossen and Walyaro 1980). The *T*-gene is also carried by the variety Catimor (Agwanda et al. 1997). A breeding strategy to enhance the stability of CBD resistance in *C. arabica* suggested by van der Vossen and Walyaro (1980) is to combine the Rume Sudan type resistance with that of Hybrido de Timor.

Agwanda et al. (1997) performed studies to identify random amplified polymorphic DNA (RAPD) markers associated with CBD resistance using SL28 and Caturra as susceptible cultivars and Rume Sudan, K7, and Catimor as resistant donors. Three RAPD markers were found to be closely associated with resistance to coffee berry disease in arabica coffee, controlled by the *T* gene found in the varieties Hibrido de Timor and Catimor (Agwanda et al. 1997). This provides an efficient way to select for the *T* gene in crosses involving Catimor and Hybrido de Timor through MAS. In another study aimed at understanding the genetic basis of host resistance and identification of molecular markers associated with coffee berry disease caused by *C. kahawae*, eight AFLP and two microsatellite markers were identified as being tightly linked to the resistant phenotypes, which were mapped to one unique chromosomal fragment introgressed from *C. canephora* (Gichuru et al. 2008). Of particular importance were the two highly repeatable and informative microsatellite markets, Sat 207 and Sat 235 (Gichuru et al. 2008).

The development of DNA markers linked to CBD resistance genes would considerably improve the efficiency of breeding programs by allowing selection at an early stage and gene stacking to increase the chances of high levels of durable resistance (Lashermes et al. 2000b; Gichuru et al. 2008). Additionally, another advantage would be preemptive breeding in countries where CBD is not yet present, but where climatic conditions are often favorable (Gichuru et al. 2008).

9.2.3 Root-Knot Nematode (*Meloidogyne* spp.)

Root-knot nematodes (*Meloidogyne* spp.) have become a major threat in all *C. arabica*-growing regions of the world (Noir et al. 2003). More than 15 species of *Meloidogyne* have been reported as pathogens of coffee, but the most damaging species reported in Central America is *M. exigua* Goeldi (Bertrand et al. 2001; Noir et al. 2003). In Guatemala, the most common species is *M. incognita*, which causes severe damage, often resulting in death of trees (Anzueto et al. 2001). In Central America, all cultivated varieties (such as Typica, Bourbon, Caturra, Catuai, Costa Rica 95, and IHCAFE90) are susceptible, with Costa Rica reporting an estimated drop in yield of 10%–20% due to general weakening of the trees (Bertrand et al. 2001).

While standard arabica cultivars are highly susceptible to *M. exigua*, several accessions of *C. canephora* have exhibited a high level of resistance, including the interspecific hybrid Timor Hybrid (as cited in Bertrand et al. 2001; Noir et al. 2003). In a study to identify molecular markers of resistant gene introgression from *C. canephora* for developing MAS, Bertrand et al. (2001) performed experiments to verify the nonexistence of *M. exigua* resistance in *C. arabica* and to confirm the presence of resistance genes in *C. canephora* and in lines derived from Timor Hybrid. The results revealed 100% resistance in *C. canephora* and susceptibility of about 98% of the *C. arabica* plants tested. Several lines derived from Timor Hybrid also exhibited high resistance similar to that observed in *C. canephora*. Study of introgressed AFLP markers indicated the existence of at least one dominant gene for resistance to *M. exigua* (Bertrand et al. 2001).

Anzueto et al. (2001) tested the response of semi-wild Ethiopian and Sudanese accessions for resistance to *M. incognita* isolate from Guatemala and Brazil. Of the Ethiopian accessions, 40% showed total resistance, with resistance dominant in F_1 and transmitted to the F_2 generations. Segregation of F_2 populations indicated the presence of a single dominant gene in some crosses and two complementary genes for other crosses (Anzueto et al. 2001). Reactions of Ethiopian semi-wild accessions to both Guatemalan and Brazilian isolates of *M. incognita* were similar, and the study confirmed the need to widen the cultivated arabica genetic base by using semiwild Ethiopian germplasm as a source of resistance to *M. incognita* (Anzueto et al. 2001).

Noir et al. (2003) performed a study to determine the mode of inheritance of *M. exigua* resistance in *C. arabica*, transferred from one of its progenitors, *C. canephora*, and to identify associated molecular markers. Using F_2 progeny derived from a cross between a root-knot nematode (*M. exigua*)-resistant introgression line, T2296, and a susceptible accession, Et6, segregation data analysis was performed, showing that resistance to *M. exigua* is controlled by a simply inherited major gene designated as the *Mex*-1 locus, with 14 AFLP markers associated with the resistance (Noir et al. 2003). A localized genetic map of the chromosome segment carrying the *Mex*-1 was constructed, which is an important starting point for MAS to enhance backcross breeding programs and to perform early selection of resistant seedlings.

9.2.4 Quantitative Traits

Of the two main cultivated coffees, robusta coffee (*C. canephora*) is characterized as a weak-flavored, neutral coffee with occasional strong and pronounced bitterness, and arabica coffee (*C. arabica*) is a milder and fruit-flavored acidulous beverage (Bertrand et al. 2003). AFLP analysis of the amount of *C. canephora* genetic material accumulated in *C. arabica* through introgressive breeding gave estimates of approximately 8%–27% presence of the *C. canephora* genome (Lashermes et al. 2000a). This could involve, in addition to desirable traits such as disease resistance, the incorporation of undesirable traits such as lower beverage quality (Bertrand et al. 2003). Breeding research in coffee has focused on incorporating disease- and pest-resistant traits of diploid coffee species into arabica coffee while preserving the beverage quality, as well as improving the beverage quality of robusta coffee. Use of species-specific markers such as microsatellites and AFLP to identify introgressed backcrossed progenies will enhance the possibility of early selection for desirable traits and speed up the breeding process (Herrera et al. 2004).

Chlorogenic acids (CGA) are phenolic compounds found in green coffee beans, and their presence modifies cup taste, with the quality of the beverage increasing when CGA content decreases (Ky et al. 1999). The three main CGA classes are caffeoylquinic acids (CQA), dicaffeoylquinic acids (diCQA), and feruloylquinic acids (FQA), each with three isomers (Ky et al. 1999). Using interspecific crosses between *C. pseudozanguebariae* Bridson and *C. liberica* var. *dewevrei* (De Wild. & T. Durand) Leburn, Ky et al. (1999) investigated CGA content in F_1 and backcross hybrids. A single major gene was identified for the 3-FQA isomer, with absence being dominant, and most other isomers showed additive patterns. For better planning of breeding programs, the authors recommend studying the relations between CGA and other interesting traits, such as morphological traits (leaf area; bean weight), phenological traits (fruitification cycle), and biochemical traits (sucrose; trigonelline; amino acids), and screening for markers near interesting QTL to develop MAS. An AFLP-based partial genetic linkage map was developed for backcross hybrids of *C. pseudozanguebariae* and *C. liberica* var. *dewevrei*, which is a starting point for identifying QTL for highly discriminant and specific traits (Ky et al. 2000).

In a study of fructification time in the interspecific cross *C. pseudozanguebariae* × *C. liberica* var. *dewevrei*, Akaffou et al. (2003) found fructification time to be an additive trait and identified a major gene, *ft1*, which was mapped on linkage group E. The main effects of the *ft1* gene were to lower caffeine content and 100-seed weight, without impacting chlorogenic acid, trigonelline, and sucrose content. Two molecular markers were found bracketing the *ft1* gene, which could be used for early MAS. Since breeding for lower caffeine content is a major aim of cup quality improvement in *C. canephora*, the identification and location of the *ft1* gene have significant implications in *C. canephora* breeding programs striving to achieve lower caffeine content (Akaffou et al. 2003).

In a more recent study of inheritance and relationships between agronomic and quality traits (fructification time, caffeine and heteroside contents, and 100-bean weight) in backcross hybrids derived from an interspecific cross between *C. pseudozanguebariae* and *C. canephora*, great variability was observed for these four traits (Akaffou et al. 2012). Short versus long fructification time was governed by one major gene with two codominant alleles, *ft1* and *ft2*. Absence versus presence of both caffeine and heteroside was also controlled by one major gene. While the allele responsible for presence of caffeine (*caf2*) dominated over the allele for absence (*caf1*), the alleles responsible for heteroside, *het1* and *het2*, were codominant (Akaffou et al. 2012). The 100-bean weight trait is controlled by multiple genes and is additive. The fructification time and heteroside content traits were additive, whereas the caffeine content was reported to be multiplicative. The two genes *ft* and *cat* were linked and were independent from the *het* gene. Significant correlation was found between fructification time and 100-bean weight and between caffeine content and 100-bean weight (Akaffou et al. 2012).

Using AFLP bands from *C. pseudozanguebariae*, Poncet et al. (2005) developed 14 SCAR primers, of which 12 successfully amplified across nine *Coffea* species. These SCAR primers will be useful in assembling a panel of PCR-based markers for comparative mapping studies and for MAS (Poncet et al. 2005).

In a study examining the phenotypic and genetic differentiation between *C. liberica* and *C. canephora* using AFLP, inter simple sequence repeat (ISSR), and SSR markers relative to 16 quantitative traits, 15 of them were found to be significantly different, with eight QTL associated with variation in petiole length, leaf area, number of flowers per inflorescence, fruit shape, fruit disc diameter, seed shape, and seed length (N'Diaye et al. 2007). This linkage map developed for the interspecific cross (*C. liberaca* × *C. canephora*) × *C. canephora* could provide a reference for any potential crosses involving the parents of these two species and for distinguishing them based on diagnostic markers (N'Diaye et al. 2007).

In one of the first studies identifying QTL combining agronomic and quality traits in coffee, Leroy et al. (2011) developed a genetic map comprising 236 molecular markers. A highly variable progeny of a *C. canephora* interspecific cross was assessed for 63 traits over 5 years and beverage quality was evaluated in relation to biochemical and cup tasting traits. Most of the QTL were consistent over the years, with the strongest QTL explaining a high percentage of variation for yield (34%–57%), bean size (25%–35%), chlorogenic acid content (22%–35%), sucrose and trigonelline content (29%–81%), and acidity and bitterness (30%–55%) (Leroy et al. 2011). Two genes coding for caffeine biosynthesis, one gene for biosynthesis of chlorogenic acids, and two genes involved in sugar metabolism were identified, making them valuable in MAS for quality improvement as well as for controlling the introgression of resistance genes from *C. canephora* to *C. arabica* without lowering quality (Leroy et al. 2011).

Production of chlorogenic acids involves the production of phenylalanine ammonia lyase (PAL) as the first enzyme entering the phenylpropanoid pathway (Lepelley et al. 2012). One gene has been characterized in *C. canephora* (*CcPAL1*) encoding the PAL gene, and Lepelley et al. (2012) identified two additional genes, *CcPAL 2* and *CcPAL3*, with similar genomic structures and encoding proteins to *CcPAL1*. As a next step to this, studying coffee *PAL* expression profiles and segregation in different *C. canephora* varieties and their offspring after crossing will be useful in advancing coffee breeding programs (Lepelley et al. 2012).

In another study for quantitative traits using *C. canephora*, Priyono and Sumirat (2012) identified 12 QTL for desirable cherry and green bean traits such as cherry maturation, harvesting period, cherry length, cherry width, cherry thickness, normal bean, weight of 100 green beans, large bean size, and extra large bean size. These were located on nine linkage groups forming clusters in 11 different chromosomal regions, providing a tool for design of a MAS program for robusta coffee improvement.

9.3 CONCLUSIONS

Efficient use of the genetic variation in wild species involves the genetic determination of the desirable trait and the ability to introgress the desirable DNA segments from wild species to the genome of the cultivated species (Prakash et al. 2004) without affecting quality traits (Herrera et al. 2002). Through MAS, DNA markers have enormous potential to improve the efficiency and precision of conventional plant breeding (Collard and Mackill 2008), especially for perennial tree crops such as coffee. The large number of QTL mapping studies has provided an abundance of DNA marker–trait associations (Collard and Mackill 2008).

In summary, progress made in coffee genetics can be utilized in breeding programs targeting MAS applications for coffee crop improvement in the following areas:

* Discrimination between parents in hybrid identification and selection of parents in breeding programs
* Breeding for enhanced resistance to pests and diseases and quality and yield improvement in arabica coffee
* Improvement of organoleptic traits of *C. canephora* and other diploid coffee species using molecular markers associated with caffeine, heterosides, flavonoids, and chlorogenic acids
* Improvement of cherry and green bean characteristics of *C. canephora* and other diploid species
* Rapid elimination of unsuitable lines after early-generation selection or tandem selection in breeding programs, thus concentrating on the most desirable materials leading to accelerated release of varieties

REFERENCES

Agwanda, C. O., P. Lashermes, P. Trouslot, M. C. Combes, and A. Charrier. 1997. Identification of RAPD markers for resistance to coffee berry disease, *Colletotrichum kahawae*, in arabica coffee. *Euphytica* 97:241–248.

Akaffou, D. S., C. L. Ky, P. Barre, S. Hamon, J. Louarn, and M. Noirot. 2003. Identification and mapping of a major gene (Ft1) involved in fructification time in the interspecific cross *Coffea pseudozanguebariae* × *C. liberica* var. Dewevrei: Impact on caffeine content and seed weight. *Theor Appl Genet* 106:1486–1490.

Akaffou, D. S., P. Hamon, S. Doulbeau et al. 2012. Inheritance and relationship between key agronomic and quality traits in an interspecific cross between *Coffea pseudozanguebariae* Bridson and *C. canephora* Pierre. *Tree Genet Genomes* 8:1149–1162.

Anthony, F., M. C. Combes, C. Astorga, B. Bertrand, G. Graziosi, and P. Lashermes. 2002. The origin of cultivated *Coffea arabica* L. varieties revealed by AFLP and SSR markers. *Theor Appl Genet* 104:894–900.

Anzueto, F., B. Bertrand, J. L. Sarah, A. B. Eskes, and B. Decazy. 2001. Resistance to *Meloidogyne incognita* in Ethiopian *Coffea arabica* accessions. *Euphytica* 118:1–8.

Bertrand, B., F. Anthony, and P. Lashermes. 2001. Breeding for resistance to *Meloidogyne exigua* in *Coffea arabica* by introgression of resistance genes of *Coffea canephora*. *Plant Pathol* 50:637–643.

Bertrand, B., B. Guyot, F. Anthony, and P. Lashermes. 2003. Impact of the *Coffea canephora* gene introgression on beverage quality of *C. arabica*. *Theor Appl Genet* 107:387–394.

Bettencourt, A. J. and C. J. Rodrigues Jr. 1988. Principles and practice of coffee breeding for resistance to rust and other disease. In R. J. Clarke and R. Macrae (eds), *Coffee, Volume 4 Agronomy*, pp. 199–234. London: Elsevier.

Collard, B. C. Y. and D. J. Mackill. 2008. Marker-assisted selection: An approach for precision plant breeding in the twenty-first century. *Phil Trans R Soc B* 363:557–572.

Cressey, D. 2013. Coffee rust gains foothold. *Nature* 494:587.

Davis, A. P. 2010. Six species of *Psilanthus* transferred to *Coffea* (Coffeeae, Rubiaceae). *Phytotaxa* 10:41–45.

Davis, A. P. 2011. *Psilanthus mannii*, the type species of *Psilanthus* transferred to *Coffea*. *Nordic J Bot* 29:471–472.

Davis, A. P., R. Govaerts, D. M. Bridson, and P. Stoffelen. 2006. An annotated taxonomic conspectus of the genus *Coffea* (Rubiaceae). *Bot J Linn Soc* 152:465–512.

Davis, A. P., J. Tosh, N. Ruch, and M. F. Fay. 2011. Growing coffee: *Psilanthus* (Rubiaceae) subsumed on the basis of molecular and morphological data: Implications for the size, morphology, distribution and evolutionary history of *Coffea*. *Bot J Linn Soc* 167:357–377.

de Brito, G. G., E. T. Caixeta, A. P. Gallina et al. 2010. Inheritance of coffee leaf rust resistance and identification of AFLP markers linked to the resistance gene. *Euphytica* 173:255–264.

Diola, V., G. G. de Brito, E. T. Caixeta, E. Maciel-Zambolim, N. S. Sakiyama, and M. E. Loureiro. 2011. High-density genetic mapping for coffee leaf rust resistance. *Tree Genet Genomes* 7:1199–1208.

Edwards, J. D. and S. R. McCouch. 2007. Molecular markers for use in plant molecular breeding and germplasm evaluation. In E. Guimaraes, J. Ruane, B. Scherf, A. Sonnino, and J. Dargie (eds), *Marker Assisted Selection: Current Status and Future Perspectives in Crops, Livestock, Forestry and Fish*, pp. 29–49. Rome: FAO.

Fazuoli, L. C., M. Perez Maluf, O. Guerreiro Filho, H. Medina Filho, and M. B. Silvarolla. 2000. Breeding and biotechnology of coffee. In T. Sera, C. R. Soccol, A. Pandey, and S. Roussos (eds), *Coffee Biotechnology and Quality*, pp. 27–45. Netherlands: Kluwer Academic Publishers.

Gichuru, E. K., C. O. Agwanda, M. C. Combes et al. 2008. Identification of molecular markers linked to a gene conferring resistance to coffee berry disease (*Colletotrichum kahawae*) in *Coffea arabica*. *Plant Pathol* 57:1117–1124.

Guerreiro Filho, O., M. B. Silvarolla and A. B. Eskes. 1999. Expressions and mode of inheritance of resistance in coffee to leaf miner *Perileucoptera coffeella*. *Euphytica* 105:7–15.

Herrera, J. C., M. C. Combes, F. Anthony, A. Charrier, and P. Lashermes. 2002. Introgression into the allotetraploid coffee (*Coffea arabica* L.): Segregation and recombination of the *C. canephora* genome in the tetraploid interspecific hybrid (*C. arabica* × *C. canephora*). *Theor Appl Genet* 104:661–668.

Herrera, J. C., M. C. Combes, H. Cortina, and P. Lashermes. 2004. Factors influencing gene introgression into the allotetraploid *Coffea arabica* L. from its diploid relatives. *Genome* 47:1053–1060.

Herrera, P. J. C., A. G. Alvarado, G. H. A. Cortina, M. C. Combes, G. G. Romero, and P. Lashermes. 2009. Genetic analysis of partial resistance to coffee leaf rust (*Hemileia vastatrix* Berk & Br.) introgressed into the cultivated *Coffea arabica* L. from the diploid *C. canephora* species. *Euphytica* 167:57–67.

International Coffee Organization. 2013a. Trade statistics. http://www.ico.org/trade_statistics.asp (accessed April 27, 2013).

International Coffee Organization. 2013b. Coffee leaf rust outbreak. http://dev.ico.org/leafrust_e.asp (accessed May 15, 2013).

Ky, C. L., P. Barre, M. Lorieux et al. 2000. Interspecific genetic linkage map, segregation distortion and genetic conversion in coffee (*Coffea* sp.). *Theor Appl Genet* 101:669–676.

Ky, C. L., J. Louarn, B. Guyot, A. Charrier, S. Hamon, and M. Noirot. 1999. Relations between and inheritance of chlorogenic acid contents in an interspecific cross between *Coffea pseudozanguebariae* and *Coffea liberica* var "dewevrei". *Theor Appl Genet* 98:628–637.

Lashermes, P., S. Andrzejewski, B. Bertrand et al. 2000a. Molecular analysis of introgressive breeding in coffee (*Coffea arabica* L.). *Theor Appl Genet* 100:139–146.

Lashermes, P., M.-C. Combes, J. Robert et al. 1999. Molecular characterization and origin of the *Coffea arabica* L. genome. *Mol Gen Genet* 261:259–266.

Lashermes, P., M. C. Combes, P. Topart, G. Graziosi, B. Bertrand, and F. Anthony. 2000b. Molecular breeding in coffee (*Coffea arabica* L.). In T. Sera, C. R. Soccol, A. Pandey, and S. Roussos. (eds), *Coffee Biotechnology and Quality*, pp. 101–112. Amsterdam: Kluwer Academic Publishers.

Lepelley, M., V. Mahesh, J. McCarthy et al. 2012. Characterization, high-resolution mapping and differential expression of three homologous *PAL* genes in *Coffea canephora* Pierre (Rubiaceae). *Planta* 236:313–326.

Leroy, T., F. De Bellis, H. Legnate et al. 2011. Improving the quality of African robustas: QTLs for yield- and quality-related traits in *Coffea canephora*. *Tree Genet Genomes* 7:781–798.

Mahe, L., M.-C. Combes, V. M. P. Varzea, C. Guilhamon, and P. Lashermes. 2008. Development of sequence characterized DNA markers linked to leaf rust (*Hemileia vastatrix*) resistance in coffee (*Coffea arabica* L.). *Mol Breed* 21:105–113.

N'Diaye, A., M. Noirot, S. Hamon, and V. Ponc. 2007. Genetic basis of species differentiation between *Coffea liberica* Hiern and *C. canephora* Pierre: Analysis of an interspecific cross. *Genet Resour Crop Evol* 54:1011–1021.

Noir, S., F. Anthony, B. Bertrand, M. C. Combes, and P. Lashermes. 2003. Identification of a major gene (*Mex*-1) from *Coffea canephora* conferring resistance to *Meloidogyne exigua* in *Coffea arabica*. *Plant Pathol* 52:97–103.

Pearl, H. M., C. Nagai, P. H. Moore, D. L. Steiger, R. V. Osgood, and R. Ming. 2004. Construction of a genetic map for arabica coffee. *Theor Appl Genet* 108:829–835.

Pinto-Maglio, C. A. F. 2006. Cytogenetics of coffee. *Braz J Plant Physiol* 18(1):37–44.

Poncet, V., P. Hamon, M.-B. Sauvage de Saint Marc, T. Bernard, S. Hamon, and M. Noiret. 2005. Base composition of *Coffea* AFLP sequences and their conversion within the genus. *J Hered* 96(1):59–65.

Prakash, N. S., M. C. Combes, N. Somanna, and P. Lashermes. 2002. AFLP analysis of introgression in coffee cultivars (*Coffea arabica* L.) derived from a natural interspecific hybrid. *Euphytica* 124:265–271.

Prakash, N. S., D. V. Marques, V. M. P. Varzea, M. C. Silva, M. C. Combes, and P. Lashermes. 2004. Introgression molecular analysis of a leaf rust resistance gene from *Coffea liberica* into *C. arabica* L. *Theor Appl Genet* 109:1311–1317.

Prakash, N. S., B. Muniswamy, B. T. Hanumantha et al. 2011. Marker assisted selection and breeding for leaf rust resistance in coffee (*Coffea arabica* L.)—Some recent leads. *Indian J Genet* 71:1–6.

Priyono, P. and U. Sumirat. 2012. Mapping of quantitative trait loci (QTLs) controlling cherry and green bean characters in the robusta coffee (*Coffea canephora* Pierre). *J Agric Sci Technol* A2:1029–1039.

Ruane, J. and A. Sonnino. 2007. Marker-assisted selection as a tool for genetic improvement of crops, livestock, forestry and fish in developing countries: An overview of the issues. In E. Guimaraes, J. Ruane, B. Scherf, A. Sonnino, and J. Dargie (eds), *Marker Assisted Selection: Current Status and Future Perspectives in Crops, Livestock, Forestry and Fish*, pp. 3–13. Rome: FAO.

Silva, M. d. C., V. Varzea, L. Guerra-Guimaraes et al. 2006. Coffee resistance to the main diseases: Leaf rust and coffee berry disease. *Braz J Plant Physiol* 18:119–147.

van der Vossen, H. A. M. and D. J. Walyaro. 1980. Breeding for resistance to coffee berry disease in *Coffea arabica* L. II. Inheritance of resistance. *Euphytica* 29:777–791.

van der Vossen, H. A. M. and D. J. Walyaro. 2009. Additional evidence for oligogenic inheritance of durable host resistance to coffee berry disease (*Colletotrichum kahawae*) in arabica coffee (*Coffea arabica* L.). *Euphytica* 165:105–111.

CHAPTER 10

Advances in Papaya Genomics

Savarni Tripathi, Luz Castro, Gustavo Fermin, and Paula Tennant

CONTENTS

10.1 INTRODUCTION

Papaya (*Carica papaya* L.) is an important fruit crop grown in tropical and subtropical regions worldwide. Described as a large, fast-growing, arborescent herb, papaya is cultivated for its climacteric, melon-like fruit. It is one of the few plant species that bears fruit continuously throughout the year within months of planting and is commonly found in small backyard gardens for household consumption and organized commercial plantings for domestic markets and export (Chan and Paull 2009). Papaya is well known for its nutritional, medicinal, and health benefits. The fruit is rich in vitamins and minerals and received a high rank on nutritional scores among 38 fruits based on the percentage of the U.S. Recommended Daily Allowance for vitamin A, vitamin C, potassium, folate, niacin, thiamine, riboflavin, iron, and calcium (Liebman 1992; OECD 2005). Consumption of the fruit is frequently recommended to help prevent and control vitamin A deficiency, which causes childhood blindness in developing countries (Chandrika et al. 2003). Despite papaya's reputation as a dessert fruit, it is of far greater importance as a staple food for subsistence farmers in many countries, where green fruit, leaves, and flowers are cooked and eaten as vegetables (Watson 1997). Papaya is also cultivated for its commercially valuable proteolytic enzyme, papain. Commonly found and extracted from green, unripe papaya fruit, papain is used as a component in meat tenderizers, the clarification of beer, the external treatment of warts, scars, and injured dermal tissue, and the development of inhibitors against animal cysteine proteases that are implicated in diseases such as muscular dystrophy (Czaplewski et al. 1999; Hewitt et al. 2000; Scheldeman et al. 2011; Starley et al. 1999). Apart from papain, other biologically active compounds such as chymopapain, cystatin, beta-tocopherol, ascorbic acid, flavonoids, and cyanogenic glucosides exhibit a wide range of pharmacological effects including antioxidation, anti-inflammation, antiplatelet, antithrombotic, and antiallergic effects as well as anticancer activity (Ching and Mohamed 2001; Mello et al. 2008; Miean and Mohammed 2001; Nguyen et al. 2013; Otsuki et al. 2010; Seigler et al. 2002; Waly et al. 2012).

Papaya belongs to the order Brassicales, which comprises 19 families including Caricaceae, which contains papaya, and Brassicaceae, the mustard or cabbage family, of which *Arabidopsis thaliana* is a member. The family Caricaceae consists of six genera. Papaya is the only species in the genus *Carica* (Badillo 1993, 2000) and economically is the most important representative of the family. The family has an amphi-Atlantic distribution with two species in tropical Africa and about 33 in Central and South America and the Caribbean (Antunes-Carvalho and Renner 2012). While the African *Cylicomorpha* species are large trees, *Horovitzia* consists of species of herbaceous plants covered in stinging hairs, members of *Jacaratia* are also tree species, and those of the genus *Jarilla* in Mexico are perennial herbs. Most of the *Vasconcellea* spp. (or highland papayas from the Andes that were formerly recognized within the genus *Carica*) are trees, but others exhibit shrub-like or climber growth habits (Badillo 1971, 1993, 2000, 2001). *C. papaya*, a soft-wooded perennial herblike pachycaul tree, is closely related to *Jarilla* and *Horovitzia*. The three presumably diverged within Caricaceae in South America during the Oligocene, 27 million years ago (MYA), around the time of the formation of the Panamanian Isthmus (Antunes-Carvalho and Renner 2012). Dioecy appears to be ancestral in the family, as hermaphroditic species do not exist. Papaya is one of two trioecious species in the family, with three basic sex forms that are morphologically distinct (Yu et al. 2008b). Sex determination is controlled by primitive sex chromosomes (Liu et al. 2004b).

Papaya has long been one of the best genetically characterized species of Caricaceae. Various aspects of genomics are relatively easy with papaya because it is diploid ($2n = 18$) and has a small genome, which initially was published as 372 Mbp (Arumuganathan and Earle 1991) and recently adjusted to 442.5 Mbp (Gschwend et al. 2013). The genome is portioned into small metacentric and homomorphic chromosomes of similar size (1.52–2.29 μm) and is highly euchromatic (genetically active region). Heterochromatic (genetically inactive condensed region) domains (30%–35%) are mostly localized in the pericentromeric regions of chromosomes. It is 65%–70% euchromatic. Similar to other dicotyledonous plants, the base composition is 64.7% AT and 35.3% GC (Araujo et al. 2010; Ming et al. 2008). In comparison with other genomes, papaya has the lowest number of genes. The genome shows 24,746 genes, with an average gene length of 2,373 bp (Ming et al. 2008). There are few genome-wide duplications (i.e., low gene redundancy) except, for example, MADS-box genes, starch synthases, and genes involved in cell expansion and the production of volatiles. Additionally, papaya is striking in its sex determination system and evolution—apparently unique for this species (Ming et al. 2011, 2012; Weingartner and Moore 2012).

Over the last decade or so, availability of genetic knowledge of papaya has accelerated with the development of molecular markers, linkage and physical maps, comparative genomics studies, and databases and the very recent release of its genome sequence. Papaya is the fifth flowering plant and first transgenic eukaryotic organism to be sequenced. The transgenic variety SunUp, which saved the Hawaiian papaya industry from devastation by the *Papaya ringspot virus* (PRSV), was developed through genetic transformation of the papaya cultivar Sunset (Fitch et al. 1992; Gonsalves 1998). Besides being the first transgenic fruit in the market since 1998, its genetic manipulation has served to uncover the basic aspects of gene silencing and the intricacies of transgenic viral resistance (Fermin et al. 2010). Papaya is also the first fleshy fruit with a climacteric ripening pattern to be sequenced. Keeping in mind the agricultural, nutritional, and medicinal value of this fruit crop, we summarize important aspects of papaya genome research. We report on the status of the papaya genome and on current molecular and genetic tools available in papaya that will contribute to the process of cultivar development as well as to the characterization of biological features relevant to the tropical environment and nutritional and medicinal profiles.

10.2 EXPERIMENTALLY BASED FUNCTIONAL GENOMICS

10.2.1 Genetic Approaches

10.2.1.1 Marker-Based Approaches

Prior to the 1990s, a classical map for papaya based on the study of cosegregation between major morphological genes was not available. Only a few linkages between morphological markers had been reported. Since then, various DNA markers have been developed. These include dominant markers such as random amplified polymorphic DNA (RAPD), simple sequence repeats (SSRs) or microsatellites, and amplified fragment length polymorphisms (AFLP). The methods have proved useful for the large-scale characterization of the papaya genome, for which genomic sequence information was not previously available.

Early attempts by Hofmeyr (1939b) to identify markers that coinherit with sex led to the discovery of a loose linkage between three morphological markers, namely, sex type, flower color, and stem color (Hofmeyr 1939b). Hofmeyr's work generated the first genetic map of papaya and covered 41 cM of the genome. Hofmeyr (1939b) proposed that sex determination in papaya is controlled by a single gene with three alleles: M, M^h, and m; males (Mm) and hermaphrodites ($M^h m$) are heterozygous, while females (mm) are homozygous recessive. Dominant combinations of MM, M^hM^h, and MM^h were reported to be lethal, resulting in a 2:1 segregation of dominant markers on the linkage group where the sex determination gene is located rather than the normal 3:1 ratio for a single autosomal dominant gene (refer to the section on sex chromosome evolution). Genes for purple versus nonpurple stem color were found to be fairly closely linked with flower color genes for yellow versus white flower color, and comparatively loosely linked with sex factors.

During the 1990s, when the utility of polymerase chain reaction (PCR)-based markers for mapping was being explored, Sondur et al. (1996) published the second genetic map of papaya. Using an F_2 progeny derived from a University of Hawaii UH breeding line 356 × Sunrise cross, Sondur et al. (1996) mapped 61 RAPD markers and one morphological marker (*SEX1*, the gene that determines plant sex in papaya) to 11 linkage groups comprising 999.3 cM. Based on the assumption that the papaya genome is about 1350 cM, Sondur et al. (1996) surmised that the map covered a significant portion of the papaya genome and was adequate for the genetic analysis of quantitative traits. Additionally, an extended model on sex expression in papaya was presented in light of their observations and the current thinking on the regulation of floral development and its control through *trans*-acting regulatory proteins. It was theorized that the three alleles controlling sex determination in papaya encode different *trans*-acting factors, which direct the expression of the different flower forms.

Further investigations into the expression of sex in papaya followed in the next few years. Sex determination in papaya is intriguing and has been a frequent subject of genetic analyses, not only from the stand point that papaya is a trioecious species with three basic sex forms that are morphologically distinct, but also because it is directly related to efficient commercial fruit production. In commercial production, the hermaphrodite papaya trees are preferred over female trees. Female trees produce round fruits that require greater container volume for shipping than the pyriform-shaped fruits from hermaphrodite trees. Moreover, fruits from female trees are variably sized, and contain large central air spaces and few seeds, generally resulting from unreliable cross-pollination (Fitch et al. 2005). Since sex determination in papaya is only possible at 6 months after germination (Gonsalves 1994), orchards are generally established using a system of overplanting. This involves the use of a number of established seedlings (6–8 weeks after germination) or a number of seeds per single planting site. At the time of flowering and depending on the variety, plants are thinned in each planting position to one bisexual plant or a ratio of eight to ten females to one male. The system is regarded as wasteful of seeds, water, and fertilizer. Moreover, less than optimum plant growth and late bearing may result given the initial intraplant competition (Fitch et al. 2005). The cost incurred during the establishment of papaya orchards could be reduced if screening of the plants, for example using sex-specific DNA markers, occurred far earlier in the establishment of orchards. On this basis, reports of molecular markers tightly linked to the sex of plants were available in the early 2000s. In a study using southern hybridization and the oligonucleotide $(GATA)_4$ as a probe, Parasnis et al. (2000) identified sex-linked DNA fragments. Urasaki et al. (2002a,b) subsequently reported on a RAPD marker specific to male and hermaphrodite plants of papaya. It was also shown that a sequence-characterized amplified region (SCAR) marker could be used to determine the sex of papaya plants at an early developmental stage. Given that the study was conducted with a small number of plants and that linkage data were not provided, Deputy et al. (2002) cloned RAPD fragments tightly linked to *SEX1* and developed a PCR-based system for rapidly and reliably determining papaya plant sex. Three RAPD products were cloned and a portion of their DNA sequenced. The sequences showed no significant similarity to sequences in public sequence databases and did not appear to encode proteins. However, SCAR primers based on these sequences generated products from hermaphrodite and male plants with an overall accuracy of 99.2%. A test population of close to 2000 plants was analyzed. Work into sex determination in papaya is further addressed in the later section on evolutionary genomics.

Although the studies up to this time together served to enhance genetic and genome work in papaya, papaya still lagged behind other crops. High-resolution genetic maps existed for the model plant *Arabidopsis* (Peters et al. 2001) and major crops such as maize (Davis et al. 1999), rice (Harushima et al. 1998; Wu et al. 2002), tomato, and potato (Haanstra et al. 1999; Tanksley et al. 1992). A high-density genetic map of papaya was sorely needed to launch into extensive genomic research on this fruit crop. This subsequently became the focus of various research groups and significant progress has since been made in recent years.

Two high-density genetic linkage maps were constructed and reported between 2004 and 2007. The first high-density genetic map of papaya was constructed with 1501 markers, included 1498 AFLP markers, 2 morphological markers (sex type and fruit flesh color), and 1 transgenic marker, namely that of the PRSV coat protein marker of the transgenic variety SunUp (Ma et al. 2004). Twelve major linkage groups covered a total length of 3294.2 cM, with an average distance of 2.2 cM between adjacent markers. However, it was the later map developed with highly informative SSRs that proved more helpful in aligning papaya genome sequences to linkage groups and integrating genetic and physical maps. The map generated by SSR markers contained 707 markers including 706 sequence-based SSR markers and one morphological marker, fruit flesh color (Chen et al. 2007), and spanned 1068.9 cM, with an average distance of 1.51 cM between adjacent markers. These SSR markers were developed from either BAC end or whole-genome shotgun (WGS) sequence reads (Chen et al. 2007). A BAC library of the transgenic papaya variety SunUp was constructed with $13.7\times$ genome equivalents (Ming et al. 2001). BAC ends of the library were sequenced (Lai et al. 2006), providing the first glimpse of the sequence composition of the papaya genome. A BAC-based physical map of papaya using high information-content fingerprinting and overgo hybridization of conserved DNA probes from *Arabidopsis* and *Brassica* was generated (Yu et al. 2009). This physical map was integrated into a sequence-tagged high-density genetic map and draft genome sequence.

Blas et al. (2009) later reported on the saturation of the papaya genetic map with AFLP markers, which generated better genome coverage. Using an F_2 mapping population derived from a cross between AU9 breeding line and SunUp papaya, a map was generated that spanned 945.2 cM, covered 14 linkage groups, and contained 712 SSR, 277 AFLP, and 1 morphological marker, fruit flesh color. The average distance between adjacent loci ranged from 0.5 to 1.6 cM. The morphological markers, flower color, and stem color of the classical map (Hofmeyr 1939b) were not integrated in this map, as there were no differences in flower or stem color in the parents or mapping population. Sex type and fruit flesh color were, however, mapped to linkage groups 1 and 5, respectively. Additionally, reductions ranging from 5% to 42% in linkage groups 1, 2, and 4 were reported with this map as well as an increase in the length of linkage group 9, by 71%. Although the authors proposed that the papaya genetic map was subsequently near saturation, a deficiency in the telomeric regions was recognized.

Maps generated after this time examined economically important commercial traits of the crop (Blas et al. 2012; Srinivasan 2004). Typically, early- and low-bearing trees are commercially preferred traits. Early flowering correlates to lower node numbers at flowering (Nakasone and Storey 1955; Subhadrabandhu and Nontaswatsri 1997). While fruit weight and flesh color depend on the market preference, there is less variation between regions on standards established for classifying and grading fruit shape for the international market. Fruits should be pyriform or pear-shaped, not rounded, and free of deforming longitudinal ridges or seams. The latter condition is referred to as catfacing or carpellody and results when stamens develop abnormally into carpel-like fleshy structures. Low night temperatures in combination with high moisture and nitrogen levels contribute to the development of carpellodic fruits (Awada and Ikeda 1957). Srinivasan (2004) analyzed two qualitative traits, phosphoglucomutase activity (PGM) and flesh color, and three quantitative traits, node number at floral conversion, carpellody, and fruit weight. Blas et al. (2012) later examined the quantitative trait loci controlling fruit size and shape in papaya.

Srinivasan (2004) generated an F_2 progeny derived from a breeding line from the Northern Mariana Islands, Saipan Red, and a commercial variety from Hawaii, Kapoho. Phenotypic data collected from trees in the field indicated two possible linkages: one between the PGM locus and one of the major QTLs controlling number of nodes to first flowering, and another between flesh color and carpellody. Marker genotyping with this F_2 progeny mapped 513 AFLP markers, an isozyme marker (phosphoglucomutase), and a morphological marker (flesh color) to 15 linkage groups covering 2066 cM, with an average interval distance of 5 cM (Srinivasan 2004). Associations observed in the field between PGM and node number at floral conversion and flesh color and carpellody were not confirmed at the molecular level. PGM was found on linkage group 5 with flanking markers 11 and 8.3 cM away, and fruit flesh color located at one end of linkage group 14 with a linked marker of 4.1 cM. Although Ma et al. (2004) identified three QTLs for fruit weight, Srinivasan (2004) reported on five markers but no QTLs, which may be a function of the population size. Markers on four linkage groups (linkage groups 2, 3, 4, 7) showed significant association with the expression of carpellody. A single QTL affecting node number at which flowering occurred was also reported.

Blas et al. (2012) generated an F_2 mapping population in order to analyze fruit size and shape in papaya. A mapping population was established by crossing Khaek Dum, a papaya variety from Thailand, and 2H94. The Khaek Dum variety produces red flesh on large cylindrical fruit weighing an average of 1.2 kg, while 2H94 is a small-fruited mutant identified among progeny from a cross between Kapoho and SunUp. 2H94 bears yellow flesh on pear-shaped fruits with weights averaging

0.2 kg. A genetic map composed of 11 dominant and 46 codominant markers spanning a total length of 672.1 cM was generated. Fifty-four SSR markers, the morphological flesh color locus, and the CPFC1 and CPFC2 SCAR markers for flesh color were mapped to 11 linkage groups corresponding to linkage groups 1–9 and 11. Fourteen QTLs with phenotypic effects on papaya fruit shape and size were found across six linkage groups, and there were clusters of two or more QTLs on linkage groups 2, 3, 4, 6, 7, and 9. One cluster of overlapping QTLs positioned at approximately 12 cM on linkage group 9 affects fruit weight, length, and shape. A second cluster located on linkage group 7 around 1.0 cM, and flanked by the SSR loci located at 16.4 and 67.0 cM, had minor effects on fruit length and shape. Putative orthologs to loci described in tomato that regulate fruit shape and size were also located. A putative *ovate* ortholog was located on linkage group 7, a *sun* ortholog on 4, and *fw2.2* on 9 (Blas et al. 2012). Identification of these putative *ovate*, *sun*, and *fw2.2* genes on linkage groups containing QTLs for papaya fruit size and shape is suggestive of parallel functions for these genes in papaya. In tomato, fruit shape is essentially determined early in flower development, and fruit size apparently correlates to the number of ovules successfully fertilized (Varga and Bruinsma 1986). The locus *ovate* is associated with tomato fruit elongation and affects early ovary development in floral organogenesis (Hedrick and Booth 1907; Ku et al. 1999, 2001). On the other hand, *sun* contributes to elongation of the ovary pericarp after fertilization (van der Knaap and Tanksley 2003), and *fw2.2* modulates fruit size (Cong et al. 2002). It is theorized that *fw2.2* links carbohydrate supply with ovary cell proliferation in tomato flowers (Nesbitt and Tanksley 2001). It has long been recognized that fruit shape in papaya is a sex-related trait. However, it may be that flower shape is also an indicator of fruit shape. Female papaya flowers tend to be shorter and broader with a bulbous base at the stem-end of the flower, whereas hermaphrodite flowers are cylindrical or peanut-shaped. Like tomato, papaya fruit size has also been linked with fertilization. Small female fruits containing few or no seeds have been documented (Nakasone and Paull 1998).

10.2.1.2 Mutagenesis

The generation of mutations has been applied to papaya more for the improvement of one or two major traits and much less for the identification of mutated genes that are fundamental to understanding gene function. Two classical approaches involving exposure to ethyl methanesulfonate (EMS) or ionizing radiation such as fast neutrons, γ-rays, and x-rays (Koornneef et al. 1982) have been successfully applied to the generation of mutants in papaya. The alkylating agent EMS induces a high density of DNA point mutations and provides an allelic series for any particular locus with different effects on gene function that can be easily isolated by screening a few thousand mutagenized plants (Østergaard and Yanofsky 2004). Compared with irradiation mutagenesis, EMS induces relatively few strand breaks that lead to inversion or deletion mutations. Ionizing radiation produces deletions, ranging from one to thousands of nucleotides, as well as chromosomal rearrangements, typically resulting in loss of function or, in most cases, null alleles (Li et al. 2001). Insertional mutagenesis using DNA sequences such as T-DNA, transposons, or retrotransposons as mutagens (Jeon et al. 2000) has not been reported with papaya.

Santosh et al. (2010) described mutants derived from EMS-mutated seeds of the papaya variety Surya. The variety Surya is a gynodioecious hybrid from the cross Sunrise Solo × Pink Flesh producing medium-sized fruits (Dinesh 2010; Santosh et al. 2010). Mutants altered in plant height, stem girth, and number of leaves were identified. Earlier studies described mutant populations generated by irradiation. γ-Ray mutagenesis studies with a local Indian papaya variety in the 1980s resulted in the release of the dwarf mutant papaya variety, PusaNanha (Dinesh 2010) in India. Trees of PusaNanha grow up to 160 cm and are male or female. Female trees bear yellow flesh fruits of medium size, round to ovate in shape, and with total soluble solutes (TSS) of about 8°Brix. The variety is well suited for high-density planting or pot cultivation and is also tolerant to water logging (Ram and Srivastava 1984). Encouraging results were also reported in the screening of putative

Eksotika papaya mutants generated by γ-ray irradiation (Chan 2009). Mutants were evaluated for traits such as plant vigor, fruiting character, yield, fruit quality, sex segregation, and resistance to PRSV and malformed top disease (MTD). The latter disease is caused by a *Cladosporium*–thrip complex and is very damaging to the foliage of young papaya plants. Mutants producing fruits with increased TSS and yellow rather than orange-red fruit flesh were identified. However, no changes in the yield parameters were obtained. Moreover, male and female trees rather than hermaphrodite and female trees were observed. Although initial inoculation tests gave encouraging levels of resistance against PRSV, fewer trees proved resistant once disease pressure was increased. Only two trees remained asymptomatic following four inoculations and the trees succumbed to virus infections on transfer to the field. More encouraging results were observed with the evaluations for resistance against MTD. Several trees were reported to be resistant to the disease and the procedure was regarded as having potential for the development of a papaya variety with MTD resistance. More recently, Mahadevamma et al. (2012) investigated the effects of gamma radiation on callus and shoot tips of the papaya variety Coorg Honey Dew. In some instances, coupling *in vitro* culture techniques with those of mutagenesis have proven effective in overcoming the limitations of mutagenesis (Ahloowalia 1998; Maluszynski et al. 1995). However, the regeneration of callus was not observed in this study, and irradiated papaya shoot tips exhibited abnormal, narrow leaves with a mosaic pattern after 4 weeks in culture. Identification of the gene mutations in these forward genetic screens has not been reported.

10.2.1.3 Transgenic Technology

Methods for validating gene function usually require the use of transformation (Dixon et al. 2007), since predictions of gene function based on sequence homology alone do not necessarily provide information on the biological role of the gene *in planta* (van Enckevort et al. 2005). Genetic transformation is used to overexpress or knock down genes in transgenic plants and facilitate assessment of the phenotypic effect.

Transformation experiments with papaya were initiated in the late 1980s (Fitch et al. 1990; Fitch and Manshardt 1990). Contrary to the approach used with other fruit crops, namely gene delivery via genetically engineered *Agrobacterium* strains, microprojectile bombardment was investigated. Initial experiments focused on the development of tissue culture conditions and identification of papaya explants that could be efficiently transformed and regenerated into plants. Embryogenic zygotic embryos, embryogenic calli, somatic embryos derived from hypocotyls, and zygotic embryos were examined as potential explants for transformation. It was established that 2,4-dichlorophenoxyacetic acid–treated zygotic embryos derived from immature seeds of 90–120-day-old green fruit had the highest transformation capacity, at 1.42% (Fitch et al. 1990; Fitch and Manshardt 1990). Since the original report in the early 1990s, an improved protocol for obtaining somatic embryos was released (Cai et al. 1999; Gonsalves et al. 1998). Two modifications proved valuable. Somatic embryos were bombarded earlier during active cell proliferation, selection and screening on antibiotics was reduced, and antibiotics were excluded from subsequent regeneration steps. Later protocols utilized somatic embryos generated from hypocotyl explants (Fitch 1993, 1995) along with transformation by *Agrobacterium* strains (Cheng et al. 1996; Fitch et al. 1993; Kung et al. 2010; Yang et al. 1996). Predictable DNA integration occurs more frequently in *Agrobacterium*-mediated transformation, without the rearrangements or deletions generally associated with biolistic-mediated transformation.

Although papaya was one of the first fruit crops to be routinely transformed and protocols have been developed for various cultivars, genetic transformation has been used mainly as a tool for genetic improvement of the crop and not for the determination of gene function. One of the primary targets has been the development of resistance against the PRSV given the commercial losses incurred by epidemics and the economic importance of the crop in a number of developing regions

(Fermin et al. 2004; Oliver et al. 2011; Suzuki et al. 2007; Tennant et al. 2005). Other target traits include resistance to mites (McCafferty et al. 2006) and *Phytophthora* (Zhu et al. 2007), tolerance to herbicides (Cabrera-Ponce et al. 1995) and aluminum toxicity (de la Fuente et al. 1997), and output traits such as delayed ripening and improved shelf life through the inhibition of ethylene production or decrease in loss of firmness (Abu Bakar et al. 2001, 2005; Daud et al. 2005; Neupane et al. 1998). The use of papaya to produce vaccines against tuberculosis and cysticercosis in animals has also been explored (Hernández et al. 2007).

10.2.2 Omics Approaches

The large-scale study of genes (genomics), proteins (proteomics), coding and noncoding RNAs (transcriptomics), and metabolites (metabolomics) has sped up the progress of uncovering the basic processes of evolution, domestication, morphogenesis, and regulatory systems and networks, among others, and paved the way to the improvement of many crop plants, including papaya. In the following sections, a succinct review of the most recent literature on diverse *omics* approaches employed in the study of papaya is presented. A few expressed sequence tag (EST) projects have examined fruit softening, aroma, color biosynthesis, and sex determination in papaya. Proteomic analysis has thus far focused mainly on fruit ripening, and there are a few metabolomic studies mostly dedicated to cataloguing groups of secondary metabolites.

10.2.2.1 Transcriptome Sequencing and Transcript Profiling

Transcriptome sequencing is fundamental to the validation of raw genomic data and the discovery of genomic variants, determination of alternative splicing forms of genes, and DNA marker development (Llaca 2012). Transcriptome analysis is primarily conducted through the sequencing of ESTs. ESTs are single-pass sequences of random cDNA clones. They are partial cDNA sequences corresponding to mRNAs generated from randomly selected cDNA library clones (Parkinson and Blaxter 2009). Expression analysis of papaya transcriptomes is at an early stage. Currently, there are 77,393 ESTs available for papaya in GenBank; 16,362 are unigenes, with 92.1% matching the assembled WGS sequences. The transcribed sequences match about 3.6% (13.4 Mb) of the whole genome and 48% of the genic region in WGS (Ming et al. 2008).

Devitt et al. (2006) published the first snapshot of genes expressed during papaya fruit development. They generated a total of 1171 ESTs from randomly selected clones of two independent fruit cDNA libraries derived from yellow- and red-flesh varieties. The most abundant sequences were those encoding enzymes that hydrolyze chitin (chitinase), control the production of ethylene (1-aminocyclopropane-1-carboxylic acid oxidase), and catalyze the decomposition of hydrogen peroxide (catalase) as well as enzymes involved in the biosynthesis of methionine (methionine synthase). DNA sequence comparisons identified ESTs with significant similarity to genes associated with fruit color biosynthesis, namely orthologs of ξ-carotene desaturase (ZDS), p-hydroxyphenyl-pyruvate dioxygenase (HPD), and β-carotene hydroxylase (β-CHX). ESTs associated with fruit aroma included isoprenoid biosynthesis and shikimic acid pathway genes and proteins associated with acyl lipid catabolism. Sequences with predicted roles in fruit softening, that is putative cell wall hydrolases, cell membrane hydrolases, and ethylene synthesis and regulation sequences, were also identified.

More recently, a cross species microarray analysis was performed to profile genes involved in papaya fruit ripening using chips from *Arabidopsis* and microarray data from tomato and grape (Fabi et al. 2012). The analysis showed a high proportion (61%) of upregulated transcription factors (TFs) that exhibited a closer resemblance to those of the other fleshy fruits than to those of *Arabidopsis*. Overall, genes related to ripening are involved in primary metabolism (genes involved in Krebs and TCA cycles; coding genes for cytochromes; genes involved in lipid metabolism, including cell membrane lipases; and genes involved in the synthesis of unsaturated fatty acids), transcriptional regulation (gene products related to nucleotide and nucleic acid binding and translation initiation, as well as ribosome constituents), cell wall modification (including genes involved in pectin hydrolysis, cellulose hydrolysis and rearrangement, and cell expansion), and response to stress and plant defense (Acyl-CoA oxidases, which have an important role in jasmonic acid biosynthesis; heat shock proteins; and leucine-rich repeat proteins). Of note, the TFs upregulated in ripening papaya included members of the MADS box, NAC, and AP2/ERF gene families, which in other fruits are master regulators of this key process of plant development.

Transcript profiling has also been used to test causal relationships between specific signals or inducers and the genes expressed upon exposure to the triggering condition or molecule. Fabi et al. (2011) analyzed the effect of ethylene on the differential expression of some 71 putative ripening genes using a cDNA-AFLP approach. Genes identified included those related to ethylene biosynthesis, transcription regulation, and plant responses to biotic and abiotic stresses, among others.

Given the importance of sex determination of papaya plants prior to the reproductive stage, Urasaki et al. (2012) used the combined Ht-SuperSAGE analysis and SOLiD sequencing technology to evaluate and compare the transcriptome of papaya flowers. RNA was extracted from two different developmental stages in male, female, and hermaphroditic flowers, and cDNA was obtained and subsequently used to extract tags from NlaIII sites. After sequencing and BLASTN analysis, the sex-chromosome tags were defined as those with perfect matches with sequences for sex chromosomes. The purpose of the analysis was the identification of candidate genes for sex determination in papaya (see later sections). The analysis showed that 312 unique tags mapped exclusively to X or Y^h chromosomes. The majority of tags corresponded to retromobile elements, most of the tags mapped to the X chromosome, and only 30/312 were common to both X and Y^h chromosomes. Some Y^h-specific tags corresponding to female specification genes were also identified. Functional analysis should allow for confirmation of sex specificity of the candidate genes identified in this work.

Porter et al. (2008) conducted a root transcriptome survey in papaya and provided the first report of miRNAs from papaya. Their study revealed cDNAs for genes associated with a complex network of interactions including defense, beneficial plant–microbe interactions, abiotic stress, and plant development. Of particular interest were a pathogen-inducible tyrosine-rich hydroxyproline-rich glycoprotein, two unique peroxidases, a homolog to the regulatory *liguleless2* maize gene, and three novel non-protein-coding RNAs (npcRNAs). One of the latter was found to show appreciable homology to the microRNA MIR162a. MIR162a regulates the miRNA pathway and its own biogenesis by inhibiting DICER-LIKE1 (DCL1), the multidomain protein that catalyzes the formation of miRNAs in plants. Later, Aryal et al. (2012) described some 60 miRNAs from papaya, 24 present in other species, along with 36 apparently specific to papaya. Of the 60 annotated miRNAs, 18 were only detected in flowers, 12 only in leaves, and 5 only in PRSV-infected leaves. Interestingly, the endogenous population of RNAs in papaya shows an asymmetric accumulation of purine-rich strands—most probably due to cellular elimination of the pyrimidine-rich strands. The extent to which these miRNAs regulate gene expression under diverse conditions of development and biotic stress needs further analysis. More on miRNAs in papaya is presented in Section 10.3 on computational genomics (genes involved in growth and development).

Similar studies have focused on proteins associated with defense responses to abiotic and biotic stresses in papaya, in particular the zinc finger proteins, C2H2 (Jiang and Pan 2012). These TFs are reportedly induced by various stresses and implicated in stress tolerance in plants. Of more than 47,000 ESTs, 87 C2H2 candidate proteins were recovered, identified, and aligned, and real-time PCR was used to analyze transcript level changes from 42 C2H2 when plants were subjected to salt stress, low temperatures, PRSV inoculation, and spermine, methyl jasmonate, and salicylic acid treatments. Responses varied according to the source of stress and tissue, and should provide clues for future cloning and expression studies.

Another fruitful field of research derived from the completion of the papaya genome is the use of the information for the *in silico* cloning of specific genes. Idrovo-Espín et al. (2012) reported on the differential expression of the TGA TF family in papaya. The TGAC MOTIF BINDING FACTOR, or TGA TFs, plays a major role in disease resistance and plant development. Using genomic information available for papaya, papaya orthologs of *Arabidopsis* TGA coding genes were obtained. Six orthologs were detected upon *in silico* identification and sequence analysis. The TGA TFs showed the highest similarity with proteins involved in flower development and plant defense. Reverse transcriptase (RT)-PCR experiments also demonstrated basal expression of TGA TFs in different tissues and differential expression in floral organs, as well as a clear response of all identified TGA TFs of papaya when plants were subjected to salicylic acid induction. Using a similar approach, Peraza-Echeverria et al. (2012) cloned the genes and characterized the structure, phylogeny, and expression of the nonexpressor of pathogenesis related gene 1 (*NPR1*) gene family of *C. papaya*. The NPR1 genes are involved in the regulation of systemic acquired resistance in dicots, and in papaya there are four genes orthologous to the *Arabidopsis* gene family that are expressed both in vegetative and reproductive tissues (refer to Section 3.1.1 on genes involved in defense and signaling).

10.2.2.2 Protein Profiling

With the avalanche of genomic information available, proteomics is becoming increasingly important in genome annotation. The proteome essentially provides a synopsis of the proteins expressed in a cell or tissue at a defined time point. In contrast to the genome, the proteome is highly dynamic and dependent on physiological functions and environmental factors. Unlike genome and transcript analysis, proteome analysis provides insight into processes fundamental to cell function such as degradation, transport, alternative splicing, and posttranslational modifications (Baginsky et al. 2010). Proteomic analysis of papaya has thus far focused on fruit ripening and more recently on the identification of lipases because of their potential industrial applications and involvement in signal transduction and plant membrane metabolism. Another study has looked at the effects of the virus disease Meleira on the regulatory network of laticifers in papaya. The major techniques employed in proteome analysis include gel-based protein separation methods: two-dimensional differential gel electrophoresis (2-D DIGE) and mass spectrophotometry (MS) (May et al. 2011).

Nogueira et al. (2012) used a differential proteomic approach to identify and distinguish proteins that accumulate during papaya fruit ripening. Proteomes of climacteric and preclimacteric papaya fruits were compared by 2-D DIGE. Following MS, 27 proteins were identified, classified, and related to ripening metabolic changes, including proteins from the cell wall, ethylene biosynthesis, climacteric respiratory burst, response to stress, synthesis of carotenoid precursors, and chromoplast differentiation, as well as a set of proteins not previously detected. A thorough comparison of the transcriptome (Fabi et al. 2012) and proteome data described by Nogueira et al. (2012) is needed to fully understand the changes in gene expression and protein activities during papaya fruit ripening, particularly when specific agents are applied, like 1-methylcyclopropene (1-MCP)—a common practice in the postharvest management of papaya fruits that might modify ripening-related expression patterns (Huerta-Ocampo et al. 2012).

A functional proteomic approach aimed at characterizing the lipase activity of *Vasconcellea heilbornii* (babaco) latex was recently applied to the analysis of the insoluble fraction of this species by MS (Dhouib et al. 2011). Twenty-eight proteins were identified, including enzymes related to plant defense, protein synthesis and processing, and polysaccharide metabolism. However, more interesting was the identification of a putative lipase in babaco that facilitated the cloning and sequencing of its ortholog in *C. papaya*. Further, EST sequences facilitated the cloning and characterization of another papaya latex protein, a new phospholipase D, which is involved in signal transduction, vesicle trafficking, and membrane metabolism (Abdelkafi et al. 2012).

Another notable proteomic analysis of papaya latex involved the characterization of its constituent proteins in plants affected by sticky disease (Rodrigues et al. 2011), caused by the dsRNA plant virus, laticifer-infecting *Papaya meleira virus* (PeMV). Proteomes of healthy and symptomatic leaves of papaya were analyzed using a combination of methods that resulted in the detection of 75 differentially expressed proteins in diseased leaves by conventional 2-DE and 79 by DIGE. It was revealed that those proteins that were downregulated were related to metabolism, while those that were upregulated were involved in stress responses. Interestingly, and following their initial findings, additional proteins were discovered in healthy leaves along with the observation of reduced proteolytic activity in sticky diseased leaves in quantitative proteomics of the papaya latex (Rodrigues et al. 2012). The authors posited that the changes observed at the proteomic level together with the accumulation of H_2O_2 and calcium in laticifers during PMeV infection are indicative of signs of a hypersensitive response against virus infection.

10.2.2.3 Metabolite Profiling

As the name suggests, one aim of metabolomics is the large-scale phytochemical analysis of plants. The components of the metabolome are generally viewed as the end products of gene expression that define the biochemical phenotype of a cell or tissue (Sumner et al. 2003). Although the technologies of metabolomics (e.g., MS) have a long history, profiling the metabolome has not been as eagerly engaged in as its omics counterparts. Metabolomic studies of papaya have examined carotenoid and volatile profiles.

Carotenoid analyses by liquid chromatography–mass spectrometry (LC–MS) revealed striking similarity of nutritionally relevant carotenoid profiles in both the red and yellow papaya varieties, but red papayas were shown to have higher levels of lycopene (51% of total carotenoids), which was associated with the presence of crystalloid structures in its chromoplasts (Schweiggert et al. 2011). Total carotenoid content in red papayas was almost double that of yellow varieties. Later chemical profiling of yellow- and red-flesh papayas from Costa Rica (Schweiggert et al. 2012) corroborated these findings, but also demonstrated that an exploitable genetic variability in carotenoid profiles in papaya germplasm, including their hybrids, might be a workable target in future breeding programs. Importantly, pre- and postharvest handling factors have been found to be important modifiers of carotenoid profiles. Ripe papayas stored at 25°C, for example, were shown to contain more carotenoids than papayas stored at 1°C (Rivera-Pastrana et al. 2010), which is an important factor that should be considered in any papaya improvement program. A missing link between the proteome involved in carotenoid biosynthesis and the genes encoding the proteins of this fundamental pathway was partially filled with the cloning and expression analysis of phytoene and ξ-desaturase genes from papaya by Yan et al. (2011). Although both genes are expressed in diverse tissues, Yan et al. (2011) showed that their highest expression occurs in maturing fruits. Higher levels of expression were obtained in red-flesh than yellow-flesh papaya fruits. In contrast, Blas et al. (2010) showed that the papaya *CYC-b* gene, which codes for a chromoplast-specific lycopene β-cyclase (*lcy*-β1 and *lcy*-β2 according to Devitt et al. 2010), is the gene that controls flesh color in papaya and that a frameshift mutant allele is always present in red-flesh papayas (see Yan et al. 2011). More work is needed to explain the genetic and biochemical data on fruit color development in papaya.

Biochemical analyses of papaya fruit aromas are well advanced, with over 40 years of investigations and the identification of about 400 fruit volatile components (Pino et al. 2003). Flavor volatiles that have been reported in all varieties include alcohols, aldehydes, esters, terpenoids, and glucosinolate derivatives, however the composition and relative abundance of each group varies among papaya varieties. For instance, various esters derived from carboxylic acids have been identified as the predominant contributors to the distinctive flavors of the Sri Lankan (MacLeod and Pieris 1983) and Cuban Maradol varieties (Pino et al. 2003), whereas terpenoids make up 81% of total volatiles in Hawaiian cultivars (Flath and Forrey 1977). The poor flavor and bitter taste of some Australian varieties have been attributed to benzyl isothiocyanate (BITC), a pungent sulfurous plant defense compound (Wills and Widjanarko 1995). More recently, headspace solid-phase microextraction (HS-SPME) procedures have been used in the characterization of papaya germplasm commercially cultivated in Portugal (Pereira et al. 2011). After isolation of metabolites, thermal desorption gas chromatography-quadrupole mass spectrometry (GC-qMS) was applied, and 23 compounds from papaya pulp were identified. The main components included BITC, linalool oxide, furfural, hydroxypropanone, linalool, and acetic acid, plus four new components: octaethyleneglycol, 1,2-cyclopentanedione, 2-furyl methyl ketone, and 3-methyl-1,2-cyclopentanedione (MCP). The latter is an agonist of the peroxisome proliferator-activated receptor γ, and it has been used in the treatment of hyperlipidemia and hyperglycemia. Additionally, MCP is considered a scavenger of reactive oxygen species.

10.3 COMPUTATIONAL GENOMICS

10.3.1 Analysis of Gene Families

In 2008, the first draft of the virus-resistant transgenic papaya genome was decoded by a team of scientists from different institutions (Ming et al. 2008, 2012). The papaya genome was obtained from a total of 2.8 million WGS sequencing reads from the transgenic cultivar SunUp. A total of 1.6 million reads were assembled into contigs of 271 Mb and supercontigs of 370 Mb including embedded gaps. Of the 16,362 unigenes derived from ESTs, 92.1% matched the protein-coding region of the assembly. Reads from 34,065 BAC clones provided alignment to the physical map. Of these 34,065, 706 BAC end–derived and WGS sequence–derived SSRs on the genetic map, 92.4% were used to anchor 167 Mb of contigs or 235 Mb of scaffolds, to the 12 papaya linkage groups in the current genetic map.

The Institute for Genomic Research (TIGR) Eukaryotic Annotation Pipeline was used to annotate the papaya genes and masking of assembled sequence was done based on similarity to known repeat elements in RepBase and the TIGR Plant Repeat Database, in addition to a *de novo* papaya repeat database. When compared with the nonredundant (NR) database from the National Center for Biotechnology Information (NCBI), 45% of the predicted papaya genes with average lengths of 1102 bp were similar to NR proteins, and about 45% of these were supported by papaya unigenes. Genes of average length of 309 bp were not similar to NR proteins, but 8% were supported by papaya unigenes. It was posited that the number of predicted genes in papaya is 24,746. This is less than that reported for other sequenced genomes published about the same time, that is, grape (19%; Jaillon et al. 2007), rice (34%; International Rice Genome Sequencing Project 2005), poplar (46%; Tuskan et al. 2006), and *Arabidopsis*, with which papaya shares a common ancestor (11%–20%; Hanada et al. 2007, The *Arabidopsis* Genome Initiative 2000). The minimal gene set of papaya remained true when compared with other recently available plant genomes, including 34,496 genes in sorghum (Paterson et al. 2009), 32,000 genes in maize (Schnable et al. 2009), 26,682 genes in cucumber (Huang et al. 2009), 57,386 genes in apple (Velasco et al. 2010), 46,430 genes in soybean (Schmutz et al. 2010), 25,050 genes in strawberry (Shulaev et al. 2011), 28,798 genes in cocoa (Argout et al. 2010), and 40,976 genes in cotton (Wang et al. 2012b).

A high percentage of repeat accounted for a large portion of the non-genic space in the papaya genome. Comparison of total intron length of orthologous gene pairs among papaya, *Arabidopsis*, and rice showed that papaya has larger introns. Additionally, the papaya genome was found to consist of approximately 52% repetitive sequences, including about 43% identifiable transposable elements. About 8.5% were reported as novel repetitive sequences. Forty percent of the genome contained retrotransposons. *Ty3-gypsy* and *Ty1-copia* were the abundant types, at 27.8% and 5.5%, respectively. However, in comparison to other sequenced genomes, far fewer DNA transposons

(0.2%) were uncovered (Ming et al. 2008). It was also revealed that genome duplication had not occurred in papaya since its divergence from *Arabidopsis* about 72 MYA (Jaillon et al. 2007; Ming et al. 2008). Apparently, *Arabidopsis* underwent at least three rounds of whole-genome duplication, the last two of which occurred after its divergence from papaya. As with papaya, the latter two duplication events did not occur in grape (Bowers et al. 2003). This is further explained in the section on evolutionary genomics.

Comparisons of the NR proteins from papaya, *Arabidopsis*, grape, rice, poplar, and papaya reveal grouping of angiosperm genes into 39,706 similarity groups or tribes. Papaya has the least members in most of the tribes compared with other genomes. A total of 13,331 grouped in 5,920 tribes are presumed to be the minimal number of genes required for angiosperm plants. Of these, 76% of the minimal tribe is present in papaya, suggesting that papaya has the most NR sets of genes (Ming et al. 2008). Details of selected gene families are summarized in the following sections.

10.3.1.1 Defense and Signaling

Porter et al. (2009) recently analyzed the draft sequence of the papaya genome for the identification of defense-associated genes. As with many tropical and subtropical crops, papaya is host to various pests and pathogens that reduce fruit yield and quality. Some of the diseases are very destructive and have great economic impact. Bioengineering is being investigated as a management strategy and in one case has been successfully applied to the control of PRSV (Fermin et al. 2010). But a number of other diseases remain, such as bunchy top and fruit and root rot, which continue to limit the production of the crop. One approach for developing plants with resistance against a range of pathogens that is receiving attention involves enhancing the natural mechanisms of defense already present in plants by using genes or components derived from the same or other plant species. As plants do not have the benefit of circulating antibodies, they maintain constant vigilance against pathogens by expressing *R* genes that subsequently activate defenses once a prospective invader is detected (McDowell and Woffenden 2003). Most of these genes encode proteins with at least three core domains, namely a C-terminal leucine-rich repeat (LRR) domain, a central nucleotide binding site (NBS) domain, and an N-terminal domain that contains homology to cytosolic domains of the *Drosophila* Toll, animal interleukin-1 receptors (TIR), or a potential coiled-coil (CC) domain (TIR-NBS-LRR or CC-NBS-LRR) (Jones 2000). The highly variable LRR domains determine recognition of the pathogen (Dodds et al. 2001; Ellis et al. 1999; Jia et al. 2000), whereas the more conserved TIR-NBS or CC-NBS regions are believed to propagate the perceived signal (Tao et al. 2000).

Porter et al. (2009) analyzed the papaya genome sequences for the identification of NBS encoding genes. Papaya appears to possess fewer NBS class resistance genes than other sequenced genomes: 28% of those in the *Arabidopsis* genome and 10% of those in the rice genome. However, papaya possesses multiple distinct classes of these proteins with both TIR and non-TIR subclasses. Seven non-TIR members with distinct motif sequence represent a novel subgroup. Some 55 NBS class resistance genes were identified. Over 50% contained more NBS-LRR domains than NBS domains. As regards to the N-terminal domain, seven genes were found to have TIR domains and six contained CC motifs. Multiple splice variants were also detected (Porter et al. 2009), suggesting that alternative splicing may contribute to the diversity of NBS-encoding genes in papaya. Presumably the "guard model" (van der Biezen and Jones 1998; Dangl and Jones 2001; DeYoung and Innes 2006) applies to papaya. This model theorizes that the mode of action of R proteins may not involve monitoring specific signals from an invading pathogen but rather monitor key physiological processes that are targeted by pathogens. This approach of targeting multiple effectors would invariably maximize the detection potential of the plant's surveillance mechanism. Interestingly, papaya NBS-encoding genes appear more closely related to those of grape than those of *Arabidopsis* (Porter et al. 2009).

Two genes with significant similarity to the transcriptional regulator WRKY-encoding genes were reported with papaya. Presumably, R proteins interact with transcriptional regulators in the nucleus, such as WRKY proteins, to start the reprogramming process and the activation of the defense response. This results in the elevated accumulation of the plant defense-potentiating molecule salicylic acid, the expression of pathogenesis-related proteins (PR), and the transport of molecular signals through the plant to trigger the *"whole plant system response"* or SAR. Homologies of other signal transduction components such as *RAR1*, *NDR1*, *EDS1*, and *PAD4* were also found in papaya, as well as homologs of *LSD1*, *SID2*, and *EDS5*. *RAR1* plays a central role in plant disease resistance triggered by a number of R proteins (Liu et al. 2004a). *NDR1*, which is associated with the activation of the CC-NBLRR family of R proteins, is also involved in two primary pathways of pathogen defense signaling, namely pathogen-associated molecular pattern (PAMP)-triggered immunity and effector-triggered immunity (ETI). EDS1 functions in the activation of TIR-NB-LRR R-proteins (Aarts et al. 1998; Knepper et al. 2011) and both *EDS1* and *PAD4* are required for the accumulation of salicylic acid. *SID2* encodes a pathogen-induced isochorismate synthase (Wildermuth et al. 2001), while *EDS5* encodes an orphan multidrug and toxin extrusion transporter that may be involved in a positive feedback loop stimulating salicylic acid accumulation (Nawrath and Métraux 1999). The TF LSD1 negatively regulates one or more signaling pathway for basal defense and cell death and likely functions either to negatively regulate a pro-death pathway component or to activate a repressor of plant cell death (Dietrich et al. 1997).

Several TFs that interact with NPR1 proteins were identified in papaya along with a large number of downstream PR genes. Signaling components involved in the jasmonic acid– and ethylene-dependent signaling pathways were also detected, including several nitric oxide (NO) synthase genes, thus implicating NO-dependent signaling in papaya. All genes known in the signal transduction were identified in papaya, thus suggesting extensive conservation of disease response signaling pathways across plant species (Ming et al. 2008; Porter et al. 2009).

10.3.1.2 Growth and Development

Relatively little is known about non-protein-coding genes of papaya, including microRNAs (miRNAs). miRNAs are implicated in diverse aspects of plant growth and development, including leaf morphology and polarity, lateral root formation, hormone signaling, transition from juvenile to adult vegetative phase and vegetative to flowering phase, flowering time, floral organ identity, and reproduction (Sunkar et al. 2007; Mallory and Vaucheret 2006). These miRNAs are noncoding RNAs that range between 20 and 24 nts in length, exist in multiple copies throughout the plant genome, and often target and bind to multiple mRNA transcripts to trigger their cleavage (Voinnet 2009). Targets include mRNAs encoding TFs involved in development, miRNA/small interfering RNA (siRNA) metabolic or effector components (DCL1, Argonaute1 [AGO1], and AGO2), components of the Skp1-Cullin-F-box protein complex involved in ubiquitin-mediated protein degradation, and metabolic and stress-related factors. More recently, several miRNA families have been reported as regulators of disease resistance and have been found to target genes encoding NBS-LRR receptors in legumes (Zhai et al. 2011) and members of the Solanaceae (Li et al. 2012; Shivaprasad et al. 2012). At least 37 miRNA families have been described, a large number of which are species specific or restricted to closely related species. Sunkar and Jagadeeswaran (2008) described homologs to five miRNA families in papaya, namely miR169, miR170/71, miR399, miR403, and miR408. miR399 is specifically upregulated in response to low nutrient conditions (Pant et al. 2008), miR169 appears to play important roles in drought- and salt-stress responses in *Arabidopsis* (Li et al. 2008) and *Medicago truncatula* (Trindade et al. 2010), and miR171 is involved in responses to cold stress in *Arabidopsis* and rice (Zhao et al. 2007). miR408, along with other miRNAs, is part of the defense mechanism in *Arabidopsis* against bacterial pathogens and is downregulated when these plants are challenged with *Pseudomonas syringae* pv. tomato DC3000 (Zhang et al. 2011).

Although papaya possesses fewer gene families than other sequenced angiosperms, it appears that some gene families have been expanded in papaya, like the expansin superfamily. Genes of this family are specifically found in the land plant lineage and are involved in biological functions of cell growth and in other instances where the movement, adhesion, and enzymatic accessibility of wall polysaccharides are important, including drought responses and the later stages of fruit ripening (Cho and Kende 1997; Rose et al. 1997). Expansins appear to disrupt the noncovalent adhesion of matrix polysaccharides to cellulose microfibrils, thereby permitting turgor-driven wall enlargement (Cosgrove 2000). The superfamily is divided into two major families, A- and B-expansins, on the basis of sequence divergence and biochemical activity and probably act on different polymers of cell walls. Papaya has a similar number of expansin A genes (24) as *Arabidopsis* (26) and poplar (27). However, a larger number of expansin B genes were described in papaya and *Arabidopsis* than in poplar (10, 6, and 3, respectively). The sequences of B-expansins are highly represented in rice and maize EST databases (Crowell 1994; Downes and Crowell 1998). In contrast to expansin-related genes, papaya has on average about 25% fewer cell wall degradation genes than *Arabidopsis* (Ming et al. 2008). For example, while *Arabidopsis* has 12, 67, and 27 endoxylanase-like genes in the glycoside hydrolase family, pectin methyl esterases, and pectin lysases, respectively, there are only 4, 29, and 15 in papaya.

More genes in the lignin biosynthesis pathway have been described for papaya than *Arabidopsis*. Lignin, the generic term for a large group of aromatic polymers resulting from the oxidative combinatorial coupling of 4-hydroxyphenylpropanoids, is the most abundant cell wall polymer after cellulose. The polymers are a part of secondary cell wall growth and makes cells rigid and impervious (Raes et al. 2003). These polymers play an important role in mechanical support, water transport, plant defense, and abiotic stress resistance (Boerjan et al. 2003). Poplar, papaya, and *Arabidopsis* have 37, 30, and 18 candidate genes for the lignin synthesis pathway, respectively. Comparatively, papaya has lower gene numbers in the cinnamoyl-CoA reductase (CCR) and cinnamyl alcohol dehydrogenase (CAD) gene families, which all mediate later steps of the lignin biosynthesis pathway. Papaya, a semiwoody, giant herb, apparently accumulates lignin at an intermediate level between the herbaceous, annual *Arabidopsis* and the woody forest tree poplar. Of note, cellulose biosynthesis gene homologs are present in larger numbers in papaya than *Arabidopsis*; poplar contains even higher numbers of cellulose synthesis–related genes. A total of 38 and 40 cellulose synthase–related genes (GT2) were identified in papaya, whereas 48 were described in poplar and 31 in *Arabidopsis* (Ming et al. 2008).

10.3.1.3 *Flower and Fruit Development*

Papaya is a trioecious species. Specific varieties of papaya are either dioecious (producing male and female plants) or gynodioecious (producing hermaphrodite and female plants). Staminate flowers consist of small sepal lobes, a long petal tube with contorted lobes, and a diplostemonous androecium. Stamens, inserted on the tube, occur in two separate whorls. The central ovary is reduced. Pistillate flowers, which are much larger than staminate flowers, possess petal lobes that are only fused at the base. Stamens are missing and five carpels are positioned opposite to the sepals. The central ovary carries five parietal placentas containing numerous ovules and is overtopped by a short style with well-developed stigmatic lobes (Badillo 1971; Ronse Decraene and Smets 1999). Female flowers develop from the suppression of stamen development at the inception stage, while male flowers develop from aborted carpels at much later stages. Sex reversal occurs in all three primary sex forms (Higgins and Holt 1914; Hofmeyr 1939a), mainly due to environmental influence. Characterization of homologous genes associated with carpel development in papaya has shown that the *AGAMOUS* gene of *Arabidopsis*, which specifies the identity of the floral organs, has a homolog in papaya named *PAG* and shares 85% identity within the MADS box and K box domain. The *AGAMOUS* positive regulator gene *LEAFY*, which integrates environmental and endogenous signals to control flower timing in *Arabidopsis*, has a papaya homolog, *PFL*, and shares 65% identity. The PFL protein of both species shares 71% identity with LFY homologs. Despite similarities in two conserved regions, proline-rich and acidic motifs differ between PFL and LFY proteins. Genomic and BAC hybridization showed that *PFL* exists as a single copy in the papaya genome. Similarly, another regulator of stamen development in *Arabidopsis*, *hua1*, has a papaya homolog, *Phua*. They share about 82% identity (Yu et al. 2005).

Comparison of ESTs and unigene builds of 15 species from The Floral Genome Project showed that more than 60% of the papaya unique sequences share similarity to floral genes of the species listed. More than 200 *Arabidopsis* genes are involved in floral development and regulation, flowering time, and floral organ-specific genes, and about 50% of these genes were represented in papaya floral libraries by unique sequences. Papaya orthologs *ZTL* of *PAS-FBOX-KELCH* genes, which control light signaling and flowering time, lack an obvious KELCH domain compared to *Arabidopsis* and poplar. In fact, the papaya genome contains fewer KELCH domains compared to *Arabidopsis* and poplar. The KELCH domains play a role in light-mediated ubiquitination. Papaya has apparently developed an alternative way of integrating light/timing information specific to neutral plants (Ming et al. 2008). More on circadian expression and regulation is presented in the next section.

The papaya genome sequence opened avenues to investigate genes potentially involved in fruit growth, development, and ripening. Similar to *Arabidopsis* and tomato, papaya has multiple genes in the same families related to cell division and growth (Ming et al. 2008). The cell wall glucosyl transferase gene family was identified, but with fewer genes than *Arabidopsis*. A similar number of symplastic and apoplastic unloading genes were found in papaya, which suggested analogous developmental requirements for carotenoid biosynthesis. Potential homologous genes to those involved in tomato fruit size and shape were found. A number of genes that may impact sugar accumulation, such as those involved in ethylene synthesis, respiration, chlorophyll degradation, and carotenoid synthesis, were predicted in the papaya genome; similar and fewer numbers of genes were previously described for *Arabidopsis* and tomato, respectively. Papaya can be a valuable and unique tool to study the evolution of fruit ripening and the complex regulatory networks in fruit ripening because of the absence of genome duplication and the presence of fewer gene families and biosynthetic pathways than its brassical relatives.

The papaya fruit is a fleshy berry with a climacteric ripening pattern that is characterized by the rise in ethylene production occurring at the same time as a respiratory increase (Paull and Chen 1983). Papaya fruit is composed of five carpels united to form a central cavity derived from the ovary, which is lined with the placenta, and numerous black seeds. Fruit shape is sex-linked, and fruits produced by female flowers range from spherical to ovoid, whereas those from hermaphrodite flowers are cylindrical or pyriform. The skin of papaya fruit, which is green when immature, turns a bright yellow on ripening (Chan and Paull 2009; Paull 1993). Ripening essentially represents the terminal stage of development at which time mature seeds are ready for release. Papaya, like all fleshy fruits, undergoes various biochemical, physiological, and structural changes with the aim of attracting seed-dispersing organisms. The changes primarily include (1) modification of color through the alteration of chlorophyll, carotenoid, and/or flavonoid accumulation; (2) textural modification via alteration of cell turgor and cell wall structure and/or metabolism; (3) modification of sugars, acids, and volatile profiles that affect nutritional quality and flavor, and (4) increased respiration and synthesis of the gaseous hormone ethylene, which likely serves to coordinate the ripening process (Giovannoni 2004).

Both *Arabidopsis* and papaya have almost the same number of genes involved in ethylene synthesis, namely *S*-adenosyl-L-methionine (S-SAM) synthase, 1-aminocyclopropane-1-carboxylate synthase (ACC) synthase, ACC oxidase, and ethylene responsive binding factors, AP2/ERF. Two genes for alternative oxidase AOX1, and not multigene families as in most plants, were predicted for papaya. Alternate oxidase (AOX) is the terminal quinol oxidase that competes for electrons with the standard cytochrome pathway in the electron transport chain of plant mitochondria. Three ORFs with similarity to other non-proton-pumping enzymes, type II dehydrogenase genes, were also described in the papaya genome. As regards to changes in color, one carotenoid cleavage dioxygenase and two 9-cis-epoxycarotenoid dioxygenases are predicted in papaya. The former group uses multiple carotenoid substrates whereas the C40-9-cis-epoxycarotenoid is the preferred substrate for the latter group of enzymes, which are presumably involved in the synthesis of abscisic acid. Carotenoid development and fruit skin degreening is apparently coordinated by the latter through either carotenoid cleavage dioxygenase or 9-cis-epoxycarotenoid. *De novo* synthesis of carotenoids in mesocarp chromoplasts is typically accompanied with the unmasking of carotenoids because of the degradation of chlorophyll. While two chlorophyllase genes are found in *Arabidopsis* and three in cabbage, only one gene was predicted in papaya. Higher numbers of β-1-3-glucanases and β-galactosidases were described and may reflect a greater need for wall degradation in ripening and wall turnover in papaya (Ming et al. 2008; Paull et al. 2008). As regards sucrose sugar accumulation in papaya fruit, which begins about 100 days after anthesis, four sucrose synthase genes and three sucrose phosphate synthase genes were predicted for papaya, and a cell wall invertase cleaves the sucrose arriving via the phloem to hexoses, thus maintaining the sucrose source to sink gradient on termination of fruit growth. Hexose transporters are implicated in the uptake of sugars into fruit cells and vacuoles. Unlike tomato, where 16 cell wall invertases, 1 acid invertase, and 3 hexose transporter genes are involved in fruit sugar accumulation, 2 cell wall invertases and 1 acid invertase as well as 5 possible hexose transporter genes appear to be active in papaya.

The major difference in ripening in papaya and *Arabidopsis* is the number of genes pertaining to volatile compounds, which presumably reflects involvement in the attraction of pollinator and seed dispersal agents. At least 162 compounds were found in papaya, far more than in *Arabidopsis*, but fewer than the number in tomato. The most commonly identified papaya volatiles are methyl butanoate, ethyl butanoate, 3-methyl-1-butanol, and 1-butanol.

10.3.1.4 Circadian Expression and Regulation

Publication of the papaya genome sequence facilitated investigations into the tropical plant circadian clock at the molecular level. Previous studies primarily focused on plants growing in temperate and subtropical climates, and little was known about the circadian transcriptional networks of plants that typically grow under tropical conditions, with relatively constant day lengths and temperatures over the year. The circadian clock is an endogenous, approximately 24-h timekeeper in living systems. In plants, the clock is crucial for adjusting growth to both time and day, as well as to the seasons. Leaf movements, stomata opening, hypocotyl elongation, and the expression of a large number of genes show circadian rhythms (Covington et al. 2008). Moreover, both jasmonate and salicylate hormone levels are circadian regulated, suggesting an underlying mechanism for clock-mediated plant defense (Goodspeed et al. 2012). The mechanism of generating rhythmicity, best described for *Arabidopsis*, consists of a central oscillator based on transcriptional negative feedback loops (Dunlap 1999). Three genes have been suggested as core components of the central oscillator: *CIRCADIAN CLOCK ASSOCIATED*1 (*CCA1*), *LATE ELONGATED HYPOCOTYL* (*LHY*), and *TIMING OF CAB EXPRESSION*1 (*TOC1*). The two morning-phased, partially redundant MYB TFs CCA1 and LHY bind directly to the promoter of the evening-phased pseudoresponse regulator *TOC1*, negatively regulating its expression. TOC1 participates in the positive regulation of the expression of the *CCA1* and *LHY* genes (Alabadí et al. 2001; Zhang and Kay 2010).

Zdepski et al. (2008) conducted a genomic and computational analysis of the circadian biology of papaya. As expected, many of the circadian clock and light-signaling gene families are smaller in papaya than in *Arabidopsis*, rice, and poplar. Papaya, like *Arabidopsis* and rice, possesses only one *TOC1* gene, and there is only one homolog of the TF *LHY* gene compared to two paralogs in *Arabidopsis*, *LHY*, and *CCA1*. Invariably, CCA1/LHY and TOC1/PRR1 were found to peak in the morning and evening, respectively, during quantitative real-time PCR expression of circadian clock orthologs and two independent 48-h time courses in mature papaya trees. Zdepski et al. (2008) noted with surprise that these findings with a tropical plant exhibiting a distinct lifestyle were consistent with those of temperate plants. They posited that circadian timing likely played a major role in the evolution of plant genomes, consistent with the selective pressure of anticipating daily environmental changes.

10.3.2 Evolutionary Genomics

10.3.2.1 Genome Evolution

Analysis of the genome information content of sequenced plant genomes, including papaya, has been fundamental in uncovering phylogenetic processes in angiosperms and estimating crucial events in the speciation process including the verification of some evolutionary dating estimates. One notable example is the phylogenomic and comparative expression analysis of the *BARREN STALK1/LAX PANICLE1* (*BA1/LAX1*) genes, whose products are essential for the formation of axillary meristems during vegetative development and all lateral structures during inflorescence development in *Arabidopsis*, papaya, medicago, rice, sorghum, and maize performed by Woods et al. (2011). The authors found that these genes reside in a syntenic region (a set of genes that co-occur in two species), and that phylogenetical analysis brings support to a clade comprising monocots and dicots with an estimated origin 125 MYA. The analysis of transposition events in rosids, on the other hand, confirmed that the lineage that gave origin to the Caricaceae family of vascular plants diverged from *Arabidopsis* 70–72 MYA (Woodhouse et al. 2010, 2011). By comparing orthologous genomic alignments of sequences from *A. thaliana* with those from two outgroup species, papaya and grape, Freeling et al. (2008) previously found that ca. 11% of *Arabidopsis* genes have "transposed" either into or out of syntenic regions shared with its common ancestor with papaya. *Carica* arose from a deep split in the order Brassicales and is distantly related to Brassicaceae, the family that contains both *Arabidopsis* and *Brassica*. According to Wang et al. (2012b), *C. papaya* and *A. thaliana* diverged from the *Gossypium raimondii–T. cacao* subclade approximately 82.3 MYA. It was also noted that selection pressure acting upon genes belonging to mobile genes families differs from that of syntenic genes (Woodhouse et al. 2010, 2011).

One of the realizations to emerge from comparative plant genomic studies is that plant gene families appear to be largely conserved over evolutionary timescales that encompass the diversification of all angiosperms and nonflowering plants (Rensing et al. 2008). On the other hand, lineage-specific fluctuations in gene family size are frequent among taxa. This suggests that the diversity and lineage-specific phenotypic variation are more likely due to the duplication and adaptive specialization of preexisting genes rather than the acquisition of a diverse set of wholly novel genes. Such duplications of genome content can occur through unequal crossing over and chromosomal anomalies as well as via mechanisms involving transposons and other reverse-transcriptase-mediated duplication events (Flagel and Wendel 2009). Phylogenomic analyses of sequenced plant genomes allowed the inference that the ancestor of all eudicots was probably hexaploid, that is, harboring a genome that was already subject to two successive rounds of whole-genome duplication (γ event). The data also support the proposition that the nuclear genome of *Arabidopsis* (an herbaceous annual) is evolving 2.11 times faster than papaya's (a semiwoody perennial) nuclear genome (Barker et al. 2009). Gene order evolution has also been analyzed using genomic data from papaya. The analysis revealed a faster rate of rearrangement in the papaya genome from a rosid ancestor than previously thought (Sankoff et al. 2009).

Analysis of fully sequenced genomes of monocots (*Oryza sativa* and *Sorghum bicolor*) and eudicots (*A. thaliana*, *C. papaya*, *Populus trichocarpa*, and *Vitis vinifera*) show evidence of at least one event of paleoploidization, or an ancient polyploid event. Interestingly, papaya has undergone only one paleohexaploidy event (that is, no recent events of WGD have occurred in the species), while the distantly related brassical *Arabidopsis* has undergone at least three ancient polyploidy events (α, β, γ events). There is nothing corresponding to the *Arabidopsis* α or β events. However, *Carica* shows evidence of γ, the early event detected in *Vitis, Populus*, and *Arabidopsis* (Jaillon et al. 2007). Since Caricaceae are estimated to have diverged from the Brassicaceae lineage ca. 72 MYA (Ming et al. 2008), the absence of recent genome duplication is in conflict with prior estimates of a much earlier age for the β genome duplication, which suggests it could correspond to the origin of the eudicots (Bowers et al. 2003). Of note, only 39.1% of genes of papaya have experienced duplications, as compared to 63.3% in *Arabidopsis*, and this gives support to the claim that the papaya genome has undergone only the most ancient WGD shared with the species examined in the study and all rosids analyzed.

Further evidence of low gene redundancy in papaya emerged from studies of papaya telomeres and some gene families. In papaya, the telomeres, as has also been found in most plants, consist of TTTAGGG repeats, but in contrast to plants of the Brassicaceae family, the telomeres are approximately 10 times longer (25–50 kb). Interestingly, several genes involved in telomere maintenance, protection, and dynamics are single copy or are in copy numbers lower than those found in other species (Shakirov et al. 2008). Comparison of telomeric sequences from papaya and *Arabidopsis* showed that three copies of the *POT1* (a protein absolutely essential for telomere protection and integrity) coding gene present in *Arabidopsis* are represented by a single copy gene in papaya (Watson and Riha 2010). The identification of only six genes coding for GH3 amino acid conjugases reflects further the lack of additional WGD in papaya (Okrent and Wildermuth 2011). Nineteen GH3 genes were found in *Arabidopsis*, 13 in poplar, and 8 in grape. From their analysis, the authors proposed that the common ancestor prior to the pre-rosid hexaploidy WGD event had three GH3 genes—leading to nine GH3 genes in the WGD hexaploidy event and a subsequent loss of two in the papaya lineage.

Although the only known genome of members in the family Caricaceae is that of *C. papaya*, an analysis of genes of *C. papaya* and other members of the family, including species of the genera *Vasconcellea* and *Jacaratia* from south America, *Jarilla* and *Horovitzia* from Mexico and Guatemala, and *Cylicomorpha* from Africa, revealed that after arrival from Africa the neotropical ancestor of the family split in the late Eocene. Papaya was found to be more closely related to the herbaceous members of the family and the diversification of the family in South America (Miocene) coincided with the collision of the subcontinent and Panama (Antunes-Carvalho and Renner 2012). Genomic data from other members of the family would help clarify these assumptions based solely on the comparison of the genomes between the brassicals *Arabidopsis thaliana* and *C. papaya* and accommodate or validate the diversification of the latter from the *Vasconcellea* members of the family 27 MYA.

10.3.2.2 Sex Chromosome Evolution

The Caricaceae family comprises 32 dioecious species, 2 trioecious species (*C. papaya* and *V. cundinamarcensis*), and 1 monoecious species (*V. monoica*). The predominance of dioecious species suggests that dioecy is an ancestral character of the family, and that trioecious and monoecious species are of recent origin (Ming et al. 2007; Gschwend et al. 2011, 2012). Most papaya varieties are dioecious or gynoecious. Male plants are characterized by the possession of long, pedunculated inflorescences with abundant flowers with atrophied pistils. Female flowers, on the other hand, have short inflorescences with few flowers with a prominent gynaeceum without stamens, while hermaphrodite plants have short inflorescences with bisexual flowers that can be sexually variable (Yu et al. 2008a). Sex determination in papaya is particularly intriguing, not only because the species has three different sexual types, but also because sexual reversion is frequent and dependent on environmental factors such as temperature and soil humidity. According to Nakasone and Lamoureux (1982), Higgings and Holt described 13 sexual forms to which Sayed added two more for a total of 15 different sexual forms. Hofmeyr (1938, 1939b) grouped the various forms into three general classes. Storey (1938) described five types, but later reclassified to eight. Prior to the advent of molecular biology techniques and their application to the study of papaya, there was limited evidence to support this diversity in sex types.

In the late 1930s, Hofmeyr (1938) and Storey (1938) independently proposed, based on the segregation ratios of flower types, that sex in papaya was controlled by a single gene with three different alleles named M_1, M_2, and m by Hofmeyer and M, M^h, and m by Storey. In this system, male individuals were heterozygous Mm plants, and hermaphrodites were heterozygous M^hm plants, while recessive homozygous mm plants were females. The dominant genotypes MM, M^hM^h, and MM^h were considered to be lethal. A year later, however, Hofmeyr proposed an alternative hypothesis. It was proposed that sexual determination in papaya was governed by a genetic balance between sexual and autosomal chromosomes despite the absence of evidence on the existence of heteromorphic sexual chromosomes. In fact, he considered the aforementioned alleles (M, M_h, and m) as "sexual chromosomes." In this hypothesis, Hofmeyr assumed that female determinant factors were located on sexual chromosomes while male factors were located in autosomal chromosomes. In addition, he proposed that M and M^h represented inactive regions of the sexual chromosome of varying lengths given that certain vital genes were eliminated. Since some vital genes were absent in these regions, any of the combinations MM, M^hM^h, and MM^h were lethal, while the Mm and M^hm genotypic plants could be viable due to the presence of the (intact) m sexual chromosome. Storey (1953) reformulated his hypothesis of a single gene with three alleles and proposed instead that sexual determination in papaya involved a group of closely linked genes located in a small, nonrecombinant region of the sexual chromosome. In this scenario, elongated peduncles were always and only associated with male flowers, and the lethal factor was associated only with the dominant genotypes of males and hermaphrodites. The genes proposed to lie in this region included Mp (elongated peduncle in male flowers), l (zygotic lethal factor), sa (androecium suppressor), sg (gyneceum suppressor), and C (hypothetical factor responsible for the suppression of recombination in the sexual determination region). The latter might be a gene, an inversion, a deletion, or a region very close to the centromere or telomere able to suppress DNA recombination. In the same fashion, the l gene might also be involved in recombination suppression if overlapped with chromosomal deletions in such a way that, as suggested by Storey, l and C were the result of a single chromosomal rearrangement instead of two different factors or genes tightly linked. Regarding the genes sa and sg, it was posited that male and hermaphrodite individuals were heterozygous due to the sexual reversion of hermaphrodite to male flowers, and the other way around, as observed in the field.

The first researchers to propose a XX–XY system of sexual determination in papaya, which involved a modified Y chromosome, were Horovitz and Jiménez (1967). This was concluded based on the analysis of intergeneric hybridization experiments between *Carica* and other *Vasconcellea* species. They proposed the combinations XX, XY, and XY_2 for female, male, and hermaphrodite plants, respectively. Additionally, the authors suggested that chromosomes Y and Y_2 harbored the lethality region.

Based on the ABCE model of flower development, Sondur et al. (1996) later suggested the existence of two dominant alleles, *SEX1-M*, which codes for a masculinizing factor that promotes stamen development while suppressing carpel development, and *SEX1-H*, with incomplete dominance since it induces stamen development, but does not completely repress carpel development. For female plants, they proposed that the presence of the recessive null allele *sex1-f* is unable to induce stamen development.

Later on, the application of DNA markers and sequencing of the sexual determination locus by diverse research groups led to the construction of linkage maps, high-density genetic maps, and physical and genomic maps of papaya, and ultimately the demonstration of primitive sexual chromosomes in the species (Gschwend et al. 2012; Liu et al. 2004b; Ma et al. 2004; Ming et al. 2007; Na et al. 2012; Wai et al. 2012; Wang et al. 2012a; Wu et al. 2010; Yu et al. 2007, 2008b,c). Sexual chromosomes in plants are very interesting since they have evolved recently, as compared with sexual chromosomes in mammals or *Drosophila*. The study of sexual chromosomes in plants allows for the identification of the initial stages of the continuous evolutionary process that led to the massive loss of genes experienced by the Y chromosome (Charlesworth 2004). In many dioecious animals and plants, the pair of sexual chromosomes can be easily observed. In mammals, the Y chromosome is smaller than the X chromosome, while in *Silene latifolia* (Caryophyllales: Caryophyllaceae) the Y chromosome is much larger than the X chromosome. Despite their differences in size and morphology, X and Y chromosomes in a species are homologous, and like the autosomal chromosomes, segregate during meiosis. The sexual chromosomes of mammals do not usually recombine and harbor a long MSY region—but also possess a small region capable of recombination called the *pseudoautosomal region*, (PAR) (Bergero and Charlesworth 2008). In this regard, although it was initially thought that *C. papaya* had heteromorphic sexual chromosomes, recent karyotype analysis in this species and two *Vasconcellea* species revealed that there were no appreciable differences in the sizes of their chromosomes (Corrêa-Damasceno et al. 2009). Furthermore, the recent comparisons of the genome sizes of the monoecious (*Vasconcellea monoica*), dioecious (*Jacaratia spinosa*, *Vasconcellea glandulosa*, *V. goudotiana*, *V. horovitziana*, *V. parviflora*, *V. pulchra*, and *V. stipulata*), and trioecious Caricaceae species (*C. papaya* and *V. cundinamarcensis*) not only allowed adjustment of the papaya genome size, but also estimation of genome size in other species in the family. Interestingly, papaya has the smallest genome of all species analyzed, but unlike other confamiliar species, their sexual chromosomes are of equal size (Gschwend et al. 2013).

Although the evolution of bisexual flowers from the unisexual cones of gymnosperms has yet to be solved, there is an apparent agreement that a bisexual flower is an ancestral or initial trait of angiosperms. The first step toward the evolution of sexual chromosomes is the occurrence of female or male sterile mutations that led to the emergence of unisexual flowers (Ming et al. 2011). Moreover, such mutants, selected randomly in a population, might involve individuals with unisexual flowers or bisexual flowers, thus allowing the manifestation of gynodioecy or androdioecy. Mutations that led to abortion of the sexual organ to generate unisexual flowers may occur in any developmental stage, from the induction of primordia formation of the stamen to pollen grain maturation in female flowers, or in the case of male flowers, from induction of primordia formation of carpels to the development of style and stigma (Ming et al. 2011). In papaya, production of unisexual flowers occurs in different stages during the floral ontogeny: male flowers pass through a bisexual stage at the beginning of their formation and then, as in *S. latifolia*, gynaeceum progress is arrested in Stage 7 of flower development. In female flowers, the bisexual stage (and staminoids) was not observed. Nonetheless, it cannot be concluded that floral primordia of gynoid plants are unisexual or sexually unipotent. Although floral primordia do not show vestiges of the inappropriate sexual organ that is observed with *Mercurialis annua* (Malpighiales: Euphorbiaceae) or *Cucumis sativa* (Cucurbitales: Cucurbitaceae), hormone treatment can revert sexuality, suggesting that they are possibly bipotent (Castro et al. 2002).

Recent advances in molecular biology have allowed for the grouping of sexual chromosomes of plants in six categories of evolution. *C. papaya* has been classified in Stage 3, where recombination suppression extended to close regions, allowing for a high number of degenerate Y-linked genes and a male-specific region on a nascent Y (MSY). Expansion of the MSY region occurred via the accumulation of retroposons, as well as by translocation and duplications of chromosomal fragments. X and Y chromosomes look homomorphic at the cytological level, but they are heteromorphic at the molecular level. Loss of genes, on the other hand, is extensive enough for YY or Y^hY^h combinations to be lethal. Liu et al. (2004b), using molecular markers, concluded that papaya has an incipient Y chromosome with one male-specific region (MSY), which corresponds to 10%

(~4.4 Mb) of the sex chromosome, showing a high percentage of polymorphic markers, suppression of recombination, and degeneration of the DNA sequence. This region, or sexual locus, has been mapped on linkage group 1 (Ma et al. 2004) and these features are consistent with the classical notion that the early stages of sex chromosome evolution involved the suppression of recombination around the sex determination locus, allowing for the gradual degeneration of the Y chromosome. The identification and sequencing of the genes located on MSY facilitated measurement of the discrepancy between the allelic sequences of X and Y chromosomes and thereby an estimation of the origin of sexual chromosomes using a molecular clock (Bergero and Charlesworth 2008). Thanks to this approach, Ma et al. (2004) estimated the divergence between the X and Y chromosomes to be around 10%–20%, which is lower than the divergence found in mammals, suggesting again the recent origin of the Y sex determination system in papaya. Another finding is that the sex chromosomes are not of the same age.

Sexual chromosomes in papaya emerged by various mutational events within a hermaphrodite population, then, a mutation that leads to male sterility emerged to give rise to a female, and one or two more mutations suppressed female fertility, giving rise to male flowers. The evolutionary process starting from a hermaphrodite apparently did not occur in a single step. Transition occurred in a genomic region where mutations arose, creating a small cluster of genes of sexual determination (MSY). Selection processes then favored new alleles via male and female recombination and would likely generate individuals with genes for sterility. Lack of recombination in this region in males led to the accumulation of repetitive sequences, loss of genes, and degeneration (Charlesworth 2004).

There are two slight differences in the papaya Y chromosome that in males is called Y and in hermaphrodites is called Y^h. Females are homogametic XX, whereas males and hermaphrodites are XY and XY^h, respectively. Using fluorescence *in situ* hybridization (FISH) and sequencing five BAC clones that harbored the MSY region, Yu et al. (2007) found that this genomic portion is located in the pericentromeric region of the Y chromosome, which has low genetic density, but a high proportion of retroelements and duplicated sequences. Low genetic density was estimated at 1/257 kb—this is higher than that found in the Y human chromosome, which evolved from an ancestral autosomal shared with birds 310 MYA. Papaya's Y chromosome is estimated to be only 0.5–2.2 million years old (Yu et al. 2008c). Comparative analysis of MSY BACs suggested evolution from a region of low gene density. Duplications, transposable elements, and plastid-derived DNA sequences contributed to the divergence between the Y and X chromosomes.

More recently, sequencing and comparisons between two homologous BAC X and Y^h pairs have increased our understanding of the characteristics and evolution of the MSY^h (Yu et al. 2008b). According to these studies, it was suggested that sexual chromosomes evolved not at the level of family but at the species level, since divergence between Caricaceae and its most closely related family, the Moringaceae family, occurred 60 MYA, while divergence between *C. papaya* and its closest relatives *Horovitzia* and *Jarilla* occurred only 25 MYA (Antunes-Carvalho and Renner 2012). Sexual chromosomes in members of the Caricaceae family emerged from the same autosomal chromosome harboring allelic forms of X, which is still found in the monoecious *V. monoica* (Wu et al. 2010).

10.4 FROM GENOME TO BREEDING AND BEYOND

As shown, the availability of map and genome sequence data of papaya has brought multiple benefits to plant biology, including the fields of omics and evolutionary biology. Undoubtedly, the data will accelerate and support research in the area of crop improvement, namely QTL analysis, to successfully unravel the genetic control of important agronomic traits and map their corresponding loci—a prerequisite for marker-assisted breeding. QTL data from papaya, if transferable, might also be applied to the genetic improvement of other members of the Caricaceae family of potential economic value. Of importance too is the anticipated advance in genome-wide association mapping in papaya when, for example, high-density single-nucleotide polymorphism (SNP) genotyping arrays (Celton et al. 2010; Ganal et al. 2009) of cultivated papaya are developed and assessed. These approaches will allow genetic selection (Morrell et al. 2012) and speed up the process of large-scale genotype–phenotype correlation in papaya, as proposed for other crop plants (Edwards and Batley 2010; Grattapaglia et al. 2012; Pérez-de-Castro et al. 2012; Troggio et al. 2012; Yang et al. 2009). Invariably, these developments will deliver new insights into pathogen-regulated host miRNAs and small interfering RNAs (siRNAs) and their roles in plant–microbe interaction and disease (Mochida and Shinozaki 2010) and, combined with physiological and reverse genetic studies, should lead to the identification of the key factors surrounding the manipulation of genes involved in abiotic stress adaptation in papaya, as demonstrated with other plant crops such as common bean and cassava (Ishitani et al. 2004).

Many recent advances in the field of genomics, proteomics, and metabolomics give support to the proposal of papaya as a workable model plant system. Although there is very limited genetic variability as estimated by SSR (Kyndt et al. 2006; Ocampo-Pérez et al. 2006), ISSR (da Costa et al. 2011), PCR-RFLP (van Droogenbroeck et al. 2004), and AFLP (Kim et al. 2002) marker analysis, a recent finding of natural populations of dioecious papayas growing in disturbed areas and within secondary lowland forests in Costa Rica may possibly serve as a reservoir of genetic and morphological diversity for this tropical fruit crop (Brown et al. 2012). Moreover, since many genes in papaya are single copy, as compared to other plant species including the plant model system *Arabidopsis*, which possesses multicopy orthologs (and paralogs), papaya could be used to better test evolutionary and phylogenetic hypotheses once its genome is completely sequenced and annotated. The genomic "simplicity" of papaya, accompanied with a comprehensive description of its coding and noncoding regions, should facilitate the correlation between the genotype and phenotype as traditionally approached by forward and reverse genetics. Certainly the application of genomic data to plant breeding and improvement will be of great help in addressing the current and future challenges of climate change, food safety, and food security; the overarching challenge is then how to convert the mass of genomic data into knowledge that can be applied in a crop breeding program.

One area of research with papaya that has not received much attention is the identification of potential allergens. Although allergies to food have afflicted populations for a long time, little is known about the specific proteins in food as well as the fundamental mechanisms by which individuals develop allergic reactions to these proteins. One study has focused on the allergenicity assessment of the transgenic papaya variety Rainbow and its heterologous gene products using conventional molecular and bioinformatics strategies (Fermin et al. 2011). With the massive amount of genetic information available, innovative strategies along with emerging proteomic technologies can be exploited to identify and characterize new allergenic proteins (Maghuly et al. 2009). This is particularly important in an era of nutrition and food safety awareness, especially with vegetable and fruit produce.

Another direction in the comprehensive analysis of papaya genetics involves the mapping of the species epigenome, that is, that group of heritable modifications that occur in the genome that change the topological structure of DNA and invariably gene expression. Although the genome defines the species and its full biological potential, deciphering the epigenome of this important tropical plant crop will surely unravel complex regulatory pathways—especially as a product of the interactions established with the plant and its environment. Mapping the epigenome of papaya will involve, as suggested by the EPIC Planning Committee (2012), full transcript profiling (including protein-coding mRNAs, and long and small noncoding RNAs), genome-wide complete methyl-cytosine modifications at single nucleotide resolution, histone modifications under varying conditions, and position (and dynamics of change) of histone variants, DNA hypersensitive sites, and nucleosomes. This knowledge will be required to advance modeling gene expression, as has been shown with other organisms (Dong et al. 2012). Furthermore, the knowledge derived from the draft of the papaya genome could also be useful in advancing studies on chemical genetics. That is, screening chemical compounds, whether natural or synthetic, that disrupt a process of interest so as to probe gene function. Because chemical genetics can mimic loss- or gain-function alleles and hence are analogous to mutational analysis, any plant process can be studied by this approach. Among the functions that can be analyzed this way are metabolism and metabolic changes (even at a single-cell scale), plant development, and tolerance to various abiotic and biotic conditions, among others (McCourt and Desveaux 2009).

Finally, despite all the information omics data can offer, a plant interacts with its environment to display a particular phenotype—which sometimes is impossible to predict based only on molecular clues (Pieruschka and Poorter 2012). Hence, an approach, most probably automated, must be put in place in order to consolidate and establish a clear phenome representation of papaya in a time when the exploitation of data for the manipulation of plant growth, development, and stress resistance, through breeding or modern genetic modification, is lagging behind the almost daily production of molecular information.

REFERENCES

Aarts, N., M. Metz, E. Holub, B. J. Staskawicz, M. J. Daniels, and J. E. Parker. 1998. Different requirements for *EDS*1 and *NDR*1 by disease resistance genes define at least two *R* gene-mediated signaling pathways in *Arabidopsis*. *Proc Natl Acad Sci USA* 95:10306–10311.

Abdelkafi, S., A. Abousalham, I. Fendri et al. 2012. Identification of a new phospholipase D in *Carica papaya* latex. *Gene* 15:243–249.

Abu Bakar, U. K., V. Pillai, M. Hashim, and H. M. Daud. 2005. Sharing Malaysian experience with the development of biotechnology-derived food crops. *Food Nutr Bull* 26:432–435.

Abu Bakar, U. K., V. Pillai, P. Muda, L. P. Fatt, C. Y. Kwok, and H. M. Daud. 2001. Molecular and biochemical characterizations of "Eksotika" papaya plants transformed with antisense ACC oxidase gene. In *Papaya Biotechnology Network of SEAsia: Technical Workshop and Coordination Meeting*, 22–26 October 2001, Vietnam.

Ahloowalia, B. S. 1998. In vitro techniques and mutagenesis for the improvement of vegetatively propagated plants. In S. M. Jain, D. S. Brar, and B. S. Ahloowalia (eds), *Somaclonal Variation and Induced Mutations in Crop Improvement*, pp. 293–309. Dordrecht: Kluwer Academic Publishers.

Alabadí, D., T. Oyama, M. J. Yanovsky, F. G. Harmon, P. Má, and S. A. Kay. 2001. Reciprocal regulation between TOC1 and LHY/CCA1 within the *Arabidopsis* circadian clock. *Science* 293:880–883.

Antunes-Carvalho, F. and S. S. Renner. 2012. A dated phylogeny of the papaya family (Caricaceae) reveals the crop's closest relatives and the family's biogeographic history. *Mol Phylogenet Evol* 65:46–53.

Araujo, F. S., C. R. Carvalho, and W. R. Clarindo. 2010. Genome size, base composition and karyotype of *Carica papaya* L. *Nucleus* 53:25–31.

Argout, X., J. Salse, J. M. Aury et al. 2010. The genome of *Theobroma cacao*. *Nat Genet* 32:101–108.

Arumuganathan, K. and E. D. Earle. 1991. Nuclear DNA content of some important plant species. *Plant Mol Biol Rep* 9:208–218.

Aryal, R., X. Yang, Q. Yu, R. Sunkar, L. Li, and R. Ming. 2012. Asymmetric purine-pyrimidine distribution in cellular small RNA population of papaya. *BMC Genomics* 13:682.

Awada, M. and W. Ikeda. 1957. Effects of water, nitrogen application on composition of sugars in fruits, yield and sex expression of the papaya plant. *Hawaii Agr Exp Sta Tech Bull* 33:16.

Badillo, V. M. 1971. *Monografía de la Familia Caricaceae*. Maracay: Asociación de Profesores, Universidad Central de Venezuela, Facultad de Agronomía.

Badillo, V. M. 1993. Caricaceae. Segundo esquema. *Rev Fac Agron Univ Cent Venezuela* 43:1–111.

Badillo, V. M. 2000. *Vasconcella* St.-Hil. (Caricaceae) con la rehabilitacion de este ultimo. *Ernstia* 10:74–79.

Badillo, V. M. 2001. Nota correctiva *Vasconcellea* St. Hill. y no *Vasconcella* (Caricaceae). *Ernstia* 11:75–76.

Baginsky, S., L. Hennig, P. Zimmermann, and W. Gruissem. 2010. Gene expression analysis, proteomics, and network discovery. *Plant Physiol* 152:402–410.

Barker, M. S., H. Vogel, and M. E. Schranz. 2009. Paleopolyploidy in the Brassicales: Analyses of the *Cleome* transcriptome elucidate the history of genome duplications in *Arabidopsis* and other Brassicales. *Genome Biol Evol* 1:391–399.

Bergero, R. and D. Charlesworth. 2008. The evolution of restricted recombination in sex chromosomes. *Trends Ecol Evol* 24:94–102.

Blas, A. L., R. Ming, Z. Liu et al. 2010. Cloning of the papaya chromoplast-specific lycopene β-cyclase, *CpCYC-b*, controlling fruit flesh color reveals conserved microsynteny and a recombination hot spot. *Plant Physiol* 152:2013–2022.

Blas, A. L., Q. Yu, C. Chen et al. 2009. Enrichment of a papaya high-density genetic map with AFLP markers. *Genome* 52:716–725.

Blas, A. L., Q. Yu, Q. J. Veatch, R. E. Paull, P. H. Moore, and R. Ming. 2012. Genetic mapping of quantitative trait loci controlling fruit size and shape in papaya. *Mol Breed* 29:457–466.

Boerjan, W., J. Ralph, and M. Baucher. 2003. Lignin biosynthesis. *Annu Rev Plant Biol* 54:519–546.

Bowers, J. E., B. A. Chapman, J. Rong, and A. H. Paterson. 2003. Unraveling angiosperm genome evolution by phylogenetic analysis of chromosomal duplication events. *Nature* 422:433–438.

Brown, J. E., J. M. Bauman, J. F. Lawrie, O. J. Rocha, and R. C. Moore. 2012. The structure of morphological and genetic diversity in natural populations of *Carica papaya* (Caricaceae) in Costa Rica. *Biotropica* 44:179–188.

Cabrera-Ponce, J., A. Vegas-García, and L. Herrera-Estrella. 1995. Herbicide resistant transgenic papaya plants produced by an efficient particle bombardment transformation method. *Plant Cell Reps* 15:1–7.

Cai, W. Q., C. Gonsalves, P. Tennant et al. 1999. A protocol for efficient transformation and regeneration of *Carica papaya* L. *In Vitro Cell Dev Biol Plant* 35:61–69.

Castro, L. T., O. Ruiz, M. Vielma, and A. J. Briceño. 2002. Determinación sexual en *Carica papaya* L. *Pittieria* 31:25–32.

Celton, J.-M., A. Christoffels, D. J. Sargent, X. Xu, and D. J. G. Rees. 2010. Genome-wide SNP identification by high throughput sequencing and selective mapping allows sequence assembly positioning using a framework genetic linkage map. *BMC Biol* 8:155.

Chan, Y.-K. 2009. Radiation-induced mutation breeding of papaya. In *Induced Mutation in Tropical Fruit Trees*, pp. 93–100. Vienna: International Atomic Energy Agency.

Chan, Y.-K. and R. E. Paull. 2009. Papaya *Carica papaya* L., Caricaceae. In R. E. Paull and J. Janick (eds), *Encyclopedia of Fruits and Nuts*, pp. 237–247. Wallington UK: CABI Publishing.

Chandrika, U. G., E. R. Jansz, S. M. D. N. Wickramasinghe, and N. D. Warnasuriya. 2003. Carotenoids in yellow- and red-fleshed papaya (*Carica papaya* L). *J Sci Food Agric* 83:1279–1282.

Charlesworth, D. 2004. Plant evolution: Modern sex chromosomes. *Curr Biol* 14:R271–R273.

Chen, C., Q. Yu, S. Hou et al. 2007. Construction of a sequence-tagged high-density genetic map of papaya for comparative structural and evolutionary genomics in Brassicales. *Genetics* 177:2481–2491.

Cheng, Y.-H., J.-S. Yang, and S.-D. Yeh. 1996. Efficient transformation of papaya by coat protein gene of *papaya ringspot virus* mediated by *Agrobacterium* following liquid-phase wounding of embryogenic tissues with carborundum. *Plant Cell Rep* 16:127–132.

Ching, L. and S. Mohamed. 2001. Alpha-tocopherol content in 62 edible tropical plants. *J Agric Food Chem* 49:3101–3105.

Cho, H. T. and H. Kende. 1997. Expression of expansin genes is correlated with growth in deep water rice. *Plant Cell* 9:1661–1671.

Cong, B., J. Liu, and S. D. Tanksley. 2002. Natural alleles at a tomato fruit size quantitative trait locus differ by heterochronic regulatory mutations. *Proc Natl Acad Sci USA* 99:13606–13611.

Corrêa-Damasceno, J. P., F. R. Costa, T. N. S. Pereira, M. F. Neto, and M. G. Pereira. 2009. Karyotype determination in three Caricaceae species emphasizing the cultivated form (*C. papaya* L). *Caryologia* 62:10–15.

Cosgrove, D. J. 2000. Loosening of plant cell walls by expansins. *Nature* 407:321–326.

Covington, M. F., J. N. Maloof, M. S. Straume, A. Kay, and S. L. Harmer. 2008. Global transcriptome analysis reveals circadian regulation of key pathways in plant growth and development. *Genome Biol* 9:R130.

Crowell, D. N. 1994. Cytokinin regulation of a soybean pollen allergen gene. *Plant Mol Biol* 25:829–835.

Czaplewski, C., Z. Grzonka, M. Jaskolski, F. Kasprzykowski, and M. Kozk. 1999. Binding modes of a new epoxysuccinyl-peptide inhibitory of cysteine proteases. Where and how do cysteine proteases express their selectivity? *Biochim Biophys Acta* 1431:290–305.

da Costa, F. R., T. N. Santana-Pereira, A. P. Candido-Gabriel, and M. Gonzaga-Pereira. 2011. ISSR markers for genetic relationships in Caricaceae and sex differentiation in papaya. *Crop Breed Appl Biot* 11:352–357.

Dangl, J. L. and J. D. Jones. 2001. Plant pathogens and integrated defence responses to infection. *Nature* 411:826–833.

Daud, H. M., U. K. Abu Bakar, V. Pillai et al. 2005. Improvement of "Eksotika" papaya through transgenic technology—Malaysia's experience. In *Proceedings of the First International Symposium on Papaya*, p. 49. 22–24 November 2005, Malaysia (Abstract).

Davis, G. L., M. D. McMullen, C. Baysdorfer et al. 1999. A maize map standard with sequenced core markers, grass genome reference points and 932 expressed sequence tagged sites (ESTs) in a 1736-locus map. *Genetics* 152:1137–1172.

de la Fuente, J. M., V. Ramirez-Rodriguez, J. L. Cabrera-Ponce, and L. Herrera-Estrella. 1997. Aluminum tolerance in transgenic plants by alteration of citrate synthesis. *Science* 276:1566–1568.

Deputy, J. C., R. Ming, H. Ma et al. 2002. Molecular markers for sex determination in papaya (*Carica papaya* L.). *Theor Appl Genet* 106:107–111.

Devitt, L. C., K. Fanning, R. G. Dietzgen, and T. A. Holton. 2010. Isolation and functional characterization of a lycopene β-cyclase gene that controls fruit colour of papaya (*Carica papaya* L.). *J Exp Bot* 61:33–39.

Devitt, L. C., T. Sawbridge, T. A. Holton, K. Mitchelson, and R. G. Dietzgen. 2006. Discovery of genes associated with fruit ripening in *Carica papaya* using expressed sequence tags. *Plant Sci* 170:356–363.

DeYoung, B. J. and R. W. Innes. 2006. Plant NBS-LRR proteins in pathogen sensing and host defense. *Nat Immunol* 7:1243–1249.

Dhouib, R., J. Laroche-Traineau, R. Shaha et al. 2011. Identification of a putative triacylglycerol lipase from papaya latex by functional proteomics. *FEBS J* 278:97–110.

Dietrich, R. A., M. H. Richberg, R. Schmidt, C. Dean, and J. L. Dangl. 1997. A novel zinc finger protein is encoded by the *Arabidopsis LSD1* gene and functions as a negative regulator of plant cell death. *Cell* 88:685–694.

Dinesh, M. R. 2010. Papaya breeding in India. *Acta Hortic* 851:69–76.

Dixon, R. A., J. H. Bouto, B. Narasimhamoorthy, M. Saha, Z.-Y. Wang, and G. D. May. 2007. Beyond structural genomics for plant science. *Adv Agron* 95:77–161.

Dodds, P. N., G. J. Lawrence, and J. G. Ellis. 2001. Six amino acid changes confined to the leucine-rich repeat β-strand/β-turn motif determine the difference between the *P* and *P2* rust resistance specificities in flax. *Plant Cell* 13:163–178.

Dong, X., M. C. Greven, A. Kundaje et al. 2012. Modeling gene expression using chromatin features in various cellular contexts. *Genome Biol* 13:R53.

Downes, B. P. and D. N. Crowell. 1998. Cytokinin regulates the expression of a soybean β-expansin gene by a post-transcription mechanism. *Plant Mol Biol* 37:437–444.

Dunlap, J. C. 1999. Molecular bases for circadian clocks. *Cell* 96:271–290.

Edwards, D. and J. Batley. 2010. Plant genome sequencing: Applications for crop improvement. *Plant Biotechnol J* 8:2–9.

Ellis, J. G., G. J. Lawrence, J. E. Luck, and P. N. Dodds. 1999. Identification of regions in alleles of the flax rust resistance gene *L* that determine differences in gene-for- gene specificity. *Plant Cell* 11:495–506.

EPIC Planning Committee. 2012. Reading the second code: Mapping epigenomes to understand plant growth, development, and adaptation to the environment. *Plant Cell* 24:2257–2261.

Fabi, J. P., L. R. B. Carelli-Mendes, F. M. Lajolo, and J. R. Oliveira-do-Nascimento. 2011. Transcript profiling of papaya fruit reveals differentially expressed genes associated with fruit ripening. *Plant Sci* 179:225–233.

Fabi, J. P., G. B. Seymour, N. S. Graham et al. 2012. Analysis of ripening-related gene expression in papaya using an *Arabidopsis*-based microarray. *BMC Plant Biol* 12:242.

Fermin, G., L. T. Castro, and P. F. Tennant. 2010. CP-Transgenic and non-transgenic approaches for the control of papaya ringspot: Current situation and challenges. *Transgenic Plant J* 4:1–15.

Fermin, G., R. C. Keith, J. Y. Suzuki et al. 2011. Allergenicity assessment of the papaya ringspot virus coat protein expressed in transgenic "Rainbow" papaya. *J Agr Food Chem* 59:1006–1012.

Fermin, G., P. Tennant, C. Gonsalves, and D. Gonsalves. 2004. Comparative development and impact of transgenic papayas in Hawaii, Jamaica and Venezuela. In L. Peña (ed.), *Methods in Molecular Biology 286 Transgenic Plants: Methods and Protocols*, pp. 399–430. New Jersey: Humana Press.

Fitch, M. M. M. 1993. High frequency somatic embryogenesis and plant regeneration from papaya hypocotyl callus. *Plant Cell Tiss Org Cult* 32:205–212.

Fitch, M. M. M. 1995. Somatic embryogenesis in papaya. In Y. P. S. Bajaj (ed.), *Biotechnology in Agriculture and Forestry*, pp. 260–79. Heidelberg: Springer Science + Business Media.

Fitch, M. M. M. and R. M. Manshardt. 1990. Somatic embryogenesis and plant regeneration from immature zygotic embryos of papaya (*Carica papaya* L.). *Plant Cell Rep* 9:320–324.

Fitch, M. M. M., R. M. Manshardt, D. Gonsalves, and J. L. Slightom. 1993. Transgenic papaya plants from *Agrobacterium* mediated transformation of somatic embryos. *Plant Cell Rep* 12:245–249.

Fitch, M. M. M., R. M. Manshardt, D. Gonsalves, J. L. Slightom, and J. C. Sanford. 1990. Stable transformation of papaya via microprojectile bombardment. *Plant Cell Rep* 9:189–194.

Fitch, M. M. M., R. M. Manshardt, D. Gonsalves, J. L. Slightom, and J. C. Sanford. 1992. Virus resistant papaya plants derived from tissues bombarded with the coat protein gene of papaya ringspot virus. *Nat Biotechnol* 10:1466–1472.

Fitch, M. M. M., P. H. Moore, and T. C. W. Leong. 2005. Clonally propagated and seed-derived papaya orchards: I. Plant production and field growth. *HortScience* 40:1283–1290.

Flagel, L. E. and J. F. Wendel. 2009. Gene duplication and evolutionary novelty in plants. *New Phytol* 183:557–564.

Flath, R. A. and R. R. Forrey. 1977. Volatile components of papaya (*Carica papaya* L., solo variety). *J Agric Food Chem* 25:103–109.

Freeling, M., E. Lyons, B. Pedersen, M. Alam, R. Ming, and D. Lisch. 2008. Many or most genes in *Arabidopsis* transposed after the origin of the order Brassicales. *Genome Res* 18:1924–1937.

Ganal, M. W., T. Altman, and M. S. Röder. 2009. SNP identification in crop plants. *Curr Opin Plant Biol* 12:1–7.

Giovannoni, J. J. 2004. Genetic regulation of fruit development and ripening. *Plant Cell* 16:S170–S180.

Gonsalves, D. 1994. papaya ringspot virus. In R. C. Ploetz, G. A. Zentmyer, W. T. Nishijima, K. G. Rohrbach, and H. D. Ohr (eds), *Compendium of Tropical Fruit Diseases*, pp. 67–68. St. Paul, MN: APS Press.

Gonsalves, D. 1998. Control of papaya ringspot virus in papaya: A case study. *Annu Rev Phytopathol* 36:415–437.

Gonsalves, C., W. Cai, P. Tennant, and D. Gonsalves. 1998. Effective development of papaya ringspot virus resistant papaya with untranslatable coat protein gene using a modified microprojectile transformation method. *Acta Hortic* 461:311–314.

Goodspeed, D., E. W. Chehab, A. Min-Venditti, J. Braam, and M. F. Covington. 2012. *Arabidopsis* synchronizes jasmonate-mediated defense with insect circadian behavior. *Proc Natl Acad Sci USA* 109:4674–4677.

Grattapaglia, D., R. E. Vaillancourt, M. Shepherd et al. 2012. Progress in Myrtaceae genetics and genomics: *Eucalyptus* as the pivotal genus. *Tree Genet Genomes* 8:463–508.

Gschwend, A. R., C.-M. Wai, F. Zee, A. K. Arumuganathan, and R. Ming. 2013. Genome size variation among sex types in dioecious and trioecious Caricaceae species. *Euphytica* 189:461–469.

Gschwend, A. R., Q. Yu, P. Moore et al. 2011. Construction of papaya male and female BAC libraries and application in physical mapping of the sex chromosomes. *J Biomed Biotechnol* 2011(929472):7.

Gschwend, A. R., Q. Yu, E. J. Tong et al. 2012. Rapid divergence and expansion of the X chromosome in papaya. *Proc Natl Acad Sci USA* 109:13716–13721.

Haanstra, J. P. W., C. Wye, H. Verbakel et al. 1999. An integrated high-density RFLP-AFLP map of tomato based on two *Lycopersicon esculentum* L. pennellii F_2 populations. *Theor Appl Genet* 99:254–271.

Hanada, K., X. Zhang, J. O. Borevitz, W.-H. Li, and S.-H. Shiu. 2007. A large number of novel coding small open reading frames in the intergenic regions of the *Arabidopsis thaliana* genome are transcribed and/or under purifying selection. *Genome Res* 17:632–640.

Harushima, Y., M. Yano, A. Shomura et al. 1998. A high-density rice genetic linkage map with 2275 markers using a single F_2 population. *Genetics* 148:479–494.

Hedrick, U. P. and N. O. Booth. 1907. Mendelian characters in tomatoes. *Proc Am Soc Hort Sci* 5:19–24.

Hernández, M., J. L. Cabrera-Ponce, G. Fragoso et al. 2007. A new highly effective anticysticercosis vaccine expressed in transgenic papaya. *Vaccine* 25:4252–4260.

Hewitt, H. H., S. Whittle, S. A. Lopez, E. Y. Bailey, and S. R. Weaver. 2000. Topical uses of papaya in chronic skin ulcer therapy in Jamaica. *West Indian Med J* 49(1):32–33.

Higgins, J. C. and V. Holt. 1914. The papaya in Hawaii. *Hawaii Agr Exp Sta Bull* 32:1–44.

Hofmeyr, J. D. J. 1938. Genetical studies of *Carica papaya* L. *S Afr Dept Agric Sci Bull* 35:300–304.

Hofmeyr, J. D. J. 1939a. Sex reversal in *Carica papaya* L. *S Afr J Sci* 26:286–287.

Hofmeyr, J. D. J. 1939b. Sex-linked inheritance in *Carica papaya* L. *S Afr J Sci* 36:283–285.

Horovitz, S. and H. Jimenéz. 1967. Cruzamientos interespecíficos e intergenéricos en caricaceas y sus implicaciones fitotécnicas. *Agron Trop* 17:323–343.

Huang, S., R. Li, Z. Zhang et al. 2009. The genome of the cucumber, *Cucumis sativus* L. *Nat Genet* 41:1275–1281.

Huerta-Ocampo, J. A., J. A. Osuna-Castro, G. J. Lino-López et al. 2012. Proteomic analysis of differentially accumulated proteins during ripening and in response to 1-MCP in papaya fruit. *J Proteomics* 75:2160–2169.

Idrovo-Espín, F. M., S. Peraza-Echeverria, G. Fuentes, and J. M. Santamaría. 2012. *In silico* cloning and characterization of the TGA (TGACG MOTIF-BINDING FACTOR) transcription factors subfamily in *Carica papaya*. *Plant Physiol Biochem* 54:113–122.

International Rice Genome Sequencing Project. 2005. The map-based sequence of the rice genome. *Nature* 436:793–800.

Ishitani, M., R. I. Rao, P. Wenzl, S. Beebe, and J. Tohme. 2004. Integration of genomics approach with traditional breeding towards improving abiotic stress adaptation: Drought and aluminium toxicity as case studies. *Field Crop Res* 90:35–45.

Jaillon, C. O., J. M. Aury, B. Noel et al. 2007. The grapevine genome sequence suggests ancestral hexaploidization in major angiosperm phyla. *Nature* 449:463–467.

Jeon, J.-S., S. Lee, K.-H. Jung et al. 2000. T-DNA insertional mutagenesis for functional genomics in rice. *Plant J* 22:561–570.

Jia, Y., S. A. McAdams, G. T. Bryan, H. P. Hershey, and B. Valent. 2000. Direct interaction of resistance gene and avirulence gene products confers rice blast resistance. *EMBO J* 19:4004–4014.

Jiang, L. and L.-J. Pan. 2012. Identification and expression of C2H2 transcription factor genes in *Carica papaya* under abiotic and biotic stresses. *Mol Biol Rep* 39:7105–7115.

Jones, D. A. 2000. Resistance genes and resistance protein function. In M. Dickinson and J. Beynon (eds), *Molecular Plant Pathology, Annual Plant Reviews*, vol. 4, pp. 108–143. Sheffield: Sheffield Academic Press.

Kim, M. S., P. H. Moore, F. Zee et al. 2002. Genetic diversity of *Carica papaya* as revealed by AFLP markers. *Genome* 45:503–512.

Knepper, C., E. A. Savory, and B. Day. 2011. The role of NDR1 in pathogen perception and plant defense signaling. *Plant Signal Behav* 6:1114–1116.

Koornneef, M., L. W. M. Dellaert, and J. H. van der Veen. 1982. EMS- and radiation-induced mutation frequencies at individual loci in *Arabidopsis thaliana* (L.) Heynh. *Mutat Res* 93:109–123.

Ku, H. M., S. Doganlar, K. Y. Chen, and S. D. Tanksley. 1999. Genetic basis of pear-shaped fruit in tomato. *Theor Appl Genet* 99:844–850.

Ku, H. M., J. Liu, S. Doganlar, and S. D. Tanksley. 2001. Exploitation of *Arabidopsis*-tomato synteny to construct a high-resolution map of the ovate-containing region in tomato chromosome 2. *Genome* 44:470–475.

Kung, Y. J., T. A. Yu, C. H. Huang et al. 2010. Generation of hermaphrodite transgenic papaya lines with virus resistance via transformation of somatic embryos derived from adventitious roots of in vitro shoots. *Transgenic Res* 19:621–635.

Kyndt, T., B. V. Droogenbroeck, A. Haegeman, I. Roldán-Ruiz, and G. Gheysen. 2006. Cross-species microsatellite amplification in *Vasconcellea* and related genera and their use in germplasm classification. *Genome* 49:786–798.

Lai, C. W., Q. Yu, S. Hou et al. 2006. Analysis of papaya BAC end sequences reveals first insights into the organization of a fruit tree genome. *Mol Genet Genomics* 276:1–12.

Li, F., D. Pignatta, C. Bendix et al. 2012. MicroRNA regulation of plant innate immune receptors. *Proc Natl Acad Sci USA* 109:1790–1795.

Li, W. X., Y. Oono, J. Zhu et al. 2008. The *Arabidopsis* NFYA5 transcription factor is regulated transcriptionally and posttranscriptionally to promote drought resistance. *Plant Cell* 20:2238–2251.

Li, X., Y. Song, K. Century et al. 2001. A fast neutron deletion mutagenesis-based reverse genetics system for plants. *Plant J* 27:235–242.

Liebman, B. 1992. Fresh fruit: A papaya a day? (nutritional aspects of fruit). Nutrition Action Healthletter, *HighBeam Research*. http://www.highbeam.com/doc/1G1-12272995.html. (Accessed: February 13, 2013).

Liu, Y., T. Burch-Smith, M. Schiff, S. D. Feng, and S. P. Dinesh-Kumar. 2004a. Molecular chaperone Hsp90 associates with resistance protein N and its signaling proteins SGT1 and Rar1 to modulate an innate immune response in plants. *J Biol Chem* 279:2101–2108.

Liu, Z., P. H. Moore, H. Ma et al. 2004b. A primitive Y chromosome in papaya marks incipient sex chromosome evolution. *Nature* 427:22–26.

Llaca, V. 2012. Sequencing technologies and their use in plant biotechnology and breeding. In A. Munshi (ed.), *DNA Sequencing—Methods and Applications*, pp. 35–60. Croatia: InTech Open.

Ma, H., P. H. Moore, Z. Liu et al. 2004. High-density linkage mapping revealed suppression of recombination at the sex determination locus in papaya. *Genetics* 166:419–436.

MacLeod, A. J. and N. M. Pieris. 1983. Volatile components of papaya (*Carica papaya* L.) with particular reference to glucosinolate products. *J Agric Food Chem* 31:1005–1008.

Maghuly, F., G. Marzban, and M. Laimer. 2009. Functional genomics of allergen gene families in fruits. *Nutrients* 1:119–132.

Mahadevamma, M., L. Sahijram, V. Kumari, and T. H. Shankarappa. 2012. In vitro mutation studies in papaya (*Carica papaya* L.). *CIBTech J Biotechnol* 1:49–55.

Mallory, A. and H. Vaucheret. 2006. Functions of microRNAs and related small RNAs in plants. *Nat Genet* 38:S31–S36.

Maluszynski, M., B. S. Ahloowalia, and B. Sigurbjornsson. 1995. Application of in vivo and in vitro mutation techniques for crop improvement. *Euphytica* 85:303–315.

May, C., F. Brosseron, P. Chartoski, C. Schumbrutzki, B. Schoenebeck, and K. Marcus. 2011. Instruments and methods in proteomics. *Meth Mol Biol* 696:3–26.

McCafferty, H. R. K., P. H. Moore, and J. Y. Zhu. 2006. Improved *Carica papaya* tolerance to carmine spider mite by the expression of *Manduca sexta* chitinase transgene. *Transgenic Res* 15:337–347.

McCourt, P. and D. Desveaux. 2009. Plant chemical genetics. *New Phytol* 185:15–26.

McDowell, J. M. and B. J. Woffenden. 2003. Plant disease resistance genes: Recent insights and potential application. *Trends Biotechnol* 21:178–183.

Mello, V., M. Gomes, F. Lemos et al. 2008. The gastric ulcer protective and healing role of cysteine proteinases from *Carica candamarcencis*. *Phytomedicine* 15:237–244.

Miean, K. H. and S. Mohammed. 2001. Flavonoid (myricetin, quercetin, kaempferol, luteolin, and apigenin) content of edible tropical plants. *J Agric Food Chem* 49:3106–3112.

Ming, R., A. Bendahmane, and S. S. Renner. 2011. Sex chromosome in land plants. *Annu Rev Plant Biol* 62:485–514.

Ming, R., S. Hou, Y. Feng et al. 2008. The draft genome of the transgenic tropical fruit papaya (*Carica papaya* Linnaeus). *Nature* 452:991–996.

Ming, R., P. H. Moore, F. Zee, C. A. Abbey, H. Ma, and A. H. Paterson. 2001. Construction and characterization of a papaya BAC library as a foundation for molecular dissection of a tree-fruit genome. *Theor Appl Genet* 102:892–899.

Ming, R., Q. Yu, and P. H. Moore. 2007. Sex determination in papaya. *Semin Cell Dev Biol* 18:401–408.

Ming, R., Q. Yu, P. H. Moore et al. 2012. Genome of papaya, a fast growing tropical fruit tree. *Tree Genet Genomes* 8:445–462.

Mochida, K. and K. Shinozaki. 2010. Genomics and bioinformatics resources for crop improvement. *Plant Cell Physiol* 51:497–523.

Morrell, P. L., E. S. Buckler, and J. Ross-Ibarra. 2012. Crop genomics: Advances and applications. *Nat Rev Genet* 13:85–96.

Na, J.-K., J. Wang, J. E. Murray et al. 2012. Construction of physical maps for the sex-specific regions of papaya sex chromosomes. *BMC Genomics* 13:176.

Nakasone, H. Y. and C. Lamoureux. 1982. Transitional form of hermaphroditic papaya flowers leading to complete maleness. *J Amer Soc Hort Sci* 1074:589–592.

Nakasone, H. Y. and R. E. Paull. 1998. Papaya. In H. Y. Nakasone and R. E. Paull (eds), *Tropical Fruit*, pp. 239–269. New York: CABI Publishing.

Nakasone, H. Y. and W. B. Storey. 1955. Studies on inheritance of fruiting height of *Carica papaya* L. *J Amer Soc Hort Sci* 66:168–182.

Nawrath, C. and J.-P. Métraux. 1999. Salicylic acid induction-deficient mutants of *Arabidopsis* express PR-2 and PR-5 and accumulate high levels of camalexin after pathogen inoculation. *Plant Cell* 11:1393–1404.

Nesbitt, T. C. and S. D. Tanksley. 2001. *fw2.2* directly affects the size of developing fruit, with secondary effects on fruit number and photosynthate partitioning. *Plant Physiol* 127:575–583.

Neupane, K. R., U. T. Mukatira, C. Kato, and J. J. Stiles. 1998. Cloning and characterization of fruit expressed ACC synthase and ACC oxidase from papaya (*Carica papaya* L.). *Acta Hortic* 461:329–337.

Nguyen, T. T., P. N. Shaw, M. O. Parat, and A. K. Hewavitharana. 2013. Anticancer activity of *Carica papaya*: A review. *Mol Nutr Food Res* 57:153–164.

Nogueira, S. B., C. A. Labate, F. C. Gozzo, E. J. Pilau, F. M. Lajolo, and J. R. Oliveira-do-Nascimento. 2012. Proteomic analysis of papaya fruit ripening using 2DE-DIGE. *J Proteomics* 75:1428–1439.

Ocampo-Pérez, J., D. Dambier, P. Ollitrault et al. 2006. Microsatellite markers in *Carica papaya* L.: Isolation, characterization and transferability to *Vasconcellea* species. *Mol Ecol Notes* 6:212–217.

OECD. 2005. Consensus document on the biology of papaya (*Carica papaya*). Organization for Economic Co-operation and Development (OECD), Series on Harmonisation of Regulatory Oversight in Biotechnology No. 33. Head of Publications Service, Paris, France: OECD.

Okrent, R. A. and M. C. Wildermuth. 2011. Evolutionary history of the GH3 family of acyl adenylases in rosids. *Plant Mol Biol* 76:489–505.

Oliver, J. E., P. F. Tennant, and M. Fuchs. 2011. Virus-resistant transgenic horticultural crops: Safety issues and lessons from risk assessment studies. In B. Mou and R. Scorza (eds), *Transgenic Horticultural Crops: Challenges and Opportunities—Essays by Experts*, pp. 263–287. Boca Raton (FL): CRC Press.

Østergaard, L. and M. F. Yanofsky. 2004. Establishing gene function by mutagenesis in *Arabidopsis thaliana*. *Plant J* 39:682–696.

Otsuki, N., N. H. Dang, E. Kumagai, A. Kondo, S. Iwata, and C. Morimoto. 2010. Aqueous extracts of *Carica papaya* leaves exhibits anti-tumor activity and immunomodulatory effects. *J Ethnopharmacol* 127:760–767.

Pant, B. D., A. Buhtz, J. Kehr, and W.-R. Scheible. 2008. MicroRNA399 is a long-distance signal for the regulation of plant phosphate homeostasis. *Plant J* 53:731–738.

Parasnis, A. S., V. S. Gupta, S. A. Tamhankar, and P. K. Ranjekar. 2000. A highly reliable sex diagnostic PCR assay for mass screening of papaya seedlings. *Mol Breed* 6:337–344.

Parkinson, J. and M. Blaxter. 2009. Expressed sequence tags: An overview. In J. Parkinson (ed.), *Expressed Sequence Tags (ESTs): Generation and Analysis, Methods in Molecular Biology*, vol. 533, pp. 1–12. New York: Humana Press Inc.

Paterson, A. H., J. E. Bowers, R. Bruggmann et al. 2009. The *Sorghum bicolor* genome and the diversification of grasses. *Nature* 457:551–556.

Paull, R. E. 1993. Pineapple and papaya. In G. Seymour, J. Taylor, and G. Tucker (eds), *Biochemistry of Fruit Ripening*, pp. 291–323. London: Chapman and Hall.

Paull, R. E. and N. J. Chen. 1983. Postharvest variation in cell wall-degrading enzymes of papaya (*Carica papaya* L.) during fruit ripening. *Plant Physiol* 72:382–385.

Paull, R. E., B. Irikura, P. Wu et al. 2008. Fruit development, ripening and quality related genes in the papaya genome. *Tropical Plant Biol* 1:246–277.

Peraza-Echeverria, S., J. M. Santamaría, G. Fuentes et al. 2012. The *NPR1* family of transcription cofactors in papaya: Insights into its structure, phylogeny and expression. *Genes Genomics* 34:379–390.

Pereira, J., J. Pereira, and J. S. Câmara. 2011. Effectiveness of different solid-phase microextraction fibres for differentiation of selected Madeira island fruits based on their volatile metabolite profile-identification of novel compounds. *Talanta* 15:899–906.

Pérez-de-Castro, A. M., S. Vilanova, J. Cañizares et al. 2012. Application of genomic tools in plant breeding. *Curr Genomics* 13:179–195.

Peters, J. L., H. Constandt, P. Neyt et al. 2001. A physical amplified fragment-length polymorphism map of *Arabidopsis*. *Plant Physiol* 127:1579–1589.

Pieruschka, R. and H. Poorter. 2012. Phenotyping plants: Genes, phenes and machines. *Funct Plant Biol* 39:813–820.

Pino, J. A., K. Almora, and R. Marbot. 2003. Volatile components of papaya (*Carica papaya* L., maradol variety) fruit. *Flavour Frag J* 18:492–496.

Porter, B. W., K. S. Aizawa, Y. J. Zhub, and D. A. Christopher. 2008. Differentially expressed and new non-protein-coding genes from a *Carica papaya* root transcriptome survey. *Plant Sci* 174:38–50.

Porter, B. W., M. Paidi, R. Ming, M. Alam, W. T. Nishijima, and Y. J. Zhu. 2009. Genome-wide analysis of *Carica papaya* reveals a small NBS resistance gene family. *Mol Genet Genomics* 281:609–626.

Raes, J., A. Rohde, J. H. Chrisrensen, Y. van der Peer, and W. Boerjan. 2003. Genome-wide characterization of the lignin toolbox in *Arabidopsis*. *Plant Physiol* 133:1051–1071.

Ram, M. and S. Srivastava. 1984. Mutagenesis in papaya. In *National Seminar on Papaya and Papain Products*, pp. 26–27. Coimbatore: Tamil Nadu Agricultural University (TNUA).

Rensing, S. A., D. Lang, A. D. Zimmer et al. 2008. The *Physcomitrella* genome reveals evolutionary insights into the conquest of land by plants. *Science* 319:64–69.

Rivera-Pastrana, D. M., E. M. Yahia, and G. A. González-Aguilar. 2010. Phenolic and carotenoid profiles of papaya fruit (*Carica papaya* L.) and their contents under low temperature storage. *J Sci Food Agr* 90:2358–2365.

Rodrigues, S. P., J. A. Ventura, C. Aguilar et al. 2011. Proteomic analysis of papaya (*Carica papaya* L.) displaying typical sticky disease symptoms. *J Proteomics* 11:2592–2602.

Rodrigues, S. P., J. A. Ventura, C. Aguilar et al. 2012. Label-free quantitative proteomics reveals differentially regulated proteins in the latex of sticky diseased *Carica papaya* L. plants. *J Proteomics* 75:3191–3198.

Ronse Decraene, L. P. and E. F. Smets. 1999. The floral development and anatomy of *Carica papaya* (Caricaceae). *Can J Bot* 77:582–598.

Rose, J. K. C., H. H. Lee, and A. B. Bennett. 1997. Expression of a divergent expansin gene is fruit-specific and ripening-regulated. *Proc Natl Acad Sci USA* 94:5955–5960.

Sankoff, D., C. Zheng, P. K. Wall, C. De Pamphilis, J. Leebens-Mack, and V. A. Albert. 2009. Towards improved reconstruction of ancestral gene order in Angiosperm phylogeny. *J Comput Biol* 16:1353–1367.

Santosh, L. C., M. R. Dinesh, and A. Rekha. 2010. Mutagenic studies in papaya (*Carica papaya* L.). *Acta Hortic* 851:109–112.

Scheldeman, X., T. Kyndt, G. C. Coppens d'Eeckenbrugge et al. 2011. *Vasconcellea*. In C. Kole (ed.), *Wild Crop Relatives*: *Genomic and Breeding Resources, Tropical and Subtropical Fruits*, pp. 213–249. Heidelberg: Springer-Verlag.

Schmutz, J., S. B. Cannon, J. Schlueter et al. 2010. Genome sequence of the palaeopolyploid soybean. *Nature* 463:178–183.

Schnable, P. S., D. Ware, R. S. Fulton et al. 2009. The B73 maize genome: Complexity, diversity, and dynamics. *Science* 326:1112–1115.

Schweiggert, R. M., C. B. Steingass, P. Esquivel, and R. Carle. 2012. Chemical and morphological characterization of Costa Rican papaya (*Carica papaya* L.) hybrids and lines with particular focus on their genuine carotenoid profiles. *J Agr Food Chem* 60:2577–2585.

Schweiggert, R. M., C. B. Steingass, A. Heller, P. Esquivel, and R. Carle. 2011. Characterization of chromoplasts and carotenoids of red- and yellow-fleshed papaya (*Carica papaya* L.). *Planta* 234:1031–1044.

Seigler, D. S., G. F. Pauli, A. Nahrstedt, and R. Leen. 2002. Cyanogenic allosides and glucosides from *Passiflora edulis* and *Carica papaya*. *Phytochemistry* 69:873–882.

Shakirov, E. V., S. L. Salzberg, M. Alam, and D. E. Shippen. 2008. Analysis of *Carica papaya* telomeres and telomere-associated proteins: Insights into the evolution of telomere maintenance in Brassicales. *Trop Plant Biol* 1:202–215.

Shivaprasad, P.V., H. M. Chen, K. Patel, D. M. Bond, B. Santos, and D. Baulcombe. 2012. A microRNA superfamily regulates nucleotide binding site–leucine-rich repeats and other mRNAs. *Plant Cell* 24:859–874.

Shulaev, V., D. J. Sargent, R. N. Crowhurst et al. 2011. The genome of woodland strawberry (*Fragaria vesca*). *Nat Genet* 43:109–116.

Sondur, S. N., R. M. Manshardt, and J. Stiles. 1996. A genetic linkage map of papaya based on randomly amplified polymorphic DNA markers. *Theor Appl Genet* 93:547–553.

Srinivasan, R. 2004. Genetic linkage mapping and QTL analysis of economic traits in papaya (*Carica papaya* L.). PhD Thesis, University of Hawaii.

Starley, I. F., P. Mohammed, G. Schneider, and S. W. Bickler. 1999. The treatment of paediatric burns using topical papaya. *Burns* 25:636–639.

Storey, W. B. 1938. Segregation of sex types in "Solo" papaya and their application to the selection of seed. *Am Soc Hortic Sci* 35:83–85.

Storey, W. B. 1953. Genetics of the papaya. *J Hered* 44:70–78.

Subhadrabandhu, S. and C. Nontaswatsri. 1997. Combining ability of some characters of introduced and local papaya cultivars. *Sci Hortic-Amsterdam* 71:203–212.

Sumner, L. W., P. Mendes, and R. A. Dixon. 2003. Plant metabolomics: Large-scale phytochemistry in the functional genomics era. *Phytochemistry* 62:817–836.

Sunkar, R., V. Chinnusamy, J. Zhu, and J. K. Zhu. 2007. Small RNAs as big players in plant abiotic stress responses and nutrient deprivation. *Trends Plant Sci* 12:301–309.

Sunkar, R. and G. Jagadeeswaran. 2008. *In silico* identification of conserved microRNAs in large number of diverse plant species. *BMC Plant Biol* 8:37.

Suzuki, J.Y., S. Tripathi, and D. Gonsalves. 2007. Virus resistant transgenic papaya: Commercial development and regulatory and environmental Issues. In Z. K. Punja, S. DeBoer, and H. Sanfacon (eds), *Biotechnology and Plant Disease Management*, pp. 436–461. Wallingford: CABI Publishing.

Tanksley, S. D., M. W. Ganal, J. P. Prince et al. 1992. High density molecular linkage maps of the tomato and potato genomes. *Genetics* 132:1141–1160.

Tao, Y., F. Yuan, R. T. Leister, F. M. Ausubel, and F. Katagiri. 2000. Mutational analysis of the *Arabidopsis* nucleotide binding site–leucine-rich repeat resistance gene *RPS2*. *Plant Cell* 12:2541–2554.

Tennant, P., M. H. Ahmad, and D. Gonsalves. 2005. Field resistance of coat protein transgenic papaya to papaya ringspot virus in Jamaica. *Plant Dis* 89:841–847.

The Arabidopsis Genome Initiative. 2000. Analysis of the genome sequence of the flowering plant *Arabidopsis thaliana*. *Nature* 408:796–815.

Trindade, I., C. Capitao, T. Dalmay, M. P. Fevereiro, and D. Metelo dos Santos. 2010. miR398 and miR408 are up-regulated in response to water deficit in *Medicago truncatula*. *Planta* 231:705–716.

Troggio, M., A. Gleave, S. Salvi et al. 2012. Apple, from genome to breeding. *Tree Genet Genomes* 8:509–529.

Tuskan, G. A., S. Difazio, S. Jansson et al. 2006. The genome of the black cottonwood, *Populus trichocarpa* (Torr. & Gray). *Science* 313:1596–1604.

Urasaki, N., K. Tarora, A. Shudo et al. 2012. Digital transcriptome analysis of putative sex-determination genes in papaya (*Carica papaya*). *PLoS ONE* 7:e40904.

Urasaki, N., K. Tarora, T. Uehara, I. Chinen, R. Terauchi, and M. Tokumoto. 2002a. Rapid and highly reliable sex diagnostic PCR assay for papaya (*Carica papaya* L.). *Breed Sci* 52:333–335.

Urasaki, N., M. Tokumoto, K. Tarora et al. 2002b. A male and hermaphrodite specific RAPD marker for papaya (*Carica papaya* L.). *Theor Appl Genet* 104:281–285.

van der Biezen, E. A. and J. D. Jones. 1998. Plant disease-resistance proteins and the gene-for-gene concept. *Trends Biochem Sci* 23:454–456.

van der Knaap, E. and S. D. Tanksley. 2003. The making of a bell pepper-shaped tomato fruit: Identification of loci controlling fruit morphology in "Yellow Stuffer" tomato. *Theor Appl Genet* 107:139–147.

van Droogenbroeck, B., T. Kyndt, I. Maertens et al. 2004. Phylogenetic analysis of the highland papayas (*Vasconcellea*) and allied genera (Caricaceae) using PCR-RFLP. *Theor Appl Genet* 108:1473–1486.

van Enckevort, L. J. G., G. Droc, P. PiVanelli et al. 2005. EUOSTID: A collection of transposon insertional mutants for functional genomics in rice. *Plant Mol Biol* 59:99–110.

Varga, A. and J. Bruinsma. 1986. Tomato. In S. P. Monselise (ed.), *CRC Handbook of Fruit Set and Development*, pp. 461–480. Boca Raton, FL: CRC Press.

Velasco, R., A. Zharkikh, J. Affourtit et al. 2010. The genome of the domesticated apple (*Malus × domestica* Borkh.). *Nat Genet* 42:833–841.

Voinnet, O. 2009. Origin, biogenesis, and activity of plant microRNAs. *Cell* 136:669–687.

Wai, C. M., P. H. Moore, R. E. Paull, R. Ming, and Q. Yu. 2012. An integrated cytogenetic and physical map reveals unevenly distributed recombination spots along the papaya sex chromosomes. *Chromosome Res* 20:753–767.

Waly, M. I., N. A. Guizani, M. S. A. Ali, and M. S. Rahman. 2012. Papaya epicarp extract protects against aluminum-induced neurotoxicity. *Exp Biol Med* 237:1018–1022.

Wang, J., J.-K. Na, Q. Yu et al. 2012a. Sequencing papaya X and Y^h chromosomes reveals molecular basis of incipient sex chromosome evolution. *Proc Natl Acad Sci USA* 109:13710–13715.

Wang, K., Z. Wang, F. Li et al. 2012b. The draft genome of a diploid cotton *Gossypium raimondii*. *Nat Genet* 44:1098–1103.

Watson, B. 1997. Agronomy/agroclimatology notes for the production of papaya. AusAid-Soil and Crop Evaluation Project. Canberra, Australia.

Watson, J. M. and K. Riha. 2010. Comparative biology of telomeres: Where plants stand. *FEBS Lett* 584:3752–3759.

Weingartner, L. A. and R. C. Moore. 2012. Contrasting patterns of X/Y polymorphism distinguish *Carica papaya* from other sex chromosome systems. *Mol Biol Evol* 29:3909–3920.

Wildermuth, M. C., J. Dewdney, G. Wu, and F. M. Ausubel. 2001. Isochorismate synthase is required to synthesize salicylic acid for plant defence. *Nature* 414:562–565.

Wills, R. B. H. and S. B. Widjanarko. 1995. Changes in physiology, composition and sensory characteristics of Australian papaya during ripening. *Aust J Exp Agric* 35:1173–1176.

Woodhouse, M. R., B. Pedersen, and M. Freeling. 2010. Transposed genes in *Arabidopsis* are often associated with flanking repeats. *PLoS Genet* 6:e1000949.

Woodhouse, M. R., H. Tang, and M. Freeling. 2011. Different gene families in *Arabidopsis thaliana* transposed in different epochs and at different frequencies throughout the Rosids. *Plant Cell* 23:4241–4253.

Woods, D. P., C. L. Hope, and S. T. Malcomber. 2011. Phylogenomic analyses of the *BARREN STALK1/LAX PANICLE1 (BA1/LAX1)* genes and evidence for their roles during axillary meristem development. *Mol Biol Evol* 28:2147–2159.

Wu, J., T. Maehara, T. Shimmokkawa et al. 2002. A comprehensive rice transcript map containing 6591 expressed sequence tag sites. *Plant Cell* 14:525–535.

Wu, X., J. Wang, J.-K. Na et al. 2010. The origin of the non-recombining region of sex chromosomes in *Carica* and *Vasconcellea*. *Plant J* 63:801–810.

Yan, P., X. Z. Gao, W. T. Shen, and P. Zhou. 2011. Cloning and expression analysis of phytoene desaturase and ξ-carotene desaturase genes in *Carica papaya*. *Mol Biol Rep* 38:785–791.

Yang, J.-S., T.-A. Yu, Y.-H. Cheng, and S.-D. Yeh. 1996. Transgenic papaya plants from *Agrobacterium*-mediated transformation of petioles of in vitro propagated multishoots. *Plant Cell Rep* 15:459–464.

Yang, X., U. C. Kalluri, S. P. Di Fazio et al. 2009. Poplar genomics: State of the science. *CRC Cr Rev Plant Sci* 28:285–308.

Yu, Q., S. Hou, F. A. Feltus et al. 2008c. Low X/Y divergence in four pairs of papaya sex-linked genes. *Plant J* 53:124–132.

Yu, Q., S. Hou, R. Hobza et al. 2007. Chromosomal location and gene paucity of the male specific region on papaya Y chromosome. *Mol Genet Genomics* 278:177–185.

Yu, Q., P. H. Moore, H. H. Albert, A. H. K. Roader, and R. Ming. 2005. Cloning and characterization of a *FLORICAULA/LEAFY* ortholog, *PFL*, in polygamous papaya. *Cell Res* 15:576–584.

Yu, Q., R. Navajas-Pérez, E. Tong et al. 2008b. Recent origin of dioecious and gynodioecious Y chromosomes in papaya. *Trop Plant Biol* 1:49–57.

Yu, Q., D. Steiger, E. M. Kramer, P. H. Moore, and R. Ming. 2008a. Floral MADS-box genes in trioecious papaya: Characterization of *AG* and *AP1* subfamily genes revealed a sex-type-specific gene. *Trop Plant Biol* 1:97–107.

Yu, Q., E. Tong, R. L. Skelton et al. 2009. A physical map of the papaya genome with integrated genetic map and genome sequence. *BMC Genomics* 10:371.

Zdepski, A., W. Wang, H. D. Priest et al. 2008. Conserved daily transcriptional programs in *Carica papaya*. *Trop Plant Biol* 1:236–245.

Zhai, J. X., D.-H. Jeong, E. De Paoli et al. 2011. MicroRNAs as master regulators of the plant NB-LRR defense gene family via the production of phased, trans-acting siRNAs. *Genes Dev* 25:2540–2553.

Zhang, E. E. and S. A. Kay. 2010. Clocks not winding down: Unraveling circadian networks. *Nat Rev Mol Cell Biol* 11:764–776.

Zhang, W., S. Gao, X. Zhou et al. 2011. Bacteria-responsive microRNAs regulate plant innate immunity by modulating plant hormone networks. *Plant Mol Biol* 75:93–105.

Zhao, B., R. Liang, L. Ge et al. 2007. Identification of drought-induced microRNAs in rice. *Biochem Biophys Res Commun* 354:585–590.

Zhu, Y. J., R. Agbayani, and P. H. Moore. 2007. Ectopic expression of *Dahlia merckii* defensin DmAMP1 improves papaya resistance to *Phytophthora palmivora* by reducing pathogen vigor. *Planta* 226:87–97.

CHAPTER 11

Advances in Omics for Improved Onion and Potato Quality

Noureddine Benkeblia

CONTENTS

11.1 INTRODUCTION

The human population is increasing, accompanied by the issues of the mounting challenge of feeding the growing population, climate and environment changes, and the low yield of wild species. It has become necessary for humanity to alleviate these issues by developing new tools and technologies to improve crops, not only for quantitative but also for qualitative purposes. These aims are to satisfy the needs of the population, cope with the changing climate, and ensure resilience, sustainability of the agrosystems, and food security (Henry 2012).

From the conceptual point of view, and regardless of modern technologies, crop improvement is the manipulation or alteration of plant genetic material to address various issues or to achieve one or more specific goals, or both.

Since prehistory, out of the thousands of plants in existence, a few have been relocated from one region to another, domesticated, and cultivated. During this transition, some members of these species have been transformed into crops through genetic alterations, via conscious and unconscious selection, resulting in some plant species becoming the main agricultural crops and thus providing the major part of the world's food supply (Mannion 1999). During this domestication process, the identification of certain useful wild species, combined with a process of selection, brought about changes in appearance, quality, and productivity. However, the exact details of the process by which the main crops were developed are not fully understood, and in many cases the genetic changes were enormous (Hancock 2005). For example, many crop plants, for example maize, have been so changed that their origins remain obscure, with seemingly little resemblance to close wild relatives.

The spectacular changes due to domestication included alteration in organ color, size, shape, and form; loss of many survival characteristics, for example, bitterness or toxicity; disarticulation of seeds in grains; seed dormancy; and changes in life span. Considering the importance and the fundamental role of plant domestication in human history and the small number of crop plants in modern societies, little is known about how these crops were first domesticated (Ross-Ibarra et al. 2007).

Surprisingly, modern humans rely on a very small number of crops: about 70% of the calories consumed are supplied by only 15 crops, while cereals, particularly rice, wheat, maize, sugarcane, and barley, supply 50% of the total calories (FAO 2012). Despite the critical importance of these crops, in most cases little is known about their domestication, and there are some obvious questions regarding the mediators responsible for the process: Who were they? How did they identify the incipient crops? What were their cultivation methods? Other questions regarding the history of the crops have also been raised: What were the wild progenitors of the modern crops? Did domestication occur more than once? If so, where? Although the application of modern methods has begun to provide partial answers, the picture remains far from complete, with many pieces missing from the puzzle (Harter et al. 2004; Heun et al. 1997).

11.2 OMICS AND CROPS IMPROVEMENT

Over the past few decades, much effort has been put into plant breeding, with the main goals being to increase the yield and enhance the resistance of crops to pests, diseases, and other attacks that cause heavy damage and losses during cropping, harvesting, handling, transportation, and storage.

Research regarding the quality attributes of horticultural crops, in particular cereals, fruits, and vegetables, was primarily designed to improve their utility for processing, for example, baking or brewing, rather than to improve their nutritional value and their postharvest physiological behavior. With the development of new analytical techniques and technologies, these approaches began to change in the early 1960s with the discovery of new varieties (Benkeblia 2012; Mertz et al. 1964; Rochfort 2005).

Therefore, the main goal of modern food-crop science is to broaden our still limited knowledge of, on the one hand, how genes can be "shaped" and consequently affect enzymes and metabolites, and, on the other hand, how gene–enzyme–metabolite interactions and effects lead to "new" crops possessing desirable agronomic, physiological, biochemical, and nutritional features. Indeed, we know that the "road ahead is still long" and the problem is huge; however, modern analytical chemistry and biochemistry techniques, bioinformatics tools, genomics, and biotechnologies are giving us hope, and researchers have started piecing together interesting clues. Thus, gene expression and function, metabolism, and, more broadly, genetic diversity within and between plants, their responses to biotic and abiotic stresses, and their limits in genetic modification, are becoming better understood and their mysteries are being revealed (Benkeblia 2011). This development and progress have resulted in the generation of new research areas, which are devoted to "looking from inside the smaller" rather than looking from the outside. Consequently, scientists have developed new terms related to this field, defining specific "smaller" research areas such as genomics, transcriptomics, proteomics, metabolomics, ionomics, foodomics, nutrigenomics, metagenomics, and so on (Powell 2007; Rezzi et al. 2007a,b; Rist et al. 2006; Subbiah 2006; Trujillo et al. 2006). On the other hand, and following on from these new emerging methodologies, as well as the newly generated knowledge, new products, and so on that have resulted from these approaches, new possibilities are arising for tailored food products that promote the health and well-being of specific groups of the population and consumers.

One good example of the application of these advanced approaches is the combination of molecular biology, chemistry, agriculture, and food science in the development of transgenic or genetically modified foods (Brown 1999; Chen and Lin 2013; Engel et al. 2002; Herrera-Estrella 2000; McGloughlin 2010; Phillips 2008; Schilter and Constable 2002; Ruth 2003).

11.3 POTATO (*SOLANUM TUBEROSUM* L.)

11.3.1 Issues with Domesticated Varieties

The domestication of crops is reported to date back more than 12,000 years. Because domesticated crops have not had time to develop differences from the wild species through mutation, the genetic diversity in these crops is expected to be much less than that found in wild populations (Mannion 1999; Olsen and Gross 2008). Although all domesticated and cultivated potatoes belong to the same botanical species, *Solanum tuberosum*, there are many thousands of varieties varying in color, shape, texture, and other features.

The potato originated in the South American Andes, and more than 200 wild species are found in this wide habitat. It is also thought that the potato was domesticated in the Andes Mountains of South America (Spooner et al. 2006). The potato is not only considered a staple of the numerous societies living in the Andes; it was also introduced into Europe, and subsequently distributed worldwide over several centuries. Nowadays, it is the most consumed vegetable (Hawkes 1990).

Great diversity in wild potato species is found in these regions, where the potato was probably domesticated between 10,000 and 7,000 years ago. Fossilized remains of possibly cultivated tubers have been found in Chilca Canyon, suggesting that the potato was cultivated at least 7,000 years ago (Ugent 1970). Most likely, *Solanum tuberosum* was domesticated from the wild diploid species *S. stenotomum*, which hybridized with *S. sparsipilum* to form the amphidiploid *S. tuberosum*, which evolved into the subspecies *S. tuberosum* subsp. *tuberosum* (Grun 1990; Hawkes 1990; Quilter et al. 1991).

Although the potato is easy to grow in many places, there are many agronomic, environmental, disease, pest, and other issues that can arise in the field and during the postharvest handling of the tubers (Struik et al. 1997). Therefore, identifying these problems is crucial, because not all conditions can be reversed. The major issues of potato production will be described.

11.3.1.1 Water Deficiency

The potato is a drought-sensitive plant. Water availability can seriously affect the morphology, production rate, and marketable yield (van Loon 1981), and adequate water supply is essential for high yields (Evans and Neild 1981). Water stress causes reduced yield, specifically reduction of leaf area, reduction of photosynthesis per unit of leaf area, or both (Epstein and Grant 1973; Jefferies 1989). Water deficit during the tuberization period also decreases yield to a larger extent than drought during other growth stages. It also causes leaf tips and margins to turn yellow and then brown and eventually die. Tubers may develop brown spots and have irregular shapes and sizes. Inconsistent watering during drought can also cause these symptoms (Mackerron and Jefferies 1988). Water-stressed potatoes can also sometimes develop high concentrations of reducing sugars due to hot weather. High concentrations of reducing sugars are undesirable at the time of harvest and when potatoes are used for processing, because they produce French fries with dark stem ends (Eldredge et al. 1996; Shock et al. 1993).

On the other hand, under excessive irrigation regimes, rapid potato growth can cause the development of hollow tubers with cavities at the center. Excessive irrigation also results in reduced root zones and yields (weight of tubers and total yield [Cappaert et al. 1992; Stark et al. 1993; Yuan et al. 2003]), as well as increased severity of potato early dying (Cappaert et al. 1992). It is therefore, recommended that irrigation is scheduled when the soil has dried to a depth of about 4 in.

11.3.1.2 Growing Issues

Commonly, tubers or tuber pieces are used to grow potato plants. However, these pieces do not always sprout, because stored potatoes are often chemically treated to prevent sprouting. Therefore, either specific tubers or true seed potatoes (TSP) should be used. If pieces are used, the cutting should have at least two eyes and should be planted in a warmer soil. Potato tubers used for cutting should also not be cut in sunlight or left in a dry place (Priestley and Woffenden 1923). The sprouting of the cutting can be accelerated if the piece is dipped in gibberellic acid (GA), especially if this treatment is applied two weeks before planting (Slomnicki and Rylski 1964). Allen (1979) investigated the effects of cutting seed tubers on the number of stems and tuber yields of several potato varieties, and found that using diced cutting reduced potato peel early growth and yield in some cultivars, while in others cutting produced a large number of stems and tubers, and increased yield of small-sized tubers. With some cultivars, however, the effect was small or no differences in growth and yield were found.

Tuberization is also one of the main issues of potato growing, and has a significant effect on potato yield and total yield (Chapman 1958). Photoperiod is a determining environmental factor for the time to flowering in potato, and is critical to promote differentiation of tubers (Rodríguez-Falcón et al. 2006).

High temperatures and gibberellin have a negative effect on tuberization by promoting haulm growth and suppressing tuber production, whereas low temperatures induce abscisic acid (ABA) production, promote tuber production, and reduce the growth of the haulms (Menzel 1979, 1985). Moreover, nitrogen nutrition, like photoperiod and temperature, also impacts tuberization, mainly by regulating the ABA/GA ratio (Krauss and Marschner 1982). Although strong evidence was reported for a role of gibberellins in the regulation of potato tuber formation (Vreugdenhil and Sergeeva 1999), synthesis is promoted by high temperatures in the buds rather than their transport to the stolons (Menzel 1983).

Potatoes should be grown completely covered by soil and stored in darkness; exposure to sun or light during the growing stage, and even during storage, results in tuber greening (Conner 1937). The green color is caused by the biosynthesis and accumulation of toxic tuber glycoalkaloids (TGA, solanines), both during growth and after harvest. These solanines appear to be more toxic to man than to other animals, and also affect the commercial quality of potato tubers (Friedman et al. 1997). However, the levels of solanines differ among varieties, growing locations (Friedman et al. 2003; Sinden and Webb 1972), and environmental conditions (Nitithamyong et al. 1999). Some investigations report that the accumulation of solanines during the harvesting intervals was markedly different depending on the cultivars (Bejarano et al. 2000; Cronk et al. 1974) and influenced by the stage of maturity of the tubers at harvest. The concentration of solanines significantly increased under drought stress conditions in most varieties (Bejarano et al. 2000).

Frost can also cause stunting or death in young potato sprouts, and damage mature plants (Sukumaran and Weiser 1972). The degree of frost tolerance varies among cultivars (Steffan and Palta 1986). Mineral nutrition also affects frost resistance of potatoes. Grewal and Singh (1980) reported an inverse relation between frost damage, potassium availability, potassium content of the soils, and its concentration in the potato leaves. They noted that potassium fertilization significantly reduced the levels of frost damage and increased tuber yields. Nevertheless, the difference in frost resistance between resistant and sensitive strains of potatoes is about 3°C–4°C (Grewal and Singh 1980). Although the potato is considered a sensitive plant, it can survive temperatures of −2.0°C in most cases. Many scientific reports have indicated that the resistance to freezing is due to the capability of the genotype to tolerate freeze-induced dehydration (Li et al. 1981).

11.3.1.3 Diseases

The potato crop is affected by several economically important bacterial, fungal, and viral pathogens. The bacterial diseases include the soft rots caused by *Erwinia* spp., the brown rot *Ralstonia solanacearum*, the ring rot *Clavibacter michiganensis* subsp. *sepedonicus,* and the scab-forming *Streptomyces* spp. (Arora and Paul Khurana 2004; Dutt 1979; Pérombelon 2002; Stead 1999). The most common fungal diseases include early blight and late blight, while the most common viral diseases are the mosaic virus, potato virus A (PVA), and the potato yellow dwarf virus (PYDV) (ICP 1990). The early blight is characterized by dark brown spots affecting the leaves, and is most common in warm, wet environments such as tropical climates. The late blight affects plants during cool, moist weather and spreads when temperatures increase. Consequently, the leaves develop a brown color, which then turns black. Potato blackleg, another disease that is very common during wet growing seasons, causes leaves to turn from pale green to yellow. The stems of the infected plant become dark brown and eventually black just above the soil, often resulting in wilting and death (Arora and Paul Khurana 2004; Dutt 1979; Rich 1983).

Even though only a few viruses are known to cause significant yield reduction, at present, about 30 different viruses, excluding strains, have been found in potato tubers. The viruses and related agents can be divided into two groups defined by their association with the host: viruses dependent on potatoes for survival and spread, and viruses not dependent on potatoes for survival and spread (Dutt 1979; ICP 1990; Rich 1983). Potato leaves affected by mosaic virus tend to curl, with light green and dark green shades. The virus is spread by aphids, and most outbreaks do not result in the death of the plants, but yields are reduced. PYDV is transmitted by leafhoppers and causes stunting of potato plants, yellowing of the leaves, and cracking and disfiguring of tubers (Dutt 1979; Rich 1983).

11.3.1.4 Pests of Potato

Potato plants can be attacked by a range of insects at any stage of their development, from seeding (planting) until tubers are harvested. Various insects, both larvae and adults, feed on tubers, seed pieces, leaves, and stems (Evans et al. 1992; Harris 1992; Morrison et al. 1967; Oerke 2006; Radcliffe 1982).

The most common pests found on potato are aphids, which are usually pink or green and live on the undersides of the leaves. The most commonly observed aphids on the potato are the green peach aphid and the potato aphid. The green peach aphid is the more abundant species, and infestation typically begins on the bottom of the plant, becoming scattered over the entire plant. Other potato diseases, such as leafroll virus, are also spread by both aphids. Although the green aphid is the more effective vector, the insect also causes damage to the plant. Populations become so abundant that their feeding weakens the plants. Aphids can also spread other viral diseases, such as cucumber mosaic and alfalfa mosaic (calico).

Other pests that affect potato plants include the flea beetles. Adult beetles chew leaves, resulting in a shot-hole appearance. However, adult potato plants can usually withstand flea beetle infestations. Wireworms are another pest of potato. These are the larvae of the click beetle, and are thin and brownish yellow. They may appear in areas that are covered with fresh sod, and they infest potato tubers before they are ripe. Cutworms, gray grubs living under the soil, attack potato plants at night by cutting young seedlings off at soil level. Mature plants are less susceptible than younger ones.

Potato plants can also be attacked by nematodes. *Globodera* spp., known as the potato cyst, are the most economically important nematodes of potatoes. Potato cyst nematodes occur worldwide, and the two species *Globodera pallida* (the white cyst nematode) and *Globodera rostochiensis* are the principal nematodes affecting potatoes in both temperate and subtropical regions. These pests are responsible for considerable yield losses (Brodie 1999; Greco 1993).

11.3.2 Breeding

Currently, cultivated potato is a vegetatively propagated autotetraploid that has been bred for different purposes, including yield improvement, resistance to stresses and diseases, fresh market requirements, and different processing goals. Breeding efforts have relied on phenotypic selection of populations developed from intra- and intermarket class crosses and introgressions of wild and cultivated *Solanum* relatives. However, breeding efforts face two classical questions: "what to breed?" and "how to breed?" (Bonnel 2008; Hanneman 1999; Hirsch et al. 2013; Thieme and Griess 2005).

Indeed, potato breeding aims to develop new varieties required for sustainable yield increases of potato production to satisfy the needs of the growing population and to cope with impending environmental change. Technically, as with other crops, both conventional and modern technologies are used to breed potato. On the one hand, conventional potato breeding aims to produce and develop cultivars from sexual crosses followed by clonal propagation and selection. Most of the new potato varieties still emerge from this process (Brown 2011). Nevertheless, conventional potato breeding strategies are not fully using the existing biodiversity within the *Solanum* genus; breeding at the 4x–4x level is slow, and the potential for breeding at the diploid and dihaploid levels is not being fully explored (Pehu 1996).

On the other hand, during the last three decades, biotechnology has been widely used to breed new potato cultivars. The efficiency of breeding has been enhanced through the use of diagnostic DNA-based markers for the selection of superior cultivars (Bradshaw 2007; Gebhardt 2013). The strategy for potato breeding using modern technologies is viewed as one of the key decisions that breeders make concerning the objectives of a breeding program, including breeding methods and the germplasm that could be used to achieve this goal (Huamán and Schmiediche 1999). A wide range of genetic resources available for the improvement of the cultivated potato, and wild and cultivated species, have been utilized in breeding of other potato cultivars; however, only a very small sample of the available biodiversity has been exploited (Bradshaw et al. 2006; Ritter et al. 2008). This strategy will also need to address the issue of whether new cultivars should preferably be propagated vegetatively or through true potato seed (TPS), whether the new cultivars should be genetically modified, and how to achieve durable disease and pest resistance (Bradshaw 2007; Simmonds 1997).

11.3.3 Transgenic Potato

Since the development and rapid expansion of the science of plant molecular biology in the last three decades, many transgenic potato plants have been generated to investigate the feasibility of transgene expression to affect different parameters such as yield, quality, stress physiology

(Watanabe et al. 2011), and pest or disease resistance (Davies et al. 2008). As described above, strategies for alternative uses of potato crops to express recombinant proteins or generate altered metabolic products are being extensively explored (Chakraborty et al. 2010; Davies 1996). Today, transgenic potatoes are among the most predominant cultivated crops, and more than 500 transgenics are undergoing field trials (Conner 2007; Dunwell 2000). For example, starch biosynthesis in potato tubers was modified using transgenes for (i) cytoplasmic expression of a protein, (ii) amyloplast targeting of a protein, and (iii) antisense expression. The altered carbohydrate metabolism resulted in (i) the production of novel carbohydrates, (ii) production of starches with altered amylose/amylopectin ratios and chain lengths, and (iii) production of more starch (increased yield) (Shewmaker and Stalker 1992).

11.3.4 Omics Improvement Goals

One of the goals of plant biologists is to achieve a significant increase in biomass and seed yields. They have always dedicated their fundamental research to the benefit of mankind, and the first green revolution likely represented a crucial step toward contemporary agriculture and the development of high-yield varieties (Den Herder et al. 2010). While research to date shows the exceptional promise of the omics technologies in crop improvement, much work remains to be done to develop and validate it, as well as to incorporate it into breeding schemes (Carreno-Quintero et al. 2013; Heffner et al. 2009).

Indeed, the rapid progress of omics technologies—for example, genomics, transcriptomics, proteomics, and metabolomics, as well as bioinformatics and molecular breeding—has provided new scientific tools to address issues encountered by farmers. These issues mainly include water stresses (both drought and flooding), biological stresses, diseases, and other environmental changes, such as global warming and CO_2 increase. Therefore, the omics technologies aim to exploit advances in these technologies to harness the rich global heritage of plant genetic resources and contribute to the development of new genetically improved varieties that are stress tolerant and high yielding, allowing crop production to keep pace with the needs of the growing world population and to cope with environmental changes and biotic and abiotic stresses (Bruskiewich et al. 2006; Mba et al. 2012; Miller et al. 2010). These emerging experimental technologies have largely driven omics science, and developments in systems biology have generated novel and quantitative hypotheses via modelling (Yin and Struik 2009).

Moreover, the responses of crops to environmental conditions are critical factors for agricultural sustainability, and these responses cannot be understood unless the system biology of the crop system is considered under different conditions. Currently, the popular view in modern science is that omics are providing better tools and deeper knowledge, allowing the characteristics of the crop to be modified, resulting in improved actual and potential crop yields, enhanced pest and disease resistance, more efficient resource use, and enhanced crop system health (Watkins et al. 2001; Yin and Struik 2007, 2009, 2010). Nevertheless, and for successful improvement of crops, the impact of omics technologies needs to be assessed at the crop level, given the potential effect of genetic alteration on phenotypes from molecular to crop level. Therefore, the success of crop systems biology will require real commitment from scientists as well as the entire knowledge chain of plant biology, from molecule or gene to crops and agroecosystems (Parry and Hawkesford 2012; Yin and Struik 2009, 2010). More recently, the Generation Challenge programme (GCP) has been launched. This programme is a global crop research consortium with the goal of crop improvement through the application of comparative biology and genetic resources characterization to plant breeding (Bruskiewich et al. 2006, 2008).

Consequently, study of the responses of potato crops to stressful conditions has resulted in the identification of a large number of genes, the expression of which is regulated by external stresses, and these stress-induced genes are either functional or regulatory genes (Leone et al. 1999).

11.3.4.1 Increasing Nitrogen-Use Efficiency

The relations between the assimilation of nitrogen (N) by crops, N-assimilation enzymes, and growth and yield are reported extensively in the literature (Fageria and Baligar 2005). However, the genetic variation in N-use efficiency (NUE) is attributed to variation in plant genotypes that may differ in the NUE to produce yield (utilization efficiency), nutrient uptake efficiency, or both (Sattelmacher et al. 1994).

The literature has also reported that manipulating the N-assimilation enzyme genes (Hirel et al. 2007; Westermann et al. 1988), or detecting gene expression markers (Li et al. 2010), could result in increased yield, increased NUE, or both. The published data suggest that, on the one hand, the levels of N-assimilation enzymes do not limit primary N assimilation and, hence, yield, and, on the other hand, the assimilation of nitrate by roots or shoots could be advantageous under some environmental conditions (Andrews et al. 2004; Hirel et al. 2007). The critical issue with N is to determine the development of new varieties that show good dose response to N input, the critical nitrogen concentration as a function of crop biomass, and the strategy of the potato plant to cope with nitrogen limitation by improving NUE (Vos 2009). Moreover, NUE characteristics of commercial potato cultivars were studied, and results reported that potato cultivars vary more in terms of nitrogen uptake capacity (NUC) than in terms of NUE (Zebarth et al. 2004). Genetic transformation of the potato showed that NUE can be increased by 16% at low nitrogen fertilization rate compared with traditional clones (Zvomuya and Rosen 2002). Different biotechnological options have been suggested (Pathak et al. 2008). For example, for an optimal NUE, plant leaves need to achieve a redistribution of nitrogen, and this achievement will increase by 30%–40% the regeneration capacities of ribulose-1,5-bisphosphate and carboxylation. Therefore, genetic interventions to improve the NUE of crops have many benefits, such as reducing pollution of water supply by nitrates. Indeed, NUE could be increased either through manipulation of the amount or properties of ribulose-1,5-bisphosphate carboxylase oxygenase (RuBiSCO) or by decreasing photorespiration. However, decreasing RuBiSCO content enhances NUE by about 5%, while a drastic decrease in photorespiration produces a change in NUE of more than 50% (Kumar et al. 2002). Attempts to improve NUE by manipulating nitrate reductase (NR) and nitrite reductase (NiR) genes have also been made, and yielded mixed results. At the transcriptional level, deregulation of NR gene expression by constitutive expression in transgenic potato caused a reduction in nitrate levels in tissues, and the NR transformants showed better performance in terms of available NR protein, rapidly restoring nitrogen assimilation (Djennane et al. 2002). Also, the enhancement of NR rate led to higher biomass production, likely through the promotion of nitrogen allocation resources (Djennane et al. 2004).

11.3.4.2 Increasing Water-Use Efficiency

The development of high-yield potato cultivars has invariably led to higher crop water consumption. Widely accepted evidence suggests that growth and yield are linearly related to transpiration of crops throughout the growing season (De Pascale et al. 2011; Hassanpanah 2010).

Due to its root system, the potato plant is sensitive to water deficits, which can significantly reduce tuber yield (Jefferies 1993; Porter et al. 1999). In order to ensure high yields, irrigation is required most of the time to maintain good soil moisture (Fabeiro et al. 2001). However, the increasing shortage of water resources requires the improvement of the water-use efficiency (WUE) of plants, including potato (Liu et al. 2006). Various biotechnological strategies, such as altering morphological, physiological, and genetic characteristics of potato, have been used to increase WUE in the potato crop to cope with water stress. Native potato and alien genes, identified using transcriptomics, might provide useful candidates for deployment against stress. Transgenic potato cultivars harboring many of these genes have been evaluated and have shown encouraging results for future release of new stress-tolerant cultivars (Levy et al. 2013). Moreover, significant variation was observed between different genotypes of the potato, and WUE was found to be positively related to carbon fractionation, but inversely related to stomatal conductance (Vos and Groenwold 1989).

Transgenic potato plants expressing phytochrome B (phyB) gene showed higher stomatal conductance, transpiration rates, and photosynthesis rates per unit leaf area, while stomata density was unaffected. These observations suggested that phyB enhanced the aperture of the stomatal pore in transgenic plants (Boccalandro et al. 2003; Thiele et al. 1999).

11.3.4.3 Tolerance to Salinity

Sensitivity of the potato crop to salinity varies with cultivars. While various reports have demonstrated the response of wild and bred potato cultivars to salinity, the mechanisms involved in the adaptation of the plant to this stress are still not clearly understood (Levy and Veilleux 2007). Overall, salinity causes a decrease of the water and the osmotic potential of tubers, but there is an increase of total soluble solids and proline (Levy et al. 1988; Patel et al. 2001).

Different strategies to overcome this sensitivity have been proposed, including biotechnological tools. By using DNA microarrays, it was demonstrated that six clones of ADP-ribosylation factor–like proteins are upregulated by salinity-imposed stress (Kim et al. 2003). In addition, the introduction of a single alien gene has been used in an attempt at improving salt stress tolerance. Potato was transformed with a glyceraldehyde-3-phosphate dehydrogenase (GDP) gene isolated from the oyster mushroom *Pleurotus sajocaju*, and six independently transformed lines were confirmed for GDP expression (Jeong et al. 2001).

The amino acid proline (Pro) is known as a compatible osmolyte accumulated by plant cells in response to salt and drought stress. It acts as an osmoprotectant, protecting cellular structures under osmotic stress (Hmida-Sayari et al. 2005a). Although the data suggest the involvement of proline in the response of potato crop to salinity stress, the role of proline still remains unclear. Salt tolerance and associated reduced proline levels under salt stress were observed in transgenic potato expressing an *Arabidopsis* osmotin-like protein (Evers et al. 1999). The results showed that plantlets of the transgenic line grown *in vitro* produced less proline under salt stress, while transgenic lines that expressed a pyrroline-5-carboxylate synthetase cDNA from *Arabidopsis* exhibited increased accumulation of proline under salt stress as well as a less severe reduction in tuber yield (Hmida-Sayari et al. 2005b).

11.3.4.4 Resistance to Pests and Diseases = Fewer Pesticides

Many potato diseases have become pandemic, highly virulent, fungicide resistant, and widely disseminated throughout the world. As a result, and because of the importance of the potato in the diet, the development of cost-effective and environmentally friendly strategies for disease control is considered a priority (Osusky et al. 2004).

The technology of insect-resistant transgenic plants has expanded rapidly during the last three decades. Although to date only expressing genes derived from the bacterium *Bacillus thuringiensis* (Bt) are commercialized, numerous other genes from higher plants have been transferred into crop cultivars, especially genes encoding inhibitors of digestive enzymes and lectins. Transfer of resistance genes from other microorganisms and animals has been limited (Grafius and Douches 2008; Schuler et al. 1998). Engineering insect resistance in potatoes is mainly based on the use of (i) proteinase inhibitors and genes to improve defenses against insects and pathogens (Ryan 1990; Valueva et al. 2003), (ii) crystal proteins of Bt and their expression in potatoes (Adang et al. 1993; Perlak et al. 1993), and (iii) biological control using the insect pests *Myzus persicae*, *Macrosiphum euphorbiae*, *Leptinotarsa decemlineata*, and *Phthorimaea operculella* (Ashouri et al. 2001; Peferoen et al. 1990). Osusky et al. (2004) described a strategy for engineering potato plants with strong protection against blight and pink rot phytopathogens without deleterious effects on plant yield or vigor. They reported that temporin A, a small, naturally occurring antimicrobial cationic peptide, when expressed in the potato, conveyed strong resistance to these two pathogens in addition to resistance to the bacterial pathogen *Erwinia carotovora*. They also reported that the transgenic tubers did not show any diseases during storage.

Indeed, Bt is considered a valuable source of insecticidal proteins for use in conventional sprayable formulations and in transgenic crops, and is regarded as the most effective biological insect control. Insect resistance is, however, a serious threat to this technology, although only one insect species has evolved significant levels of resistance in the field so far (Ferré and Van Rie 2002). Other experience has shown that insect adaptation to the Bt toxins expressed in transgenic potato cultivars is a high risk for rapid adaptation if these cultivars are misused (Gould 1998).

11.3.4.5 Enhancing Nutritional Attributes

During the last few decades, tremendous advances in analytical chemistry and molecular biology have been achieved, and have led to increased potential for further improvements in nutritional compounds of potato, such as vitamin content and protein/carbohydrate ratio (Bajaj 1987). By using biotechnology and engineering the plant's functional properties, new potatoes can be obtained that contain nutritional, industrial, and even therapeutic compounds. Indeed, two achievements have been crucial to expanding the role of the potato: progress in functional potato genomics and the ability to integrate genes of interest into the potato genome (Bajaj 1987; Mullins et al. 2006).

Chakraborty et al. (2000) cloned the seed albumin gene AmA1 (a nonallergenic protein rich in all essential amino acids) from *Amaranthus hypochondriacus*, and successfully introduced and expressed this gene in a tuber-specific and constitutive manner. The transgenic potato obtained had a high total protein content, and an increase in almost all essential amino acids was observed.

The mineral composition of potato has also been improved by genetic transformation. Park et al. (2005) reported that potato tubers expressing the *Arabidopsis* H^+/Ca_2^+ transporter sCAX1 contained ca. three times more calcium than wild-type tubers. The increase in calcium levels in the sCAX1-expressing tubers did not appear to alter tuber morphology or yield.

Novel methods have been identified to enhance the content of β-carotene in potato. For example, RNA interference (RNAi) was used to silence the β-carotene hydroxylase gene (bch) responsible for the conversion of β-carotene to zeaxanthin. *Agrobacterium tumefaciens*-mediated transformation was employed to introduce two RNAi constructs into different potato lines. The results showed that β-carotene content increased from trace amounts in wild-type tubers to ca. 331 µg 100 g^{-1} fresh weight, while some transformants exhibited a significant decrease in zeaxanthin content, an increase in lutein, or both (Van Eck et al. 2007). On the other hand, screening of hundreds of potato somaclones has led to the selection of different advanced lines. The levels of several phenolic compounds increased from ten to 100 times in some selected lines (Nassar et al. 2013).

Enhanced ascorbic acid (AsA) transgenic potato overexpressing either l-gulono-γ-lactone oxidase or strawberry GalUR gene was also obtained. The AsA content of the transgenic tubers was enhanced by 141% and 160%–200%, respectively, and this transformation conferred tolerance to various abiotic stresses (Upadhyaya et al. 2009, 2010). Another transgenic potato overexpressed the cytosolic dehydroascorbate reductase (DHAR) gene and chloroplastic DHAR, and AsA content was enhanced through recycling of ascorbate via DHAR overexpression (Qin et al. 2011).

11.3.4.6 Low Acrylamide-Forming Potential

Acrylamide is a chemical compound found in some foods that have been cooked at temperatures above 120°C, such as potato chips and French fries. Recent studies have shown that the acrylamide content in these two foods is very high in comparison to other foods (FAO/WHO

2008; Friedman 2003; Mottram et al. 2002; Stadler et al. 2002). According to the FAO/WHO joint committee, the levels of acrylamide in foods are considered to be a "major concern," and more research is needed to assess the risk of exposure to this dietary compound (FAO/WHO 2008). Recommendations have been made to decrease acrylamide levels in some food products by, for example decreasing cooking time, blanching before frying, and postdrying (Kita et al. 2004; Skog et al. 2008). From the chemical point of view, asparagine is one of the main precursors of acrylamide formation, and its level varies considerably with potato variety, growing year, storage temperature, storage time (Olsson et al. 2004), soil type, underwater weight, and tuber size (De Wilde et al. 2006). A subset of clonal progeny from potato has been assessed, and results showed that the clone with the lowest acrylamide-forming susceptibility had low reducing sugars and asparagine content. Presumably, both asparagine and reducing sugar levels are implicated in variation in acrylamide level, and both metabolites should be targeted for potato improvement (Shepherd et al. 2010).

Consequently, strategies for reducing acrylamide formation in potato products have been developed, for example by breeding varieties containing lower concentrations of free asparagine and/or reducing sugars, and establishing best agronomic practice (BAP) to keep the concentrations of these two compounds as low as possible (Halford et al. 2012). Biotechnological alternatives have also been investigated as a means of reducing the acrylamide content in food, by reducing the accumulation of the precursors of acrylamide formation. Metabolic change was accomplished by silencing two asparagine synthetase genes through "all-native DNA" transformation, and transformant tubers contained ca. 20-fold reduced levels of free asparagine. This metabolic change did not affect the shape or yield of the tubers (Rommens et al. 2008).

On the other hand, the application of constitutive expression of an antisense copy of the cold-inducible acid invertase (Inv) gene led to 33% lower hexose content, while tuber-specific expression of an invertase-derived inverted repeat led to significantly lower reducing sugars (only ca. 2%) in the transgenic potatoes (Ye et al. 2010). Rommens et al. (2006) also reported that lowering the expression of tuber-expressed polyphenol oxidase (Ppo), starch-associated R1, and phosphorylase-L (PhL) genes without inserting any foreign DNA into the potato genome. French fries made from these intragenic potatoes contained less reduced acrylamide.

11.4 PERSPECTIVES AND CONCLUSIONS

Although interesting work has been done to improve crops, breeders are still aiming to produce new cultivars that adapt better to the changing environment and nutritional needs. These new varieties should also have better growing, storage, and processing characteristics. Transgenic plants are now grown commercially in many countries, and the potato is one of the most predominant ones, especially in North America. Although new potato cultivars were, and still are, produced by sexual hybridization, genetic modification of wild or existing cultivars by genome transformation is offering the best alternative way to improve the potato. The two techniques are complementary because they have different outcomes.

Besides the attempts to transform crops using all-native DNA, transformation by introducing "nonpotato genes" to give desirable novel traits is considered to be potentially of greater benefit, although controversial. The development of omics technologies would be an asset for the progress of potato breeding and discovery of other cultivars, using new genes, possibly from different sources, which are required for the appropriate genetic improvement of potatoes.

In conclusion, improvement of the potato will have many positive impacts, such as (i) environmental benefits, by reducing water use and applications of fertilizers and pesticides, (ii) consumer benefits, by developing new and improved nutritional crops and health benefits as well as novel products, and (iii) economic benefits, with higher yield, reduced losses, and processing variants.

ACKNOWLEDGMENTS

The author thanks Prof. Paula Tennant for her critical reading of the manuscript and her comments.

REFERENCES

Adang, M. J., M. S. Brody, G. Cardineau et al. 1993. The reconstruction and expression of a *Bacillus thuringiensis* cryIIIA gene in protoplasts and potato plants. *Plant Mol Biol* 21:1131–1145.

Allen, E. J. 1979. Effects of cutting seed tubers on number of stems and tubers and tuber yields of several potato varieties. *J Agric Sci* 93:121–128.

Andrews, M., P. J. Lea, J. A. Raven, and K. Lindsey. 2004. Can genetic manipulation of plant nitrogen assimilation enzymes result in increased crop yield and greater N-use efficiency? An assessment. *Ann Appl Biol* 145:25–40.

Arora, R. K. and S. M. Paul Khurana. 2004. Major fungal and bacterial diseases of potato and their management. *Fruit Veg Dis Dis Manage Fruits Veg* 1:189–231.

Ashouri, A., D. Michaud, and C. Cloutier. 2001. Recombinant and classically selected factors of potato plant resistance to the Colorado potato beetle, *Leptinotarsa decemlineata*, variously affect the potato aphid parasitoid *Aphidius nigripes*. *BioControl* 46:401–418.

Bajaj, S. 1987. Biotechnology of nutritional improvement of potato. In S. Bajaj (ed.), *Potato*, pp. 136–154. Berlin Heidelberg: Springer.

Bejarano, L., E. Mignolet, A. Devaux, N. Espinola, E. Carrasco, and Y. Larondelle. 2000. Glycoalkaloids in potato tubers: The effect of variety and drought stress on the α-solanine and α-chaconine contents of potatoes. *J Sci Food Agric* 80:2096–2100.

Benkeblia, N. 2011. *Sustainable Agriculture and New Biotechnologies*. Boca Raton: CRC Press.

Benkeblia, N. 2012. Metabolomics and food science: Concepts and serviceability in plant foods and nutrition. In N. Benkeblia (ed.), *Omics Technologies: Tools for Food Science*, pp. 59–75. Boca Raton: CRC Press.

Boccalandro, H. E., E. L. Ploschuk, M. J. Yanovsky, R. A. Sanchez, C. Gatz, and J. J. Casal. 2003. Increased phytochrome B alleviates density effects on tuber yield of field potato crops. *Plant Physiol* 133:1539–1546.

Bonnel, E. 2008. Potato breeding: A challenge, as ever! *Potato Res* 51:327–332.

Bradshaw, J. E. 2007. Potato-breeding strategy. In D. Vreugdenhil, J. Bradshaw, C. Gebhardt et al. (eds), *Potato Biology and Biotechnology. Advances and Perspectives*, pp. 157–177. Amsterdam: Elsevier.

Bradshaw, J. E., G. J. Bryan, and G. Ramsay. 2006. Genetic resources (including wild and cultivated *Solanum* species) and progress in their utilisation in potato breeding. *Potato Res* 49:49–65.

Brodie, B. B. 1999. Classical and molecular approaches for managing nematodes affecting potato. *Can J Plant Pathol* 21:222–231.

Brown, C. R. 2011. The contribution of traditional potato breeding to scientific potato improvement. *Potato Res* 54:287–300.

Brown, T. A. 1999. How ancient DNA may help in understanding the origin and spread of agriculture. *Philos Trans R Soc Lond B Biol Sci* 354:89–98.

Bruskiewich, R., G. Davenport, T. Hazekamp et al. 2006. Generation challenge programme (GCP): Standards for crop data. *OMICS* 10:215–219.

Bruskiewich, R., M. Senger, G. Davenport et al. 2008. The generation challenge programme platform: Semantic standards and workbench for crop science. *Int J Plant Genomics* 2008:6.

Cappaert, M. R., M. L. Powelson, N. W. Christensen, and F. J. Crowe. 1992. Influence of irrigation on severity of potato early dying and tuber yield. *Phytopathology* 82:1448–1453.

Carreno-Quintero, N., H. J. Bouwmeester, and J. J. B. Keurentjes. 2013. Genetic analysis of metabolome–phenotype interactions: From model to crop species. *Trends Genet* 29:41–50.

Chakraborty, S., N. Chakraborty, L. Agrawal et al. 2010. Next-generation protein-rich potato expressing the seed protein gene AmA1 is a result of proteome rebalancing in transgenic tuber. *Proc Natl Acad Sci USA* 107:17533–17538.

Chakraborty, S., N. Chakraborty, and A. Datta. 2000. Increased nutritive value of transgenic potato by expressing a nonallergenic seed albumin gene from *Amaranthus hypochondriacus*. *Proc Natl Acad Sci USA* 97:3724–3729.

Chapman, H. W. 1958. Tuberization in the potato plant. *Physiol Plantarum* 11:215–224.

Chen, H. and Lin, Y. 2013. Promise and issues of genetically modified crops. *Curr Opin Plant Biol* 16:255–260.

Conner, A. J. 2007. Field-testing of transgenic potatoes. In D. Vreugdenhil, J. Bradshaw, C. Gebhardt et al. (eds), *Potato Biology and Biotechnology. Advances and Perspectives*, pp. 687–703. Amsterdam: Elsevier.

Conner, W. H. 1937. Effect of light on solanine synthesis in the potato tuber. *Plant Physiol* 12:79–98.

Cronk, T. C., G. D. Kuhn, and F. J. McArdle. 1974. The influence of stage of maturity, level of nitrogen fertilization and storage on the concentration of solanine in tubers of three potato cultivars. *Bull Environ Contam Tox* 11:163–168.

Davies, H., G. J. Bryan, and M. Taylor. 2008. Advances in functional genomics and genetic modification of potato. *Potato Res* 51:283–299.

Davies, H. V. 1996. Recent developments in our knowledge of potato transgenic biology. *Potato Res* 39:411–427.

Den Herder, G., G. Van Isterdael, T. Beeckman, and I. De Smet. 2010. The roots of a new green revolution. *Trends Plant Sci* 15:600–607.

De Pascale, S., L. Dalla Costa, S. Vallone, G. Barbieri, and A. Maggio. 2011. Increasing water use efficiency in vegetable crop production: From plant to irrigation systems efficiency. *HortTechnology* 21:301–308.

De Wilde, T., B. De Meulenaer, F. Mestdagh et al. 2006. Selection criteria for potato tubers to minimize acrylamide formation during frying. *J Agric Food Chem* 54:2199–2205.

Djennane, S., J. E. Chauvin, and C. Meyer. 2002. Glasshouse behaviour of eight transgenic potato clones with a modified nitrate reductase expression under two fertilization regimes. *J Exp Bot* 53:1037–1045.

Djennane, S., I. Quilleré, M. T. Leydecker, C. Meyer, and J. E. Chauvin. 2004. Expression of a deregulated tobacco nitrate reductase gene in potato increases biomass production and decreases nitrate concentration in all organs. *Planta* 219:884–893.

Dunwell, J. M. 2000. Transgenic approaches to crop improvement. *J Exp Bot* 51:487–496.

Dutt, B. L. 1979. *Bacterial and Fungal Diseases of Potato*. Wallingford: CAB International.

Eldredge, E. P., Z. A. Holmes, A. R. Mosley, C. C. Shock, and T. D. Stieber. 1996. Effects of transitory water stress on potato tuber stem-end reducing sugar and fry color. *Am Potato J* 73:517–530.

Engel, K. H., T. Frenzel, and A. Miller. 2002. Current and future benefits from the use of GM technology in food production. *Toxicol Lett* 127:329–336.

Epstein, E. and W. J. Grant. 1973. Water stress relations of the potato plant under field conditions. *Agron J* 65:400–404.

Evans, K., D. L. Trudgill, K. V. Raman, and E. B. Radcliffe. 1992. Pest aspects of potato production. In P. M. Harris (ed.), *The Potato Crop. The Scientific Basis for Improvement*, pp. 438–506. Amsterdam: Springer.

Evans, S. A. and J. R. A. Neild. 1981. The achievement of very high yields of potatoes in the U.K. *J Agric Sci Camb* 97:391–396.

Evers, D., S. Overney, P. Simon, H. Greppin, and J. F. Hausman. 1999. Salt tolerance of *Solanum tuberosum* L.: Overexpressing an heterologous osmotin-like protein. *Biol Plant* 42:105–112.

Fabeiro, C., F. Martin de Santa Olalla, and J. A. de Juan. 2001. Yield and size of deficit irrigated potatoes. *Agric Water Manage* 48:255–266.

Fageria, N. K. and V. C. Baligar. 2005. Enhancing nitrogen use efficiency in crop plants. *Adv Agron* 88:97–185.

FAO (Food and Agriculture Organization). 2012. Statistical data of crops production. http://faostat.fao.org (accessed: December 20, 2013).

FAO/WHO (Food and Agriculture Organization/World Health Organization). 2008. Summary report of the sixty-fourth meeting of the Joint FAO/WHO Expert Committee on Food Additives (JECFA). http://www.who.int/entity/ipcs/food/jecfa/summaries/summary_report_64_final.pdf (accessed: December 17, 2013).

Ferré, J. and J. Van Rie. 2002. Biochemistry and genetics of insect resistance to *Bacillus thuringiensis*. *Annu Rev Entomol* 47:501–533.

Friedman, M. 2003. Chemistry, biochemistry, and safety of acrylamide. A review. *J Agric Food Chem* 51:4504–4526.

Friedman, M., G. M. McDonald, and M. A. Filadelfi-Keszi. 1997. Potato glycoalkaloids: Chemistry, analysis, safety, and plant physiology. *Crit Rev Plant Sci* 16:55–112.

Friedman, M., J. N. Roitman, and N. Kozukue. 2003. Glycoalkaloid and calystegine contents of eight potato cultivars. *J Agric Food Chem* 51:2964–2973.

Gebhardt, C. 2013. Bridging the gap between genome analysis and precision breeding in potato. *Trends Genet* 29:248–256.

Gould, F. 1998. Sustainability of transgenic insecticidal cultivars: Integrating pest genetics and ecology. *Annu Rev Entomol* 43:701–726.

Grafius, E. J. and D. S. Douches. 2008. The present and future role of insect-resistant genetically modified potato cultivars in IPM. In J. Romeis, A. M. Shelton, and G. G. Kennedy (eds), *Integration of Insect-Resistant Genetically Modified Crops within IPM Programs*, pp. 195–221. Amsterdam: Springer.

Greco, N. 1993. Nematode problems affecting potato production in subtropical climates. *Nematropica* 23:213–220

Grewal, J. S. and S. N. Singh. 1980. Effect of potassium nutrition on frost damage and yield of potato plants on alluvial soils of the Punjab (India). *Plant Soil* 57:105–110.

Grun, P. 1990. The evolution of cultivated potatoes. *Econ Bot* 44:39–55.

Halford, N. G., T. Y. Curtis, N. Muttucumaru, J. Postles, J. Stephen-Elmore, and D. S. Mottram. 2012. The acrylamide problem: A plant and agronomic science issue. *J Exp Bot* 63:2841–2851.

Hancock, J. F. 2005. Contributions of domesticated plant studies to our understanding of plant evolution. *Ann Bot* 96:953–963.

Hanneman, Jr. R. E. 1999. The reproductive biology of the potato and its implication for breeding. *Potato Res* 42:283–312.

Harris, P. M. 1992. *The Potato Crop: The Scientific Basis for Improvement*. Wallingford: CAB International.

Harter, A. V., K. A. Gardner, D. Falush, D. L. Lentz, R. A. Bye, and L. H. Rieseberg. 2004. Origin of extant domesticated sunflowers in eastern North America. *Nature* 430:201–205.

Hassanpanah, D. 2010. Evaluation of potato cultivars for resistance against water deficit stress under in vivo conditions. *Potato Res* 53:383–392.

Hawkes, J. G. 1990. *The Potato: Evolution, Biodiversity and Genetic Resources*. London: Bellhaven Press.

Heffner, E. L., M. E. Sorrells, and J. L. Jannink. 2009. Genomic selection for crop improvement. *Crop Sci* 49:1–12.

Henry, R. J. 2012. Next-generation sequencing for understanding and accelerating crop domestication. *Brief Func Genomics* 11:51–56.

Herrera-Estrella, L. R. 2000. Genetically modified crops and developing countries. *Plant Physiol* 124:923–926.

Heun, M., R. Schafer-Pregl, D. Klawan et al. 1997. Site of einkorn wheat domestication identified by DNA fingerprinting. *Science* 278:1312–1314.

Hirel, B., J. Le Gouis, B. Ney, and A. Gallais. 2007. The challenge of improving nitrogen use efficiency in crop plants: Towards a more central role for genetic variability and quantitative genetics within integrated approaches. *J Exp Bot* 58:2369–2387.

Hirsch, C. N., C. D. Hirsch, K. Felcher et al. 2013. Retrospective view of North American potato (*Solanum tuberosum* L.) breeding in the 20th and 21st centuries. *G3 Genes Genomes Genetics* 3:1003–1013.

Hmida-Sayari, A., R. Gargouri-Bouzid, A. Bidani, L. Jaoua, A. Savouré, and S. Jaoua. 2005a. Overexpression of Δ1-pyrroline-5-carboxylate synthetase increases proline production and confers salt tolerance in transgenic potato plants. *Plant Sci* 169:746–752.

Hmida-Sayari, A., R. Gargouri-Bouzid, A. Bidani, L. Jaoua, A. Savouré, and S. Jaoua. 2005b. Overexpression of A'-pyrroline-5-carboxylate synthetase increases proline production and confers salt tolerance in transgenic potato plants. *Plant Sci* 169:746–752.

Huamán, Z. and P. Schmiediche. 1999. The potato genetic resources held in trust by the International Potato Center (CIP) in Peru. *Potato Res* 42:413–426.

ICP (International Potato Center). 1990. Control of virus and virus-like diseases of potato and sweet potato. Report of the 3rd Planning Conference, Lima, Peru.

Jefferies, R. A. 1989. Water-stress and leaf growth in field-grown crops of potato (*Solanum tuberosum* L.). *J Exp Bot* 40:1375–1381.

Jefferies, R. A. 1993. Responses of potato genotypes to drought. I. Expansion of individual leaves and osmotic adjustment. *Ann App Biol* 122:93–104

Jeong, M. J., S. C. Park, and N. O. Byun. 2001. Improvement of salt tolerance in transgenic potato plants by glyceraldehyde-3 phosphate dehydrogenase gene transfer. *Mol Cell* 12:185–189.

Kim, D. Y., H. E. Lee, K. W. Yi et al. 2003. Expression pattern of potato (*Solanum tuberosum*) genes under cold stress by using cDNA microarray. *Korean J Genet* 25:345–352.

Kita, A., E. Brathen, S. H. Knutsen, and T. Wicklund. 2004. Effective ways of decreasing acrylamide content in potato crisps during processing. *J Agric Food Chem* 52:7011–7016.

Krauss, A. and H. Marschner. 1982. Influence of nitrogen nutrition, daylength and temperature on contents of gibberellic and abscisic acid and on tuberization in potato plants. *Potato Res* 25:13–21.

Kumar, P. A., M. A. J. Parry, R. A. C. Mitchell, A. Ahmad, and Y. P. Abrol. 2002. Photosynthesis and nitrogen-use efficiency. In C. H. Foyer and G. Noctor (eds), *Photosynthetic Nitrogen Assimilation and Associated Carbon and Respiratory Metabolism*, pp. 23–34. Amsterdam: Springer.

Leone, A., A. Costa, F. Consiglio et al. 1999. Tolerance to abiotic stresses in potato plants: A molecular approach. *Potato Res* 42:333–351.

Levy, D., W. K. Coleman, and R. E. Veilleux. 2013. Adaptation of potato to water shortage: Irrigation management and enhancement of tolerance to drought and salinity. *Am J Potato Res* 90:186–206.

Levy, D., E. Fogelman, and Y. Itzhak. 1988. The effect of water salinity on potatoes (*Solanum tuberosum* L.): Physiological indices and yielding capacity. *Potato Res* 31:601–610.

Levy, D. and R. E. Veilleux. 2007. Adaptation of potato to high temperatures and salinity: A review. *Am J Potato Res* 84:487–506.

Li, P. H., N. P. A. Huner, M. Toivio-Kinnucan, H. H. Chen, and. J. P. Palta. 1981. Potato freezing injury and survival, and their relationships to other stress. *Am Potato J* 58:15–29.

Li, X.-Q., D. Sveshnikov, B. J. Zebarth et al. 2010. Detection of nitrogen sufficiency in potato plants using gene expression markers. *Am J Potato Res* 87:50–59.

Liu, F., A. Shahnazari, M. N. Andersen, S. E. Jacobsen, and C. R. Jensen. 2006. Physiological responses of potato (*Solanum tuberosum* L.) to partial root-zone drying: ABA signalling, leaf gas exchange, and water use efficiency. *J Exp Bot* 57:3727–3735.

Mackerron, D. K. L. and R. A. Jefferies. 1988. The distributions of tuber sizes in droughted and irrigated crops of potato. I. Observations on the effect of water stress on graded yields from differing cultivars. *Potato Res* 31:269–278.

Mannion, A. M. 1999. Domestication and the origins of agriculture: An appraisal. *Prog Phys Geog* 23:37–56.

Mba, C., E. P. Guimaraes, and K. Ghosh. 2012. Re-orienting crop improvement for the changing climatic conditions of the 21st century. *Agr Food Security* 1:7. http://www.agricultureandfoodsecurity.com/content/1/1/7.

McGloughlin, M. N. 2010. Modifying agricultural crops for improved nutrition. *New Biotechnol* 27:494–504.

Menzel, C. M. 1979. Tuberization in potato at high temperatures: Responses to gibberellin and growth inhibitors. *Ann Bot* 46:259–265.

Menzel, C. M. 1983. Tuberization in potato at high temperatures: Gibberellin content and transport from buds. *Ann Bot* 52:697–702.

Menzel, C. M. 1985. Tuberization in potato at high temperatures: Interaction between temperature and irradiance. *Ann Bot* 55:35–39.

Mertz, E. T., L. S. Bates, and O. E. Nelson. 1964. Mutant maize that changes the protein composition and increases the lysine content of maize endosperm. *Science* 145:279–280.

Miller, J. K., E. M. Herman, M. Jahn, and K. J. Bradford. 2010. Strategic research, education and policy goals for seed science and crop improvement. *Plant Sci* 179:645–652.

Morrison, H. E., L. G. Gentner, R. F. Koontz, and R. W. Every. 1967. The changing role of soil pests attacking potato tubers. *Am Potato J* 44:137–144.

Mottram, D. S., B. L. Wedzicha, and A. T. Dodson. 2002. Acrylamide is formed in the Maillard reaction. *Nature* 419:448–449.

Mullins, E., D. Milbourne, C. Petti, B. M. Doyle-Prestwich, and C. Meade. 2006. Potato in the age of biotechnology. *Trends Plant Sci* 11:254–260.

Nassar, A. M. K., S. Kubow, Y. N. Leclerc, and D. J. Donnelly. 2013. Somatic mining for phytonutrient improvement of "russet burbank" potato. *Am J Potato Res* 91:89–100.

Nitithamyong, A., J. H. Vonelbe, R. M. Wheeler, and T. W. Tibbitts. 1999. Glycoalkaloids in potato tubers grown under controlled environments. *Am J Potato Res* 76:337–343.

Oerke, E. C. 2006. Crop losses to pests. *J Agric Sci* 144:31–43.

Olsen, K. M. and B. L. Gross. 2008. Detecting multiple origins of domesticated crops. *Proc Natl Acad Sci USA* 105:13701–13702.

Olsson, K., R. Svensson, and C. A. Roslun. 2004. Tuber components affecting acrylamide formation and colour in fried potato: Variation by variety, year, storage temperature and storage time. *J Sci Food Agric* 84:447–458.

Osusky, M., L. Osuska, R. E. Hancock, W. W. Kay, and S. Misra. 2004. Transgenic potatoes expressing a novel cationic peptide are resistant to late blight and pink rot. *Transgenic Res* 13:181–190.

Park, S., T. S. Kang, C. K. Kim et al. 2005. Genetic manipulation for enhancing calcium content in potato tuber. *J Agric Food Chem* 53:5598–5603.

Parry, M. A. J. and M. J. Hawkesford. 2012. An integrated approach to crop genetic improvement. *J Integr Plant Biol* 54:250–259.

Patel, R. M., S. O. Prasher, D. Donnelly, and R. B. Bonnel. 2001. Effect of initial soil salinity and subirrigation water salinity on potato tuber yield and size. *Agr Water Manage* 46:231–239.

Pathak, R. R., A. Ahmad, S. Lochab, and N. Raghuram. 2008. Molecular physiology of plant nitrogen use efficiency and biotechnological options for its enhancement. *Curr Sci* 94:1394–1403.

Peferoen, M., S. Jansens, A. Reynaerts, and J. Leemans. 1990. Potato plants with engineered resistance against insect attack. In M. E. Vayda and W. D. Park (eds), *The Molecular and Cellular Biology of the Potato*, pp. 193–204. Wallingford: CABI Publishing.

Pehu, E. 1996. The current status of knowledge on the cellular biology of potato. *Potato Res* 39:429–435.

Perlak, F. J., T. B. Stone, Y. M. Muskopf et al. 1993. Genetically improved potatoes: Protection from damage by Colorado potato beetles. *Plant Mol Biol* 22:313–321.

Pérombelon, M. C. M. 2002. Potato diseases caused by soft rot erwinias: An overview of pathogenesis. *Plant Pathol* 51:1–12.

Phillips, T. 2008. Genetically modified organisms (GMOs): Transgenic crops and recombinant DNA technology. *Nat Edu* 1:213.

Porter, G. A., G. B. Opena, W. B. Bradbury, J. C. McBurnie, and J. A. Sisson. 1999. Soil management and supplemental irrigation effects on potato. I. Soil properties, tuber yield and quality. *Agron J* 91:416–425.

Powell, K. 2007. Functional foods from biotech-an unappetizing prospect? *Nature* 25:525–531.

Priestley, J. H. and L. M. Woffenden. 1923. The healing of wounds in potato tubers and their propagation by cut sets. *Ann Appl Biol* 10:96–115.

Qin, A., Q. Shi, and X. Yu. 2011. Ascorbic acid contents in transgenic potato plants overexpressing two dehydroascorbate reductase genes. *Mol Biol Rep* 38:1557–1566.

Quilter, J., B. Ojeda E., D. Pearsall et al. 1991. Subsistence economy of El Paraiso, an early Peruvian site. *Science* 251:277–283.

Radcliffe, E. B. 1982. Insect pests of potato. *Ann Rev Entomol* 27:173–204.

Rezzi, S., Z. Ramadan, L. B. Fay, and S. Kochhar. 2007a. Nutritional metabonomics: Applications and perspectives. *J Proteome Res* 6:513–525.

Rezzi, S., Z. Ramadan, F. P. Martin et al. 2007b. Human metabolic phenotypes link directly to specific dietary preferences in healthy individuals. *J Proteome Res* 6:4469–4477.

Rich, A. E. 1983. *Potato Diseases*. Wallingford: Cab International.

Rist, M. J., U. Wenzel, and H. Daniel. 2006. Nutrition and food science go genomic. *Trends Biotechnol* 24:1–7.

Ritter, R., L. Barandalla, R. López, and J. I. Ruiz de Galarreta. 2008. Exploitation of exotic, cultivated *Solanum* germplasm for breeding and commercial purposes. *Potato Res* 51:301–311.

Rochfort, S. 2005. Metabolomics reviewed: A new "omics" platform technology for systems biology and implications for natural products research. *J Nat Prod* 68:1813–1820.

Rodríguez-Falcón, M., J. Bou, and S. Prat. 2006. Seasonal control of tuberization in potato: Conserved elements with the flowering response. *Ann Rev Plant Biol* 57:151–180.

Rommens, C. M., H. Yan, K. Swords, C. Richael, and J. Ye. 2008. Low-acrylamide French fries and potato chips. *Plant Biotechnol J* 6:843–853.

Rommens, C. M., J. Ye, C. Richael, and K. Swords. 2006. Improving potato storage and processing characteristics through all-native DNA transformation. *J Agric Food Chem* 54:9882–9887.

Ross-Ibarra, J., P. L. Morrell, and B. S. Gaut. 2007. Plant domestication, a unique opportunity to identify the genetic basis of adaptation. *Proc Natl Acad Sci USA* 104:8641–8648.

Ruth, L. 2003. Tailoring thresholds for GMO testing. Social and economic factors shape new regulations that in turn drive the technology. *Anal Chem* 9:392–396.

Ryan, C. A. 1990. Protease inhibitors in plants: Genes for improving defenses against insects and pathogens. *Annu Rev Phytopathol* 28:425–449.

Sattelmacher, B., W. J. Horst, and H. C. Becker. 1994. Factors that contribute to genetic variation for nutrient efficiency of crop plants. *Z Pflanz Bodenkunde* 157:215–224.

Schilter, B. and A. Constable. 2002. Regulatory control of genetically modified (GM) foods: Likely developments. *Toxicol Lett* 127:341–349.

Schuler, T. H., G. M. Poppy, B. R. Kerry, and I. Denholm. 1998. Insect-resistant transgenic plants. *Trends Biotechnol* 16:168–175.

Shepherd, L. V. T., J. E. Bradshaw, M. F. B. Dale et al. 2010. Variation in acrylamide producing potential in potato: Segregation of the trait in a breeding population. *Food Chem* 123:568–573.

Shewmaker, C. K. and D. M. Stalker. 1992. Modifying starch biosynthesis with transgenes in potatoes. *Plant Physiol* 100:1083–1086.

Shock, C. C., Z. A. Holmes, T. D. Stieber, E. P. Eldredge, and P. Zhang. 1993. The effect of timed water stress on quality, total solids and reducing sugar content of potatoes. *Am Potato J* 70:227–241.

Simmonds, N. W. 1997. A review of potato propagation by means of seed, as distinct from clonal propagation by tubers. *Potato Res* 40:191–214.

Sinden, S. L. and R. E. Webb. 1972. Effect of variety and location on the glycoalkaloid content of potatoes. *Am Potato J* 49:334–338.

Skog, K., G. Viklund, K. Olsson, and I. Sjoholm. 2008. Acrylamide in home-prepared roasted potatoes. *Mol Nutr Food Res* 52:307–312.

Slomnicki, I. and I. Rylski. 1964. Effect of cutting and gibberellin treatment on autumn-grown seed potatoes for spring planting. *Eur Potato J* 7:184–192.

Spooner, D. M., K. McLean, G. Ramsay, R. Waugh, and G. J. Bryan. 2006. A single domestication for potato based on multilocus amplified fragment length polymorphism genotyping. *Proc Natl Acad Sci USA* 102:14694–14699.

Stadler, R. H., I. Blank, N. Varga et al. 2002. Acrylamide from Maillard reaction products. *Nature* 419:449–450.

Stark, J. C., I. R. McCann, D. T. Westermann, B. Izadi, and T. A. Tindall. 1993. Potato response to split nitrogen timing with varying amounts of excessive irrigation. *Am Potato J* 70:765–777.

Stead, D. 1999. Bacterial diseases of potato: Relevance to *in vitro* potato seed production. *Potato Res* 42:449–456.

Steffan, K. L. and J. P. Palta. 1986. Effect of light on photosynthetic capacity during cold acclimation in a cold-sensitive and a cold-tolerant potato species. *Physiol Plantarum* 66:353–359.

Struik, P. C., M. F. Askew, A. Sonnino et al. 1997. Forty years of potato research: Highlights, achievements and prospects. *Potato Res* 40:5–18.

Subbiah, M. T. R. 2006. Nutrigenetics and nutraceuticals: The next wave riding on personalized medicine. *Transl Res* 149:55–61.

Sukumaran, N. P. and C. J. Weiser. 1972. Freezing injury in potato leaves. *Plant Physiol* 50:564–567.

Thiele, A., M. Herold, I. Lenk, P. H. Quail, and C. Gatz. 1999. Heterologous expression of *Arabidopsis* phytochrome B in transgenic potato influences photosynthetic performance and tuber development. *Plant Physiol* 120:73–82.

Thieme, R. and H. Griess. 2005. Somaclonal variation in tuber traits of potato. *Potato Res* 48:153–165.

Trujillo, E., C. Davis, and J. Milner. 2006. Nutrigenomics, proteomics, metabolomics and the practice of diets. *J Am Diet Assoc* 106:403–414.

Ugent, D. 1970. The potato. *Science* 170:1161–1166.

Upadhyaya, H. C. P., N. Akula, K. E. Young, S. C. Chun, D. H. Kim, and S. W. Park. 2010. Enhanced ascorbic acid accumulation in transgenic potato confers tolerance to various abiotic stresses. *Biotechnol Lett* 32:321–330.

Upadhyaya, H. C. P., K. E. Young, N. Akula et al. 2009. Over-expression of strawberry d-galacturonic acid reductase in potato leads to accumulation of vitamin C with enhanced abiotic stress tolerance. *Plant Sci* 177:659–667.

Valueva, T. A., T. A. Revina, E. L. Gvozdeva, N. G. Gerasimova, and O. L. Ozeretskovskaia. 2003. Role of proteinase inhibitors in potato protection. *Bioorg Khim* 29:499–504.

Van Eck, J., B. Conlin, D. F. Garvin, H. Mason, D. A. Navarre, and C. R. Brown. 2007. Enhancing β-carotene content in potato by RNAi-mediated silencing of the β-carotene hydroxylase gene. *Am J Potato Res* 84:331–342.

van Loon, C. D. 1981. The effect of water stress on potato growth, development, and yield. *Am Potato J* 58:51–69.

Vos, J. 2009. Nitrogen responses and nitrogen management in potato. *Potato Res* 52:305–317.

Vos, J. and J. Groenwold. 1989. Genetic differences in water-use efficiency, stomatal conductance and carbon isotope fractionation in potato. *Potato Res* 32:113–121.

Vreugdenhil, V. and L. I. Sergeeva. 1999. Gibberellins and tuberization in potato. *Potato Res* 42:471–481.

Watanabe, K. N., A. Kikuchi, T. Shimazaki, and M. Asahina. 2011. Salt and drought stress tolerances in transgenic potatoes and wild species. *Potato Res* 54:319–324.

Watkins, S. M., B. D. Hammock, J. W. Newman, and J. B. German. 2001. Individual metabolism should guide agriculture toward foods for improved health and nutrition. *Am J Clin Nutr* 74:283–286.

Westermann, D. T., G. E. Kleinkopf, and L. K. Porter. 1988. Nitrogen fertilizer efficiencies on potatoes. *Am Potato J* 65:377–386.

Ye, J., R. Shakya, P. Shrestha, and C. M. Rommens. 2010. Tuber-specific silencing of the acid invertase gene substantially lowers the acrylamide-forming potential of potato. *J Agric Food Chem* 58:12162–12167.

Yin, X. and P. C. Struik. 2007. Crop system biology. An approach to connect functional genomics with crop modelling. In J. H. J. Spiertz, P. C. Struik, and H. H. van Laar (eds), *Scale and Complexity in Plant Systems Research: Gene-Plant-Crop Relations*, pp. 63–73. Berlin: Springer.

Yin, X. and P. C. Struik. 2009. Applying modelling experiences from the past to shape crop systems biology: The need to converge crop physiology and functional genomics. *New Phytol* 179:629–642.

Yin, X. and P. C. Struik. 2010. Modelling the crop: From system dynamics to systems biology. *J Exp Bot* 61:2171–2183.

Yuan, B. Z., S. Nishiyama, and Y. Kang. 2003. Effects of different irrigation regimes on the growth and yield of drip-irrigated potato. *Agr Water Manage* 63:153–167.

Zebarth, B. J., G. Tai, R. Tarn, H. de Jong, and P. H. Milburn. 2004. Nitrogen use efficiency characteristics of commercial potato cultivars. *Can J Plant Sci* 84:589–598.

Zvomuya, F. and C. J. Rosen. 2002. Biomass partitioning and nitrogen use efficiency of "Superior" potato following genetic transformation for resistance to Colorado potato beetle. *J Am Soc Hort Sci* 127:703–709.

CHAPTER 12

Omics-Based Approaches for Improvement of the Common Bean

Sajad Majeed Zargar, Chumki Bhattacharjee, Rashmi Rai, Muslima Nazir
Yoichiro Fukao, Ganesh Kumar Agrawal, and Randeep Rakwal

CONTENTS

12.1 COMMON BEAN: IMPORTANCE AND NEED FOR IMPROVEMENT

Legumes belong to the family Fabaceae (or Leguminosea), subfamily Papilionoideae. The common bean (*Phaseolus vulgaris* L.) belongs to its section *Phaseoli*. The Fabaceae family includes about 650 genera and 18,000 species, thus making it the third largest family within the plant kingdom (Doyle 2001). Legumes are podded plants that have been part of the human diet since the dawn of agriculture and are cultivated worldwide. The major legumes include soybean, ground nut, common bean, pea, fava bean, lentil, chickpea, cowpea, pigeonpea, and so on (Burstin et al. n.d.). Legumes are generally rich in protein, with low saturated fat and high complex carbohydrates, fiber, and micronutrients. They also contain two of the most critical nutrients for humans, that is, folic acid, which essential for normal cell division, immune response, and correct development of the fetus in the womb, and thiamine (vitamin B1), which is essential for the process of metabolism. Legumes in general also possess high iron and B vitamins, particularly B6. Besides its nutritional value, the common bean has been shown to have medicinal properties due to the presence of a number of phytochemicals including polyphenols and flavonoids (Akond et al. 2011). Different plant parts are used to treat various disorders such as ulcers, joint pain, cough, measles, eye infection, gonorrhea, diabetes, and so on (Sharma et al. 2011). Most members of the Fabaceae family contain significant quantities of isoflavones that reduce the risk of cancer, heart disease, and osteoporosis, and also help relieve menopausal symptoms (Messina 1999). The nutraceutical properties of grain legume proteins have been reviewed by Duranti (2006).

Regarding food security, the common bean is the most important food legume, representing 50% of grain legumes consumed worldwide (McClean et al. 2004). Variation in the principal constituents among different grain legumes, including the common bean, is presented in Table 12.1.

Beans are an important source of protein (~22%), vitamins (folate), and minerals (Ca, Cu, Fe, Mg, Mn, Zn) in the human diet, especially in the developing countries (Broughton et al. 2003). The bioavailability of micronutrients in the diet is very important, as their deficiency causes various diseases and abnormalities. Compared with meat-based diets, plant-based diets have a limited content of these essential micronutrients, which is mainly due to the presence of various toxins and antinutritional elements (Ologhobo 1980). Iron and zinc deficiencies are the prevalent micronutrient problems in the world. According to a report by Welch and Graham (1999), more than 2 billion people globally are affected by Fe deficiency, and another report (Brown and Peerson 2001) estimates that more than 49% of the world population is at risk from low zinc intake. As the minerals in beans are readily available, various health-related problems can be potentially solved, including hypertension (Appel et al. 1997).

Referring to a report published by Commodity Online on December 1, 2010, the Indian government is considering the import of around 900,000 t of the legume (Commodity Online 2010). These data ring alarm bells, showing the magnitude of the crisis and the need to work for the improvement of this crop. The common bean is afflicted by various environmental stress factors that have direct impact on its yield and quality. These may be biotic, principally from pathogenic fungi, bacteria, and viruses that cause diseases such as root rot, anthracnose, angular leaf spot (ALS), bean common mosaic virus (BCMV), and bacterial blight. Other factors include abiotic stresses such as drought, heat, cold, low soil fertility, soil salinity, and heavy metal. Anticipating the increasing demand for protein and micronutrient-rich food sources, the improvement of the common bean is a critically important area of research with the aim of enhancing its yield potential.

12.2 GENOMICS APPROACH

For improvement of any crop, understanding its genetics is essential. Recent advances in molecular genetics/availability of genomics-based tools have made it easy to understand the genetics of any trait. *P. vulgaris* is a true diploid with 11 pairs of chromosomes, and it is estimated to have a genome size between 588 and 637 mega base pairs (Mbp) (Arumuganthan and Earle 1991; Bennett and Leitch 1995; Thibivilliers et al. 2009). Its sequencing will help serve as a model for understanding the ~1100-Mbp soybean genome (McConnell et al. 2010). These features are the reason why initiatives have been taken to complete the sequencing of the common bean genome. Its immediate applications will help in developing fine maps that can be used in identification and introgression of various biotic and abiotic stress tolerance genes in desirable backgrounds through a molecular breeding approach. It should be

Table 12.1 Variation in Principle Constituents among Different Grain Legumes

Bean	Latin Name	Protein and Energy	Dietary fiber	Fat	Thiamin (vitamin B1)	Riboflavin (vitamin B2)	Pyridoxine (vitamin B6)	Folate (vitamin B9)	Mo	Ca	Zn	Fe	K	Se	References
Red bean	*Phaseolus vulgaris*	8.67 g, 127 calories	7.3 g	1 g	0.216 mg	0.058 mg	0.12 mg	130 mcg	132 mcg	28 mg	1.07 mg	2.94 mg	403 mg	1.2 mcg	Dr. Decuypere (n.d.), Nutrition-and-You (n.d.), Laberge (n.d.)
Navy bean	*Phaseolus vulgaris*	8.23 g, 140 calories	10.5 g	1.5 g	0.237 mg	0.066 mg	0.138 mg	140 mcg	100 mcg	69 mg	1.03 mg	2.36 mg	389 mg	2.9 mcg	Dr. Decuypere (n.d.), Nutrition-and-You (n.d.), Laberge (n.d.)
Pinto bean	*Phaseolus vulgaris*	9.01 g, 143 calories	9 g	0.5 gm	0.193 mg	0.062 mg	0.229 mg	172 µg	74 µg	46 mg	0.98 mg	2.09 mg	436 mg	6.2 µg	Dr. Decuypere (n.d.), Martinac (n.d.)
Lima bean, butter bean	*Vigna lunatus*	7.80 g, 115 calories	7.0 g	0.7 g	0.161 mg	0.055 mg	0.161 mg	83 µg	72 µg	17 mg	0.95 mg	2.39 mg	508 mg	4.5 µg	Dr. Decuypere (n.d.), Laberge (n.d.), Messina (1999)
Adzuki bean	*Vigna angularis*	7.52 g, 128 calories	7.3 g	0.53 g	0.115 mg	0.064 mg	0.096 mg	121 µg	210 µg	28 mg	1.77 mg	2 mg	532 mg	1.2 µg	Dr. Decuypere (n.d.), Fattree Archive (n.d.)

(continued)

Table 12.1 (Continued) Variation in Principle Constituents among Different Grain Legumes

Bean	Latin Name	Protein and Energy	Dietary fiber	Fat	Thiamin (vitamin B1)	Riboflavin (vitamin B2)	Pyridoxine (vitamin B6)	Folate (vitamin B9)	Mo	Ca	Zn	Fe	K	Se	References
Mung bean	Vigna radiata	7.02 g of protein, 105 calories	7.6 g	1.2 g	0.164 mg	0.061 mg	0.067 mg	159 µg	410 µg	27 mg	0.84 mg	1.4 mg	266 mg	2.5 µg	Dr. Decuypere (n.d.)
Black bean (urad dal or kaali dal)	Vigna mungo	8.86 g protein, 132 calories	8.7 g	1.4 g	0.244 mg	0.059 mg	0.069 mg	149 µg	130 µg	27 mg	1.12 mg	2.1 mg	355 mg	1.2 µg	Dr. Decuypere (n.d.), Laberge (n.d.) Messina (1999)
Broad bean or fava bean	Vicia faba	7.6 g of protein, 110 calories	5.4 g	1.5 g	0.202 mg	0.055 mg	0.1 mg	208 µg	260 µg	24 mg	1.29 mg	2.51 mg	278 mg	2.5 µg	Dr. Decuypere (n.d.), Laberge (n.d.)
Cow pea	Vigna unguiculata	7.73 g, 116 calories	6.5 g	2 gm	0.202 mg	0.055 mg	0.1 mg	208 µg	92 µg	24 mg	1.29 mg	2.51 mg	278 mg	2.5 µg	Dr. Decuypere (n.d.), Laberge (n.d.)
Pigeon pea (arhar dal or toor dal)	Cajanus cajan	6.76 g, 121 calories	6.7 g	1 gm	0.146 mg	0.059 mg	0.05 mg	111 µg	–	43 mg	0.9 mg	1.11 mg	384 mg	2.9 µg	Dr. Decuypere (n.d.)

															References
Winged bean	*Psophocarpus teragonolobus*	10.62 g, 147 calories.	–	18%	0.295 mg	0.129 mg	0.047 mg	10 µg	–	142 mg	1.44 mg	4.33 mg	280 mg	2.9 µg	Dr. Decuypere (n.d.), the full wiki (2009)
Chickpea	*Cicer arietinum*	8.86 g, 164 calories	7.6 g	15%	0.116 mg	0.063 mg	0.139 mg	172 µg	150 µg	49 mg	1.53 mg	2.89 mg	291 mg	3.7 µg	Dr. Decuypere (n.d.), Indobase (n.d.)
Lentil	*Lens culinaris*	25.8 g, protein, 353 calories	1.06 g	1198.8 ug	83.33 mg	0.52 mg	479 µg	76 µg	21.11 mg	1.38 mg	3.66 mg	955 mg	8.3 µg		Dr. Decuypere (n.d.), Laberge (n.d.), the full wiki (2009), Commodity Online (2010)
Groundnut or peanut	*Arachis hypogaea*	25.80 g, 567 Kcal	8.5 g	49.24 g	0.640 mg	0.135 mg	0.348 mg	240 µg	22 cg	92 mg	3.27 mg	4.58 mg	705 mg	7.2 µg	Dr. Decuypere (n.d.), Nutrition-and-You (n.d.), Laberge (n.d.)
Soybean	*Glycine max*	35.22 g, 471 calories	17.7 g	47%	0.1 mg	0.145 mg	0.208 mg	211 µg	6.5 µg	138 mg	3.14 mg	3.9 mg	1470 mg	19.1 µg	Dr. Decuypere (n.d.), Laberge (n.d.)

Note: Nutritional values are per 100 g.

mentioned that the first large-scale identification of the common bean transcriptome derived by 454 pyrosequencing has been recently published (Kalavacharla et al. 2011). This database along with the whole genome sequence will help in candidate gene discovery not only in the common bean but also other legume crops, including soybean. In order to understand the potential of the common bean germplasm, the immediate requirement is to characterize the available germplasm.

12.2.1 Need for Genetic Diversity Studies

The best way to understand the potential of the available germplasm is by analysis of the genetic diversity and subsequent characterization. Genetic diversity can be analyzed by various means. But the main objective is to understand the genetic makeup of each genotype. The United Nations General Assembly declared 2010 as the International Year of Biodiversity (resolution 61/203) with the slogan, "Biodiversity is life. Biodiversity is our life." Devoting a year to biodiversity indicates its importance in human development and survival. The immense genetic diversity of landraces of crops is the most directly useful and economically valuable part of biodiversity. With the increase in human population, there is now more focus on biodiversity for the sole reason of choosing important candidates for crop improvement. The control of diversity is necessary to avoid loss of some important landraces/genotypes due to changed patterns of crop cultivation. Unlike high-yielding varieties, the landraces maintained by farmers are endowed with tremendous genetic variability, as they are not subjected to subtle selection over a long period of time.

For determination of genetic diversity, the first and most important step is collection and cataloguing of the germplasm. Zargar's Laboratory has collected common bean germplasm from different regions of Jammu and Kashmir (India) (Zargar et al. 2014). The collected germplasm is unique and is comprised mostly of the landraces. Genetic diversity can be assessed by diverse methods, however as molecular markers generate greater polymorphism it is better to use these markers. These molecular markers are selected based on their availability, reproducibility, and accuracy. In a number of studies detailed later in the chapter, random amplified polymorphic DNA (RAPD) and simple sequence repeat (SSR) have been extensively used for genetic diversity analysis in the case of the common bean. For determination of genetic diversity in crop plants, previous researchers have used various morphological and biochemical markers. However, due to their limited polymorphism, molecular markers are currently being employed for characterizing germplasm diversity. Molecular markers have demonstrated the potential to detect genetic diversity and to aid in the management of plant genetic resources (Virk et al. 2000; Song et al. 2003). In contrast to morphological traits, molecular markers can reveal differences among genotypes at DNA level, providing a more direct, reliable, and efficient tool for germplasm characterization, conservation, and management. Several types of molecular markers are available today, including those based on restriction fragment length polymorphism (RFLP) (Botstein et al. 1980), RAPD (Welsh and McClelland 1990; Williams et al. 1990), amplified fragment length polymorphism (AFLP) (Vos et al. 1995), and SSR (Tautz 1989). The advantages of RAPDs are their improved reproducibility, suitability in large numbers, and validation (Malviya and Yadav 2010). RAPD markers have been used to investigate the genetic diversity and relationships in the common bean gene pool, not only in the center of origin but also outside the centers (McClean et al. 2004). Metais et al. (2000) used seven RAPDs that generated 120 bands and RFLP for exploring the genetic diversity among commercial bean lines. They observed that RAPDs and RFLP lead to the same clustering of the bean lines according to their geographical origins. Maras et al. (2008) used 10 AFLPs and 14 SSRs to generate the polymorphism for gene pool classification of common beans. They observed that both marker systems showed comparable accuracy in grouping the genotypes according to their gene pool of origin. Biswas et al. (2010) used RAPDs for evaluation of 14 French bean varieties of three ecogeographical regions of Bangladesh. They concluded that simplicity and efficiency of RAPD analysis will even facilitate the construction of genetic maps in French bean. Jose et al. (2009)

used 13 RAPD primers to evaluate an intraspecific genetic variation that prevails in landraces of common bean in the Nilgiris biosphere reserve. A dendrogram was constructed based on Jaccard's coefficients of 102 RAPD markers using the average distance method (UPGMA), separating the accessions into two major clusters, A and B, with Mesoamerican and Andean gene pools, respectively. Kumar et al. (2009) utilized 17 microsatellite markers to examine genetic diversity at the molecular level, which showed polymorphic information content (PIC) in the range of 0.00–0.684. Dendrograms based on Euclidean distances and UPGMA analysis showed the presence of the majority of released varieties in a single cluster, which pointed toward their low genetic base in comparison to indigenous landraces and exotic germplasm.

In addition to the above-mentioned gel-based molecular markers, new-generation molecular markers, called single-nucleotide polymorphisms (SNPs) do not always need these gel-based assays (Zhou et al. 2004). They are the most abundant of all marker systems known so far. There is some evidence that the stability of SNPs and, therefore, the fidelity of their inheritance are higher than those of other molecular systems such as SSRs and AFLPs. SNPs at a particular site in the DNA molecule should in principle involve four possible nucleotides, but in actual practice, only two of these four possibilities have been observed at a specific site in a population (Gupta et al. 2001). Consequently, SNPs are biallelic. According to recent estimates, one SNP occurs every 100–300 bp in any genome, thus making SNPs the most abundant molecular markers known so far. Cortex et al. (2011) developed 94 SNPs and tested them across the common bean germplasm. The SNP diversity was accessed at 84 gene-based and 10 non-genic loci, using KASPar technology in a panel of 70 genotypes. SNPs exhibited high levels of genetic diversity. The authors concluded that both SNP and SSR markers are ideal genetic marker combinations to carry out genetic diversity, mapping, and association studies in the common bean. Blair et al. (2013) used the Illumina Golden Gate assay for the evaluation of parental polymorphism and genetic diversity among a minicore set of common bean accessions. They scored 736 SNPs from 236 diverse common bean genotypes and recorded low levels of missing data. Those authors observed that Illumina SNPs were effective in distinguishing between gene pools and therefore are most useful in saturation of inter–gene pool genetic maps. The other high-throughput technique, referred to as diversity arrays technology (DArT), is an upcoming technology that does not require sequence information. DArT discovers a large number of markers in parallel and does not require further development of an assay once markers are discovered. DArT was developed to provide a practical and cost-effective whole-genome fingerprinting tool (Jaccoud et al. 2001). Development of DArT starts with assembling a group of DNA samples representative of the germplasm anticipated to be analyzed with the technology. This group of samples usually corresponds to the primary gene pool of a crop species, but can be restricted to the two parents of a cross or expanded to secondary or even tertiary gene pools. The DNA mixture representing the gene pool of interest is processed using a complexity reduction method—a process by which reproducibility selects a defined fraction of genomic fragments, called representation, is then used to construct a library in *Escherichia coli*. The inserts from individual clones are then amplified and used as molecular probes on DArT assays. DArT markers are biallelic markers that are either dominant or hemidominant. Rice was used for the initial proof-of-concept work of DArT (Jaccoud et al. 2001). Brinz et al. (2012) applied the DArT system for the first time to the common bean and tested on 89 accessions. A total of 2501 polymorphic markers were found, and in further diversity analysis it was evident that DArT technology was accurate for studying the genetic diversity in the common bean and efficient for the large-scale detection of polymorphism. Taken together, all these marker technologies can be used to access the genetic diversity of the common bean.

12.2.2 Genomics Approaches for Identification and Introgression of Stress Resistance Genes

Biotic and abiotic stresses are the major concerns for crop improvement. Here we discuss the major biotic and abiotic stresses and their impact on the common bean. In general, a number of biotic (bacteria, fungi, nematodes, viruses, insects, etc.) and abiotic (drought, salinity, phosphate deficiency, etc.) stresses severely affect yield. In order to enhance resistance to these stresses, understanding the genetics behind resistance is essential. Genetic resistance is considered the most effective strategy to protect the bean crop against such stresses. Although extensive literature on resistance to biotic/abiotic stresses is available, bean breeders still struggle with the decision as to which gene(s) are better utilized/deployed in resistance breeding programs. Genes for resistance are reported either as a single locus or in gene clusters at complex loci. Given the information of identification, mapping, and gene tagging, new information on the location of most of the major genes including those controlling resistance to various stresses is now available on the Bean Improvement Cooperation (BIC) website. Similarly, a list of SCAR markers is available for genes of the common bean. Readers are referred to these sites for detailed information. Although conventional breeding has been practiced for many decades and has helped breeders to develop new varieties, the molecular breeding approach is more precise and advantageous. Hence, practicing molecular breeding for the incorporation of resistant genes in locally adopted varieties is a better option. Figures 12.1 and 12.2 provide details about the precision of molecular breeding and also the strategy that can be followed to obtain resistant cultivars. Table 12.2 provides information on disease-resistance genes in beans. Additionally, the details on some disease-resistance genes are presented in the following subsections.

12.2.2.1 Gene Discovery: Identification of Resistance Gene(s)

Anthracnose (ANT) in the common bean is caused by *Colletotrichum lindemuthianum* and exhibits a high level of pathogenic diversity. This disease causes huge yield loss in almost all bean-growing areas of world. Resistance against anthracnose is conditioned primarily by 11 major independent genes: *Co-1, Co-2, Co-3, Co-4, Co-5, Co-6, Co-7, Co-10, Co-11, Co-12,* Co-13, and one

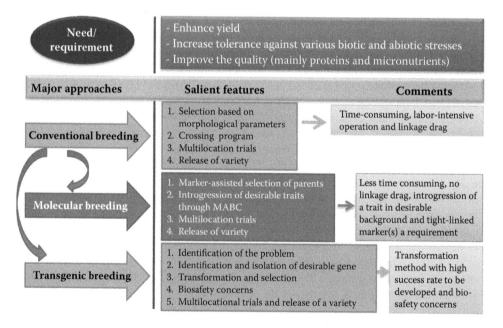

Figure 12.1 Approaches for improvement of the common bean (*Phaseolus vulgaris* L.). The three main breeding approaches for improvement of the common bean are Conventional, molecular, and transgenic breeding. Each breeding method has its own advantages and limitations. However, molecular breeding is preferred as it has high precision and does not require any consideration for biosafety regulations.

Table 12.2 List of Biotic Stress Resistance Genes in Common Bean along with Linked Markers

S.No.	Pest/Pathogen	Gene	Linked/Flanking Molecular Markers	Linkage Group	References for Genes/Markers/LG
Fungal					
1.	Anthracnose: *Colletotrichum lindemuthianum* (Sacc. & Magnus) Lams.-Scrib	*Co-1*	–	1	McRostie (1919) and Kelly et al. (2003)
		Co-1²	SEACT/MCCA	1	Melotto and Kelly (2000) and Vallejo and Kelly (2008)
		Co-1³	–	1	Melotto and Kelly (2000)
		Co-1⁴	–	1	Alzate-Marin et al. (2003a)
		Co-1⁵	–	1	Gonçalves-Vidigal and Kelly (2006)
		Co-2	SCAreoli, H20, SQ4, OQ4	11	Mastenbroek (1960), Geffroy et al. (1998), Adam-Blondon et al. (1994) and Awale et al. (2008)
		Co-3	SW12, W12	–	Bannerot (1965), Miklas et al. (2000b), Singh et al. (2000) and Rodríguez-Suárez et al. (2008)
		Co-3²	–	–	Fouilloux (1979)
		Co-3³	–	–	Geffroy et al. (1999)
		Co-4	SY20, Y20 SC80, C80	8	Bannerot (1965), Queiroz et al. (2004b) and Kelly et al. (2003)
		Co-4²	SAS13, AS13 SH18, H18, SBB14, BB14	8	Young et al. (1998), Kelly et al. (2003) and Awale and Kelly (2001)
		Co-4³	–	–	Alzate-Marin et al. (2002)
		Co-5	SAB3, AB-3	7	Vallejo and Kelly (2001) and Campa et al. (2005)
		Co-5²	–	–	Vallejo and Kelly (2009)
		Co-6	Z20, SZ04₅₆₀, SZ04, Z04	7	Schwartz et al. (1982), Kelly et al. (2003)
		Co-7	–	–	Young et al. (1998)
		co-8	–	–	Alzate-Marin et al. (1997)
		Co-9	SB12, B12	4	Geffroy et al. (1999)
		Co-10	F10, SF10	4	Alzate-Marin et al. (2003b), Corrêa et al. (2000) and Alzate-Marin et al. (2003c)
		Co-11	–	–	Gonçalves-Vidigal et al. (2007)
		Co-12	–	–	Gonçalves-Vidigal et al. (2008)
		Co-13	–	–	Gonçalves-Vidigal et al. (2009)

(continued)

Table 12.2 (Continued) List of Biotic Stress Resistance Genes in Common Bean along with Linked Markers

S.No.	Pest/Pathogen	Gene	Linked/Flanking Molecular Markers	Linkage Group	References for Genes/Markers/LG
2.	Angular Leaf Spot: *Pseudocercospora griseola* pv. *phaseolicola* (Psp)	*Phg-1*	H13, SH13	1	Queiroz et al. (2004a)
		Phg-2	N02, SN02,	8	Nietsche et al. (2000)
			E-ACA/MCTT₃₉₀	–	Mahuku et al. (2004)
		ON	AA19, SAA19, BA16, SBA16, M02, SM02	–	Queiroz et al. (2004a)
3.	Rust: *Uromyces appendiculatus* (Pers.) Unger var. *appendiculatus*	*Ur-1*	–	–	Ballantyne (1978)
		Ur-2	–	–	Ballantyne (1978)
		Ur-2²	–	–	Ballantyne (1978)
		Ur-3	–	11	Ballantyne (1978), Haley et al. (1994), Nemchinova and Stavely (1998) and Miklas et al. (2002)
		Ur-4	K14, SK14	6	Ballantyne (1978), Miklas et al. (1993, 2002) and Meine et al. (2004)
		Ur-5 (Ur5A to Ur5H)	A14, SA14	4	Stavely (1984), Haley et al. (1994), Melotto and Kelly (1998) and Miklas et al. (2000b)
		Ur-6	BC06, SBC06	11	Ballantyne (1978), Park et al. (2003b, 2004b) and Miklas et al. (2002)
		Ur-7	AD12, SAD12	11	Augustin et al. (1972) and Park et al. (2003a, 2004a, 2008)
		Ur-8	–	–	Christ and Groth (1982)
		Ur-9	–	–	Finke et al. (1986)
		Ur-10	–	1	Webster and Ainsworth (1988)
		Ur-11	AE19, SAE19, GT02, UR11-GT02	11	Stavely (1990), Boone et al. (1999), Johnson et al. (1995), Queiroz et al. (2004c) and Miklas et al. (2002)
		Ur-12	–	7	Jung et al. (1998)
		Ur-13	E-AAC/M-AAC, KB126	8	Liebenberg and Pretorius (2004) and Mienie et al. (2005)
		Ur-NO	F10, SF10, BA8, SBA8	4	Corrêa et al. (2000) and Miklas et al. (2002)
		Ur-14	–	–	Souza et al. (2011)
4.	Wilt: *Fusarium oxysporum* f. sp. *phaseoli*	*Fop-1*	–	–	Ribeiro and Hagedorn (1979) and Woo et al. (1996)
		Fop-2	–	–	

	Disease	Gene	Markers		References
Bacterial					
1.	Halo blight: *Pseudomonas syringae*	*Pse-1*	R13, SR13, T8, ST8, H11, SH11	10	Teverson (1991) and Miklas et al. (2009)
		Pse-2	–		Teverson (1991)
		Pse-3	W13, SW13	2	Haley et al. (1994), Melotto et al. (1996), Fourie et al. (2004) and Teverson (1991)
		Pse-4	–	–	Teverson (1991)
		pse-5	–	–	Teverson (1991)
		Un-named	B10, SB10	4	Fourie et al. (2004)
Viral					
1.	Alfalfa mosaic virus	*Amv-1*	–	–	Wade and Zaumeyer (1940)
		Amv-2	–	–	
2.	Bean common mosaic virus	$bc\text{-}1^{1}$	–	–	Drijfhout (1978)
		$bc\text{-}1^{2}$	BD5, SBD5	3	Drijfhout (1978) and Miklas et al. (2000a)
		$bc\text{-}2^{1}$	–	–	Drijfhout (1978)
		$bc\text{-}2^{2}$	–	–	Drijfhout (1978)
		$bc\text{-}3$	C11, OC11, OG6, SG6, eIF4E	6	Drijfhout (1978), Johnson et al. (1997) and Naderpour et al. (2010)
3.	Blackeye cowpea mosaic virus (confers temperature-sensitive resistance)	*Bcm*	*I* gene	–	Provvidenti et al. (1983) and Kyle and Provvidenti (1987)
4.	Beet curly top virus	*Bct*	A08, SAS8	6	Larsen and Miklas (2004) and Mukeshimana et al. (2005)
5.	Bean dwarf mosaic virus	*Bdm*	–	–	Seo et al. (2004)
6.	Bean golden yellow mosaic virus	*Bgm*	$R_2\ SR_2$	3	Velez et al. (1998), Urrea et al. (1996) and Blair et al. (2007)
		bgm-2	–	–	Velez et al. (1998)
		bgm-3			Osorno et al. (2007)
		bgp-2	–	–	Osorno et al. (2007)

(continued)

Table 12.2   (Continued) List of Biotic Stress Resistance Genes in Common Bean along with Linked Markers

S.No.	Pest/Pathogen	Gene	Linked/Flanking Molecular Markers	Linkage Group	References for Genes/Markers/LG
7.	Bean pod mottle virus	Bpm	–	–	Thomas and Zaumeyer (1950)
8.	Bean southern mosaic virus	Bsm	–	–	Zaumeyer and Harter (1943)
9.	Bean yellow mosaic virus	By-1	–	–	Schroeder and Provvidenti (1968)
		By-2	–	–	Dickson and Natti (1968)
10.	Cowpea aphid-borne mosaic virus (temperature-sensitive resistance)	Cam	–	–	Provvidenti et al. (1983) and Kyle and Provvidenti (1987)
11.	Clover yellow vein virus	cvt-1, cvt-2 Cyv	–	–	Schultz and Dean (1947) and Provvidenti and Schroeder (1973)
Nematode					
1.	Meloidogyne incognita	Mel	–	–	Omwega et al. (1990)
2.	M.javanica and M. arenaria	Mel-2	–	–	Omwega and Roberts (1992)
3.	Mosaicorugoso del frijol (bean rusgo mosaic virus)	Mrf	–	–	Machado and Pinchinat (1975)
		Mrf2	–	–	

recessive gene, *co-8* (reviewed by Kelly and Vallejo 2004; Gonçalves-Vidigal et al. 2007, 2008, 2009). Although Geffroy et al. (2000) detected colocalization of resistance genes, they also found resistance gene analogs and defense response genes along with 10 quantitative trait loci for partial resistance against ANT in beans. Among these resistance genes, *Co-2–Co-10* are of Middle American origin and four resistance loci, *Co-1*, *Co-11*, *Co-12*, and *Co-13*, are from the Andean gene pool (Kelly and Vallejo 2004; Gonçalves-Vidigal et al. 2007, 2008, 2009). The gene *Co-9* identified earlier is now identified as an allele of the *Co-3* gene (Mendez de Vigo et al. 2002). *Co-1* is not only the first described disease-resistance gene, but for the first time, genotype by pathogen (different strains of anthracnose) interaction was confirmed for this gene (Kelly and Vallejo 2004). Some of these loci have multiple alleles. The locus *Co-1* has four alleles: *Co-1*, *Co-1²*, *Co-1³*, and *Co-1⁴*. Similarly, *Co-3* has two and *Co-4* has three alleles. The seven resistance loci have been mapped on five linkage groups of the integrated bean linkage map published by Freyre et al. (1998); *Co-1* on B1, *Co-2* on B11, *Co-3* and *Co-10* on B4, *Co-4* on B8, and *Co-5* and *Co-6* on B8 are where most of them are tagged with suitable molecular markers.

Wilt in common bean is another disease caused by *Fusarium oxysporum* f. sp. *Phaseoli*. The genes for two races (Brazilian/race 2 and U.S./race 1) of wilt disease have been identified and named as *Fop-1* and *Fop-2*, respectively (Woo et al. 1996). ALS disease in bean is caused by a biotrophic fungus *Pseudocercospora griseola* (Sacc.) Crous & Braun (sin. *Phaeoisariopsis griseola* (Sacc.)). The pathogen causes necrotic lesions on the aerial parts of the plant, reducing the productivity as well as quality of the bean seeds. This pathogen is highly variable with multiple known races belonging to two gene pools: Mesoamerican and Andean. The isolates of the Mesoamerican gene pool infect the bean cultivars of both Mesoamerican and Andean origin, whereas the Andean isolates infect beans plants of the same origin (Wagara et al. 2004). For ALS, two dominant resistance genes have been described, *Phg-1* and *Phg-2*, which are mapped on LG B1 and B8, respectively (Gonçalves-Vidigal et al. 2011). The *Phg-1* is of Andean origin, identified in the AND 277 variety, and *Phg-2* is of Mesoamerican origin, identified in the Mexico 54 variety. In varieties Ouro Negro and G10474, dominant monogenic inheritance for ALS has been described, but their relation to *Phg-1* and *Phg-2* is not known (Mahuku et al. 2004). Later, by allelism test, three more genes, *Phg*-3, *Phg*-4, and *Phg*-5, with two alleles of each controlling resistance in four bean cultivars, AND 277, Mexico 54, MAR 2, and Cornell 49-242, were identified (Caixeta et al. 2005; Mahuku et al. 2004). In addition to qualitative resistance genes, there are also reports of QTLs controlling resistance to ALS. Bean rust, another major disease, is caused by a highly variable seed-borne macrocyclic fungus, *Uromyces appendiculatus* (Pers.) Unger var. *appendiculatus*. Fourteen genes (*Ur-1–Ur-14*) are known for providing resistance against this disease (Souza et al. 2011). *Ur-2²* is a resistance allele located on the *Ur-2* locus. The resistance locus *Ur-5* is a cluster of eight tightly linked genes, *Ur-5A–Ur-5H*, detected in a rust differential variety, Mexico 309, which is of Middle American origin (Stavely 1984). All of these are specific resistance genes, except *Ur-12*, which is identified as an adult plant resistance gene by Jung et al. (1998). The resistance genes *Ur-1*, *Ur-2*, *Ur-3*, *Ur-5*, *Ur-7*, *Ur-11*, *Ur-13*, and *Ur-14* are of Middle American origin, and the rest are from the Andean gene pool. The resistance genes of Andean origin are susceptible to Andean races of pathogen but very effective against many Middle American races. Meanwhile, the resistance genes of Middle American origin have broad resistance particularly to most Andean races, but are susceptible to many Middle American races (Pastor-Corrales 2006). *Ur-11* is considered the most effective gene, resistant to almost all known races of rust pathogen, except race 108, which is of Middle American origin. *Ur-3* and *Ur-5* rust resistance genes are susceptible to several Middle American races of the rust fungus, but together they show resistance to most of the Andean races (Beaver et al. 2003). Therefore, combining selected rust-resistance genes (gene pyramiding) from Andean and Middle American gene pools could result in bean cultivars with effective and durable rust resistance throughout the world. Halo blight (HB) is a serious seed-borne bacterial disease of bean caused by *Pseudomonas syringae* pv. *phaseolicola* (Psp) (Saettler 1991; Young et al. 1998).

Five monogenic genes (*Pse-1*, *Pse-2*, *Pse-3*, *Pse-4*, and *pse-5*) conditioned resistance against nine known races of HB (Teverson 1991; Taylor et al. 1996). The *Pse-1* and *Pse-4* genes are supposed to be from cultivar UI-3, are located on LG B4, and conditioned resistance against Races 1, 7, and 9 and 5, respectively. *Pse-2* is located on LG B10 and has confirmed resistance for Races 2, 3, 4, 5, 7, and 9. The genes *Pse-3* and *Pse-5* both conditioned resistance to Races 3 and 4. The molecular markers have been identified for three of these genes: *Pse-1*, *Pse-2*, and *Pse-3* (Fourie et al. 2004; Haley et al. 1994; Melotto et al. 1996; Miklas et al. 2009).

The virus diseases caused by BCMV and bean common mosaic necrotic virus (BCMNV) are very serious, causing considerable yield loss worldwide (McKern et al. 1992). A major gene, *I*, along with a series of independent multiallelic loci, is involved in resistance to both of these potyviruses. The three strain-specific recessive *bc* genes, *bc1* (LG3), *bc2¹* or *bc2²*, and *bc3* (LG6), with the strain nonspecific *bc-u* gene, give resistance to different strains of BCMV (Drijhout 1978). The gene *bc-3* provide resistance to all strains of BCMV (Drijfhout 1978). The three recessive loci that prevent chlorotic responses caused by BCMNV are *bgm*, *bgm1*, and *bgm2* (Velez et al. 1998; Osorno et al. 2007). Another gene, *Bgp-2*, prevents pod deformation in the presence of BGYMV (Osorno et al. 2007). Many other genes or loci that have been identified for resistance against different viruses are listed in Table 12.2. The *I* gene, or the complex I region located at the terminal position on LG2, conditions resistance to a set of nine potyviruses: BCMV, WMV, BlCMV, CAbMV, AzMV, ThPV, SMV, PWV-K, and ZYMV (Fisher and Kyle 1994; Vallejos et al. 2000).

12.2.2.2 Identification of Linked Markers and Molecular Breeding for Introgression of Resistance Gene(s)

Most of the genes identified for different stresses are tightly linked with molecular markers. Many of them are tagged with RAPD markers from which SCAR markers are developed. Miklas et al. (1993) identified RAPD marker $A14_{(1100)}$ linked with rust resistance gene *Ur-4* for the first time, and this was later used to pyramid gene *Ur-4* with gene *Ur-11*n (Stavely et al. 1994). Since then, many of the resistance genes have been tagged, as shown in Table 12.2. Some of the genes are also tagged with other markers such as *Ur-13* and *bc-c* tagged with an AFPL marker and an ALS resistance allele tagged with a SSR marker PV-$act001_{282C}$ (Mienie et al. 2005; Mukeshimana et al. 2005; Silva et al. 2003). Molecular markers linked to genes controlling resistance to different races of a pathogen or to different pathogens can be used to pyramid the different genes to enhance disease resistance. Oliveria et al. (2002) used a RAPD marker and introgressed ALS resistance genes into bean cultivars. Miklas et al. (2003) successfully introgressed *Co-4⁴* from the unadapted SEL1308 breeding line into the adapted pinto bean breeding line USPTANT-1 through marker-assisted back crossing. Similarly, Ragagnin et al. (2003) enhanced resistance to anthracnose in the cultivar Perola via marker-assisted selection (MAS). Souza et al. (2009) developed bean lines with durable resistance spectrum to rust (*Ur-5*, *Ur-11*, and *Ur-14*) along with high productivity to recurrent parent through a gene pyramiding approach assisted by molecular markers. With the help of molecular marker–assisted breeding, pyramided lines have been obtained with resistance alleles to several pathogens (Ragagnin et al. 2005, 2009; Costa et al. 2010; Rocha et al. 2012). Figure 12.2 represents the strategy of marker-assisted backcrossing for introgression of the desirable genes in cultivars of choice.

In the common bean genome, the disease-resistance genes are clustered at many locations, and the appropriate linkage phase is present between marker and gene loci. Therefore, simultaneous selection for different disease resistances using the same markers is possible. In such locations, the majority of the markers originated from the resistant parent and are linked in the coupling phase with QTL or major genes for disease resistance. There is some information available on linkages among genes controlling resistance to different pathogens in the common bean. The ANT resistance *Co*-genes mapped on linkage groups B1, B4, B7, and B11 are clustered with the *Ur*-genes for rust

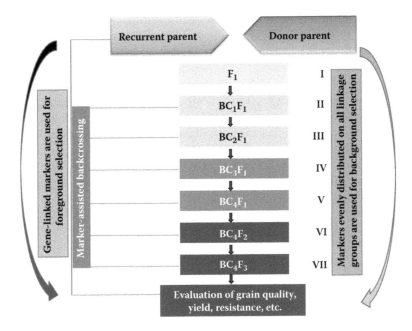

Figure 12.2 Marker-assisted backcrossing (MABC) for introgression of desirable gene(s) in the common bean. Through MABC, the desirable genes can be transferred from the same gene pool. This ensures introgression of the relevant gene without any linkage drag by using molecular markers for both background as well as foreground selection.

resistance (Kelly et al. 2003; Miklas et al. 2002). The colocalization of QTL on LG B4 along with partial resistance provided by the *Co* gene on B4 leads to specific resistance against ANT (Geffroy et al. 2000). The cluster on LG B4 has genes and QTLs responsible for conditioning resistance to ANT, ALS, BGYMV, rust, and so on. Similarly, the resistance gene cluster on LG B7 conditions resistance to anthracnose, bean golden yellow mosaic virus (BGYMV), common bacterial blight (CBB), macrophomina, white mold, and so on. The *I* gene, as mentioned earlier, is a complex loci conditioning resistance to multiple potyviruses.

Since synteny exists between genomes of different legumes (Choi et al. 2004; Hougaard et al. 2008; McClean et al. 2010), this can be used for bean improvement programs. Recently, Gonçalves-Vidigal et al. (2011) revealed the cosegregation of the resistances to ANT and ALS simultaneously with Races 73 of *C. lindemuthianum* and 63-23 of *P. griseola*. Those authors identified two new molecular markers, CV542014[450] and TGA1.1[570], tagging the *Co-1[4]* and *Phg-1* loci, which are physically linked in *cis* configuration. These two markers are basically discovered through synteny mapping between the common bean and soybean. Therefore, synteny mapping illustrates the possibility of pyramiding of these two disease-resistance genes within a short time and at reduced cost. In some cases, where molecular breeding is not possible, transferring of gene(s) from other sources to provide resistance can be considered. In case of abiotic stress resistance, transgenic breeding is an option.

12.2.3 Transgenics and Transgenic Breeding: An Option

Different breeding approaches have significantly contributed to the improvement of the common bean for a range of desirable traits. But through conventional breeding it is not possible to introgress those genes that do not exist naturally in the bean gene pool, that are tightly linked with some undesirable trait, or that have a desirable trait with an unknown underlying physiology. In such cases, genetic transformation comes into play. Although many countries are still against the cultivation of genetically modified crops, the fact is that the number of biotech crop hectares has increased by an unprecedented hundredfold from 1.7 million ha in 1996 to over 170 million ha in 2012 (http://isaaa. org/). The basic strategies behind transgenic development are introduction of modified endogenous genes or foreign genes and overexpression or suppression (antisense or cosuppression) of endogenous genes. It is now possible to develop genetically engineered common bean plants. However, it is cumbersome, and the rate of recovery of transgenic lines is still limited, because of the lack of availability of efficient protocols for regeneration of transformed undifferentiated tissue.

The potential benefits for transgenic beans range from resistance to pesticides or herbicides (Aragao et al. 2002; Kwapata et al. 2012) to enhanced tolerance to biotic/abiotic stresses (Aragao et al. 1998; Faria et al. 2006; Kwapata et al. 2012; Liu et al. 2005). Through transgenic breeding of *P. vulgaris*, nutritional properties have also been enhanced. The methionine content in transgenic beans has been improved and some antinutritional factors like phytate and raffinose-family oligosaccharides have been reduced (Aragao et al. 1998). Aragao et al. (1998) introduced 2S-albumin genes (*be2s1*) to Brazil nuts (*Bertholletia excels* H.B.K.), which increased the methionine content of transgenic plants to 23%. In most studies, *Agrobacterium tumefaciens*-mediated and gene gun microprojectile bombardment/biolistic transformation methods have been used for development of transgenics. Recently, an ultrasonic and vacuum infiltration–assisted *Agrobacterium*-mediated transformation technique has been used to investigate transgenic efficiency and the function of the introduced foreign gene in *P. vulgaris* (Liu et al. 2005). This method is comparatively cheap and increased the transformation efficiency. The *Agrobacterium rhizogenes*–mediated root transformation for development of composite *P. vulgaris* can be used as a powerful transformation tool to obtain valuable information on gene expression and putative gene function, particularly functional genomics based on root biology and root–microbe interaction (Colpaert et al. 2008). The first attempt at stable genetic transformation in the common bean was carried out by Russell et al. (1993) via an electric discharge–mediated particle acceleration device. The method was not only time consuming and variety specific, but gave a low frequency of transformation of only 0.03%. These authors introduced the *bar* gene and coat protein gene from the bean golden mosaic gemini viruses (BGMV). The bar gene encodes phosphinothricin acetyl transferase (PAT), which confers resistance against phosphinothricin and glufosinate ammonia in glasshouse conditions; however, coat protein did not exhibit virus resistance. Later, Aragao et al. (2002) developed and field tested transgenic bean lines having the *bar* gene, giving resistance to glufosinate ammonium. This is considered the first field-release transgenic line of *P. vulgaris*. Aragao et al. (1998) used antisense RNA strategy to generate transgenic bean lines via a biolistic method that showed delayed and attenuated symptoms to the bean golden mosaic gemini virus (BGMV-BR). In that study, a construct was developed with the genes *Rep-TrAP-REn* (coat protein [*AV1*], *Rep* [*AC1*], *TrAP* [*AC2*], and *REn* [*AC3*]) and *BC1* (*BV1* and *BC1*) from the Brazilian isolate bean golden mosaic gemini virus (BGMV-BR), cloned in antisense orientation under the transcriptional control of the CaMV 35S promoter. Similarly, Faria et al. (2006) transformed common beans with a construct of the *rep* gene (replication-associated protein, REP) from BGMV mutated at amino acid 262, through the replacement of an aspartic acid by an arginine (D262R) to evaluate its resistance to BGMV. The RNAi-hairpin construct was used for silencing of the *AC1* region of the viral genome of the BGMV by Bonfim et al. (2007).

Among abiotic stresses, drought is a damaging factor and is responsible for substantial yield loss of *P. vulgaris*. Under drought conditions, plants lose their cellular turbidity, causing protein aggregation and misfolding (Zhu 2002). The stress-tolerant genes encode late embryogenesis abundant proteins (LEA), a class of heat shock proteins. Through an ultrasonic plus vacuum infiltration–assisted *Agrobacterium*-mediated transformation technique, the *Brassica napus* group 3 *lea* gene was introduced into a local cultivar of Japan common bean cv. Green Light (Liu et al. 2005). The transgenic lines showed enhanced growth ability under salt and water deficit stress conditions. The stress tolerance power was more highly correlated at expression level than gene integration level because transgenic lines with high levels of *lea* gene expression showed higher tolerance than lines with lower expression level. Kwapata et al. (2012) transferred the *Hordeum vulgare* LEA protein type III encoding gene (*hva1*) into different varieties of *P. vulgaris* and reported development of drought-tolerant transgenic plants. Hence, transgenics and transgenic breeding are an option for improvement of the common bean.

12.3 PROTEOMICS APPROACH

Proteomics is an analytical and technical approach to study the structural and functional determinants of cells. The word "proteome" is a fusion of "protein" and "genome" and was coined by Marc Wilkins in 1994 (Wilkins et al. 1995). The proteome is the entire complement of proteins, which are fundamental components of the physiological and metabolic pathways that are vital for the growth, development, and interaction of living organisms with their environment. It varies with time, among tissues, and in response to stimuli or stresses that a cell or organism undergoes. The genome sequence does not show how proteins function or biological processes occur. In general, proteomic approaches can be used (1) for proteome profiling to understand how proteins execute biological processes, (2) for comparative expression analysis of two or more protein samples, where proteins may fold into specific 3-D structures that determine function, (3) for understanding of posttranslational modifications (PTMs), which are regulated after synthesis and are major regulatory mechanisms controlling many basic cellular processes, (4) for the study of protein–protein interactions, and (5) for identifying new biomarkers to detect and monitor specific stress manifestations (Chandramouli and Qian 2009). With great effort, proteome coverage has steadily increased over the past decade, especially at the technical level, in the plant proteomics field (readers are referred to further study in Agrawal and Rakwal 2008; Agrawal et al. 2011).

12.3.1 Need for Proteomics Study: Major Focus on Signaling against Stress

As discussed above, genomics-based investigation can only predict the expression of a protein; information about its location, function, and so on cannot be revealed. Proteomics technology has been applied to study the cellular and subcellular components of plant tissues (Agrawal and Rakwal 2008). Moreover, due to a large number of PTMs, the product of a gene may be totally different from the predictions based on genomics studies. Hence, proteomics studies are important to resolve this issue. It is worth mentioning that crops are exposed to the elements of nature, many of them unfavorable, and it is the protein that can induce tolerance in plants against a stress. The sensing of stress signals and their transduction into appropriate responses is crucial for the adaptation and survival of plants. Plants have evolved signaling networks that rely on PTMs of their components and these PTMs and protein–protein interactions can be studied by proteomics. Keeping this in mind, we discuss here the role of proteomics in elucidating signaling pathways in plants, so that similar approaches can be adopted in common beans for improvement against various stresses. A proteomics study of the mung bean epicotyl was carried out by Huang et al. (2006). The regulatory relationship between the steroid hormone brassinosteroid (BR) and chilling stress was investigated and possible mechanisms were elucidated. Protein phosphorylation has a prominent role in cellular processes and much research in plants has focused on both protein kinases and phosphatases, which, respectively, catalyze phosphorylation and dephosphorylation of specific substrates (Kersten et al. 2009). Protein phosphorylation has been shown to play an important role in BR signal transduction in *Arabidopsis*, where signal-induced phosphorylation and dephosphorylation of two known BR signaling proteins, BAK1 and BZR1, were identified using prefractionation and two-dimensional difference in gel electrophoresis (2-D DIGE) (Tang et al. 2008). Kinase cascades of the mitogen-activated protein kinase (MAPK) class play a remarkably important role in plant signaling of a variety of abiotic and biotic stresses. A few kinase-mediated signaling pathways have been elucidated in the model plants *Arabidopsis thaliana* (van Bentem and Hirt 2007; Pitzschke et al. 2009) and rice (Chen and Ronald 2011). Kinase pathways have been demonstrated to become active within minutes—in some cases seconds—of a stimulus in crop plants (Nirmala et al. 2010; Schulze et al. 2010), and, furthermore, in most cases the signal is transient and weak, presenting challenges to detection and quantification. Phosphorylation cross-talk with PTMs and hormone signals also occurs, which further increases the complexity of understanding and ultimately manipulates this signal. Several research groups have deciphered the involvement of different kinases such as cyclin-dependent protein kinase (CDPKs) (Wan et al. 2009), CBL-interacting protein kinase (CIPKs) (Batistic and Kudla 2009; Luan 2009), and MAPKs (Lee et al. 2008; Popescu et al. 2009) in abiotic and biotic stress signaling networks. Phosphoproteins were reviewed in detail to understand their role in transmitting stress signals from the cell surface to the nucleus, so that an appropriate response is elicited (Rampitsch and Bykova 2012a). With the advent of proteomics, it has become evident that when plants are stressed there are changes in their proteome as they attempt to overcome the stress and maintain homeostasis (Kosová et al. 2011; Rampitsch and Bykova 2012b). In a recent study, an approach integrating genetics with phosphoproteomics was used to identify multiple components of the ABA-responsive protein phosphorylation network in *Arabidopsis* (Umezawa et al. 2013). Turning this knowledge into a tool to create improved crop varieties (i.e., a selectable marker) is the ultimate goal.

12.3.2 Plant Proteomics: Technological Update

Proteome analyses have become a powerful tool for the investigation of complex cellular processes (Lottspeich 1999; Görg et al. 2000) and have also been successfully used for genetic and physiological studies in plants (Yun et al. 2012; Ren et al. 2009; Thiellement et al. 1999) including the common bean (De La Fuente et al. 2011). The key techniques of proteomics include gel-based and gel-free approaches, which have been reviewed by (Chandramouli and Qian 2009). Gel-based applications include 1-D and 2-D polyacrylamide gel electrophoresis (2-DGE) and gel-free high-throughput screening technologies include multidimensional protein identification technology (MudPit), isotope-coded affinity tag (ICAT), stable isotope labeling by amino acids in cell culture (SILAC), and isobaric tagging for relative and absolute quantitation (iTRAQ). Shotgun proteomics and 2-D DIGE, as well as protein microarrays, can be applied to obtain overviews of protein expression in cells, tissues, and organelles. Large-scale Western blot assays, multiple reaction monitoring assay (MRM), and label-free quantification of high-mass-resolution LC-MS data are being explored for high-throughput analysis (Chandramouli and Qian 2009). 3-DGE has also been introduced (Ventzki and Stegemann 2003). Other techniques include reverse-phase high-performance liquid chromatography (RP-HPLC), surface-enhanced laser desorption/ionization (SELDI) protein chip, and tandem affinity purification (TAP) (Ruan et al. 2006). Regardless of the proteomic separation technique, gel-based or gel-free, mass spectrometry (MS) is always the primary tool for protein identification. Mass spectrometry, such as matrix-assisted laser desorption/ionization-time-of-flight-mass spectrometry (MALDI-TOF-MS) and electrospray ionization tandem mass spectrometry ESI-(MS/MS) are routinely used.

2-DGE with immobilized pH gradients (IPGs) combined with protein identification by MS has been used for proteome analysis. 2-DGE allows for the separation of highly complex mixtures of proteins according to isoelectric point (pI), molecular mass (Mr), solubility, and relative abundance and delivers a map of intact proteins, which reflects changes in protein expression level, isoforms, and PTMs. 2-DGE can resolve more than 5000 proteins simultaneously (~2000 proteins routinely) and can detect and quantify <1 ng of protein per spot (Weiss and Görg 2007; Görg et al. 2009). Recently, new methods have been developed for separation of peptides that is helpful in the identification of low-abundance proteins. These methods include OFFGEL fractionation, 2-D liquid chromatography (2D-LC), and long-column method. In a recent study by Fukao et al. (2013), these three different methods (OFFGEL electrophoresis, 2D-LC, and the long monolithic silica-C18 capillary column method) of peptide separation allowed the identification of 1132, 836,and 795 proteins using 5–10 µg of proteins from protoplasts, respectively. On the other hand, only 140 proteins were identified without any pretreatments prior to LC-MS analysis. Therefore, these results indicate that peptide fractionation prior to LC-MS analysis is highly effective for the identification of a much higher number of proteins. One major challenge, however, remains, and that is the need for advanced automated peptide purification systems with high precision and greater reproducibility (Zargar et al. forthcoming).

12.3.3 Proteomics Studies in Beans: A Few Examples

To the best of our knowledge, some selected proteome-wide studies have been done on the common bean. Considering the vast potential of plant proteomics, this section will help to understand the gaps of using this approach in common beans and to formulate strategies to bridge them.

Lum et al. (2005) investigated the downstream signaling pathways of nitric oxide (NO) in plants using a proteomic approach in *Phaseolus aureus* (mung bean). Comparative 2-DGE analysis revealed seven downregulated and two upregulated proteins identified after treatment with 0.5 mM sodium nitroprusside (SNP) for 6 h. The identities of these proteins were analyzed by a combination of peptide mass fingerprinting and postsource decay using a MALDI-TOF mass spectrometer. Six out of these nine proteins found were involved in either photosynthesis or cellular metabolism. NO^+ rapidly and drastically decrease the amount of glucose in mung bean leaves, suggesting chloroplasts might be one of the main subcellular targets of NO in plants.

In order to improve the understanding of the complex mechanisms involved in the response of the common bean to drought stress, a proteomic approach was used (Zadražnik et al. 2012) to identify drought-responsive proteins in leaves of two cultivars differing in their response to drought. 2-D DIGE was used to compare differences in protein abundance between control and stressed plants. The majority of identified proteins were classified into functional categories that include energy metabolism, photosynthesis, ATP interconversion, protein synthesis and proteolysis, and stress- and defense-related proteins. Marsolais et al. (2010) studied the nutritional quality of the common bean using a proteomic approach. The common bean seed with storage protein deficiency revealed upregulation of sulfur-rich proteins and starch and raffinose metabolic enzymes and downregulation of the secretory pathway. This provided information on the pleiotropic phenotype associated with storage protein deficiency in a dicotyledonous seed. Torres et al. (2007) used a proteomics approach to examine responses of cultivated bean (*P. vulgaris* L. cv. IDIAP R-3) and maize (*Zea mays* L. cv. Guarare 8128) plants to ozone (O_3) stress. Gel-based proteomics revealed a clear modulation of oxidative stress, heat shock, and secondary metabolism-related proteins by O_3. Potential novel protein markers of ozone stress in leaves of cultivated bean and maize species of Panama were identified. Mensack et al. (2010) used three different omics approaches, transcriptomics, proteomics, and metabolomics, to qualitatively evaluate the diversity of the common bean from two centers of domestication (COD). It was observed that all three approaches were able to classify the common bean according to their COD. They concluded that omics platforms significantly contribute not only to the rapid identification of traits of agronomic and nutritional importance, but also in characterization of genetic diversity.

Proteomics provides a much-needed platform to investigate the various regulatory pathways that are involved in stress tolerance, nutritional fortification, and yield improvements in the common bean. Figure 12.3 depicts the proteomics approach involved in identification of novel proteins induced due to tolerance/resistance against a specific stress in crops.

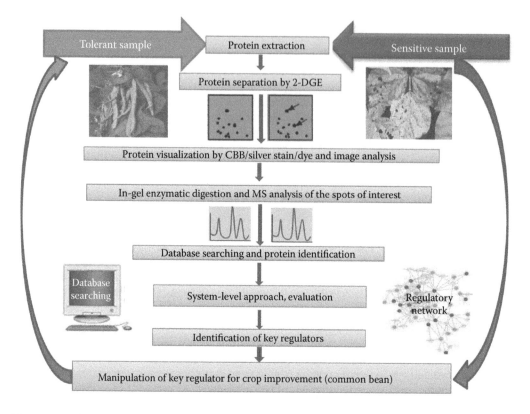

Figure 12.3   (See color insert) Proteomics approach for identification of novel proteins in the common bean. A gel-based proteomics approach can be used to identify the differentially expressed proteins among tolerant and susceptible cultivars. Identification of these novel proteins will help in revealing the metabolic pathways involved in inducing tolerance.

12.3.4 Subproteomic Analyses

Although most of the proteomic studies are based on the separation of protein mixtures and subsequent identification of the resolved proteins by sequencing or mass spectrometry, the resolution capacity of 2-DE is still insufficient to monitor entire protein sets of eukaryotic cells. Hence, proteome research often is based on a subset of proteins of eukaryotic cells called the *subproteome* (Cordewell et al. 2000; Jung et al. 2000). Revealing the subcellular protein localizations is an important step to understand protein function and cell behaviors. Subcellular proteomics has more potential to discover protein functions systematically from spatial and time scales. Salavati et al. (2011) studied the responses of mitochondrial proteins of *P. vulgaris* L. root cells in the early stages of symbiotic interaction with *Rhizobium etli*. The understanding of the subcellular compartment in which a protein is likely to reside can facilitate proteomics analysis and protein isolation experiments (Cui et al. 2011). In plant biology, very successful subproteomic analyses were carried out for the cell wall of *Arabidopsis*, carrot, French bean, tomato, and tobacco (Robertson et al. 1997), the plasma membrane of *Arabidopsis*, *Populus*, and barley embryo (Santoni et al. 1998, 2000; Prime et al. 2000; Nilsson et al. 2010; Hynek et al. 2009), the thylakoids of pea (Peltier et al. 2000; van Wijk 2000), the peroxisome of *Arabidopsis* and spinach (Reumann et al. 2009; Babujee et al. 2010), the mitochondria of *Arabidopsis* and the common bean (Millar et al. 2005), the endoplasmic reticulum of castor bean (Maltman et al. 2002), and the tonoplast of *Arabidopsis* (Shimaoka et al. 2004). During these investigations, hundreds of proteins were separated, several of which were identified for the first time. For a comprehensive review of organelle proteomics, readers are referred to Agrawal et al. (2011). With the improvements in biological mass spectrometry and the rapidly growing protein and DNA databases, the field of proteomics has been revolutionized.

12.4 BEYOND PROTEOMICS APPROACH: NEED FOR ADOPTION OF METABOLOMICS- AND IONOMICS-BASED APPROACHES

Metabolomics plays a major role in the field of plant biology and its application is expanding at a fast pace in crop biotechnology. Metabolomics refers to the comprehensive and simultaneous systematic determination of the complete set of metabolites in a biological system and their changes over time as a consequence of stimuli (F). The term metabolomics was coined in the late 1990s (Fiehn 2002). Various separation techniques that are being used include gas chromatography (GC), capillary electrophoresis (CE), high-performance liquid chromatography (HPLC), and ultraperformance liquid chromatography (UPLC) in combination with the detection techniques that include nuclear magnetic resonance spectroscopy (NMR), and MS. Omics approaches including transcriptomics and metabolomics have been successfully applied to crop systems (Howarth et al. 2008; Wan et al. 2009). A detailed review of the metabolomic approach for crop improvement is available in Saito and Matsuda (2010) and Kusano and Saito (2012).

Nutrient minerals are essential yet potentially toxic, and homeostatic mechanisms are required to regulate their intracellular levels. A network of gene products regulates the overall mechanisms of uptake, binding, transportation, and sequestration of these minerals. Lahner and colleagues first described the ionome to include all the metals, metalloids, and nonmetals present in an organism (Lahner et al. 2003). Ionomics, the study of the ionome, involves the quantitative measurement of the ionome and changes in this composition in response to physiological stimuli, developmental state, and genetic modifications of the living organisms. Ionomic analysis has the potential to provide a powerful approach to the functional analysis of not only the genes and gene networks that directly control the ionome, but also to the more extended gene networks that control developmental and physiological processes that affect the ionome indirectly (Salt et al. 2008). Ionomics provides a rapid way to identify genes that control elemental accumulation in plants, and therefore, this approach has been applied extensively and has been reviewed by Ivan Baxter (2010).

12.5 FUTURE PERSPECTIVES

Living systems are supported and sustained by the genome through the action of the transcriptome, proteome, metabolome, and ionome, the four basic biochemical pillars of functional genomics. These pillars represent the sum of all the expressed genes, proteins, metabolites, and elements within an organism. The dynamic response and interaction of these biochemical "omes" defines how a living system functions, and its study, systems biology, is now one of the biggest challenges in the life sciences. As the majority of genes and gene networks involved in its regulation are still unknown, the study of omics promises to yield new and significant biological insight. The use of these high-throughput technologies in understanding regulatory mechanisms governing various traits is possible. It is therefore vital that scientists bring these technologies into practice for the improvement of the common bean.

12.6 SUMMARY

The common bean (*P. vulgaris* L.) is undoubtedly one of the world's most important food legumes. The bean has high protein content in addition to fiber, carbohydrates, folic acid, iron, magnesium, copper, potassium, and zinc. Today, when more of the world's population is relying on beans to satisfy their basic nutritional needs, per capita consumption has decreased in some countries. This is partly due to the low availability/production of the common bean. Due to a decrease in bean consumption, people in developing countries face an increased likelihood of malnutrition of proteins and minerals like iron and zinc. Hence, there is an urgent need to enhance the yield potential of beans. However, numerous biotic and abiotic factors affect the yield and quality of the common bean. These include some important diseases such as anthracnose, ALS, BCMV, bacterial blight, and root rot. Abiotic factors include drought, low soil fertility, high temperature, and heavy metal stress. Increased yields can be accomplished by analyzing the genetic diversity of the common bean to gain a picture of the broader genetic base, which may help in identifying novel genes that can address these biotic and abiotic stresses. New bean cultivars with higher yields, better quality, multiple disease resistance, and greater tolerance to drought and low soil fertility need to be developed in order to achieve greater yield stability. To do so, sustained, comprehensive, and integrated genetic improvement is required, in which favorable alleles from cultivated and wild relatives are accumulated in superior cultivars. Genetic diversity can very likely provide genes for yield and resistance that are not in the current narrow pool. Molecular markers generating DNA polymorphisms have become the choice for assessing genetic variation and are being employed to carry out phylogenetic analysis. The level of genetic diversity present in the genotypes can be assessed by these methods and the information obtained will aid selection and more efficient utilization of germplasm in its breeding programs. Already markers for some important diseases in the bean are known and it will be practical to introgress these genes into adopted cultivars through marker-assisted backcrossing. In addition, proteomics analysis of the common bean will also be required to examine the proteome profile under stress. 2-DGE coupled with tandem MS as well as the gel-free proteomic approaches can be used to identify novel proteins and consequently screen candidate genes responsible for regulating a particular metabolic pathway under specific conditions. Molecular breeding coupled with high-throughput omics approaches like proteomics will play a vital role in the improvement of the common bean with the aim of catering to the increased global demand for the common bean.

ACKNOWLEDGMENTS

SMZ acknowledges the financial support of SERB-DST, New Delhi (Government of India) for carrying out genetic diversity studies in the common bean. SMZ also acknowledges the DBT, New Delhi (Government of India), for the award of a CREST overseas fellowship. SMZ is grateful to the Honorable Vice Chancellor, SKUAST-Jammu (Dr. D. K. Arrora), for his unselfish encouragement and provision of necessary facilities. Randeep Rakwal (RR) acknowledges the great support and unselfish encouragement of Professor Yoshihiro Shiraiwa (Provost, Faculty of Life and Environmental Sciences), Professor Koji Nomura (Organization for Educational Initiatives, University of Tsukuba), and Professor Seiji Shioda and Dr. Tetsuo Ogawa (Department of Anatomy I, Showa University School of Medicine) in promoting interdisciplinary research. The authors acknowledge the INPPO (www.inppo.com) platform for the initiative in bringing together scientists of different disciplines to construct this review. Finally, given the vast amount of research in these disciplines and space limitations, many relevant works could not be cited and discussed in this chapter.

REFERENCES

Adam-Blondon, A. F., M. Sevignac, M. Dron et al. 1994. A genetic map of common bean to localize specific resistance genes against anthracnose. *Genome* 37:915–924.

Agrawal, G. K., J. Bourguignon, N. Rolland et al. 2011. Plant organelle proteomics: Collaborating for optimal cell function. *Mass Spec Revs* 30:772–853.

Agrawal, G. K. and R. Rakwal. 2008. *Plant Proteomics: Technologies, Strategies, and Applications*. Wiley: Hoboken, NJ.

Akond, A. S. M. G. M., L. Khandaker, J. Berthold et al. 2011. Anthocyanin, total polyphenols and anti-oxidant activity of common bean. *Am J Food Tech* 6:385–394.

Alzate-Marin, A. L., K. M. Arruda, E. G. De Barros et al. 2003a. Allelism studies for anthracnose resistance genes of common bean cultivar AND277. *Annu Rep Bean Improv Coop* 46:173–174.

Alzate-Marin, A. L., G. S. Baia, T. J. De Paula et al. 1997. Inheritance of anthracnose resistance in common bean differential cultivar AB136. *Plant Dis* 81:996–998.

Alzate-Marin, A. L., M. R. Costa, K. M. Arruda et al. 2003b. Characterization of the anthracnose resistance gene present in Ouro Negro (Honduras 35) common bean cultivar. *Euphytica* 133:165–169.

Alzate-Marin, A. L., M. G. D. Moraissilva, M. A. Moreira et al. 2002. Inheritance of anthracnose resistance genes of common bean cultivar PI 207262. *Annu Rep Bean Improv Coop* 45:112–113.

Alzate-Marin, A. L., M. G. D. Moraissilva, E. J. D. Oliveria et al. 2003c. Identification of the second anthracnose resistant gene present in the common bean cultivar PI 207262. *Annu Rep Bean Improv Coop* 46:177–178.

Appel, L. J., T. J. Moore, E. Obaranek et al. 1997. A clinical trial of the effects of dietary patterns on blood pressure. *N Engl J Med* 336:1117–1124.

Aragao, F. J. L., S. G. Ribeiro, and L. M. G. Barros. 1998. Transgenic beans (*Phaseolus vulgaris*) engineered to express viral antisense RNAs show delayed and attenuated symptoms to bean golden mosaic geminivirus. *Mol Breed* 4:491–499.

Aragao, F. J. L., G. R. Vianna, M. M. C. Albino et al. 2002. Transgenic dry bean tolerant to the herbicide glufosinate ammonium. *Crop Sci* 42:1298–1302.

Arumuganthan, K. and E. Earle. 1991. Nuclear DNA content of some important plant species. *Plant Mol Biol Rep* 9:208–218.

Augustin, E., D. P. Coyne, and M. L. Schuster. 1972. Inheritance of resistance in *Phaseolus vulgaris* to *Uromyces phaseoli typica* Brazilian rust race B11 and of plant habit. *J Am Soc Hort Sci* 97:526–529.

Awale, H., S. M. Ismail, V. A. Vallejo et al. 2008. SQ4 SCAR marker linked to the *Co-2* gene on B11 appears to be linked to the *Ur-11* gene. *Annu Rep Bean Improv Coop* 51:174–175.

Awale, H. E. and J. D. Kelly. 2001. Development of SCAR markers linked to *Co-4²* gene in common bean. *Annu Rep Bean Improv Coop* 44:119–120.

Babujee, L., V. Wurtz, C. Ma et al. 2010. The proteome map of spinach leaf peroxisomes indicates partial compartmentalization of phylloquinone (vitamin K1) biosynthesis in plant peroxisomes. *J Exp Bot* 61:1441–1453.

Ballantyne, B. J. 1978. The genetic basis of resistance to rust caused by *Uromyces appendiculatus* in bean (*Phaseolus vulgaris*). PhD. Thesis, University of Sydney, Australia.

Bannerot, H. 1965. Résults de l'infection d'une collection de haricots par six races physiologiques d'anthracnose. *Ann de Amêlior des Plantes* 15:201–202.

Batistic, O. and J. Kudla. 2009. Plant calcineurin B-like proteins and their interacting protein kinases. *Biochim Biophys Acta* 1793:985–992.

Baxter, I. 2010. Ionomics: The functional genomics of elements. *Brief Funct Genomics* 9:149–156.

Beaver, J. S., J. C. Rosas, J. Myera et al. 2003. Contributions of the bean/cowpea CRSP to cultivar and germplasm development in common bean. *Field Crops Res* 82:87–102.

Bennett, M. D. and I. J. Leitch. 1995. Nuclear DNA amounts in angiosperms. *Ann Bot* 76:113–116.

Biswas, M. S., J. Hassan, and M. M. Hossain. 2010. Assessment of genetic diversity in French bean (*Phaseolus vulgaris* L) based on RAPD marker. *Afr J Biotech* 9:5073–5077.

Blair, M. W., A. J. Cortes, R. V. Penmetsa et al. 2013. A high-throughput SNP marker system for parental polymorphism screening and diversity analysis in common bean (*Phaseolus vulgaris* L.). *Theor Appl Genet* 126:535–548.

Blair, M. W., L. M. Rodriguez, F. Pedraza et al. 2007. Genetic mapping of the bean golden yellow mosaic geminivirus resistance gene *bgm-1* and linkage with potyvirus resistance in common bean (*Phaseolus vulgaris* L.). *Theor Appl Genet* 114:261–271.

Bonfim, K., J. C. Faria, E. O. P. L. Nogueira et al. 2007. RNAi-mediated resistance to bean golden mosaic virus in genetically engineered common bean (*Phaseolus vulgaris*). *Mol Plant Microbe Interact* 20:717–726.

Boone, W. E., J. R. Stavely, and N. F. Weeden. 1999. Development of a sequence-tagged site (STS) marker for *Ur-11*, a gene conferring resistance to the bean rust fungus, *Uromyces appendiculatus*. *Annu Rep Bean Improv Coop* 42:33–34.

Botstein, D., R. L. White, M. Skolnick et al. 1980. Construction of a genetic linkage map in man using restriction fragment length polymorphism. *Am J Hum Genet* 32:314–331.

Brinz, B., M. W. Blair, A. Kilian et al. 2012. A whole genome DArT assay to access germplasm collection diversity in common beans. *Mol Breed* 30:181–193.

Broughton, W. J., G. Hernandez, M. Blair et al. 2003. Bean (*Phaseolus* spp.)-model food legumes. *Plant Soil* 252:55–128.

Brown, K. H. and J. M. Peerson. 2001. The importance of zinc in human nutrition and estimation of global prevalence of zinc deficiency. *Food Nut Bull* 22:113–125.

Burstin, J., K. Gallardo, R. R. Mir, R. K. Varshney, and G. Duc (n.d.). Improving protein content and nutrition quality. ICRISAT. http://oar.icrisat.org/61/1/ChapFood-v6__2_.pdf (accessed 9th December 2012).

Caixeta, E. T., A. Borém, A. L. Azate-Marin et al. 2005. Allelic relationships for genes that confer resistance to angular leaf spot in common bean. *Euphytica* 145:237–245.

Campa, A., C. Rodríguez-Suárez, A. Pañeda et al. 2005. The bean anthracnose resistance gene *Co-5*, is located in linkage group B7. *Annu Rep Bean Improv Coop* 48:68–69.

Chandramouli, K. and P. Y. Qian. 2009. Proteomics: Challenges, techniques and possibilities to overcome biological sample complexity. *Hum Genomics Proteomics* 8:23920.

Chen, X. and P. C. Ronald. 2011. Innate immunity in rice. *Trends Plant Sci* 16:451–459.

Choi, H. K., J. H. Mun, D. J. Kim et al. 2004. Estimating genome conservation between crop and model legume species. *Proc Natl Acad Sci USA* 101:15289–15294.

Christ, B. J. and J. V. Groth. 1982. Inheritance of resistance in three cultivars of beans to the bean rust pathogen and the interaction of virulence and resistance genes. *Phytopathology* 72:771–773.

Colpaert, N., S. Tilleman, M. van Montagu et al. 2008. Composite *Phaseoulus vulgaris* plants with transgenic roots as research tool. *Afr J Biotechnol* 7:404–408.

Commodity Online (2010) With demand for pulses exceeding 18-19 million tons and supply short falls of 4-5 million tons, the Indian government is mulling to import around 9,00,000 tons of the legume. http://www.commodityonline.com/news/Indian-export-embargo-on-pulses-may-continue-33976-3-1.html (accessed 8th January 2013).

Cordewell, S. J., A. S. Nouwens, N. M. Verrills et al. 2000. Sub proteomics based upon protein cellular location and relative solubilities in conjunction with composite two-dimensional electrophoresis. *Electrophoresis* 21:1094–1103.

Corrêa, R. X., M. R. Costa, P. I. Good-Godra et al. 2000. Sequence characterized amplified regions linked to rust resistance genes in the common bean. *Crop Sci* 40:804–807.

Cortex, A. J., M. C. Chavarro, and M. W. Blair. 2011. SNP marker diversity in common bean (*Phaseolus vulgaris* L.). *Theor Appl Genet* 123:827–845.

Costa, M. R., J. P. M. Tanure, K. M. A. Arruda et al. 2010. Development and characterization of common black bean lines resistant to anthracnose, rust and angular leaf spot in Brazil. *Euphytica* 176:149–156.

Cui, J., J. Liu, Y. Li et al. 2011. Integrative identification of *Arabidopsis* mitochondrial proteome and its function exploitation through protein interaction network. *PLoS ONE* 6:e16022.

De La Fuente, M., A. Borrajo, J. Bermúdez et al. 2011. 2-DE-based proteomic analysis of common bean (*Phaseolus vulgaris* L.) seeds. *J Proteomics* 74:262–267.

Dickson, M. H. and J. J. Natti. 1968. Inheritance of resistance of *Phaseolus vulgaris* to bean yellow mosaic virus. *Phytopathology* 58:1450.

Doyle, J. J. 2001. Leguminosae. In Brenner, S. and Miller, J. H. (eds) *Encyclopedia of Genetics*, pp. 1081–1085. Academic Press: San Diego.

Dr. Decuypere. n.d. Dr. Decuypere's nutrient charts: Legumes chart. http://www.health-alternatives.com/legumes-nutrition-chart.html (accessed 20th November 2012).

Drijfhout, E. 1978. *Genetic Interaction between Phaseolus vulgaris and Bean Common Mosaic Virus with Implications for Strain Identification and Breeding for Resistance*. Centre for Agricultural Publishing and Documentation. Issue 872 of Agricultural Research Reports.

Duranti, M. 2006. Grain legume proteins and nutraceutical properties. *Fitoterapia* 77:67–82.

Faria, J. C., M. M. C. Albino, B. B. A. Dias et al. 2006. Partial resistance to bean golden mosaic virus in a transgenic common bean (*Phaseolus vulgaris*) line expressing a mutant *rep* gene. *Plant Sci* 171:565–571.

Fatfree Archive. n.d. Nutritional information for adzuki. http://www.fatfree.com/foodweb/nutrition/nadzuki.html (accessed 26th December 2012).

Fiehn, O. 2002. Metabolomics—The link between genotypes and phenotypes. *Plant Mol Biol* 48:155–171.

Finke, M. L., D. P. Coney, and J. R. Steadman. 1986. The inheritance and association of resistance to rust, common bacterial blight, plant habit and foliar abnormalities in *Phaseolus vulgaris* L. *Euphytica* 35:969–982.

Fisher, M. L. and M. M. Kyle. 1994. Inheritance of resistance to potyviruses in *Phaseolus vulgaris* L. III. Cosegregation of phenotypically similar dominant responses to nine potyviruses. *Theor Appl Genet* 89:818–823.

Fouilloux, G. 1979. New races of bean anthracnose and consequences on our breeding programs. In H. Maraite and J. A. Meyer (eds), *Diseases of Tropical Food Crops*, pp. 221–235. Universite Catholique de Louvain-la-Neuve: Belgium.

Fourie, D., P. N. Miklas, and H. M. Ariyarathne. 2004. Genes conditioning halo blight resistance to races 1, 7, and 9 occur in a tight cluster. *Annu Rep Bean Improv Coop* 47:103–104.

Freyre, R., P. W. Skroch, V. Geffroy et al. 1998. Towards an integrated linkage map of common bean. 4. Development of a core linkage map and alignment of RFLP maps. *Theor Appl Genet* 97:847–856.

Fukao, Y., M. Yoshida, R. Kurata et al. 2013. Peptide separation methodologies for in-depth proteomics in Arabidopsis. *Plant Cell Physiol* 54:808–815.

Geffroy, V., F. Creusot, J. Falquet et al. 1998. A family of LRR sequences at the *Co-2* locus for anthracnose resistance in *Phaseolus vulgaris* and its potential use in marker-assisted selection. *Theor Appl Genet* 96:494–502.

Geffroy, V., S. Delphine, J. C. F. D. Oliveira et al. 1999. Identification of an ancestral resistance gene cluster involved in the coevolution process between *Phaseolus vulgaris* and its fungal pathogen *Colletotrichum lindemuthianum*. *Mol Plant Microbe Interact* 12:774–784.

Geffroy, V., M. Sévignac, J. De Oliveira et al. 2000. Inheritance of partial resistance against *Colletotrichum lindemuthianum* in *Phaseolus vulgaris* and co-localization of QTL with genes involved in specific resistance. *Mol Plant Microbe Interact* 13:287–296.

Gonçalves-Vidigal, M. C., A. S. Cruz, A. Garcia et al. 2011. Linkage mapping of the *Phg-1* and *Co-14* genes for resistance to angular leaf spot and anthracnose in the common bean cultivar AND 277. *Theor Appl Genet* 122:893–903.

Gonçalves-Vidigal, M. C. and J. D. Kelly. 2006. Inheritance of anthracnose resistance in the common bean cultivar Widusa. *Euphytica* 151:411–419.

Gonçalves-Vidigal, M. C., G. F. Lacanallo, and P. S. Vidigal Filho. 2008. A new gene conferring resistance to anthracnose in Andean common bean (*Phaseolus vulgaris* L.) cultivar Jalo in specific resistance. *Mol Plant Microbe Interact* 13:287–296.

Gonçalves-Vidigal, M. C., C. R. D. Silva, P. S. Vidigal-Filho et al. 2007. Allelic relationships of anthracnose (*Colletotrichum lindemuthianum*) resistance in the common bean (*Phaseolus vulgaris* L.) cultivar Michelite and the proposal of a new anthracnose resistance gene, *Co-11*. *Genet Mol Biol* 30:589–593.

Gonçalves-Vidigal, M. C., P. S. Vidigal Filho, A. F. Medeiros et al. 2009. Common bean landrace Jalo Listras Pretas is the source of a new Andean anthracnose resistance gene. *Crop Sci* 49:133–138.

Görg, A., O. Drews, C. Lück et al. 2009. 2-DE with IPGs. *Electrophoresis* 1:122–132.

Görg, A., C. Obermaier, G. Boguth et al. 2000. The current state of two-dimensional electrophoresis with immobilized pH gradients. *Electrophoresis* 21:1037–1105.

Gupta, P. K., J. K. Roy, and M. Prasad. 2001. Single nucleotide polymorphisms: A new paradigm for molecular marker technology and DNA polymorphism detection with emphasis on their use in plants. *Curr Sci* 80:524–535.

Haley, S. D., L. Afanador, and J. D. Kelly. 1994. Identification and application of a random amplified polymorphic DNA marker for the *I* gene (potyvirus resistance) in common bean. *Phytopathology* 84:157–160.

Hougaard, B. K., L. H. Madsen, N. Sandal et al. 2008. Legume anchor markers link syntenic regions between *Phaseolus vulgaris*, *Lotus japonicus*, *Medicago truncatula* and *Arachis*. *Genetics* 179:2299–2312.

Howarth, J. R., S. Parmar, and J. Jones. 2008. Co-ordinated expression of amino acid metabolism in response to N and S deficiency during wheat grain filling. *J Exp Bot* 59:3675–3689.

Huang, B., C. H. Chu, S. L. Chen et al. 2006. A proteomics study of the mung bean epicotyl regulated by brassinosteroids under conditions of chilling stress. *Cell Mol Biol Lett* 11:264–278.

Hynek, R., B. Svensson, O. N. Jensen et al. 2009. The plasma membrane proteome of germinating barley embryos. *Proteomics* 9:3787–3794.

Indobase. n.d. http://nutrition.indobase.com/articles/lentils-nutrition.php (accessed 26th December 2012).

Jaccoud, D., K. Peng, D. Feinstein, and A. Kilian. 2001. Diversity arrays: A solid state technology for sequence information independent genotyping. *Nucleic Acids Res* 29:e25.

Johnson, E., P. N. Miklas, J. R. Stavely et al. 1995. Coupling- and repulsion-phase RAPDs for marker-assisted selection of PI 181996 rust resistance in common bean. *Theor Appl Genet* 90:659–664.

Johnson, W. C., P. Guzman, D. Mandala et al. 1997. Molecular tagging of the *bc-3* gene for introgression into Andean common bean. *Crop Sci* 37:248–254.

Jose, F. C., M. M. S. Mohammed, G. Thomas et al. 2009. Genetic diversity and conservation of common bean (*Phaseolus vulgaris* L., Fabaceae) landraces in Nilgiris. *Curr Sci* 97:227–237.

Jung, E., M. Heller, J. C. Sanchez et al. 2000. Proteomics meets cell biology: The establishment of subcellular proteomes. *Electrophoresis* 21:3369–3377.

Jung, G., D. P. Coyne, J. M. Bokosi et al. 1998. Mapping genes for specific and adult plant resistance to rust and abaxial leaf pubescence and their genetic relationship using random amplified polymorphic DNA (RAPD) markers in common bean. *J Am Soc Hort Sci* 123:859–886.

Kalavacharla, V., L. Zhanji, B. C. Meyers et al. 2011. Identification and analysis of common bean (*Phaseolus vulgaris* L.) transcriptomes by massively parallel pyrosequencing. *BMC Plant Biol* 11:135.

Kelly, J. D., P. Gepts, P. N. Miklas et al. 2003. Tagging and mapping of genes and QTL and molecular marker-assisted selection for traits of economic importance in bean and cowpea. *Field Crops Res* 82:135–154.

Kelly, J. D. and P. N. Miklas. 1999. Marker-assisted selection. In S. Singh (ed.), *Common Bean Improvement in the Twenty-First Century*, pp. 93–123. Kluwer Academic Publishers: The Netherlands.

Kelly, J. D. and V. A. Vallejo. 2004. A comprehensive review of the major genes conditioning resistance to anthracnose in common bean. *Hort Sci* 39:1196–1207.

Kersten, B., G. Agrawal, P. Durek et al. 2009. Plant phosphoproteomics: An update. *Proteomics* 9:964–988.

Kosová, K., P. Vitámvás, I. T. Prášil et al. 2011. Plant proteome changes under abiotic stress contribution of proteomics studies to understanding plant stress response. *J Proteomics* 12:1301–1322.

Kumar, V., S. Sharma, A. K. Sharma et al. 2009. Comparative analysis of diversity based on morpho-agronomic traits and microsatellite markers in common bean. *Euphytica* 170:249–262.

Kusano, M. and K. Saito. 2012. Role of metabolomics in crop improvement. *J Plant Biochem Biotechnol* 21:24–31.

Kwapata, K., T. Nguyen, and M. Sticklen. 2012. Genetic transformation of common bean (*Phaseolus vulgaris* L.) with the gus colormarker, the bar herbicide resistance, and the barley (*Hordeum vulgare*) HVA1 drought tolerance genes. *Int J Agron* 2012:8.

Kyle, M. M. and R. Provvidenti. 1987. Inheritance of resistance to potato y viruses in *Phaseolus vulgaris* L. *Theor Appl Genet* 74:595–600.

Laberge, M. n.d. Molybdenum. http://www.diet.com/g/molybdenum (accessed 26th December 2012).

Lahner, B., J. Gong, M. Mahmoudian et al. 2003. Genomic scale profiling of nutrient and trace elements in *Arabidopsis thaliana*. *Nat Biotechnol* 21:1215–1221.

Larsen, R. C. and P. N. Miklas. 2004. Generation and molecular mapping of a SCAR marker linked with the Bct gene for resistance to beet curly top virus in common bean. *Phytopathology* 94:320–325.

Lee, M. O., K. Cho, S. H. Kim et al. 2008. Novel rice OsSIPK is a multiple stress responsive MAPK family member showing rhythmic expression at mRNA level. *Planta* 227:981–990.

Liebenberg, M. M. and Z. A. Pretorius. 2004. Proposal for designation of a rust resistance gene in the large-seeded cultivar Kranskop. *Annu Rep Bean Improve Coop* 47:255–256.

Liu, Z. C., B. J. Park, A. Kanno et al. 2005. The novel use of a combination of sonication and vacuum infiltration in Agrobacterium-mediated transformation of kidney bean (*Phaseolus vulgaris* L.) with lea gene. *Mol Breed* 16:189–197.

Lottspeich, F. 1999. Proteome analysis: A pathway to the functional analysis of proteins. *Angew Chem Int Ed* 38:2476–2492.

Luan, S. 2009. The CBL-CIPK network in plant calcium signalling. *Trends Plant Sci* 14:37–42.

Lum, H. K., C. H. Lee, Y. K. Butt et al. 2005. Sodium nitroprusside affects the level of photosynthetic enzymes and glucose metabolism in *Phaseolus aureus* (mung bean). *Nitric Oxide* 12:220–230.

Machado, P. F. R. and A. M. Pinchinat. 1975. Herencia de la reacción del frijol común a la infección por el virus del mosaico rugoso. *Turrialba* 25:418–419.

Mahuku, G., C. Montoya, M. A. Henriquez et al. 2004. Inheritance and characterization of angular leaf spot resistance gene present in common bean accession G 10474 and identification of an AFLP marker linked to the resistance gene. *Crop Sci* 44:1817–1824.

Maltman, D. J., W. J. Simon, C. H. Wheeler et al. 2002. Proteomic analysis of the endoplasmic reticulum from developing and germinating seed of castor (*Ricinus communis*). *Electrophoresis* 23:626–639.

Malviya, N. and D. Yadav. 2010. RAPD analysis among pigeon pea [*Cajanus cajan* (L.) Mill sp] cultivars for their genetic diversity. *Genet Eng Biotechnol J* 1:1–9.

Maras, M., J. Sustar-Vozlic, B. Javornik et al. 2008. The efficiency of AFLP and SSR markers in genetic diversity estimation and gene pool classification of common bean (*Phaseolus vulgaris* L.). *Acta Agric Slov* 91:87–96.

Marsolais, F., A. Pajak, F. Yin et al. 2010. Proteomic analysis of common bean seed with storage protein deficiency reveals up-regulation of sulfur-rich proteins and starch and raffinose metabolic enzymes, and down-regulation of the secretory pathway. *J Proteomics* 73:1587–1600.

Martinac, P. n.d. Pinto beans nutrition. SFGate. http://healthyeating.sfgate.com/pinto-beans-nutrition-1547. html (accessed 26th December 2012).

Mastenbroek, C. 1960. A breeding programme for resistance to anthracnose in dry shell haricot beans, based on new gene. *Euphytica* 9:177–184.

McClean, P., P. Kami, and P. Gepts. 2004. Genomics and genetic diversity in common beans. In R. F. Wilson, H. T. Stalker, and E. C. Brumer (eds), *Legumes Crop Genomics*, pp. 61–81. AOCS Press: Champaign, IL.

McClean, P. E., S. Mamidi, M. McConnell et al. 2010. Synteny mapping between common bean and soybean reveals extensive blocks of shared loci. *BMC Genomics* 11:184.

McConnell, M., S. Mamidi, R. Lee et al. 2010. Syntenic relationships among legumes revealed using a gene-based genetic linkage map of common bean (*Phaseolus vulgaris* L.). *Theor Appl Genet* 121:1103–1116.

McKern, N. M., G. I. Mink, O. W. Barnett et al. 1992. Isolates of bean common mosaic virus comprising two distinct potyviruses. *Phytopathology* 82:923–928.

McRostie, G. P. 1919. Inheritance of anthracnose resistance as directed by a cross between a resistant and a susceptible bean. *Pathology* 9:144–148.

Meine, C. M. S., R. Naidoo, and M. M. Liebenberg. 2004. Conversion of the RAPD marker for *Ur-4* to a co-dominant SCAR marker SA141079/800. *Annu Rep Bean Improv Coop* 47:261–262.

Melotto, M., L. Afanador, and J. D. Kelly. 1996. Development of a SCAR marker linked to the I gene in common bean. *Genome* 39:1216–1219.

Melotto, M. and J. D. Kelly. 1998. SCAR markers linked to major disease resistance genes in common bean. *Annu Rept Bean Improv Coop* 41:64–65.

Melotto, M. and J. D. Kelly. 2000. An allelic series at the *Co-1* locus conditioning resistance to anthracnose in common bean of Andean origin. *Euphytica* 116:143–149.

Mendez de Vigo, B., C. Rodriguez, A. Paneda et al. 2002. Development of a SCAR marker linked to *Co-9* in common bean. *Annu Rep Bean Improv Coop* 45:116–117.

Mensack, M. M., V. K. Fitzgerald, E. P. Ryan et al. 2010. Evaluation of diversity among common beans (*Phaseolus vulgaris* L.) from two centers of domestication using "omics" technologies. *BMC Genomics* 11:686.

Messina, M. J. 1999. Legumes and soybeans: Overview of their nutritional profiles and health effects. *Am J Clin Nutr* 70:439s–450s.

Metais, I., C. Aubry, B. Hamon et al. 2000. Description and analysis of genetic diversity between commercial bean lines (*Phaseolus vulgaris* L.). *Theor Appl Genet* 101:1207–1214.

Mienie, C. M. S., M. M. Liebenberg, Z. A. Pretorius et al. 2005. SCAR markers linked to the *Phaseolus vulgaris* rust resistance gene *Ur-13*. *Theor Appl Genet* 111:972–979.

Miklas, P. N., D. Fourie, J. Wagner et al. 2009. Tagging and mapping *Pse-1* gene for resistance to halo blight in common bean host differential cultivar UI-3. *Crop Sci* 49:41–48.

Miklas, P. N., J. D. Kelly, and S. P. Singh. 2003. Registration of anthracnose-resistant pinto bean germplasm line USPT-ANT-1. *Crop Sci* 43:1889–1890.

Miklas, P. N., R. Larsen, K. Victry et al. 2000a. Marker-assisted selection for the *bc-1²* gene for resistance to BCMV and BCMNV in common bean. *Euphytica* 116:211–219.

Miklas, P. N., M. A. Pastor-Corrales, G. Jung et al. 2002. Comprehensive linkage map of bean rust resistance genes. *Annu Rep Bean Improv Coop* 45:125–129.

Miklas, P. N., J. R. Stavely, and J. D. Kelly. 1993. Identification and potential use of a molecular marker for rust resistance in common bean. *Theor Appl Genet* 85:745–749.

Miklas, P. N., V. Stone, M. J. Daly et al. 2000b. Bacterial, fungal, and viral disease resistance loci mapped in a recombinant inbred common bean population ('Dorado'/XAN 176). *J Am Soc Hort Sci* 125:476–481.

Millar, A. H., J. L. Heazlewood, B. K. Kristensen et al. 2005. The plant mitochondrial proteome. *Trends Plant Sci* 10:36–43.

Mukeshimana, G., A. Paneda, C. Rodriguez et al. 2005. Markers linked to the *bc-3* gene conditioning resistance to bean common mosaic potyviruses in common bean. *Euphytica* 144:291–299.

Naderpour, M., O. S. Lund, R. Larsen et al. 2010. Potyviral resistance derived from cultivars of Phaseolus vulgaris carrying *bc-3* is associated with the homozygotic presence of a mutated eIF4E allele. *Mol Plant Pathol* 11:255–263.

Nemchinova, Y. P. and J. R. Stavely. 1998. Development of SCAR primers for the *Ur-3* rust resistance gene in common bean. *Phytopathology* 88:S67.

Nietsche, S., A. Borem, G. A. Carvalho et al. 2000. RAPD and SCAR markers linked to a gene conferring resistance to angular leaf spot in common bean. *J Phytopathol* 148:117–121.

Nilsson, R., K. Bernfur, N. Gustavsson et al. 2010. Proteomics of plasma membranes from poplar trees reveals tissue distribution of transporters, receptors, and proteins in cell wall formation. *Mol Cell Proteomics* 9:368–387.

Nirmala, J., T. Drader, X. Chen et al. 2010. Stem rust spores elicit rapid RPG1 phosphorylation. *Mol Plant Microbe Interact* 23:1635–1642.

Nutrition-and-You. n.d. Peanuts nutrition facts. http://www.nutrition-and-you.com/peanuts.html (accessed 20th November 2012).

Oliveria, E., J. de, A. L. Alzate-Marin, and C. L. P. de Melo. 2002. Backcross assisted by RAPD markers for the introgression of angular leaf spot resistance genes in common bean cultivars. *Annu Rep Bean Improv Coop* 45:142–143.

Ologhobo, A. D. 1980. Biochemical and nutritional studies of cowpea and limabean with particular reference to some inherent antinutritional components. PhD. Thesis, University of Ibadan, Ibadan, Nigeria.

Omwega, C. O. and P. A. Roberts. 1992. Inheritance of resistance to *Meloidogyne* spp. in common bean and the genetic basis of its sensitivity to temperature. *Theor Appl Genet* 83:720–726.

Omwega, C. O., I. J. Thomason, and P. A. Roberts. 1990. A single dominant gene in common bean conferring resistance to three root-knot nematode species. *Phytopathology* 80:745–748.

Osorno, J. M., C. G. Munoz, J. S. Beaver et al. 2007. Two genes from *Phaseolus coccineus* confer resistance to bean golden yellow mosaic virus in common bean. *J Am Soc Hort Sci* 132:530–533.

Park, S. O., D. P. Coyne, J. R. Steadman et al. 2003a. Mapping of the *Ur-7* gene for specific resistance to rust in common bean. *Crop Sci* 43:1470–1476.

Park, S. O., D. P. Coyne, and J. R. Steadman. 2004a. Development of a SCAR marker linked to the *Ur-7* gene in common bean. *Annu Rep Bean Improv Coop* 47:269–270.

Park, S. O., D. P. Coyne, J. R. Steadman et al. 2004b. RAPD and SCAR markers linked to the *Ur-6* Andean gene controlling specific rust resistance in common bean. *Crop Sci* 44:1799–1807.

Park, S. O., K. M. Crosby, D. P. Coyne et al. 2003b. Development of a SCAR marker linked to the *Ur-6* gene for specific rust resistance in common bean. *Annu Rep Bean Improv Coop* 46:189–190.

Park, S. O., J. R. Steadman, D. P. Coyne et al. 2008. Development of a coupling phase SCAR marker linked to the *Ur-7* rust resistance gene and its occurrence in diverse common bean lines. *Crop Sci* 48:357–363.

Pastor-Corrales, M. A. 2006. Diversity of the rust pathogen and common bean guide gene deployment for development of bean cultivars with durable rust resistance. *Annu Rep Bean Improv Coop* 49:51–52.

Peltier, J. B., G. Friso, D. E. Kalume et al. 2000. Proteomics of the chloroplast: Systematic identification and targeting analysis of lumenal and peripheral thylakoid proteins. *Plant Cell Rep* 12:319–341.

Pitzschke, A., A. Schikora, and H. Hirt. 2009. MAPK cascade signalling networks in plant defence. *Curr Opin Plant Biol* 12:421–426.

Popescu, S. C., G. V. Popescu, S. Bachan et al. 2009. MAPK target networks in *Arabidopsis thaliana* revealed using functional protein microarrays. *Genes Dev* 23:80–92.

Prime, T. A., D. J. Sherrier, P. Mahon et al. 2000. A proteomic analysis of organelles from *Arabidopsis thaliana*. *Electrophoresis* 21:3488–3499.

Provvidenti, R., D. Gonsalves, and M. A. Taiwo. 1983. Inheritance of resistance to blackeye cowpea mosaic virus and cowpea aphid-borne mosaic virus in *Phaseolus vulgaris*. *J Hered* 74:60–61.

Provvidenti, R. and W. T. Schroeder. 1973. Resistance in *Phaseolus vulgaris* to the severe strain of bean yellow mosaic virus. *Phytopathology* 63:196–197.

Queiroz, V. T., C. S. Sousa, M. R. Costa et al. 2004a. Development of SCAR markers linked to common bean angular leaf spot resistance genes. *Annu Rep Bean Improv Coop* 47:237–238.

Queiroz, V. T., C. S. Sousa, M. R. Costa et al. 2004b. Development of SCAR markers linked to common bean anthracnose resistance genes *Co-4* and *Co-6*. *Annu Rep Bean Improv Coop* 47:249–250.

Queiroz, V. T., C. S. Sousa, T. L. P. O. Souza et al. 2004c. SCAR marker linked to the common bean rust resistance gene *Ur-11*. *Annu Rep Bean Improv Coop* 47:271–272.

Ragagnin, V. A., A. L. Alzate-Marin, T. L. P. O. Souza et al. 2005. Use of molecular markers to pyramiding multiple genes for resistance to rust, anthracnose and angular leaf spot in the common bean. *Annu Rep Bean Improv Coop* 48:94–95.

Ragagnin, V. A., D. A. Sanglard., T. L. P. O. Souza et al. 2003. Simultaneous transfer of resistance genes for rust, anthracnose, and angular leaf spot to cultivar Perola assisted by molecular markers. *Annu Rev Bean Improv Coop* 46:159–160.

Ragagnin, V. A., T. L. P. O. Souza, D. A. Sanglard et al. 2009. Development and agronomic performance of common bean lines simultaneously resistant to anthracnose, angular leaf spot and rust. *Plant Breed* 128:156–163.

Rampitsch, C. and N. V. Bykova. 2012a. Proteomics and plant disease: Advances in combating a major threat to the global food supply. *Proteomics* 12:673–690.

Rampitsch, C. and N. V. Bykova. 2012b. The beginnings of crop phosphoproteomics: Exploring early warning systems of stress. *Front Plant Sci* 3:144.

Ren, Y., J. Lv, H. Wang et al. 2009. A comparative proteomics approach to detect unintended effects in transgenic *Arabidopsis*. *J Genet Genomics* 36:629–639.

Reumann, S., S. Quan, K. Aung et al. 2009. In-depth proteome analysis of *Arabidopsis* leaf peroxisomes combined with *in vivo* subcellular targeting verification indicates novel metabolic and regulatory functions of peroxisomes. *Plant Physiol* 150:125–143.

Ribeiro, R. L. D. and D. J. Hagedorn. 1979. Inheritance and nature of resistance in beans to *Fusarium oxysporum* f. sp. *phaseoli*. *Phytopathology* 69:859–861.

Robertson, D., G. P. Mitchell, J. S. Gilroy et al. 1997. Differential extraction and protein sequencing reveals major differences in patterns of primary cell wall proteins from higher plants. *J Biol Chem* 272:15841–15848.

Rocha, G. S., L. P. L. Pereira, P. C. S. Carneiro et al. 2012. Common bean breeding for resistance to anthracnose and angular leaf spot assisted by SCAR molecular markers. *Crop Breed Appl Biot* 12:34–42.

Rodríguez-Suárez, C., J. J. Ferreira, A. Campa et al. 2008. Molecular mapping and intra-cluster recombination between anthracnose race-specific resistance genes in the common bean differential cultivars Mexico 222 and Widusa. *Theor Appl Genet* 116:807–814.

Ruan, S. L., H. S. Ma, S. H. Wan et al. 2006. Advances in plant proteomics-I. Key techniques of proteome. *Yi Chuan* 28:1472–1486.

Russell, D., K. Wallace, J. Bathe et al. 1993. Stable transformation of *Phaseolus vulgaris via* electric-discharge mediated particle acceleration. *Plant Cell Rep* 12:165–169.

Saettler, A. W. 1991. Angular leaf spot. In R. Hall (ed.), *Compendium of Bean Diseases*, pp. 15–16. APS Press: St. Paul, MN.

Saito, K. and F. Matsuda. 2010. Metabolomics for functional genomics, systems biology, and biotechnology. *Annu Rev Plant Biol* 61:463–489.

Salavati, A., A. Taleei, A. A. S. Bushehri et al. 2011. Comprehensive expression analysis of time-dependent responses of mitochondrial proteins in *Phaseolus vulgaris* L. root cells to infection with *Rhizobium etli*. *J Plant Sci* 6:155–164.

Salt, D. E., I. Baxter, and B. Lahner. 2008. Ionomics and the study of the plant ionome. *Ann Rev Plant Bio* 59:709–733.

Santoni, V., S. Kieffer, D. Desclaux et al. 2000. Membrane proteomics: Use of additive main effects with multiplicative interaction model to classify plasma membrane proteins according to their solubility and electrophoretic properties. *Electrophoresis* 21:3329–3344.

Santoni, V., D. Rouquié, P. Doumas et al. 1998. Use of a proteome strategy for tagging proteins present at the plasma membrane. *Plant J* 16:633–641.

Schroeder, W. T. and R. Provvidenti. 1968. Resistance of bean (*Phaseolus vulgaris*) to the PV2 strain of bean yellow mosaic virus conditioned by a single dominant gene *By*. *Phytopathology* 58:1710.

Schultz, H. K. and L. L. Dean. 1947. Inheritance of curly top disease reaction in the bean, *Phaseolus vulgaris*. *J Amer Soc Agron* 39:47–51.

Schulze, B., T. Menzel, A. K. Jehle et al. 2010. Rapid heteromerization and phosphorylation of ligand-activated plant transmembrane receptors and their associated kinase BAK1. *J Biol Chem* 285:9444–9451.

Schwartz, H. F., M. A. Pastor-Corrales, and S. P. Singh. 1982. New sources of resistance to anthracnose and angular leaf spot of beans (*Phaseolus vulgaris* L.). *Euphytica* 31:741–754.

Seo, Y. S., P. Gepts, and R. L. Gilbertson. 2004. Genetics of resistance to the geminivirus, bean dwarf mosaic virus, and the role of the hypersensitive response in common bean. *Theor Appl Genet* 108:786–793.

Sharma, S., N. Agarwal, and P. Verma. 2011. Pigeon pea (*Cajanus cajan* L): A hidden treasure of regime nutrition. *J Funct Environ Bot* 1:91–101.

Shimaoka, T., M. Ohnishi, T. Sazuka et al. 2004. Isolation of intact vacuoles and proteomic analysis of tonoplast from suspension-cultured cells of Arabidopsis thaliana. *Plant Cell Physiol* 45:672–683.

Silva, G. F. D., J. B. D. Santos, and M. A. P. Ramalho. 2003. Identification of SSR and RAPD markers linked to a resistance allele for angular leaf spot in the common bean (*Phaseolus vulgaris*) line ESAL 550. *Genet Mol Biol* 26:459–463.

Singh, S. P., F. J. Morales, P. N. Miklas et al. 2000. Selection for bean golden mosaic resistance in intra- and inter-racial bean populations. *Crop Sci* 40:1565–1572.

Song, Z. P., X. Xu, B. Wang et al. 2003. Genetic diversity in the northern most *Oryza rufipogon* populations estimated by SSR markers. *Theor Appl Genet* 107:1492–1499.

Souza, T. L. P. O., S. N. Dessaune, V. A. Ragagnin et al. 2009. Rust resistance gene pyramiding in the common bean assisted by molecular markers. In Plant and Animal Genomes XVII Conference, January 10–14, 2009. San Diego, CA.

Souza, T. L. P. O., S. N. Dessaune, D. A. Sanglard et al. 2011. Characterization of the rust resistance gene present in the common bean cultivar Ouro Negro, the main rust resistance source used in Brazil. *Plant Pathol* 60:839–845.

Stavely, J. R. 1984. Genetics of resistance to *Uromyces phaseoli* in a *Phaseolus vulgaris* line resistant to most races of the pathogen. *Phytopathology* 74:339–344.

Stavely, J. R. 1990. Genetics of rust resistance in *Phaseolus vulgaris* plant introduction PI 181996. *Phytopathology* 80:1056.

Stavely, J. R., J. D. Kelly, and K. F. Grafton. 1994. BelMiDak-rust-resistant navy dry beans germplasm lines. *Hort Sci* 29:709–710.

Tang, W., Z. Deng, J. A. Oses-Prieto et al. 2008. Proteomics studies of brassinosteroid signal transduction using prefractionation and two-dimensional DIGE. *Mol Cell Proteomics* 7:728–738.

Tautz, D. 1989. Hypervariability of simple sequences as a general source for polymorphic DNA markers. *Nucleic Acids Res* 17:6463–6471.

Taylor, J. D., D. M. Teverson, and J. H. C. Davis. 1996. Sources of resistance to *Pseudomonas syringae* pv. *phaseolicola* races in *Phaseolus vulgaris*. *Plant Pathol* 45:479–485.

Teverson, D. M. 1991. Genetics of pathogenicity and resistance in the halo-blight disease of beans in Africa. PhD. Thesis, University of Birmingham, Birmingham.

the full wiki. 2009. Winged bean: Wikis. http://www.thefullwiki.org/Winged_bean (accessed 29th December 2012).

Thibivilliers, S., T. Joshi, K. B. Campbell et al. 2009. Generation of *Phaseolus vulgaris* ESTs and investigation of their regulation upon *Uromycesm pendiculatus* infection. *BMC Plant Biol* 9:46.

Thiellement, H., N. Bahrman, C. Damerval et al. 1999. Proteomics for genetic and physiological studies in plants. *Electrophoresis* 20:2013–2026.

Thomas, H. R. and W. J. Zaumeyer. 1950. Inheritance of symptom expression of pod mottle virus. *Phytopathology* 40:1007–1010.

Torres, N. L., K. Cho, J. Shibato et al. 2007. Gel-based proteomics reveals potential novel protein markers of ozone stress in leaves of cultivated bean and maize species of Panama. *Electrophoresis* 28:4369–4381.

Umezawa, T., N. Sugiyama, F. Takahashi et al. 2013. Genetics and phosphoproteomics reveal a protein phosphorylation network in the abscisic acid signaling pathway in *Arabidopsis thaliana*. *Sci Signal* 6:1–13.

Urrea, C. A., P. N. Miklas, J. S. Beaver et al. 1996. A codominant randomly amplified polymorphic DNA (RAPD) marker useful for indirect selection of BGMV resistance in common bean. *J Am Soc Hort Sci* 121:1035–1039.

Vallejo, V. and J. D. Kelly. 2001. Development of a SCAR marker linked to *Co-5* gene in common bean. *Annu Rep Bean Improv Coop* 44:121–122.

Vallejo, V. A. and J. D. Kelly. 2008. Molecular tagging and genetic characterization of alleles at the *Co-1* anthracnose resistance locus in common bean ICFAI University. *J Genet Evol* 1:7–20.

Vallejo, V. and J. D. Kelly. 2009. New insights into the anthracnose resistance of common bean landrace G 2333. *Open Hort J* 2:29–33.

Vallejos, C. E., J. J. Malandro, K. Sheehy et al. 2000. Detection and cloning of expressed sequences linked to a target gene. *Theor Appl Genet* 101:1109–1113.

van Bentem, S. and H. Hirt. 2007. Using phosphoproteomics to reveal signalling dynamics in plants. *Trends Plant Sci* 12:404–411.

van Wijk, K. J. 2000. Proteomics of the chloroplast: Experimentation and prediction. *Trends Plant Sci* 5:420–425.

Velez, J. J., M. J. Bassett, J. S. Beaver et al. 1998. Inheritance of resistance to bean golden mosaic virus in common bean. *J Am Soc Hort Sci* 123:628–631.

Ventzki, R. and J. Stegemann. 2003. High-throughput separation of DNA and proteins by three-dimensional geometry gel electrophoresis: Feasibility studies. *Electrophoresis* 24:4153–4160.

Virk, P. S., H. J. Newbury, M. T. Jackson et al. 2000. Are mapped or anonymous markers more useful for assessing genetic diversity? *Theor Appl Genet* 100:607–613.

Vos, P., R. Hogers, M. Bleeker et al. 1995. AFLP: A new technique for DNA fingerprinting. *Nucleic Acids Res* 23:4407–4414.

Wade, B. L. and W. J. Zaumeyer. 1940. Genetic studies of resistance to alfalfa mosaic virus and stringiness in *Phaseolus vulgaris*. *J Am Soc Agron* 32:127–134.

Wagara, N., A. W. Mwangombe, J. W. Kimenju et al. 2004. Genetic diversity of *Phaeoisariopsis griseola* in Kenya as revealed by AFLP and group-specific primers. *J Phytopathol* 152:235–242.

Wan, Y., C. Underwood, and G. Toole. 2009. A novel transcriptomic approach to identify candidate genes for grain quality traits in wheat. *Plant Biotechnol J* 7:401–410.

Webster, D. M. N. and P. M. Ainsworth. 1988. Inheritance and stability of a small pustule reaction of snap beans to *Uromyces appendiculatus*. *J Am Soc Hort Sci* 113:938–940.

Weiss, W. and A. Görg. 2007. Two-dimensional electrophoresis for plant proteomics. *Methods Mol Biol* 355:121–143.

Welch, R. M. and R. D. Graham. 1999. A new paradigm for world agriculture: Meeting human needs productive, sustainable, nutritious. *Field Crops Res* 60:1–10.

Welsh, J. and M. McClelland. 1990. Fingerprinting genomes using PCR with arbitrary primers. *Nucleic Acids Res* 18:7213–7218.

Wilkins, M. R., J. C. Sanchez, A. A. Gooley et al. 1995. Progress with proteome projects: Why all proteins expressed by a genome should be identified and how to do it. *Biotechnol Genet Eng Rev* 13:S19–S50.

Williams, J. G. K., A. R. Kubelik, K. J. Livak et al. 1990. DNA polymorphism amplified by arbitrary primers as useful genetic markers. *Nucleic Acids Res* 18:6531–6535.

Woo, S. L., A. Zonia, S. G. Gel et al. 1996. Characterization of *Fusarium oxysporum* f. sp. *phaseoli* by pathogenic races, VCGs, RFLPs, and RAPD. *Phytopathology* 86:966–973.

Young, R. A., M. Melotto, R. O. Nodari et al. 1998. Marker assisted dissection of oligogenic anthracnose resistance in the common bean cultivar, G2333. *Theor Appl Genet* 96:87–94.

Yun, Z., S. Jin, Y. Ding et al. 2012. Comparative transcriptomics and proteomics analysis of citrus fruit, to improve understanding of the effect of low temperature on maintaining fruit quality during lengthy postharvest storage. *J Exp Bot* 63:2873–2893.

Zadražnik, T., K. Hollung, W. Egge-Jacobsen et al. 2012. Differential proteomic analysis of drought stress response in leaves of common bean (*Phaseolus vulgaris* L.). *J Proteomics* 78:254–272.

Zargar, S. M., R. Kurata, R. Rakwal, and Y. Fukao. forthcoming. Peptide separation methodologies for in-depth proteomics. In *Plant Cell Growth and Expansion—Methods and Protocols*. Humana Press, a part of Springer Science + Business Media (in press).

Zargar, S., M. Sharma, A., Sadhu, G. K. Agrawal, and R. Rakwal. 2014. Exploring genetic diversity in common bean from unexploited regions of Jammu and Kashmir, India. *Mol Plant Breeding* 5(2):5–9.

Zaumeyer, W. J. and L. L. Harter. 1943. Inheritance of symptom expression of bean mosaic virus. *J Agric Res* 67:295–300.

Zhou, G. H., H. Shirakura, M. Kamahori et al. 2004. A gel-free SNP genotyping method: Bioluminometric assay coupled with modified primer extension reactions (BAMPER) directly from double-stranded PCR products. *Hum Mutat* 24:155–163.

Zhu, J. K. 2002. Salt and drought stress signal transduction in plants. *Ann Rev Plant Biology* 53:247–273.

CHAPTER 13

Genomics, Transcriptomics, and Molecular Breeding for Improving Cereals

Arvind H. Hirani, Muhammad Asif, Manorma Sharma,
Saikat K. Basu, Muhammad Iqbal, and Muhammad Sajad

CONTENTS

ABBREVIATIONS

BAC	Bacterial artificial chromosome
CBC	Clone by clone
De-TILLING	Deletion TILLING
DGE	Digital gene expression
dsRNA	Double-stranded RNA
EPSPS	5-Enolpyruvylshikimate-3-phosphate synthase
EST	Expressed sequence tags
HSE	Heat shock element
HSF	Heat shock transcription factor
IWGSC	International wheat genome sequencing consortium
MAPK	Mitogen-activated protein-kinase
NGS	Next generation sequencing
PAC	P1-derived artificial chromosome
QTL	Quantitative trait loci
RNAi	RNA interference
ROI	Reactive oxygen intermediates
SAGE	Serial analysis of gene expression
SOD	Superoxide dismutase
TILLING	Targeting-induced local lesions in genomes
WGS	Whole-genome shotgun selective gene sequencing
YAC	Yeast artificial chromosome

13.1 INTRODUCTION

Cereal grains are the most important source of food for humans. Among the cereals, rice, maize, barley, and wheat form more than 87% of all the food grains produced globally. Conventional breeding approaches have long been used for increasing cereal production to mitigate the rapid rise in the demand of food globally. However, the human population is increasing at a faster pace than that of food production. This situation demands a further increase in global cereal production through the application of modern technologies. Currently, most research studies are focused on generating molecular information for improving cereal crops and to improve our understanding of various aspects of cereal production and function. This review focuses on the improvement of cereal crops through the application of genomics, transcriptomics, and molecular breeding approaches. The review highlights the role of gene mapping, whole-genome sequencing, and molecular-marker-assisted selection in accelerating cereal breeding. Areas of omics and transcriptomics with focus on functional genomics and transcriptional profiling of whole genomes are also discussed in detail to bridge the gap between conventional and molecular breeding for cereal improvement. We propose that genomics- and proteomics-based integrated approaches and methodologies hold great promise for enhancing the global agricultural productivity of cereals on a sustainable basis.

Plants, being sessile, have to face a variety of stress situations that adversely affect their growth and productivity (Kumar and Sharma 2011). These stresses affect plants on transcriptome, cellular, and physiological levels (Atkinson and Urwin 2012). Recent advances in genetics and molecular biology have paved the path to find the link between biotic and abiotic stress signaling in plants. Some of the research findings suggest that ROS signaling pathways, reversible ion fluxes (Ca^{2+}, K^+, H^+), transcription factors, promoter elements, and hormone signaling pathways play key roles in regulation of responses to various stresses in plants (Fujita et al. 2006).

On the basis of origin, these stress factors are divided into two types: biotic and abiotic. As the names suggest, biotic stresses are caused by biotic factors or living organisms, whereas abiotic stresses result from environmental or physical factors. Both types adversely affect plant growth and development. A list of various biotic and abiotic stresses includes nematodes, insects, mites, viruses, fungi, bacteria, and animals (biotic); and temperature changes, chemical drift, nutrient deficiency, sunscald, improper planting practices, and radiation (abiotic).

13.2 DROUGHT OR WATER DEFICIT STRESS

Drought or water deficit is a situation when the demand of water exceeds the supply in plants. Water potential and relative water content are the two parameters used to measure this stress. Drought may be caused by high salt conditions, low temperature, or high temperature.

Salt stress interferes with various physiological, biochemical, cellular, metabolic, and photosynthetic activities (Chen et al. 2007). It increases the toxins in a plant and results in loss of the protective hydration shell around vulnerable molecules. In addition to a plant's own defense system to cope with salt stress, some soil bacteria also facilitate plant growth even under stress conditions. In one study, the bacterium *Achromobacter piechaudii* ARV8 was found to be effective in tomato and pepper plants to cope with drought stress conditions (Mayak et al. 2004). With high salt concentration, low water potential is generated, thus making it difficult for plants to absorb nutrients and water from the soil. In salt stress conditions, cell homeostasis is regulated by Na^+, K^+, H^+, and Ca^{2+} ions, which are also involved in salt stress signaling.

A specific temperature is required for the growth and development of each plant species. Rapid fluctuations in temperature cause serious tissue damage in plants. The threshold of temperature tolerance varies from species to species and even among different varieties within the same species. The major symptoms in cold stress are chlorosis, wilting, reduced leaf expansion, and sometimes necrosis (Mahajan and Tuteja 2005). These symptoms arise due to freezing dehydration or ice crystal formation that result in expansion-induced cell lyses and fracture lesions causing solute leakage, impairment of photosynthesis, loss of compartmentalization, and reduction in protein assembly and general metabolic processes (Uemura and Steponkus 1997). Likewise, an increase in temperature can also be fatal for plants. Generally, a 10°C–15°C increase in temperature may cause heat stress or heat shock (Wahid and Close 2007). Heat stress results in anatomical, physiological, and metabolic changes in plants: sunburn on plant parts, leaf senescence and abscission, inhibition of root and shoot growth, fruit damage, reduced cell size, increased density of stomata, and stomata closure (Bañon et al. 2004; Vollenweider and Günthardt-Goerg 2005).

13.3 PLANT MECHANISMS TO COPE WITH THESE STRESSES AND RECENT DEVELOPMENTS

Plant cells respond to osmotic stress by producing nontoxic compounds called *compatible solutes* or *osmolites* or *osmoprotectants*: mannitol, pinitol, proline, and so on. Most of them are sugar–alcohol and zwitterionic compounds. Glycine betaine is a quaternary ammonium osmolite that stabilizes protein and membrane via its interaction as an osmolite. The glycine betaine pathway is dependent on two important enzymes of chloroplast stroma, choline monooxygenase and betaine aldehyde dehydrogenase, which are synthesized under stress conditions. These enzymes are introduced in plants to induce stress tolerance.

A combination of modern tools and conventional breeding techniques may accelerate the development of stress-tolerant varieties of cereal crops (Singh et al. 2010). Among various genetic approaches to producing stress-tolerant plants, activation of the stress signaling pathway is of great importance. Genetic manipulation of mitogen-activated protein kinase (MAPK) increases stress tolerance in *Arabidopsis* and cereals (Šamajová et al. 2013). In a study, overexpression of the cDNA encoding DREB1A along with promoter stress-inducible rd29A increased stress tolerance in the Columbia ecotype of *Arabidopsis* (Kasuga et al. 1999). DREB1A is a transcription factor that specifically interacts with cis-acting promoter element, the dehydration response element (DRE), to increase tolerance to drought, salt, and freezing.

Salt tolerance can be induced by a vascular Na/H⁺ antiport system. Halophytes excrete extra salt through stomata on the leaf surface or store it in the vacuole to avoid its accumulation in cytoplasm. Antiport systems contain antiport protein and antiport H^+ pumps, which work against the concentration gradient. Antiport H^+ pumps act as energy suppliers and antiport protein acts as a revolving door for the transport of ions. Increased expression of antiport protein or antiport H^+ pump leads to an increase in salt tolerance. In wheat, expression of a HKT1 (Na^+ transporter) was reduced by antisense to increase NaCl tolerance (Laurie et al. 2002; Munns et al. 2003; Tester and Bacic 2005). For cold tolerance, some plant species produce specific types of proteins, known as cold-induced proteins. In *Arabidopsis*, the COR 15a protein is produced, which increases cold tolerance in chloroplast. To protect the whole plant during cold stress, a single promoter containing the CRT/DRE regulatory element that is the same for different cold tolerance related genes is introduced. Transcription factor CBF1 binds to the CRT/DRE regulatory element, which activates the core regulon group of genes (Figure 13.1) to combat cold stress in plant.

Similar to cold stress, plants produce five types of heat shock proteins, chaperones that help to cope with heat stress. These heat shock proteins include HSP100, HSP90, HSP70, HSP60, and a fifth one that is abundant in nature but whose function is unknown. These proteins are responsible for protein refolding during stress but reestablish normal protein conformation in the absence of heat shock. In higher plants, there are 20–40 kinds of HSP (Bañon et al. 2004). Heat shock transcription factor (HSF) binds to HSP70 as a monomer, but during heat stress, these dissociate and HSF forms a trimer to attach to a heat shock element (HSE), which positively regulates expression of heat shock proteins, thereby inducing heat tolerance in plants. In wheat, transfer of the CtHSR1 gene from *Candida tropicalis* increased its tolerance against both heat and salt stress (Blumwald and Arif 2007).

Reactive oxygen intermediates (ROI) are produced in plants in normal conditions through photosynthesis and respiration and also in stress conditions (Knight and Knight 2001). In stress

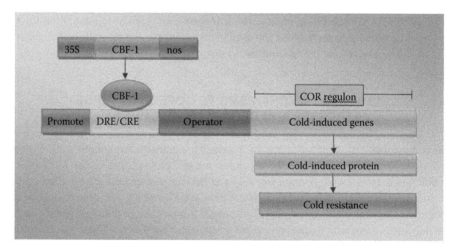

Figure 13.1 CBF-1 gene from *Arabidopsis* transformed in tobacco plant; it binds to DRE regulator and positively expresses COR regulon to produce cold-induced protein. COR regulon: an operon-containing group of cold-responsive genes.

situations, the amount of ROI is higher than normal (240–720 μMS^{-1} O_2^-). ROIs work as indicators triggering various stress-responsive pathways (Mittler 2002). Superoxide dismutase (SOD) plays an important role in protecting plants against oxidative stress. On the basis of the metal cofactor, SODs are of three types: Cu/Zn-SOD, Mn-SOD, and Fe-SOD isoenzymes. Transformation of these SODs in plants increases tolerance against oxidative stresses. One example of the development of a transgenic plant is described in Figure 13.2.

Historically, cereal grains were the first plant species to be grown, replacing the hunting and gathering habits of humans during the Neolithic age (Vasil 1994). They are widely grown edible grains of grasses that belong to monocot families Poaceae and Gramineae. For the last 50 years, the main focus of conventional and modern biotechnology research was improvement of these food grains to alleviate hunger of the ever-growing population (Chavan et al. 1989). Among all the cereals, wheat, rice, maize, and barley are the most important; together they account for 87% of grain production worldwide.

Wheat is a cereal of temperate regions and is consumed worldwide. This grass species originated from the Fertile Crescent region of the Near East (Kumar and Sharma 2011). Wheat is classified into spring and winter (based on growth habit), hard and soft (based on grain texture), and white and red (based on grain color). According to U.S. Department of Agriculture, wheat was planted on 56.53 million acres and its yield was 46.2 bushels/acre and its price was US$6.5–7.5/bushel (www.ers. usda.gov). The wheat genome is 10–20 times larger than that of cotton or rice, hence transformation is more complex and slow. The physical map of wheat contains 3025 loci including 252 phenotypically characterized genes and 17 quantitative trait loci (QTLs) relative to 334 deletion breakpoints (Kumar and Sharma 2011). The spring wheat variety Bob is the most responsive to genetic transformation and has been successfully transformed with several genes of importance. Roundup Ready™ wheat is the first genetically modified herbicide-tolerant wheat, developed by Monsanto. It contains a 5-enolpyruvylshikimate-3-phosphate synthase (EPSPS) protein from *Agrobacterium* sp. strain CP4 (CP4 EPSPS) (Obert et al. 2004).

Rice is a crop of tropical and temperate regions and is a staple food of India, China, Korea, Japan, Thailand, Brazil, and the Philippines. In the 1960s, the father of hybrid rice, Yuan Longping, developed the first genetic male sterile rice. The second most important revolution in the breeding history of rice was the development of golden rice. It is genetically engineered rice that biosynthesizes β-carotene, a precursor of vitamin A, in the endosperm (Ye et al. 2000).

Maize is a staple food of America and Africa. America is the largest producer of maize, growing about 332 million metric tons annually. Roundup Ready Corn is a glyphosate-resistant variety, resistant to glyphosate herbicide, which was commercialized in 1996 by Monsanto (Tan et al. 2005). Another genetically modified crop is Bt corn, which produces a poisonous protein against certain

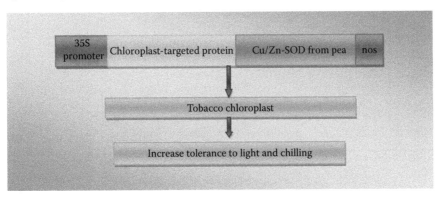

Figure 13.2 Cu/Zn-SOD from pea along with chloroplast-targeted protein insert in tobacco chloroplast to increase light and chilling tolerance.

insects. Besides food, maize is also grown for many other industrial purposes. Recently, the U.S. Department of Agriculture approved commercial growth of genetically engineered corn for ethanol production (Pollack 2011).

Barley's precise origin is not known, but fossils found in the pits and pyramids of Egypt indicate that it originated 5000 years ago. It is an annual crop cultivated for food and feed production and also used commercially to make beer, malt syrup, malted milk, and breakfast foods and as a model organism in molecular research (Magness et al. 1971).

13.4 OMICS APPROACHES

13.4.1 What Is Omics?

In biology, omics informally or formally refers to a field of study ending with the suffix - omics. The word omic(s) commonly addresses the objects of studies that employ molecular techniques or approaches in the areas of the genome, transcriptome, proteome, metabolome, and phenome (Figure 13.3). These disciplines are known as genomics, transcriptomics, proteomics, metabolomics, and phenomics, respectively. All the omics fields are directly or indirectly connected to a specific biological function of cell, tissue, organ, or whole organism. The study of omics includes genomics (DNA data), transcriptomics (RNA data), proteomics (protein data), metabolomics (chemical reaction data), and phenomics (trait data). Each omic area of studies has its advantages and challenges in terms of basic and applied research.

In agriculture, all the omics studies have been engaged one way or another in improving the production of crops in terms of yield, quality, biotic and abiotic stress tolerance, and agronomic performance. Substantial progress has been made in some of the omics areas of research in cereals and other crops. Genomics is the area where the most progress has been made, followed by transcriptomics. Proteomics, metabolomics, and phenomics need to be explored more for better understanding of the biological mechanisms in crop plants.

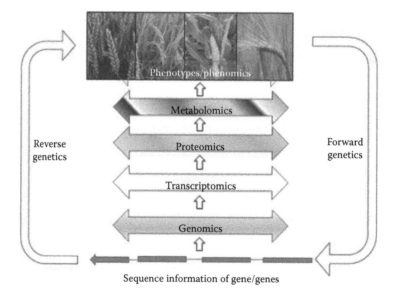

Figure 13.3 (See color insert) Flowchart of omic tools and strategies in forward and reverse genetic approaches.

13.5 GENOMICS/TRANSCRIPTOMICS/PROTEOMICS/ METABOLOMICS/PHENOMICS

According to the central dogma of molecular biology, genetic information contained in the DNA is transcribed into RNA, which is translated into proteins, which are involved in chemical reactions (metabolome) responsible for phenotypes (phenome). This is a hierarchy of biological events with complex interactions. These biological functional events have been investigated through omics research to enhance our knowledge and to improve food production with novel applications in agriculture.

13.5.1 Genomics

Genomics is the study of physical location, biological function, and expression pattern of genes in the plant genome. Broadly, genomics is the study of genes and genomes, which includes determining the locations of genes and regulatory function in plant growth and development. Recent advances in genomic approaches inevitably redefined plant breeding as molecular plant breeding or genome-assisted breeding (Varshney et al. 2005). Large numbers of genomic tools and strategies have been employed for structural and functional characterization of plant genomes. Some of the most attractive strategies being used for cereal improvement are discussed here.

13.5.1.1 Structural Genomics

The structure of a plant genome is best described by genome size, number of diploid chromosomes, gene content, repetitive sequences, and polyploidy or duplication events. Several genomic tools have been used for determining structural features of the plant genome. The most common genomic tools are genetic mapping, physical mapping, cytogenetic mapping, bacterial artificial chromosome (BAC) library construction, genome-wide scanning, whole-genome sequencing, and genome annotation. Several studies have reported genetic mapping in wheat (Peng et al. 2000; Poland et al. 2012; Quarrie et al. 2005; Somers et al. 2004; Zhang et al. 2012b), barley (Druka et al. 2002; IBGSC 2012; Poland et al. 2012; Wenzl et al. 2006), maize (Cone et al. 2002; Helentjaris 1987; Lee et al. 2002b; Wei et al. 2009), and rice (Chen et al. 2002; Harushima et al. 1998), using different molecular marker systems and population types.

Whole-genome sequencing projects have been supported by fingerprinting of BAC-, yeast artificial chromosomes (YAC)-, or P1-derived artificial chromosomes (PAC)-based physical mapping. In rice, 62,509 fingerprinted BAC clones have been organized into 458 contigs, which contain 363 Mb of the rice genome physical map, which was correlated to the genetic map (Chen et al. 2002). Hexaploid wheat (*Triticum aestivum* L.) has a large genome size of 17 Gb because of two allopolypoidization events and a very high repetitive DNA content. Each chromosome of wheat is on average twice the size (809 Mb) of the whole rice genome (390 Mb). In the International Wheat Genome Sequencing Consortium (IWGSC), different research groups performed physical mapping of individual wheat chromosomes or chromosome arms to integrate the whole wheat genome in the genetic map (Berkman et al. 2011; Paux et al. 2008; Sehgal et al. 2012; Vitulo et al. 2011).

13.5.1.2 Whole-Genome Sequencing

Decoding of the genetic code by DNA sequencing has been unraveling numerous genetic bases of complex biological systems. Knowing the sequence of the genes in cereals opens an avenue to understanding functional properties that have better production prospects (Table 13.1). Earlier DNA sequencing technology advanced slowly with advances in the di-deoxy or Sanger method. Two whole-genome sequencing projects of plant species, *A. thaliana* and *O. sativa*,

Table 13.1   Whole-Genome Sequencing Projects in Cereals

Common Name	Species Name	Genome Size (Mbp)	Predicted Genes	Databases	References
Rice	*Oryza sativa (japonica)*	420	40,600	http://rgp.dna.affrc.go.jp/IRGSP/index.html	IRGSP (2005)
Rice	*Oryza sativa (indica)*	466	40,600	http://rice.genomics.org.cn/rice/index2.jsp	IRGSP (2005)
Wheat	*Triticum aestivum*	16,500	50,000	http://www.wheatgenome.org/	Choulet et al. (2010) and Gill et al. (2004)
Maize	*Zea mays*	2,500	300,000	http://www.maizegdb.org/	Schnable et al. (2009)
Barley	*Hordeum vulgare*	5,500	32,000	1. http://www.public.iastate.edu/~imagefpc/IBSC%20Webpage/IBSC%20Template-home.html; 2. http://webblast.ipk-gatersleben.de/barley/index.php	Mayer et al. (2011)
Sorghum	*Sorghum bicolor*	659	34,500	http://www.plantgdb.org/SbGDB/	Paterson et al. (2009)
Brachypodium	*Brachypodium distachyon*	273	25,500	http://www.brachypodium.org/	IBI (2010)

were accomplished exclusively by Sanger sequencing technology (AGI 2000; IRGSP 2005). Next-generation sequencing (NGS) technology has dramatically reduced the cost of DNA sequencing and significantly increased the amount of data collection. It has also initiated whole-genome sequencing of several plant species. It has also been applied in *de novo* sequencing and assembly, whole-genome sequencing or targeted resequencing, RNA sequencing, epigenomic analysis and small RNA analysis, genome-wide scanning, and SNP genotyping in cereals (Berkman et al. 2012; Eckardt 2009; Elling and Deng 2009; Huang et al. 2009; Mayer et al. 2011; Schulte et al. 2009; Subbaiyan et al. 2012; Trick et al. 2012).

Whole-genome sequencing can be performed by three different approaches or their combinations: clone by clone (CBC), whole-genome shotgun (WGS), and selective gene sequencing. The IWGSC took a chromosome-based approach and constructed a BAC library and a gene filtration library by considering the large genome size and multinational support (Safar et al. 2004). The rice genome (*Oryza sativa* L. ssp. *japonica*), however, was sequenced by WGS sequencing method (Goff et al. 2002). Similarly, inbred line B73 of maize was used to sequence the maize genome using the CBC approach from the BAC-derived physical map (Nelson et al. 2005; Schnable et al. 2009). Similarly, 5.1 Gb haploid genome of barley has been sequenced by a combination of CBC, WGS, and NGS technology (IBGSC 2012).

13.5.1.3 Comparative Genomics

Expanding technologies in biological research, especially in genomics, bioinformatics, and molecular biology, have generated large datasets that are being compared through comparative genomics in crop plant species. Comparative genomics has been used for within-genome (that is, between chromosomes), between genomes within polyploid species, and between diploid species to estimate structural and functional similarities of biological systems. Comparative analysis is used for gene sequence (DNA), transcript sequence (RNA), and translate sequence (protein) comparison for divergent accessions of plant species that vary for traits or growth pattern, for example, cold tolerance, drought tolerance, and so on. Comparison of genetic maps developed from different populations in the same species or different species within the same genus provides opportunities to determine macro- and microcolinearity for genes, chromosome regions, and whole chromosomes. Microcolinearity has been reported in some regions between wheat and rice (Yan et al. 2003) and barley and rice (Dunford et al. 1995). Comparative findings reveals that about 71% of predicted rice proteins have a homolog in *Arabidopsis*. On the other hand, about 2859 genes appear to be unique to rice and other cereals (Paterson et al. 2005).

13.5.1.4 Functional Genomics

In the postgenomic era of plant biology, evaluation of the functions of all the coding genes in the plant genome is most challenging, especially in cereals, due to the existence of polyploidy and large gene redundancy. Determination of functional properties of genes is called functional genomics and can be achieved by two strategies: forward genetics and reverse genetics.

13.5.2 Forward versus Reverse Genetics in Cereals

Forward genetics is a functional genomics approach that refers to the process of determining a phenotype and subsequently characterizing the genetic region responsible for that phenotype. Forward genetics begins with the phenotype and then reveals the genetic basis of that phenotype. Mutagenesis and mapping populations are considered forward genetic approaches. On the other hand, reverse genetics begins with a known gene and subsequently determines the function of that gene in the phenotype by producing an altered phenotype. In the reverse genetic approach, genome or gene sequence information is prerequisite. Antisense- or RNA interference (RNAi)-based gene silencing, T-DNA insertional mutants, virus-induced gene silencing, and TILLING are reverse genetic approaches that have been employed in different crop species.

13.5.2.1 Gene Knockout and Mutagenesis

Crop plant improvements to enhance quality, production, and productivity with better biotic and abiotic stress-tolerance mechanisms are most crucial in agricultural industries. Genetic strategies such as capturing spontaneous mutations or creating beneficial mutations by artificial means such as chemical or physical mutagenesis and insertion or deletion of small DNA fragments in coding sequences have been commonly practiced to create novel genetic variation for important traits in economically important crops. In waxy wheat, three waxy alleles, *Wx-A1h*, *Wx-B1b*, and *Wx-D1b* were characterized by mutations (Vrinten et al. 1999). Similarly, genetic analysis of an F_2 population revealed a mutation detected in common wheat that produced three pistils; segregation ratios suggested that a single dominant gene was involved in conferring three-pistils bread wheat (Peng 2003). Likewise, mutant wheat displayed an altered resistance phenotype to yellow rust caused by *Puccinia striiformis* (Boyd and Minchin 2001). In rice, various useful mutations have been reviewed for their applications in rice breeding (Kadam 1932).

13.5.2.2 Target Gene or Trait Introgression by Recombinations

In traditional breeding, gene introgression for important traits is commonly utilized with phenotype evaluation for subsequent progeny selection. Recent advances in genomic tools and technologies such as molecular markers, genetic and physical maps, whole or partial genome sequences, and expressed sequence tags (EST) databases have indefinitely hastened gene introgression by homologous or homeologous recombinations. In Chinese spring wheat mutant lines, *Ph1*, *Ph1b*, and *Ph2* demonstrated that these genes are involved in regulating homologous pairing at Metaphase-I

(Martinez et al. 2001; Mello-Sampayo 1971; Riley and Chapman 1958; Sears 1977). These mutant lines have been widely employed for interspecific hybridization and subsequent trait introgression in spring and winter wheat. Several resistance genes for leaf rust, green bug, powdery mildew, and stem rust have been transferred to bread wheat using *Ph1* mutant lines from *Ae. speltoides* (Dubcovsky et al. 1998; Friebe et al. 1996). More recently, numerous studies reported the use of Chinese spring *Ph1* mutant lines for homeologous recombination and subsequent trait integration in various wheat breeding programs (Table 13.2).

In the genus *Oryza*, genetic variation is available due to the existence of different genomes in independent wild relative species such as AA, BB, CC, BBCC, CCDD, EE, FF, GG, and HHJJ (Brar and Khush 1997). Interspecific wide hybridization and subsequent embryo rescue have been performed in several studies to introgress important traits from wild distant relatives to cultivated rice species (Brar and Khush 1997; Multani et al. 2003). In rice, interspecific hybridization and subsequent alien chromosome addition lines were developed from a cross between *O. latifolia* ($2n = 48$, CCDD) and *O. sativa* ($2n = 24$, AA) to introgress important biotic traits conferring resistance against organisms such as bacterial blight, brown plant hopper, and white-backed plant hopper (Multani et al. 2003). The introgression of alien genes from wild distant relatives to the A-genome of cultivated rice was achieved for cytoplasmic male sterility, and resistance to grassy stunt virus, bacterial blight, and brown plant hopper (Brar and Khush 1997).

Similarly, in maize, studies reported the spontaneous hybridization between these taxa and the subsequent introgression of "Teosinte" (the closest wild relatives of maize) genomic fragments into the cultivated maize genetic background, which played a crucial role in the evolution of modern maize (Ellstrand et al. 2007). Alien genetic germplasm introgression was reported through a cross of *Z. mays* ssp. *mexicana* and cultivated maize inbred line Ye515; the results revealed improved agronomic performance and combining ability of the maize hybrid, which had alien introgressed genetic regions. (Wang et al. 2008a).

Table 13.2 Target Gene or Trait Introgression by Wide Hybridization in Wheat Using Chinese Spring Wheat *ph1* Mutant Stock

Crop Species	Wild Species	Trait Introgressed	Chromosome Segments	References
T. aestivum	*Agropyron elongatum*	Water-stress adaptation and high biomass	7DL	Placido et al. (2013)
T. aestivum	*Aegilops speltoides*	Resistance to multiple stem rust races	2B	Mago et al. (2009) and Niu et al. (2011)
T. aestivum	*Leymus racemosus*	Fusarium head blight	6B	Chen et al. (2005)
T. aestivum	*Aegilops speltoides*	Endosperm structure	5DS	Pshenichnikova et al. (2010)
T. aestivum	*Psathyrostachys huashanica*	Stripe rust	1B↔3Ns and 3NsS introgressed in 3BL	Kang et al. (2011)
T. aestivum	*Psathyrostachys huashanica*	Storage protein genes (HMW-GS, LMW-GS, gliadin)	A pair of wheat chromosomes replaced	Zhao et al. (2010)
T. aestivum	*Aegilops speltoides*	Spike length and number of spikelets	5A	Simonov et al. (2009)
T. aestivum	*Aegilops speltoides*	Leaf hairiness, whole plant nonglaucousness, spike glume color	7B, 2B, 1BS	Pshenichnikova et al. (2007)
T. durum	*Aegilops speltoides*	Seven races of stem rust	2BL	Faris et al. (2008)

13.5.2.3 Gene Silencing by RNAi or Antisense

Functional characteristics of genes can be determined by silencing of genes through RNAi or antisense reverse genetic approaches. The whole-genome sequence with annotation information is required to investigate the functional properties of each gene in biological systems of plants. In genomics, gene silencing by RNAi or antisense is often used in reverse genetic approaches in crop species for functional characterization of genes involved in biosynthesis pathways or regulatory mechanisms. The RNAi technique is based on targeted sequence-specific RNA degradation initiated by formation of double-stranded RNA (dsRNA), which is homologous to the targeted gene (Baulcombe 2004). Gene silencing through RNAi is most likely deployed in polyploid species to silence multigene families, all homologs, or homeologous copies of the gene(s) (Travella et al. 2006). The downregulation of the *VRN1* gene in hexaploid wheat through RNAi delayed flowering times (Loukoianov et al. 2005). Similarly, reduced transcript abundance of the *VRN2* gene by RNAi accelerated flowering time by more than a month in transgenic winter wheat plants (Yan et al. 2004). Regina et al. (2006) employed RNAi to develop high-amylose transgenic wheat lines for improved seed quality. Similarly, in rice, RNAi was used for viral disease–resistant transgenic rice development for rice strip disease (Zhou et al. 2012) and rice tungro (Tyagi et al. 2008). Likewise, the RNAi approach was successfully used for plant height (Qiao et al. 2007) and aroma (2-acetyl-1-pyrroline) production in rice seeds (Niu et al. 2008). Similarly, in maize, high-lysine transgenic maize plants were developed by suppressing lysine catabolism in the seed endosperm to improve feed quality of seeds (Houmard et al. 2007). Maize dwarf mosaic virus–resistant transgenic maize plants were developed through RNAi-based gene silencing of P1 protein (protease gene of MDMV virus) (Zhang et al. 2010).

13.5.2.4 TILLING and Its Modifications

Targeting-induced local lesions in genomes (TILLING) is a powerful reverse genetic approach initially employed in the model plant *Arabidopsis* (McCallum et al. 2000). Subsequently, due to feasibility of the technique, it has been successfully applied in cereal and oilseed crops including wheat (Slade et al. 2005, 2012; Tsai et al. 2011), rice (Till et al. 2007; Wu et al. 2005), maize (Till et al. 2004; Weil and Monde 2007), barley (Caldwell et al. 2004; Gottwald et al. 2009; Lababidi et al. 2009), and oats (Chawade et al. 2010). Availability of large genome sequencing data has opened up new possibilities of employing conventional mutation techniques in alternate yet more useful reverse genetic ways. Recently, TILLING has become the method of choice for functional characterization of genes because of its broad applications, which can be applied to any species for any kind of target gene. In addition to that, achievement of mutation in TILLING is more stable than other transgenic approaches such as insertional mutagenesis and antisense or RNAi gene silencing (Alonso and Ecker 2006). Most recently, genome editing through TILLING has become possible for improved phenomics as TILLING produces a broad range of alleles that includes nonsense, missense, splicing, and cis-regulatory alleles (Till et al. 2007).

The first modification of the TILLING strategy was proposed by Comai et al. (2004), and this modification was designated as ecoTILLING. In ecoTILLING, instead of mutagenized plants as templates, the DNA of several cultivars or germplasm lines each mixed with a reference DNA can be used. In *Arabidopsis*, 55 haplotypes in 5 genes in more than 150 individuals were reported, which differed by a SNP (Comai et al. 2004). Since its inception, many ecoTILLed genes have been screened in natural populations of different species including *Alk*, which encodes soluble starch synthase IIa in *O. sativa* (Kadaru et al. 2006), and *Pina-D1* and *Pinb-D1*, which are involved in conditioning of kernel hardness through allelic variation in these genes in *T. aestivum* (Wang et al. 2008b).

Recently, attractive and cost-effective modifications in TILLING have been reported and found to be useful for application in crop species. An alternate TILLING method that reduces cost and time required for screening mutations was developed in *Arabidopsis*, and it is called individualized TILLING (iTILLING) (Bush and Krysan 2010). In iTILLING, seeds from M_1 are collected in bulk and one to two M_2 plants are grown per well on 96-well spin plates on agar plugs, and the tissue for DNA isolation is harvested directly from the plates using the Ice-Cap method. The roots of seedlings grow through the agar toward a second 96-well plate at the bottom of the first plate that is filled with water. After about 3 weeks, a second plate is filled with roots to serve as the source of tissue for DNA isolation and to maintain seedlings of plants intact. The mutations can be revealed without using enzymatic cleavage and gel electrophoresis by a high-resolution melt curve analysis of amplified fragments. Since one or two plants per well are grown so the isolated DNA is either individual or pooled twofold, the identification of mutation is both heterozygous and homozygous in case of two plants and heterozygous in case of one plant. Corresponding seedlings with mutation can be transferred to soil and grown for phenotypic characterization. The iTILLING strategy, however, has not been employed in crop species so far.

An additional modification of the TILLING approach called Deletion TILLING (De-TILLING) was reported by Rogers et al. (2009). In the TILLING method, approximately 5% null mutations have been detected; in contrast to that, De-TILLING exclusively detects knockout mutations. The Delete-a-gene® approach is based on suppressing the wild-type sequence amplification that can provide preferential detection of deletion alleles (Li et al. 2002).

13.5.3 Transcriptomics

Besides the advancement of functional genomics, transcriptomic analyses of large gene sets in different crop species have been underway to investigate functional features of a few to large numbers of genes within genomes. Various genomic techniques such as EST screening and sequencing, microarray hybridization, serial analysis of gene expression (SAGE), quantitative PCR (qPCR), digital gene expression (DGE), and NGS Solexa/Illumina RNA-sequencing have employed for large-scale transcriptome profiling to investigate the roles of existing genes in plant growth and development. Here, transcriptomic studies in major cereals have been discussed with their potential applicability in overall production and productivity improvement in cereals.

13.5.3.1 Transcriptomics in Rice

Rice whole-genome sequencing and annotation information was supported by large-scale transcriptome analysis for different growing conditions and treatments. Several studies reported global transcriptome analysis for cold tolerance (Zhang et al. 2012a), drought-stress response (Degenkolbe et al. 2009; Fu et al. 2007; Lenka et al. 2011), root response to potassium deficiency (Ma et al. 2012), salt stress (Chao et al. 2005), and pollen development and germination after pollination (Wei et al. 2010). Transcriptome profiling of rice during embryogenesis revealed a large number of genes significantly varying in expression during embryo development (Xu et al. 2012). Transcript abundance in rice endosperm development revealed over 10,371 differentially expressed genes 3, 6, and 10 days after pollination by high-throughput sequencing (RNA-Seq) (Gao et al. 2013). Interestingly, this study reported a set of genes related to ribosomes: the spliceosome and oxidative phosphorylation were expressed in the early and middle stages, whereas genes related to defense mechanisms and starch/sucrose metabolism pathways were strongly expressed during later stages of endosperm development (Gao et al. 2013).

13.5.3.2 Transcriptomics in Wheat

In bread wheat, transcriptome analysis has been carried out for numerous regulatory and trait-specific genes for better understanding of global expression patterns during plant growth and development. Studies on transcriptomic profiling permit functional involvement of the genes in specific growth conditions regardless of gene structure and sequence. Jia et al. (2009) revealed transcript accumulation variation for 14 genes for Fusarium head blight–resistant and Fusarium head blight–susceptible phenotypes in near-isogenic lines of wheat. Similarly, transcriptional response of wheat to salt stress was carried out through 22,000 oligo-DNA microarray and twofold expression was observed for 1,811 genes. Among them, some of these genes were involved in stress-tolerance-related pathways (Kawaura et al. 2006). Likewise, the effect of hydrogen peroxide (H_2O_2) on global transcript variation in wheat was reported and results suggested that 3000–6000 transcripts were differentially expressed within them; 260 genes showed consistent expression in both powdery mildew–resistant and powdery mildew–susceptible wheat lines (Li et al. 2011). Likewise, transcriptomic studies have been reported for grain development (Wan et al. 2008), terminal drought stress (Aprile et al. 2009), salinity tolerance (Liu et al. 2012), starch biosynthesis during grain development (Stamova et al. 2009), and grain quality traits in wheat (Wan et al. 2009). Abundance of transcripts in specific growth conditions or traits can provide avenues to track the responsible gene(s) that can be useful to enhance our knowledge in systems biology as well as molecular breeding for crop improvements.

13.5.3.3 Transcriptomics in Barley

Barley transcriptome profiling was performed to examine gene expression during the early stages of germination (Potokina et al. 2002; Watson and Henry 2005) and malting quality (Munoz-Amatriain et al. 2010a,b; Potokina et al. 2004; White et al. 2006). Gene expression in barley seedlings during short or prolonged cold treatment was done using the Barley 1 Affimetrix chip. The results revealed altered expression of many genes, including known genes *VRN1*, *ODDSOC2*, *PR17d*, and serine acetyltransferase (Greenup et al. 2011). Similar to transcript expression, variation in different treatments or growth stages are not caused by various biotic stresses. Boddu et al. (2006) reported 467 differentially accumulated gene transcripts in the *F. graminearum*–treated barley plants compared to the water-treated control plants. Annotation of differentially expressed transcripts revealed host genes encoding defense-responsive proteins, oxidative burst–related enzymes, and enzymes involved in the phenylpropanoid pathway. The global transcript analysis can be used to navigate useful genes for functional application in crop improvements in cereals having complex genomes such as wheat, barley, and maize.

13.5.3.4 Transcriptomics in Maize

Maize is one of the most important crops in the world for food, feedstock, and biofuel production. Increasing world population has increased the demands on food and feed production of major crops including maize. In addition to enhancing knowledge of the whole genome, it is important to understand the transcriptomic profile of expressing genes that have crucial roles in biological enzymatic production, which are eventually key players for final traits. Numerous gene expression studies through transcript analysis have been reported for regulation of seed development and maturation, DNA replication, storage protein accumulation, induction of morphological changes, and pattern of cell division through microRNA genes during seed development (Alexandrov et al. 2009; Lee et al. 2002a; Zhang et al. 2009). Li et al. (2010) reported a maize leaf transcriptome using Illumina sequencing to determine gene structure and alternative splicing events and to quantify transcript accumulation during leaf developmental stages. They found 64% and 21% of genes were differentially expressed along the developmental stages between bundle sheath and mesophyll cells, respectively. Similarly, three omics—transcriptomics, proteomics, and metabolomics—were employed for maize nitrogen use efficiency determination through long-term nitrogen deficiency and its impact on leaf gene transcript, protein, and metabolite accumulation. The study revealed that several genes were up- or downregulated by nitrogen deficiency (Amiour et al. 2012). The role of gene transcripts in heterosis was evaluated in highly heterotic maize hybrid Zhengdan 958 and its parents during spikelet and floscule differentiation using the GeneChip β Maize Genome Array. The study found that the differentially expressed genes were involved in lipid metabolism, signal transduction, transport, and catabolism and could be involved in heterosis (Li et al. 2012).

13.6 CONCLUSION

This review clearly indicates that the technologies and developments observed in the last few decades demonstrate an ascending curve in the field of molecular breeding. However, much needs to be explored and exploited in terms of technological progress from the perspectives of cost efficiency, ease of handling and operation, high-throughput nature, and reproducibility. The genomics, transcriptomics, and phenomics technologies are progressing rapidly, but there is more to expect in the near future. We believe that such technological developments will certainly have strong impacts in shaping the future of molecular breeding techniques for major cereals grown all over the world. Significant contributions have already been made in the fundamental understanding of omics technology. What we now expect to see is the development and progress of the applied aspect of this powerful omics tool in solving many of the agronomic and breeding-related challenges of different cereal crops grown under widely divergent agroclimatic conditions. The progress will have a tremendous impact on average annual global cereal production in the coming years.

REFERENCES

AGI. 2000. Analysis of the genome sequence of the flowering plant *Arabidopsis thaliana*. *Nature* 408:796–815.
Alexandrov, N. N., V. V. Brover, S. Freidin et al. 2009. Insights into corn genes derived from large-scale cDNA sequencing. *Plant Mol Biol* 69:179–194.
Alonso, J. M. and J. R. Ecker. 2006. Moving forward in reverse: Genetic technologies to enable genome-wide phenomic screens in *Arabidopsis*. *Nat Rev Genet* 7:524–536.
Amiour, N., S. Imbaud, G. Clément et al. 2012. The use of metabolomics integrated with transcriptomic and proteomic studies for identifying key steps involved in the control of nitrogen metabolism in crops such as maize. *J Exp Bot* 63:5017–5033.
Aprile, A., A. M. Mastrangelo, A. M. De Leonardis et al. 2009. Transcriptional profiling in response to terminal drought stress reveals differential responses along the wheat genome. *BMC Genomics* 10:279.
Atkinson, N. J. and P. E. Urwin. 2012. The interaction of plant biotic and abiotic stresses: From genes to the field. *J Exp Bot* 63:3523–3543.
Bañon, S., J. A. Fernandez, J. A. Franco, A. Torrecillas, J. J. Alarcón, and M. J. Sánchez-Blanco. 2004. Effects of water stress and night temperature preconditioning on water relations and morphological and anatomical changes of *Lotus creticus* plants. *Sci Hortic* 101:333–342.
Baulcombe, D. 2004. RNA silencing in plants. *Nature* 431:356–363.
Berkman, P. J., K. Lai, M. T. Lorenc, and D. Edwards. 2012. Next-generation sequencing applications for wheat crop improvement. *Am J Bot* 99:365–371.
Berkman, P. J., A. Skarshewski, M. T. Lorenc et al. 2011. Sequencing and assembly of low copy and genic regions of isolated *Triticum aestivum* chromosome arm 7DS. *Plant Biotechnol J* 9:768–775.
Blumwald, E. and A. Arif. 2007. Gene pyramiding through genetic engineering for increase salt tolerance in wheat. National Academy of Sciences. http://sites.nationalacademies.org/PGA/dsc/pakistan/PGA_052680. Accessed: October 23, 2013.
Boddu, J., S. Cho, W. M. Kruger, and G. J. Muehlbauer. 2006. Transcriptome analysis of the barley–*Fusarium graminearum* interaction. *Mol Plant-Microbe Interact: MPMI* 19:407–417.
Boyd, L. and P. Minchin. 2001. Wheat mutants showing altered adult plant disease resistance. *Euphytica* 122:361–368.

Brar, D. S. and G. S. Khush. 1997. Alien introgression in rice. *Plant Mol Biol* 35:35–47.

Bush, S. M. and P. J. Krysan. 2010. iTILLING: A personalized approach to the identification of induced mutations in *Arabidopsis*. *Plant Physiol* 154:25–35.

Caldwell, D. G., N. McCallum, P. Shaw, G. J. Muehlbauer, D. F. Marshall, and R. Waugh. 2004. A structured mutant population for forward and reverse genetics in Barley (*Hordeum vulgare* L.). *Plant J* 40:143–150.

Chao, D. Y., Y. H. Luo, M. Shi, D. Luo, and H. X. Lin. 2005. Salt-responsive genes in rice revealed by cDNA microarray analysis. *Cell Res* 15:796–810.

Chavan, J. K., S. S. Kadam, and L. R. Beuchat. 1989. Nutritional improvement of cereals by fermentation. *Crit Rev Food Sci Nutr* 28:349–400.

Chawade, A., P. Sikora, M. Brautigam et al. 2010. Development and characterization of an oat TILLING-population and identification of mutations in lignin and β-glucan biosynthesis genes. *BMC Plant Biol* 10:86.

Chen, C., C. Tao, H. Peng, and Y. Ding. 2007. Genetic analysis of salt stress responses in asparagus bean (*Vigna unguiculata* (L.) ssp. sesquipedalis Verdc.). *J Hered* 98:655–665.

Chen, M., G. Presting, W. B. Barbazuk et al. 2002. An integrated physical and genetic map of the rice genome. *Plant Cell Online* 14:537–545.

Chen, P., W. Liu, J. Yuan et al. 2005. Development and characterization of wheat–*Leymus racemosus* translocation lines with resistance to fusarium head blight. *Theor Appl Genet* 111:941–948.

Choulet, F., T. Wicker, C. Rustenholz et al. 2010. Megabase level sequencing reveals contrasted organization and evolution patterns of the wheat gene and transposable element spaces. *Plant Cell* 22:1686–1701.

Comai, L., K. Young, B. J. Till et al. 2004. Efficient discovery of DNA polymorphisms in natural populations by Ecotilling. *Plant J* 37:778–786.

Cone, K. C., M. D. McMullen, I. V. Bi et al. 2002. Genetic, physical, and informatics resources for maize. On the road to an integrated map. *Plant Physiol* 130:1598–1605.

Degenkolbe, T., P. T. Do, E. Zuther et al. 2009. Expression profiling of rice cultivars differing in their tolerance to long-term drought stress. *Plant Mol Biol* 69:133–153.

Druka, A., D. Kudrna, C. G. Kannangara, D. von Wettstein, and A. Kleinhofs. 2002. Physical and genetic mapping of barley (*Hordeum vulgare*) germin-like cDNAs. *Proc Natl Acad Sci* 99:850–855.

Dubcovsky, J., A. Lukaszewski, M. Echaide, E. Antonelli, and D. Porter. 1998. Molecular characterization of two Triticum speltoides interstitial translocations carrying leaf rust and greenbug resistance genes. *Crop Sci* 38:1655–1660.

Dunford, R. P., N. Kurata, D. A. Laurie, T. A. Money, Y. Minobe, and G. Moore. 1995. Conservation of fine-scale DNA marker order in the genomes of rice and the Triticeae. *Nucleic Acids Res* 23:2724–2728.

Eckardt, N. A. 2009. Deep sequencing maps the maize epigenomic landscape. *Plant Cell Online* 21:1024–1026.

Elling, A. A. and X. W. Deng. 2009. Next-generation sequencing reveals complex relationships between the epigenome and transcriptome in maize. *Plant Signal Behav* 4:760–762.

Ellstrand, N. C., L. C. Garner, S. Hegde, R. Guadagnuolo, and L. Blancas. 2007. Spontaneous hybridization between maize and teosinte. *J Hered* 98:183–187.

Faris, J. D., S. S. Xu, X. Cai, T. L. Friesen, and Y. Jin. 2008. Molecular and cytogenetic characterization of a durum wheat-Aegilops speltoides chromosome translocation conferring resistance to stem rust. *Chromosome Res* 16:1097–1105.

Friebe, B., J. Jiang, W. J. Raupp, R. A. McIntosh, and B. S. Gill. 1996. Characterization of wheat-alien translocations conferring resistance to diseases and pests: Current status. *Euphytica* 91:59–87.

Fu, B. Y., J. H. Xiong, L. H. Zhu et al. 2007. Identification of functional candidate genes for drought tolerance in rice. *Mol Genet Genomics* 278:599–609.

Fujita, M., Y. Fujita, Y. Noutoshi et al. 2006. Crosstalk between abiotic and biotic stress responses: A current view from the points of convergence in the stress signaling networks. *Curr Opin Plant Biol* 9:436–442.

Gao, Y., H. Xu, Y. Shen, and J. Wang. 2013. Transcriptomic analysis of rice (*Oryza sativa*) endosperm using the RNA-Seq technique. *Plant Mol Biol* 81:363–378.

Gill, B. S., R. Appels, A. M. Botha-Oberholster et al. 2004. A workshop report on wheat genome sequencing: International Genome Research on Wheat Consortium. *Genetics* 168:1087–1096.

Goff, S. A., D. Ricke, T. H. Lan et al. 2002. A draft sequence of the rice genome (*Oryza sativa* L. ssp. japonica). *Science* 296:92–100.

Gottwald, S., P. Bauer, T. Komatsuda, U. Lundqvist, and N. Stein. 2009. TILLING in the two-rowed barley cultivar "Barke" reveals preferred sites of functional diversity in the gene *HvHox1*. *BMC Res Notes* 2:258.

Greenup, A. G., S. Sasani, S. N. Oliver, S. A. Walford, A. A. Millar, and B. Trevaskis. 2011. Transcriptome analysis of the vernalization response in barley (*Hordeum vulgare*) seedlings. *PLoS ONE* 6:e17900.

Harushima, Y., M. Yano, A. Shomura et al. 1998. A high-density rice genetic linkage map with 2275 markers using a single F_2 population. *Genetics* 148:479–494.

Helentjaris, T. 1987. A genetic linkage map for maize based on RFLPs. *Trends Genet: TIG* 3:217–221.

Houmard, N. M., J. L. Mainville, C. P. Bonin, S. Huang, M. H. Luethy, and T. M. Malvar. 2007. High-lysine corn generated by endosperm-specific suppression of lysine catabolism using RNAi. *Plant Biotechnol J* 5:605–614.

Huang, X., Q. Feng, Q. Qian et al. 2009. High-throughput genotyping by whole-genome resequencing. *Genome Res* 19:1068–1076.

IBGSC (International Barley Genome Sequencing Consortium). 2012. A physical, genetic and functional sequence assembly of the barley genome. *Nature* 491:711–716.

IBI (International-Brachypodium-Initiative). 2010. Genome sequencing and analysis of the model grass *Brachypodium distachyon*. 2010. *Nature* 463:763–768.

IRGSP (International Rice Genome Sequencing Project). 2005. The map-based sequence of the rice genome. *Nature* 436:793–800.

Jia, H., S. Cho, and G. J. Muehlbauer. 2009. Transcriptome analysis of a wheat near-isogenic line pair carrying Fusarium head blight-resistant and -susceptible alleles. *Mol Plant Microbe Interact: MPMI* 22:1366–1378.

Kadam, B. S. 1932. Mutation in rice. *Nature* 129:616–617.

Kadaru, S., A. Yadav, R. Fjellstrom, and J. Oard. 2006. Alternative ecotilling protocol for rapid, cost-effective single-nucleotide polymorphism discovery and genotyping in rice (*Oryza sativa* L.). *Plant Mol Biol Rep* 24:3–22.

Kang, H., Y. Wang, G. Fedak et al. 2011. Introgression of chromosome 3Ns from *Psathyrostachys huashanica* into wheat specifying resistance to stripe rust. *PLoS ONE* 6:e21802.

Kasuga, M., Q. Liu, S. Miura, K. Yamaguchi-Shinozaki, and K. Shinozaki. 1999. Improving plant drought, salt, and freezing tolerance by gene transfer of a single stress-inducible transcription factor. *Nat Biotechnol* 17:287–291.

Kawaura, K., K. Mochida, Y. Yamazaki, and Y. Ogihara. 2006. Transcriptome analysis of salinity stress responses in common wheat using a 22k oligo-DNA microarray. *Func Integr Genomics* 6:132–142.

Knight, H. and M. R. Knight. 2001. Abiotic stress signalling pathways: Specificity and cross-talk. *Trends Plant Sci* 6:262–267.

Kumar, A. and M. Sharma. 2011. Wheat genome phylogeny and improvement. *Aust J Crop Sci* 5:1120–1126.

Lababidi, S., N. Mejlhede, S. K. Rasmussen et al. 2009. Identification of barley mutants in the cultivar "Lux" at the *Dhn* loci through TILLING. *Plant Breed* 128:332–336.

Laurie, S., K. A. Feeney, F. J. Maathuis, P. J. Heard, S. J. Brown, and R. A. Leigh. 2002. A role for HKT1 in sodium uptake by wheat roots. *Plant J* 32:139–149.

Lee, J. M., M. E. Williams, S. V. Tingey, and J. A. Rafalski. 2002a. DNA array profiling of gene expression changes during maize embryo development. *Funct Integr Genomics* 2:13–27.

Lee, M., N. Sharopova, W. D. Beavis et al. 2002b. Expanding the genetic map of maize with the intermated B73×Mo17 (IBM) population. *Plant Mol Biol* 48:453–461.

Lenka, S. K., A. Katiyar, V. Chinnusamy, and K. C. Bansal. 2011. Comparative analysis of drought-responsive transcriptome in Indica rice genotypes with contrasting drought tolerance. *Plant Biotechnol J* 9:315–327.

Li, A., R. Zhang, L. Pan et al. 2011. Transcriptome analysis of H_2O_2-treated wheat seedlings reveals a H_2O_2-responsive fatty acid desaturase gene participating in powdery mildew resistance. *PLoS One* 6:e28810.

Li, P., L. Ponnala, N. Gandotra et al. 2010. The developmental dynamics of the maize leaf transcriptome. *Nat Genet* 42:1060–1067.

Li, X., M. Lassner, and Y. Zhang. 2002. Deleteagene: A fast neutron deletion mutagenesis-based gene knockout system for plants. *Comp Funct Genomics* 3:158–160.

Li, Z. Y., T. Zhang, and S. C. Wang. 2012. Transcriptomic analysis of the highly heterotic maize hybrid zhengdan 958 and its parents during spikelet and floscule differentiation. *J Integr Agric* 11:1783–1793.

Liu, C., S. Li, M. Wang, and G. Xia. 2012. A transcriptomic analysis reveals the nature of salinity tolerance of a wheat introgression line. *Plant Mol Biol* 78:159–169.

Loukoianov, A., L. Yan, A. Blechl, A. Sanchez, and J. Dubcovsky. 2005. Regulation of *VRN-1* vernalization genes in normal and transgenic polyploid wheat. *Plant Physiol* 138:2364–2373.

Ma, T. L., W. H. Wu, and Y. Wang. 2012. Transcriptome analysis of rice root responses to potassium deficiency. *BMC Plant Biol* 12:161.

Magness, J. R., G. M. Markle, and C. C. Compton. 1971. Food and feed crops of the United States. *Interregional Research Project* IR-4, IR Bul. 1, (Bul. 828 New Jersey Agr. Expt. Sta.). Accessed: July 10, 2012.

Mago, R., P. Zhang, H. S. Bariana et al. 2009. Development of wheat lines carrying stem rust resistance gene Sr39 with reduced Aegilops speltoides chromatin and simple PCR markers for marker-assisted selection. *Theor Appl Genet* 119:1441–1450.

Mahajan, S. and N. Tuteja. 2005. Cold, salinity and drought stresses: An overview. *Arch Biochem Biophy* 444:139–158.

Martinez, M., N. Cuñado, N. Carcelén, and C. Romero. 2001. The *Ph1* and *Ph2* loci play different roles in the synaptic behaviour of hexaploid wheat *Triticum aestivum*. *Theor Appl Genet* 103:398–405.

Mayak, S., T. Tirosh, and B. R. Glick. 2004. Plant growth-promoting bacteria that confer resistance to water stress in tomatoes and peppers. *Plant Sci* 166:525–530.

Mayer, K. F. X., M. Martis, P. E. Hedley et al. 2011. Unlocking the barley genome by chromosomal and comparative genomics. *Plant Cell Online* 23(4):1249–1263.

McCallum, C. M., L. Comai, E. A. Greene, and S. Henikoff. 2000. Targeted screening for induced mutations. *Nat Biotechnol* 18:455–457.

Mello-Sampayo, T. 1971. Promotion of homoeologous pairing in hybrids of *Triticum aestivum × Aegilops longissima*. *Genetic Iber* 23:1–9.

Mittler, R. 2002. Oxidative stress, antioxidants and stress tolerance. *Trends Plant Sci* 7:405–410.

Multani, D. S., G. S. Khush, B. G. delos Reyes, and D. S. Brar. 2003. Alien genes introgression and development of monosomic alien addition lines from *Oryza latifolia* Desv. to rice, *Oryza sativa* L. *Theor Appl Genet* 107:395–405.

Munns, R., G. J. Rebetzke, S. Husain, R. A. James, and R. A. Hare. 2003. Genetic control of sodium exclusion in durum wheat. *Aust J Agric Res* 54:627–635.

Munoz-Amatriain, M., L. Cistue, Y. Xiong et al. 2010a. Structural and functional characterization of a winter malting barley. *Theor Appl Genet* 120:971–984.

Munoz-Amatriain, M., Y. Xiong, M. Schmitt et al. 2010b. Transcriptome analysis of a barley breeding program examines gene expression diversity and reveals target genes for malting quality improvement. *BMC Genomics* 11:653.

Nelson, W. M., A. K. Bharti, E. Butler et al. 2005. Whole-genome validation of high-information-content fingerprinting. *Plant Physiol* 139:27–38.

Niu, X., W. Tang, W. Huang et al. 2008. RNAi-directed downregulation of *OsBADH2* results in aroma (2-acetyl-1-pyrroline) production in rice (*Oryza sativa* L.). *BMC Plant Biol* 8:100.

Niu, Z., D. L. Klindworth, T. L. Friesen et al. 2011. Targeted introgression of a wheat stem rust resistance gene by DNA marker-assisted chromosome engineering. *Genetics* 187:1011–1021.

Obert, J. C., W. P. Ridley, R. W. Schneider et al. 2004. The composition of grain and forage from glyphosate tolerant wheat MON 71800 is equivalent to that of conventional wheat (*Triticum aestivum* L.). *J Agric Food Chem* 52:1375–1384.

Paterson, A. H., J. E. Bowers, R. Bruggmann et al. 2009. The sorghum bicolor genome and the diversification of grasses. *Nature* 457:551–556.

Paterson, A. H., M. Freeling, and T. Sasaki. 2005. Grains of knowledge: Genomics of model cereals. *Genome Res* 15:1643–1650.

Paux, E., P. Sourdille, J. Salse et al. 2008. A physical map of the 1-gigabase bread wheat chromosome 3B. *Science* 322:101–104.

Peng, J., A. B. Korol, T. Fahima et al. 2000. Molecular genetic maps in wild emmer wheat, *Triticum dicoccoides*: Genome-wide coverage, massive negative interference, and putative quasi-linkage. *Genome Res* 10:1509–1531.

Peng, Z. S. 2003. A new mutation in wheat producing three pistils in a floret. *J Agron Crop Sci* 189:270–272.

Placido, D. F., M. T. Campbell, J. J. Folsom et al. 2013. Introgression of novel traits from a wild wheat relative improves drought adaptation in wheat. *Plant Physiol* 161:1806–1819.

Poland, J. A., P. J. Brown, M. E. Sorrells, and J. L. Jannink. 2012. Development of high-density genetic maps for barley and wheat using a novel two-enzyme genotyping-by-sequencing approach. *PLoS ONE* 7(2):e32253.

Pollack, A. 2011. U.S. approves corn modified for ethanol. Business Day, *New York Times*. Accessed: February 11, 2011.

Potokina, E., M. Caspers, M. Prasad et al. 2004. Functional association between malting quality trait components and cDNA array based expression patterns in barley (*Hordeum vulgare* L.). *Mol Breed* 14:153–170.

Potokina, E., N. Sreenivasulu, L. Altschmied, W. Michalek, and A. Graner. 2002. Differential gene expression during seed germination in barley (*Hordeum vulgare* L.). *Funct Integr Genomics* 2:28–39.

Pshenichnikova, T. A., I. F. Lapochkina, and L. V. Shchukina. 2007. The inheritance of morphological and biochemical traits introgressed into common wheat (*Triticum aestivum* L.) from *Aegilops speltoides* Tausch. *Genet Resour Crop Evol* 54:287–293.

Pshenichnikova, T. A., A. Simonov, M. Ermakova, A. Chistyakova, L. Shchukina, and E. Morozova. 2010. The effects on grain endosperm structure of an introgression from *Aegilops speltoides* Tausch. into chromosome 5A of bread wheat. *Euphytica* 175:315–322.

Qiao, F., Q. Yang, C. L. Wang, Y. L. Fan, X. F. Wu, and K. J. Zhao. 2007. Modification of plant height via RNAi suppression of *OsGA20ox2* gene in rice. *Euphytica* 158:35–45.

Quarrie, S. A., A. Steed, C. Calestani et al. 2005. A high-density genetic map of hexaploid wheat (*Triticum aestivum* L.) from the cross Chinese Spring × SQ1 and its use to compare QTLs for grain yield across a range of environments. *Theor Appl Genet* 110:865–880.

Regina, A., A. Bird, D. Topping et al. 2006. High-amylose wheat generated by RNA interference improves indices of large-bowel health in rats. *Proc Natl Acad Sci USA* 103:3546–3551.

Riley, R. and V. Chapman. 1958. Genetic control of the cytologically diploid behaviour of hexaploid wheat. *Nature* 182:713–715.

Rogers, C., J. Wen, R. Chen, and G. Oldroyd. 2009. Deletion-based reverse genetics in *Medicago truncatula*. *Plant Physiol* 151:1077–1086.

Safar, J., J. Bartos, J. Janda et al. 2004. Dissecting large and complex genomes: Flow sorting and BAC cloning of individual chromosomes from bread wheat. *Plant J* 39:960–968.

Šamajová, O., O. Plíhal, H. Al-Yousif, H. Hirt, and J. Šamaj. 2013. Improvement of stress tolerance in plants by genetic manipulation of mitogen-activated protein kinases. *Biotechnol Adv* 31:118–128.

Schnable, P. S., D. Ware, R. S. Fulton et al. 2009. The B73 maize genome: Complexity, diversity, and dynamics. *Science* 326:1112–1115.

Schulte, D., T. J. Close, A. Graner et al. 2009. The International Barley Sequencing Consortium—At the threshold of efficient access to the barley genome. *Plant Physiol* 149:142–147.

Sears, E. R. 1977. An induced mutant with homoeologous pairing in common wheat. *Can J Genet Cytol* 19:585–593.

Sehgal, S., W. Li, P. Rabinowicz et al. 2012. Chromosome arm-specific BAC end sequences permit comparative analysis of homoeologous chromosomes and genomes of polyploid wheat. *BMC Plant Biol* 12:64.

Simonov, A. V., T. A. Pshenichnikova, and I. F. Lapochkina. 2009. Genetic analysis of the traits introgressed from Aegilops speltoides Tausch. to bread wheat and determined by chromosome 5A genes. *Russ J Genet* 45:799–804.

Singh, R., E. Redoña, and L. Refuerzo. 2010. Varietal improvement for abiotic stress tolerance in crop plants: Special reference to salinity in rice. In A. Pareek, S. K. Sopory, and H. J. Bohnert (eds), *Abiotic Stress Adaptation in Plants*, pp. 387–415. New York: Springer.

Slade, A. J., S. I. Fuerstenberg, D. Loeffler, M. N. Steine, and D. Facciotti. 2005. A reverse genetic, nontransgenic approach to wheat crop improvement by TILLING. *Nat Biotechnol* 23:75–81.

Slade, A. J., C. McGuire, D. Loeffler et al. 2012. Development of high amylose wheat through TILLING. *BMC Plant Biol* 12:69.

Somers, D. J., P. Isaac, and K. Edwards. 2004. A high-density microsatellite consensus map for bread wheat (*Triticum aestivum* L.). *Theor Appl Genet* 109:1105–1114.

Stamova, B. S., D. Laudencia-Chingcuanco, and D. M. Beckles. 2009. Transcriptomic analysis of starch biosynthesis in the developing grain of hexaploid wheat. *Int J Plant Genomics* 2009:23.

Subbaiyan, G. K., D. L. E. Waters, S. K. Katiyar, A. R. Sadananda, S. Vaddadi, and R. J. Henry. 2012. Genome-wide DNA polymorphisms in elite indica rice inbreds discovered by whole-genome sequencing. *Plant Biotechnol J* 10:623–634.

Tan, S., R. R. Evans, M. L. Dahmer, B. K. Singh, and D. L. Shaner. 2005. Imidazolinone-tolerant crops: History, current status and future. *Pest Manage Sci* 61:246–257.

Tester, M. and A. Bacic. 2005. Abiotic stress tolerance in grasses. From model plants to crop plants. *Plant Physiol* 137:791–793.

Till, B. J., J. Cooper, T. H. Tai et al. 2007. Discovery of chemically induced mutations in rice by TILLING. *BMC Plant Biol* 7:19.

Till, B. J., S. H. Reynolds, C. Weil et al. 2004. Discovery of induced point mutations in maize genes by TILLING. *BMC Plant Biol* 4:12.

Travella, S., T. E. Klimm, and B. Keller. 2006. RNA interference-based gene silencing as an efficient tool for functional genomics in hexaploid bread wheat. *Plant Physiol* 142:6–20.

Trick, M., N. Adamski, S. Mugford, C.-C. Jiang, M. Febrer, and C. Uauy. 2012. Combining SNP discovery from next-generation sequencing data with bulked segregant analysis (BSA) to fine-map genes in polyploid wheat. *BMC Plant Biol* 12:14.

Tsai, H., T. Howell, R. Nitcher et al. 2011. Discovery of rare mutations in populations: TILLING by sequencing. *Plant Physiol* 156:1257–1268.

Tyagi, H., S. Rajasubramaniam, M. V. Rajam, and I. Dasgupta. 2008. RNA-interference in rice against Rice tungro bacilliform virus results in its decreased accumulation in inoculated rice plants. *Transgenic Res* 17:897–904.

Uemura, M. and P. L. Steponkus. 1997. Effect of cold acclimation on membrane lipid composition and freeze induced membrane destabilization. In P. H. Li and T. H. H. Chen (eds), *Plant Cold Hardiness. Molecular Biology, Biochemistry and Physiology*, pp. 171–179. New York: Plenum Press.

Varshney, R. K., A. Graner, and M. E. Sorrells. 2005. Genomics-assisted breeding for crop improvement. *Trends Plant Sci* 10:621–630.

Vasil, I. 1994. Molecular improvement of cereals. *Plant Mol Biol* 25:925–937.

Vitulo, N., A. Albiero, C. Forcato et al. 2011. First survey of the wheat chromosome 5A composition through a next generation sequencing approach. *PLoS One* 6:e26421.

Vollenweider, P. and M. S. Günthardt-Goerg. 2005. Diagnosis of abiotic and biotic stress factors using the visible symptoms in foliage. *Environ Pollut* 137:455–465.

Vrinten, P., T. Nakamura, and M. Yamamori. 1999. Molecular characterization of waxy mutations in wheat. *Mol Gen Genet* 261:463–471.

Wahid, A. and T. J. Close. 2007. Expression of dehydrins under heat stress and their relationship with water relations of sugarcane leaves. *Biol Plant* 51:104–109.

Wan, Y. F., R. Poole, A. Huttly et al. 2008. Transcriptome analysis of grain development in hexaploid wheat. *BMC Genomics* 9:121.

Wan, Y. F., C. Underwood, G. Toole et al. 2009. A novel transcriptomic approach to identify candidate genes for grain quality traits in wheat. *Plant Biotechnol J* 7:401–410.

Wang, L., A. Yang, C. He, M. Qu, and J. Zhang. 2008a. Creation of new maize germplasm using alien introgression from *Zea mays* ssp. mexicana. *Euphytica* 164:789–801.

Wang, N., Y. Wang, F. Tian et al. 2008b. A functional genomics resource for Brassica napus: Development of an EMS mutagenized population and discovery of *FAE1* point mutations by TILLING. *New Phytol* 180:751–765.

Watson, L. and R. J. Henry. 2005. Microarray analysis of gene expression in germinating barley embryos (*Hordeum vulgare* L.). *Funct Integr Genomics* 5:155–162.

Wei, F., J. Zhang, S. Zhou et al. 2009. The physical and genetic framework of the maize B73 genome. *PLoS Genet* 5:e1000715.

Wei, L. Q., W. Y. Xu, Z. Y. Deng, Z. Su, Y. Xue, and T. Wang. 2010. Genome-scale analysis and comparison of gene expression profiles in developing and germinated pollen in *Oryza sativa*. *BMC Genomics* 11:338.

Weil, C. F. and R. A. Monde. 2007. Getting the point—Mutations in maize abbreviations: EMS, ethyl methonyl sulfonate; TILLING, maize targeting induced local lesions in genomes. *Crop Sci* 47(Suppl. 1):S-60–S-67.

Wenzl, P., H. Li, J. Carling et al. 2006. A high-density consensus map of barley linking DArT markers to SSR, RFLP and STS loci and agricultural traits. *BMC Genomics* 7:206.

White, J., T. Pacey-Miller, A. Crawford et al. 2006. Abundant transcripts of malting barley identified by serial analysis of gene expression (SAGE). *Plant Biotechnol J* 4:289–301.

Wu, J. L., C. Wu, C. Lei et al. 2005. Chemical- and irradiation-induced mutants of indica rice *IR64* for forward and reverse genetics. *Plant Mol Biol* 59:85–97.

Xu, H., Y. Gao, and J. Wang. 2012. Transcriptomic analysis of rice (*Oryza sativa*) developing embryos using the RNA-Seq technique. *PLoS ONE* 7:e30646.

Yan, L., A. Loukoianov, A. Blechl et al. 2004. The wheat VRN2 gene is a flowering repressor down-regulated by vernalization. *Science* 303:1640–1644.

Yan, L., A. Loukoianov, G. Tranquilli, M. Helguera, T. Fahima, and J. Dubcovsky. 2003. Positional cloning of the wheat vernalization gene *VRN1*. *Proc Natl Acad Sci USA* 100:6263–6268.

Ye, X., S. Al-Babili, A. Kloti et al. 2000. Engineering the provitamin A (β-carotene) biosynthetic pathway into (carotenoid-free) rice endosperm. *Science* 287:303–305.

Zhang, F. D., L. M. Huang, W. D. Wang et al. 2012a. Genome-wide gene expression profiling of introgressed indica rice alleles associated with seedling cold tolerance improvement in a japonica rice background. *BMC Genomics* 13:461.

Zhang, L., J. M. Chia, S. Kumari et al. 2009. A genome-wide characterization of microRNA genes in maize. *PLoS Genet* 5:e1000716.

Zhang, L., J.-T. Luo, M. Hao et al. 2012b. Genetic map of *Triticum turgidum* based on a hexaploid wheat population without genetic recombination for D genome. *BMC Genet* 13:69.

Zhang, Z. Y., F. L. Fu, L. Gou, H. G. Wang, and W. C. Li. 2010. RNA interference-based transgenic maize resistant to maize dwarf mosaic virus. *J Plant Biol* 53:297–305.

Zhao, J.-X., W.-Q. Ji, J. Wu et al. 2010. Development and identification of a wheat–*Psathyrostachys huashanica* addition line carrying HMW-GS, LMW-GS and gliadin genes. *Genet Resour Crop Evol* 57:387–394.

Zhou, Y., Y. Yuan, F. Yuan et al. 2012. RNAi-directed down-regulation of *RSV* results in increased resistance in rice (*Oryza sativa* L.). *Biotechnol Lett* 34:965–972.

CHAPTER 14

Next-Generation Sequencing
Principle and Applications to Crops

Pradeep K. Jain, Pooja Choudhary Taxak, Prasanta K. Dash,
Kishor Gaikwad, Rekha Kansal, and Vijay K. Gupta

CONTENTS

14.1 INTRODUCTION

DNA sequencing has undergone a tremendous technological shift from the small-scale state-of-the-art Sanger sequencing (Sanger et al. 1977) to a large-scale venture comprising several of the latest technologies. This shift is marked by a large increase in throughput; greatly reduced per-base cost of raw sequence; specialized infrastructure of robotics, bioinformatics, databases, and instrumentation; and an accompanying requirement for extensive investment in equipment for proper utilization of the technologies. Prior to this metamorphosis in DNA sequencing technology, the landmark achievements by Sanger's and Gilbert's groups (Maxam and Gilbert 1977; Sanger and Coulson 1975) and the development of the chain termination method provided the framework for sequence-based research for decades. The original sequencing technology, the dideoxy chain termination method also commonly referred to as Sanger sequencing, dominated the sequencing industry for almost two decades, and was used to complete the human genome sequencing initiatives led by the International Genome Sequencing Consortium and Celera Genomics (International Human Genome Consortium 2004; Lander et al. 2001; Venter et al. 2001), along with other colossal accomplishments. The automated Sanger method pioneered by Sanger and Coulson (1975) and Sanger et al. (1977) is considered a *first-generation* technology and the newer methods are referred to as *next-generation sequencing* (NGS). These new technologies constitute various strategies that are based on a combination of one of the many protocols for template preparation, sequencing, imaging, genome alignment, and assembly. The commercial introduction of NGS technologies has changed our perspective on various scientific approaches in basic, applied, and clinical research.

14.2 HISTORY AND ADVANCES IN SEQUENCING TECHNOLOGIES

Since its initial introduction in 1977, Sanger sequencing remained conceptually unchanged. The method involved the DNA polymerase-dependent synthesis of a complementary DNA strand using natural 2′-deoxynucleotides (dNTPs) and termination of synthesis using 2′,3′-dideoxynucleotides (ddNTPs) that serve as nonreversible synthesis terminators. The introduction of a ddNTP into the growing oligonucleotide chain terminates the DNA synthesis reaction, resulting in a set of truncated fragments of varying lengths with an appropriate ddNTP at their 3′ terminus. These truncated fragments are then separated by size using high-resolution polyacrylamide gel electrophoresis and analyzed to reveal the DNA sequence. Ever-increasing advances in fluorescence detection technology (Prober et al. 1987; Smith et al. 1986), enzymology (Tabor and Richardson 1989), fluorescent dyes (Ju et al. 1995; Lee et al. 1997; Metzker et al. 1996), polymer biochemistry (Carrilho et al. 1996; Guttman 2002a,b; Madabhushi et al. 1996; Ruiz-Martinez et al. 1993; Salas-Solano et al. 1998), and capillary-array electrophoresis (Kheterpal et al. 1996; Takahashi et al. 1994) have helped DNA sequencing to reach its current status.

Automated Sanger sequencing involves tagging of either the primer or the terminating ddNTP with a specific fluorescent dye. Excitation and detection of the fluorophores produces fluorescence of four different colors as they pass through the detection region in the sequencers. Initially four different reactions, each containing a different ddNTP terminator (ddATP, ddCTP, ddTTP, or ddGTP) were required per template. However, advances in fluorescence detection allowed combination of all the four terminators into one reaction by labeling them with differently colored fluorescent dyes. Additionally, advances in electrophoresis systems replaced the original slab gel electrophoresis with capillary gel electrophoresis, allowing much higher electric fields to be applied to the separation matrix, enhancing the rate of separation of the fragments. The introduction of capillary arrays further increased the throughput of capillary electrophoresis (CE), as it allowed the analysis of a number of samples in parallel. Further, advances in polymer biochemistry, including the development of linear polyacrylamide and polydimethylacrylamide, allowed the reuse of capillaries in multiple electrophoretic runs, further increasing the sequencing efficiency. These advances, along with other major breakthroughs in sequencing technology, contributed to the relatively low error rate and long read lengths of the modern Sanger sequencers. The most advanced version of the automated high-throughput Sanger sequencer, with a 96-capillary-array format, is capable of sequencing up to 1 kb for 96 individual samples at a time. Sanger sequencing was used to obtain the first consensus sequence of the human genome in 2001 (Lander et al. 2001; Venter et al. 2001) and the first individual human diploid sequence of J. Craig Venter in 2007 (Levy et al. 2007). For the Human Genome Project (HGP), the core technologies of Sanger sequencing were coupled with fluorescence detection. Additionally, the HGP could not have been a success without major advances in recombinant protein engineering, fluorescent dye development, CE, automation, robotics, informatics, and process management. Despite many technical improvements and the robust performance of some of the high-end versions of automated Sanger sequencers (e.g., the 3730xl from Applied Biosystems), the application of this relatively expensive technique to large-scale sequencing projects has remained beyond the realm of a typical investigator and necessitated the need for new and improved technologies for sequencing a large number of human genomes.

Since 2004, efforts have been directed toward the development of new technologies called *next-generation* or *massively parallel* sequencing technologies, leaving Sanger sequencing with fewer reported advances. The NGS technologies have the ability to potentially generate enormous amounts of data cheaply—in some cases in excess of one billion sequence reads per instrument per run. These massively parallel high-throughput sequencers are capable of generating sequence reads from fragmented libraries of a particular genome (genome sequencing); from cDNA library fragments generated through reverse transcription (i.e., RNA sequence or transcriptome sequencing); and from a pool of PCR-amplified products (amplicon sequencing). Additionally, next-generation sequences are generated without the use of conventional vector-based cloning and *E. coli*-based cloning procedures, which form an integral part of capillary sequencing. As a result, some of the cloning bias issues that affect genome representation in sequencing projects may be avoided, although each platform may have its own associated limitations.

The NGS technologies face several limitations or challenges. There have been several attempts to address these, and they are still undergoing improvement. The first limitation of the next-generation sequencers is the shorter read lengths (35–250 bp, depending on the platform) compared with capillary sequencers (650–800 bp), which affects the utility of data for various applications. However, the developing NGS technologies, such as single-molecule sequencing, which has the potential to read several continuous kilo base pairs (kbp), may surpass Sanger sequencing in the near future. The second challenge is associated with the accuracy of their sequencing reads and quality values, although efforts are under way to benchmark them relative to Sanger sequencing. Third, the amplification step prior to sequencing includes different sources of polymerase chain reaction (PCR) bias, formation of chimeric sequences, and secondary structure-related issues (Mardis 2008a; Shendure and Ji 2008). A timeline comparison of different NGS technologies in terms of read length, accuracy, and total output reveals rapid progress in the sequencing abilities of NGS platforms (Table 14.1).

Today, several NGS platforms coexist in the market, each striking a different balance between cost, read length, data volume, and rate of data generation, with some having advantages over the other for particular applications. Although the currently available NGS platforms utilize quite different chemistry and base incorporation and detection tools, they have two main steps in common: template library preparation and detection of incorporated nucleotides (Glenn 2011;

Table 14.1 Comparative Profile of Different Next-Generation Sequencing (NGS) Technologies

Generation	Company	Platform	Template Preparation Method	Sequencing Method	Detection Method	Approximate Read Length (Bases)	Approximate Sequencing Output/Run	Run Time (Days)
Second	Roche	Roche/454GS FLX system	Fragment, mate pair library/ emPCR	Pyrosequencing	Optical	350–450	0.45 GB	0.35
Second	Illumina	Illumina GA IIx	Fragment, mate pair library/ bridge amplification	Reversible terminator/SBS	Fluorescence/ optical	50–75	≤95 GB	7–14
Second	ABI	ABI/SOLiD 5500xl system	Fragment, mate pair library/ emPCR	Cleavage probe/ sequencing by ligation	Fluorescence/ optical	35–75	~250 GB	7–8
Second	Helicos	HeliScope	Fragment, mate pair library/ single molecule	Single molecule/SBS	Fluorescence/ optical	25–35	~20–28 GB	≤1
Third	Pacific biosciences	PacBio RS system	Fragment library/ single molecule	Real-time SMS	Fluorescence/ optical	1000	~60–75 Mb	0.02
Third	Life Technologies/ Ion Torrent	Personal genome machine (with Ion Torrent 314, 316, and 318 chip)	Single molecule	SBS	Change in pH detected by ion-sensitive field effect transistors (ISFETs)	100–200	≥10 Mb for 314, ≥100 Mb for 316 and ≥1 GB for 318 chip	0.15 for 314, 0.2 for 316 and 0.23 for 318 chip

Zhang et al. 2011). Currently, six NGS platforms available commercially are classified into two groups. The first group incorporates PCR-based technologies and is comprised of four platforms: Roche GS FLX 454 sequencer (Roche Diagnostics Corp., Branford, CT, USA), Illumina genome analyzer (Illumina Inc., San Diego, CA, USA), ABI SOLiD System (Life Technologies Corp., Carlsbad, CA, USA), and Ion Personal Genome Machine (Life Technologies, South San Francisco, CA, USA). The second group, including the HeliScope (Helicos BioScience Corp., Cambridge, MA, USA) and PacBio RS single-molecule real-time (SMRT) system (Pacific Biosciences, Menlo Park, CA, USA), is based on single-molecule sequencing technologies and hence does not require an amplification step prior to sequencing. Among the six commercially available platforms, the Illumina/Solexa Genome Analyzer, Roche 454 GS FLX sequencer, Applied Biosystems SOLiD Analyzer, and HeliScope (second-generation sequencing technologies) currently dominate the market, whereas the Pacific Biosciences PacBio RS SMRT system and Ion Personal Genome Machine by Life Technologies (third-generation sequencing technologies) have been introduced recently and are hence not in wide use. With the final goal of bringing the cost of human genome sequencing to under $1,000, alternative approaches to improve second- and third-generation sequencing, as well as novel approaches to sequencing, are in development. These include the use of scanning tunneling microscope (TEM), fluorescence resonance energy transfer (FRET), single-molecule detector, and protein nanopores. With the NGS technologies, investigations that were unreachable luxuries just a few years ago are being increasingly enabled at a rapid pace. Large-scale sequencing centers are now switching to NGS. On the other hand, until small-scale next-generation sequencers can outperform CE on cost per accurate base called as well as read length, CE systems will likely remain in heavy use at the bench-top scale for targeted sequencing for directed investigations.

14.3 NEXT-GENERATION SEQUENCING METHODS

Template preparation, sequencing, imaging, and data analysis are the general steps included in all the NGS technologies, and the unique combination of different steps differentiates one technology from another. In the following text, different methods of template preparation, sequencing, and imaging that are used by the current and near-future NGS platforms are discussed.

Clonally amplified templates and single DNA molecule templates are the two methods used for template preparation, whereas different methods of sequencing are classified as cyclic reversible termination (CRT), single-nucleotide addition (SNA), sequencing by ligation (SBL), and real-time sequencing. Different methods of imaging used by NGS platforms include measuring bioluminescent signals, pH changes, and temperature changes, and one- to four-color imaging of single molecular events. Additionally, the massive scale of sequencing requires a matching scale of computational analysis that includes image analysis, signal processing, background subtraction, base calling, and quality control to produce the final sequence data from each run. In other words, these analyses place substantial demands on the information technology (IT), data storage, packing, and library information management system (LIMS) infrastructures.

14.3.1 Template Preparation

The standard methodology followed by different NGS technologies for template preparation involves randomly breaking genomic DNA into small fragments, which are then used to generate either a fragment library or a mate pair library. The template fragments in libraries are then immobilized on a solid surface or support to allow thousands to billions of sequencing reactions to be undertaken simultaneously.

14.3.1.1 Clonally Amplified Templates

Amplification of templates is required to generate enough signal to be detected by the imaging systems, which are incapable of detecting single fluorescent signals. The two common methods used to prepare clonally amplified templates are emulsion PCR (emPCR) (Dressman et al. 2003) and solid-phase amplification (Fedurco et al. 2006).

14.3.1.1.1 Emulsion PCR

emPCR allows cloned amplification of the templates without the use of the bacterial cloning method, which usually results in the loss of genomic sequences. In emPCR, a fragment or mate pair library is generated, following which adapter sequences carrying universal priming sites are attached to the library templates to allow their PCR amplification using universal sequencing primers. Adapter-ligated DNA templates are made single stranded and then captured onto agarose beads whose surfaces are decorated with oligomers complementary to adapter sequences under conditions which favor the hybridization of one DNA molecule per bead. Each of the beads, carrying a single DNA fragment hybridized to the oligodecorated surface, is then isolated into the individual oil–water micelles containing PCR reagents by mixing bead–DNA complex in water–oil emulsion and vortexing. The micelles are then subjected to emPCR to generate millions of copies of a single template molecule present on the surface of each bead. After the successful amplification and enrichment of emPCR beads, millions of copies of them can be chemically cross-linked to an amino-coated glass surface (Life/APG; Polonator), immobilized on polyacrylamide gel on a standard microscope slide (Polonator; Shendure et al. 2005), or deposited into individual PicoTiter plate (PTP) wells (Roche/454 Genome analyzer; Leamon et al. 2003), in which NGS reactions can be performed.

14.3.1.1.2 Solid-Phase Amplification

In the solid phase, DNA amplification is achieved by attaching adapter-ligated single-stranded library fragments to a solid surface called a single-molecule array or flow cell and then conducting solid-phase bridge amplification of these fragments. In bridge amplification, one end of the single-stranded DNA fragment is immobilized to a solid surface through an adapter. The fragments subsequently bend over and hybridize (creating the bridge) to complementary primers covalently attached to the solid surface in high density, thereby forming the template for the synthesis of their complementary strands. After amplification, a flow cell with 100–200 million spatially separated clusters is produced, wherein each cluster is composed of millions of copies of a single template molecule, which provide free ends to which a universal sequencing primer can be hybridized to initiate the NGS reaction.

14.3.1.2 Single-Molecule Templates

Although clonally amplified templates carry certain advantages over bacterial cloning methods, including increase in the levels of signals obtained, the protocols have certain disadvantages also: (i) clonal amplification necessarily introduces amplification bias, such that the distribution of sequence reads on the template sequence is neither uniform nor random, finally resulting in generation of "hot spots" and "cold spots" of artificially deep or shallow coverage, respectively; (ii) the stepwise sequencing of a population of template molecules leads to inaccurate reads if the individual molecules fall out of synchronization or become "out of phase;" (iii) the protocols for clonal amplification require a large amount (3–20 µg) of genomic DNA material.

Single-molecule template preparation methods deliver consistently low error rates by avoiding PCR-associated bias and problems with intensity averaging and phasing or synchronization, and also require less starting material (<1 μg). Before setting up a NGS reaction, single-molecule templates are generally immobilized on a solid support using any one of at least three different approaches. In the first approach, individual primer molecules complementary to adapters ligated to library fragments are covalently attached to the solid support (Harris et al. 2008), following which the template molecules prepared by randomly cleaving the starting material into smaller fragments and adding common adapters to the ends are hybridized to the immobilized primer. In the second approach, spatially distributed primer molecules are immobilized on the solid support, following which single-stranded adapter-ligated single-molecule templates are hybridized to the immobilized primer and the primer is extended, resulting in covalently attached single-stranded single-molecule templates. A common primer is then hybridized to the template. In both of the above approaches, DNA polymerase binds to the immobilized primed template configuration to initiate the NGS reaction (Harris et al. 2008). In the third approach, spatially distributed single polymerase molecules are attached to the solid support (Eid et al. 2009), to which a single primed template molecule is attached. This approach can be used with larger DNA molecules and with real-time methods, resulting in longer read sequences. Clonal amplification generates a population of identical templates, each of which has undergone the sequencing reaction, and the signal observed upon imaging represents the consensus of the nucleotides or probes added to the identical templates for a given cycle. Dephasing (both lagging and leading strand) is the major problem with clonally amplified templates, occurring when individual molecules move out of synchronicity, and results in increased fluorescence noise, base-called errors, and shorter reads (Erlich et al. 2008). Dephasing is not a problem with single-molecule templates; however, they are susceptible to multiple nucleotide or probe additions. Additionally, deletion errors due to quenching effects between adjacent dye molecules or no signal detection due to the incorporation of dark nucleotides also limit the approach of single-molecule template preparation.

14.3.2 Sequencing and Imaging

Different methods of sequencing can be broadly classified as CRT, SBL, SNA, and real-time sequencing. They are discussed in the following sections.

14.3.2.1 Cyclic Reversible Termination

As the name suggests, CRT is a cyclic method that comprises incorporation of modified nucleotides, also called reversible terminators, imaging of the fluorescence generated by fluorescent dye attached to the nucleotide, and cleavage of the fluorescent dye and terminating or inhibiting group. In this method, DNA polymerase binds to the primed template and incorporates a single fluorescently modified nucleotide or reversible terminator, which is complementary to the template base. This is followed by washing of the unincorporated nucleotides. The incorporated nucleotides are then identified by imaging, following which the terminating or inhibiting groups and the fluorescent dye are removed by a cleavage step. An additional washing step is also included before switching to the next incorporation step.

Two types of reversible terminators are utilized by the CRT method: 3′-blocked terminator and 3′-unblocked terminator. 3′-blocked terminators carry a cleavable group attached to the 3′-oxygen of the 2′-deoxyribose sugar. Blocking groups such as 3′-O-allyl-2′-deoxyribonucleoside triphosphates (dNTPs) (Ju 2006) and 3′-O-azidomethyl-dNTPs (Guo et al. 2008; Bentley et al. 2008) have been successfully utilized in the CRT method. Mutant DNA polymerase is used to facilitate the incorporation of the 3′-blocked terminators. Additionally, while using 3′-blocked terminators, two chemical bonds are cleaved, one to remove the fluorophore from the nucleotide and the other to restore the 3′-OH group.

The need to screen large libraries of mutant DNA polymerases to incorporate 3′-blocked terminators led to the development of 3′-unblocked reversible terminators. The 3′-unblocked terminators show more favorable enzymatic incorporation and can be incorporated using wild-type DNA polymerase. A small terminating group attached to the base of a 3′-unblocked nucleotide can act as an effective reversible terminator (Lightning Terminators™ (LaserGen Inc.); Wu et al. 2009), whereas a second nucleoside analog attached to the base of a 3′-unblocked nucleotide can act as an inhibitor (Virtual Terminators™ (Helicos BioSciences); Bowers et al. 2009). The terminating (Lightning Terminators) or inhibiting (Virtual Terminators) groups of the 3′-unblocked terminators need appropriate modifications so that they can terminate DNA synthesis once a single nucleotide is added. This is essential because the 3′-unblocked terminators contain a free 3′-OH group, which is a natural substrate for incorporating the next incoming nucleotide. In the case of 3′-unblocked terminators, cleavage of a single bond releases both the terminating or inhibiting group and the fluorophore group from the base.

14.3.2.2 Sequencing by Ligation (SBL)

As the name implies, SBL uses DNA ligase (Tomkinson et al. 2006) instead of DNA polymerase in a cyclic method that comprises probe hybridization, ligation of probe to the primer, fluorescence imaging, and cleavage. The method uses either a one-base encoded probe, which is an oligonucleotide sequence in which one interrogation base is associated with a particular dye, or a two-base encoded probe, in which two interrogation bases are associated with a particular dye in an oligonucleotide sequence. The first step involves the hybridization of labeled probe to its complementary sequence adjacent to the primer, followed by a DNA ligase-mediated joining of the dye-labeled probe to the primer. A washing step is then incorporated to wash away the nonligated probes, following which fluorescence is imaged to determine the identity of the ligated probe (Landegren et al. 1988). The next cycle can be initiated either by using cleavable probes to remove the fluorescent dye and regenerate a 5′-PO$_4$ group for subsequent ligation cycles or by removing the dye and hybridizing a new primer to the template.

14.3.2.3 Single-Nucleotide Addition: Pyrosequencing

In pyrosequencing, each nucleotide incorporated by DNA polymerase results in the release of a pyrophosphate molecule that initiates a series of downstream enzymatic reactions to produce light by the action of the enzyme luciferase. The amount of light generated is directly proportional to the number of nucleotides incorporated (Margulies et al. 2005; Ronaghi et al. 1998). In the first step, DNA beads amplified by emPCR are loaded into the individual PTP well in such a manner that each PTP well carries a single DNA bead. Smaller magnetic beads containing enzymes, sulfurylase, and luciferase are also loaded into each well to surround the DNA bead. The slide is then mounted in a flow chamber, following which the sequencing reagents, along with a single type of 2′-dNTP, are added to each well. Following the incorporation of complementary dNTP, DNA polymerase extends the primer and pauses. The addition of the next complementary dNTP then reinitiates the DNA synthesis. The order and the intensity of light generated from each PTP well undergoing the pyrosequencing reaction are then recorded as a series of peaks or flowgrams of high resolution using a charge-coupled device (CCD) camera planted below the fiber-optic slide, and the DNA sequence data are revealed.

14.3.2.4 Real-Time Sequencing

Real-time sequencing technology is the third generation of sequencing technology to be launched commercially, and is currently led by Pacific Biosciences. The method of real-time sequencing involves recording the fluorescence emitted as the phosphate chain is cleaved by the continuous incorporation of dye-labeled nucleotides during DNA synthesis by DNA polymerase

(Metzker 2009). In the Pacific Biosciences platform, sequence information of template DNA is obtained when the single DNA polymerase molecules deposited at the bottom of individual zero-mode waveguide detectors (ZMW) incorporate phospholinked nucleotides into the growing primer strand.

14.4 PCR-BASED NEXT-GENERATION SEQUENCING PLATFORMS

14.4.1 Roche/454 FLX Pyrosequencer

Introduced in 2004, the Roche 454 Genome Sequencer was the first NGS technology to gain commercial relevance, and is based on the sequencing-by-synthesis (SBS) pyrosequencing technology. The pyrosequencing approach makes use of the pyrophosphate molecule released on each incorporation of a nucleotide by DNA polymerase to fuel a downstream enzyme cascade that finally produces light from the cleavage of oxyluciferin by luciferase. The amount of light emitted is directly proportional to the number of a particular nucleotide incorporated, up to the level of detector saturation. In this approach, instead of sequencing in PCR tubes or microtiter plate wells, the library templates are amplified *en masse* following the technique of emPCR (Dressman et al. 2003) on the surface of agarose beads. In emulsion PCR, the library fragments are mixed with agarose beads with millions of oligomers attached to their surfaces that are complementary to the 454-specific adapter sequences ligated or PCR generated on both ends of the fragments during library construction. Each of these beads, carrying a single unique DNA fragment, then hybridizes to the oligodecorated surface and is separated into the individual oil: aqueous micelles containing PCR reagents. The DNA beads are then subjected to emPCR to generate millions of copies of the same fragment covering the surface of each bead. The amplified beads are recovered from the emulsion, followed by an enrichment step that retains only the amplified beads, discarding the failed ones. The beads (each containing a unique amplified fragment) are then arrayed into the several hundred thousand single wells on the surface of the PTP, with each well holding a single bead and providing a fixed location for each sequencing reaction to be monitored. Subsequently, much smaller magnetic and latex beads 1 μm in diameter, containing active enzymes (sulfurylase and luciferase), are added and centrifuged to surround the DNA-containing agarose beads in the PTP wells. The PTP is then placed in the sequencer, where it acts as a flow cell into which each of the nucleotides and other pyrosequencing reagents are delivered in a sequential fashion. Each incorporation step is then followed by imaging, in which a CCD camera placed opposite the PTP records the light emitted from each bead due to luciferase activity. A defined single-nucleotide pattern in the adapter sequence adjacent to the universal sequencing primer, which corresponds to the sequence of the first four sequences added, enables the 454-analysis software to calibrate the level of light emitted from single-nucleotide incorporation for the downstream base-calling analysis that occurs after the run is completed. For homopolymeric repeats of up to six nucleotides, the number of dNTPs incorporated is directly proportional to the intensity of light. However, long stretches of homopolymers (more than six) cannot be properly interpreted by the calibrated base calling; hence, for runs of multiple nucleotides (homopolymers) the linearity of response can exceed the detector sensitivity, making these areas (reads) prone to indel errors during base calling. On the other hand, the sequential flow of nucleotides entirely eliminates the occurrence of substitution errors in the Roche/454 sequence reads. The current GS FLX system provides 200 flow cycles of nucleotides, giving, an average read length of 800 bp during a 7 h run. These raw signals are processed by 454 pyrosequencing analysis software and then screened by various quality filters to remove poor-quality sequences (Mardis 2008a), resulting in a combined throughput of 100 Mb of high-quality sequence data. After the processing of the FLX sequences, they are assembled using the assembly algorithm (Neobler). The raw base accuracy reported by Roche is over 99%. Roche 454 genome sequencers are currently available in two versions: GS FLX+ system (1 Mb sequence read capacity) and the recently introduced GS Junior system (100 kb sequence read capacity).

14.4.2 Illumina Genome Analyzer

Genome Analyzer, introduced by Illumina (formerly known as Solexa) in 2007, currently dominates the NGS market. The Illumina platform is based on the concept of SBS coupled with bridge amplification on the surface of a flow cell. Single-stranded adapter-ligated DNA fragments are attached to a solid surface, known as a single-molecule array or flow cell, using a microfluidic cluster station. Each flow cell is an eight-channel sealed glass microfabricated device, on the interior containing covalently attached oligos complementary to the specific adapters ligated onto the library fragments. The DNA fragments are hybridized to the oligos using active heating and cooling steps, followed by subsequent incubation with the amplification reagents and an isothermal polymerase that results in the generation of discrete areas or clusters of the library fragments.

The flow cell is then placed in the fluidics cassette within the sequencer. Each cluster is supplied with all four reversible terminators (modified nucleotides) with removable fluorescent moieties and a special DNA polymerase that is capable of incorporating the terminators into the growing oligonucleotide chains. Terminators are differentially labeled fluorescent nucleotides with their 3' OH groups chemically blocked. This modification of 3'OH ensures that only a single base is incorporated per cycle. Each nucleotide incorporation step is then followed by an imaging step to identify the incorporated nucleotides on each cluster, following which a chemical treatment cleaves the fluorescent group and deblocks the 3' end, preparing each strand for incorporation of the next base in the next flow cycle. This series of steps is continued for a specific number of cycles, as determined by user-defined instrument settings: generally a read length of 25–35 bases. At the end of the sequencing run, the sequence of each cluster is computed and subjected to quality filtering to remove poor-quality sequences. A typical Illumina genome analyzer yields ~35 bp reads, producing at least 1 GB of sequence per run of two to three days, with raw base accuracy greater than 99.5%. The Illumina approach is incapable of resolving short sequence repeats in spite of being more effective at sequencing homopolymeric stretches than pyrosequencing (Bentley 2006). Due to the use of modified DNA polymerases and reversible terminators, substitution errors are the most common error types noted in Illumina sequencing data, with a higher proportion of errors occurring when G is the previous incorporated nucleotide (Dohm et al. 2008). Additionally, an underrepresentation of AT-rich (Dohm et al. 2008; Harismendy et al. 2009) and GC-rich regions (Harismendy et al. 2009; Hillier et al. 2008) was revealed by the genome analysis of Illumina data, probably due to amplification bias during the template preparation. The Illumina Genome Analyzer is the most adaptable and easy-to-use sequencing platform. Superior data quality, proper read length, and high capacity make it a system of choice for whole-genome sequencing applications, including human and model organisms. At present, four versions of Illumina sequencers dominate the commercial market: the HiSeq 2000, HiSeq 1000, and Genome Analyzer IIx have sequencing outputs of up to 600, 300, and 95 GB, respectively. The recently introduced MiSeq platform is capable of generating up to 150 bp sequencing reads with a combined throughput of 1.5–2 GB per run. In 2012, Illumina introduced the HiSeq 2500 platform as an upgraded form of HiSeq 2000. It is capable of generating up to 120 GB of data in 27 h, resulting in the sequencing of the entire genome in 24 h (i.e., genome in a day).

14.4.3 Applied Biosystems SOLiD Analyzer

Applied Biosystems (Life Technologies) introduced SOLiD (support oligonucleotide ligation detection) technology in 2007 and commercialized it as their NGS platform. SOLiD uses a unique SBL approach catalyzed by DNA ligase. The process used in this platform involves attaching the adapter-ligated library templates to 1 μm magnetic beads whose surfaces are covered with the oligos complementary to the SOLiD-specific adapter sequences, and then amplifying each of the DNA–bead complexes by emPCR. Following amplification, the beads are covalently attached to the surface of a chemically treated glass slide that is placed into a fluidics cassette within the sequencer. Initiation of ligation-based sequencing is marked by the hybridization of a universal sequencing primer complementary to the SOLiD-specific adapters ligated to the library templates amplified by emPCR, following which the semidegenerate 8-mer fluorescent oligos and the DNA ligase are added in an automated manner within the instrument. DNA ligase seals the phosphate backbone as soon as the matching 8-mer oligo hybridizes to the DNA fragment sequence adjacent to the attached universal sequencing primer at the 3' end. The ligation step is followed by an imaging step, in which a fluorescent readout identifies the ligated 8-mer oligo that corresponds to one of the four possible nucleotides. Subsequently, the linkage between the fifth and the sixth base of the ligated 8-mer is cleaved chemically to remove the fluorescent group, enabling the subsequent round of ligation. The probe hybridization, ligation, imaging, and cleavage cycle is repeated 10 times to yield 10 color calls spaced at five-base intervals, following which the extended primer (synthesized fragment) is stripped from the bound templates by denaturation. The second round of sequencing starts with the hybridization of the $n-1$ positioned universal primer and subsequent cycles of ligation-mediated sequencing. This round resets the interrogation bases and the corresponding color calls one position to the left. The same process is repeated with $n-2$, $n-3$ and $n-4$ positioned universal primers. The fluorescence obtained from the five ligation rounds is then decoded with a two-base calling processing software to generate the color calls, which are ordered into a linear sequence and aligned to a reference genome to decode the DNA sequence. The use of two base-encoded probes enables an extra quality check of the accuracy of reads in color calling and SNV calling. Additionally, the two base-encoding schemes enable the distinction between a sequencing error and a sequence polymorphism: an error would be detected in only one particular ligation reaction, whereas a polymorphism would be detected in both. In SOLiD system, two slides can be processed per instrument run, one slide receiving sequencing reagents as the second is being imaged (Mardis 2008b), and each slide can be divided to contain different libraries in four or eight quadrants. With the SOLiD data, substitutions are the most common error types, along with underrepresentation of AT-rich and GC-rich regions, as is the case with the genome analysis of Illumina/Solexa reads. The read length for the SOLiD Analyzer is user defined between 25 and 35 bp, with a combined throughput of 2–4 Gbp per sequencing run. Today, Applied Biosystems SOLiD sequencers are available in two versions, the 5500 systems and the 5500xl system, with up to 100 and 250 GB sequencing capacity, respectively, and a raw base accuracy of 99.94%.

14.4.4 Life Technologies Ion Torrent

This NGS platform can be regarded as the world's smallest solid-state pH meter. Ion Personal Genomic Machine (Ion Torrent), a start-up introduced by Life Technologies in 2010, marked significant progress in bringing to the world market a next-generation (third-generation) sequencing system that utilizes pH changes to detect base incorporation events. The system is based on the real-time detection of hydrogen ions, a by-product of nucleotide incorporation into a growing DNA strand by DNA polymerase. To achieve high throughput, Ion Torrent makes use of sequencing chips consisting of a high-density array of microwells, where each well acts as an individual DNA polymerization chamber containing a DNA polymerase and the sequencing template. Beneath the layer of microwells is an ion-sensitive layer, and beneath this is a highly dense field effect transistor (FET) array (acting as ion sensor) aligned with the array of microwells. As in the pyrosequencing scheme, the four nucleotides are added to the microwell in sequential order, and the pH change created during nucleotide incorporation is detected by the FET sensors, which convert this signal to a recordable voltage change, thereby revealing the primary sequence. The change in voltage is directly proportional to the number of nucleotides added at each step.

At present, Ion Torrent offers three different sequencing chips: Ion 314, Ion 316, and Ion 318. Ion 314 chips carry 1.2 million microwells, generating roughly 10 Mb of sequence data with an average read length of 100 bases. The second-generation sequencing chip Ion 316 carries 6.2 million microwells, generating 100 Mbp of sequence information with an average read length of 100 bases, while the third-generation sequencing chip 318 is built with 11.1 million microwells to produce 1 GB of sequencing data with an average read length of 200 bases. In 2012, Life Technologies "democratized" sequencing by introducing a further new generation of Ion semiconductor sequencers called the Ion Proton bench-top sequencers, which offer a reasonable price, bench-top scale, and high-throughput sequencing, and can potentially decipher the human genome or the human exome in just a few hours. There will be two versions of Ion Proton chips: the Ion Proton I chip, built with 165 million wells (about a hundredfold more than the Ion 314 chip), and the Ion Proton II chip, having 660 million wells (about a thousandfold more than the Ion 314 chip). These chips will be based on complementary metal oxide semiconductor (CMOS) technology to record chemistry changes instead of light and translate these changes into digital data.

This newly introduced method of SBS based on ion sensing greatly reduces the sequencing cost but has several limitations as far as sequencing the complete genome is considered. The first limitation is posed by short read lengths, which limit the assembly of *de novo* sequencing projects due to inability to read long repetitive regions in the genome. Second, due to sequential addition of nucleotides, error accumulation can occur if reaction wells are not properly rinsed between reaction steps. Thirdly, as is the case with pyrosequencing, sequencing through smaller repetitive regions of the same nucleotide (monopolymeric) regions of 5–10 bases can be challenging.

14.5 SINGLE-MOLECULE DNA SEQUENCING PLATFORMS

14.5.1 Pacific Biosciences SMRT DNA Sequencer

In 2010, Pacific Biosciences (PacBio) introduced a reliable third-generation sequencing platform based on the SMRT DNA synthesis technology. The technology directly measures the fluorescence emitted by the cleavage of the phosphate chain of fluorescently labeled nucleotides incorporated by DNA polymerase onto a complementary sequencing template. The most important part of this technology is a dense array of nanostructures called ZMW, which allow optical interrogation of single fluorescent molecules using electron beam lithography and ultraviolet photolithography. PacBio efficiently packed ZMW nanostructures onto a surface and were also successful in developing a parallel confocal imaging system that revealed high sensitivity and resolution of fluorescent nucleotides in each of the ZMW nanostructures. After the development of the ZMW array fabrication and detection scheme, the major technical challenge for this technology was to immobilize a single DNA polymerase molecule at the base of each ZMW that can incorporate fluorescently labeled nucleotides efficiently. As a first step toward this, a set of fluorescently labeled deoxyribonucleoside pentaphosphate (dN5P) substrates was synthesized to enable the spectrum differentiation of each base without decreasing the processivity of the DNA polymerase (Korlach et al. 2008). In the second step, the surface of each ZMW nanostructure, which was composed of a fused silicon bottom layer and an aluminum top layer, was chemically treated to allow the selective localization of the DNA polymerase. The derivation of the aluminum surface with polyvinylphosphonic acid (PVPA) significantly decreased the protein adsorption to the aluminum layer without compromising protein adsorption to the bottom glass layer (Korlach et al. 2008). SMRT bell templates were generated by PacBio for SMRT sequencing technology, which allows consecutive sequencing of both the sense and the antisense strand of double-stranded DNA fragments by ligating universal hairpin loops to the ends of the fragment. This technology does not require any amplification step for the template preparation, thus reducing the time needed for sample preparation. Additionally, DNA fragments over a broad size range can be used to generate SMRT bell templates. The accuracy of sequencing and single-nucleotide polymorphism (SNP) detection also increases with the use of bell templates.

Once the ZMW array fabrication, immobilization of the DNA polymerase, and preparation of the SMRT bell templates are accomplished, the complementary DNA strand is synthesized from a single-stranded template by the action of DNA polymerase attached at the bottom of each waveguide. In this technology, the fluorescent label is attached to the terminal phosphate group rather than the nucleotide base, leading to the release of a different colored fluorescent moiety with nucleotide incorporation (Flusberg et al. 2010; Pushpendra 2008). The technology eliminates the need for a washing step between nucleotide flows, accelerating the speed of nucleotide incorporation and improving sequence quality. Additionally, the natural capacity of DNA polymerase to incorporate 10 or more nucleotides per second in several thousand parallel ZMWs (Eid et al. 2009; Zhou et al. 2010) is utilized in this approach. The PacBio RS system, the commercially available sequencing system introduced by Pacific Biosciences, includes single-use ZMW arrays (called SMART cells) that contain 150,000 ZMWs and kits for SMRT bell template preparation. The robustness of the genetic data generated by the PacBio's single-nucleotide sequencing arrays was improved by correlating polymerase kinetic data to DNA methylation pattern during DNA sequencing (Flusberg et al. 2010). Additionally, to sequence mRNA strands using this technology, the DNA polymerase attached to the bottom of each ZMW can be replaced with a ribosome and the incorporation of fluorescently labeled tRNAs can be monitored.

14.5.2 Helicos BioSciences HeliScope

The HeliScope, introduced in 2008 by Helicos BioSciences, was the first commercially available single-molecule sequencing (SMS) platform to rely upon a highly sensitive fluorescence detection system to directly record each nucleotide as it is incorporated. The system utilizes SBS using a one-color CRT method on a single DNA molecule template. Using the HeliScope method, template fragments 100–200 bases in size were first attached to a substrate within a microfluidic flow cell. During sequencing, nucleotides carrying fluorescent dye (which increases detectability and removes the need for amplification of the template DNA) are introduced one species at a time and incorporated by DNA polymerase to the growing complementary strand. The fluorescent nucleotides carry appropriate modifications to stop the polymerase extension until the fluorescence of the incorporated nucleotides is captured, recorded, and analyzed to identify which nucleotide was incorporated into which growing strand with the help of a highly sensitive CCD camera connected to the fluorescent microscope. Unincorporated nucleotides and the by-products of the previous cycles are then washed off, following which the fluorescent labels on the extended strands are cleaved by chemical treatment and removed. Another cycle of addition of different species of nucleotide, label cleaving, and imaging then follows (Ewing et al. 1998; Harris et al. 2008; Zhang et al. 2011).

Harris et al. (2008) used Cy-5-1255-dNTPs, the earliest versions of their Virtual Terminators, lacking the inhibiting group and reported that the deletion errors in homopolymeric repeat regions were found to be the most common error types when the primer immobilization strategy was used to generate single-molecule templates. This may be due to incorporation of two or more Cy5-1255-dNTPs in a given cycle. These errors can be reduced to a great extent using two-pass sequencing, which gives ~25 base consensus reads using the immobilized template strategy. The read length obtained using HeliScope ranges from 30 to 35 bp, with 20 to 28 Gbp of potential sequence reads per run and a raw base accuracy >9.

14.6 APPLICATIONS OF THE CURRENT SEQUENCING TECHNOLOGIES TO CROPS

The simultaneous development of new sequencing technologies over the past 7 years or so has initiated a new dawn in the era of structural and functional genomics. Each of these technologies is based on a different principle and has a distinct role and suitability as per the experiment and the organism to be sequenced. However, there is one common thread in all these technologies: they produce data on a scale never before imagined and bypass the tedious cloning procedures. The cost and time required for sequencing have come down to levels where individual users can now execute their own experiments rather than having huge consortiums and genome centers. The science of genomics is now entering a phase where sequencing of model organisms is not a priority, as these technologies have led to the concept of personal genomics. This means that no organism remains untouched just because of its genome size and complexity, and this is a boon for solving complex biological problems in fields as varied as agriculture, environment, and medicine (Deschamps and Campbell 2010; Egan et al. 2012; Marguerat et al. 2008; Morozova and Marra 2008; Pareek et al. 2011). Some of the typical applications have addressed both RNA and DNA, and are profiled below.

14.6.1 Sequencing of Whole Genomes

Genome sequencing has come a long way over the past few decades, from sequencing partial genes to a set of genes, then large chunks of chromosomes, and finally the whole genome itself. All of this was possible due to periodic innovations in the application of the Sanger sequencing methodology. Each organism is complex in its own way. This was observed as the sequencing march began from the most simple to the most complex of organisms. Hence, sequencing of a model organism became a compulsion rather than a choice, spawning a promising era of comparative genomics. With the sequencing of *Arabidopsis*, human, and rice genomes, rapid advances were made in respect to understanding of the gene function, cloning of major quantitative trait loci (QTLs), identification of disease-resistant loci, and so on, in many species. Most of the sequenced organisms were simple in nature and chromosome constitution. It was not possible to sequence a polyploid species like wheat or sugarcane with the existing technologies, as the genome size inflated the cost and time required for complete sequencing, which thus remained an elusive dream. With the advent of commercial NGS technology in the early part of the last decade, it has become clear that all the major organisms of importance need to be sequenced in an effort to rapidly understand their adaptation and evolution and thereby accelerate their improvement for the sake of society. Today there are many genome projects, and data are easily available in databases. Individual laboratories are deciphering the genomes of species that have local importance rather than regional or national importance. Truly, genome sequencing has now acquired global proportions, making inroads into sequencing newer organisms, and with rapid advances it is likely that all major species will be sequenced before the end of this decade.

14.6.2 Metagenomics

Metagenomics refers to the sequencing of the diverse microbial flora present in any environmental location: soil, seawater, hot springs, coal mines, and so on. Microbes are probably the most diverse of all species on this planet. They inhabit almost all terrestrial and aquatic ecosystems and often thrive in conditions that are very hostile to any kind of life. Most of the species often remain elusive, and characterizing them becomes impossible, as these mini-ecosystems are constantly changing. It is also not possible to catalog the entire population, as isolation of individual species remains a constraint. NGS provides an excellent opportunity to sequence the diverse species in a sample, and helps in unraveling their composition. Metagenomics has immense applications in agriculture, where 1 g soil containing a heavy load of microbes could yield as much as 1 GB sequence data, helping to better understand the plant–microbe interaction. It also has immense potential for the human microbiome, as data from different individuals may lead to better insights into the effect of these microbes on human health. Structural and functional metagenomics also finds applications in biofuels (for effective biomass conversion) and environmental remediation (cleaning of contaminated soil and water, industrial wastes, etc.).

14.6.3 Development of Markers

The sequencing of the human and rice genomes led to a race for identification of important genes required for both abiotic and biotic stress tolerance in plants and disease management in humans. Once a reference genome is available, it becomes easy to genotype other members of the same species and then carry out comparative genomics. Molecular markers are an integral part of any breeding program, and availability of sequenced genomes provides a platform for marker discovery. SSRs are the most commonly used markers for mapping purposes and provide cross-transferability within a species. Thus, sequencing one member of a species can lead to development of markers for that species with wide usage. Among the most polymorphic and abundant markers for this cause are the SNPs, which can be used to arrive at fine linkage maps. Once these maps are available they can be used for high-throughput genotyping, allele mining, association genetics- and gene discovery.

14.6.4 Deep Transcriptome Investigations

14.6.4.1 Study of Whole Transcriptome

Before the advent of NGS technologies, transcriptome analysis was limited to the powers of the classical cDNA synthesis methods. This led to development of expressed sequence tags (ESTs) in diverse species and in different stages of development, environmental interactions, tissue, and so on. Transcript information is available for many genomes and has been useful in cloning of genes and promoters. EST databases are a must for any kind of functional genomics. However, the limiting factor remains the poor quality of the reads and low representation of the expressed genome. With NGS intervention, it is now possible to sequence the total RNA content of the cell. RNA sequencing (RNA-seq) does not require a reference genome, and hence in nonmodel organism it remains a tool of choice. RNA-seq is slowly taking over other methods of transcriptome analysis, including microarrays, serial analysis of gene expression (SAGE), and so on. Some of the advantages of NGS RNA-seq include identification of alternately spliced transcript and capture of low-abundance messages. RNA-seq by NGS thus provides a snapshot of the expressed messages directly, without any bias, and accurately identifies the differentially expressed sequences, expressed SSRs, and SNPs for the purpose of allele mining. NGS sequencing using different platforms also results in cloning of full-length cDNAs, which have far greater significance in functional genomics.

14.6.4.2 Study of Small/MicroRNAs

Small RNAs constitute a very important component of the cellular machinery. It is now widely accepted that these microRNAs (miRNAs) regulate a number of cellular processes and are important in developmental pathways. Mutations in these genes often lead to impairment in the normal function of the cell. Historically, miRNAs were identified based on cloning of individual miRNAs, and required tedious laboratory procedures, leading to several errors. It is also difficult to separate and analyze miRNAs and small interfering RNAs (siRNAs) using traditional procedures, although they have distinct functional roles. Using NGS, it is now possible to sequence the small RNA component of the cell and dissect the different si- and miRNAs present in a species. Due to deep and unbiased coverage, NGS remains the tool of choice for profiling the small RNA composition in any organism, and thus aids in identifying novel small RNAs that operate in a particular organism and understanding their regulatory role.

14.6.5 Mapping the Epigenome

Higher eukaryotes are very complex and dynamic in nature. They are subjected to many forms of regulation both within and outside the cell. The cellular form of regulation is well defined and studied in different organisms, which provides an insight into its functioning. However, there exists another form of regulation that remains elusive and hidden from the naked eye, as the manifestations are often not seen in the phenotype. This consists of epigenetic variation, arising from many intrinsic and extrinsic factors. These factors can lead to methylation of the genome or modifications in the histone proteins. These lead to a combined effect on the expression of the genome in both positive and negative ways. These variations, known as epigenetic variations, also keep the genome in a state of flux with heterochromatinization of DNA, which may or may not be reversible. These variations are of prime importance in cancer biology, and the Human Epigenome Project (HEP) aims to catalog all the DNA methylation that occurs in the human genome and discover its association with gene regulation during oncogenesis. In plants, too, many genes are governed by these factors, and the major effects are seen in the hybrids that arise from contrasting genotypes. Such hybrids are reported to carry many epigenetic variations, which could be responsible for their hybrid vigor. In contrast to classical ways of finding or mapping these variations, NGS now has the power to map the entire methylation pattern in the genome and develop an epigenome map for different species. These maps provide a tool for researchers to dissect the role of epigenetic variations like CpG methylation in the regulatory pathways that exist in a particular genotype.

14.6.6 Study of Protein–DNA Interactions

Chromosomal DNA in higher eukaryotes is generally bound with histone proteins and condensed into chromatin, which leads to chromosome packing and then remodeling to allow expression of the genes. Historically, these studies have been carried out using chip assays in combination with hybridization, PCR, and, lately, microarrays. Limitations related to the presence of restriction sites narrow down the scope of analysis. The advent of NGS gave rise to a new procedure called Chip-Seq, which sequences the immunoprecipitated DNA on a genomic scale. This technique has revealed in detail the histone modifications in the human genome. Other studies that came out of Chip-Seq included mapping of nucleosome binding sites, transcription factor binding sites, and ribosome binding sites, and the chromosome conformation capture (3C) method to detect higher orders of chromosome structure.

Rapid advances in sequencing technologies and different chemistries will make NGS a major force in both structural and functional genomics. Compared with the Sanger method, the next-generation techniques will provide a holistic approach to researchers, with the dual advantage of cost and time savings. There are certain issues with small read lengths, but the accuracy and volume of data generated somewhat make up for these limitations. For complex organisms such as plants, many of which are polyploids harboring big genomes, resolving the repeat regions is still a challenge, and there is a need to improve upon the NGS techniques and the computational tools. NGS is likely to provide highly valuable application(s) in genome sequencing, epigenome mapping, transcriptome analysis, and discovery and profiling of small RNAs. With the availability of the latest software with simpler algorithms, it will be increasingly easy for researchers to address the postsequencing data issues to expand the scope of research and generate novel ideas.

ACKNOWLEDGMENTS

The authors are highly thankful to Project Director, NRC on Plant Biotechnology for providing the necessary facilities and encouragement.

REFERENCES

Bentley, D. R. 2006. Whole-genome re-sequencing. *Curr Opin Genet Dev* 16:545–552.

Bentley, D. R., S. Balasubramanium, H. P. Swerdlow et al. 2008. Accurate whole human genome sequencing using reversible terminator chemistry. *Nature* 456:53–59.

Bowers, J., J. Mitchell, E. Beer et al. 2009. Virtual terminator nucleotides for next-generation DNA sequencing. *Nat Methods* 6:593–595.

Carrilho, E., M. C. Ruiz-Martinez, J. Berka et al. 1996. Rapid DNA sequencing of more than 1000 bases per run by capillary electrophoresis using replaceable linear polyacrylamide solutions. *Anal Chem* 68:3305–3313.

Deschamps, S. and M. Campbell. 2010. Utilization of next-generation sequencing platforms in plant genomics and genetic variant discovery. *Mol Breed* 25:553–570.

Dohm, J. C., C. Lottaz, T. Borodina, and H. Himmelbauer. 2008. Substantial biases in ultra-short read data sets from high-throughput DNA sequencing. *Nucleic Acids Res* 36:e105.

Dressman, D., H. Yan, G. Traverso, K. W. Kinzler, and B. Vogelstein. 2003. Transforming single DNA molecules into fluorescent magnetic particles for detection and enumeration of genetic variations. *Proc Natl Acad Sci USA* 100:8817–8822.

Egan, A. N., J. Schlueter, and D. M. Spooner. 2012. Applications of next-generation sequencing in plant biology. *Am J Bot* 99:175–185.

Eid, J., A. Fehr, J. Gray et al. 2009. Real-time DNA sequencing from single polymerase molecules. *Science* 323:133–138.

Erlich, Y., P. P. Mitra, M. delaBastide, W. R. McCombie, and G. J. Hannon. 2008. Alta-Cyclic: A self-optimizing base caller for next-generation sequencing. *Nat Methods* 5:679–682.

Ewing, B., L. Hillier, M. C. Wendl, and P. Green. 1998. Base-calling of automated sequencer traces using phred. I. Accuracy assessment. *Genome Res* 8:175–185.

Fedurco, M., A. Romieu, S. Williams, I. Lawrence, and G. Turcatti. 2006. BTA, a novel reagent for DNA attachment on glass and efficient generation of solid-phase amplified DNA colonies. *Nucleic Acids Res* 34:e22.

Flusberg, B. A., D. R. Webster, J. H. Lee et al. 2010. Direct detection of DNA methylation during single-molecule, real-time sequencing. *Nat Methods* 7:461–465.

Glenn, T. C. 2011. Field guide to next-generation DNA sequencers. *Mol Ecol Resour* 11:759–769.

Guo, J., N. Xu, Z. Li et al. 2008. Four-color DNA sequencing with 3′-*O*-modified nucleotide reversible terminators and chemically cleavable fluorescent dideoxynucleotides. *Proc Natl Acad Sci USA* 105:9145–9150.

Guttman, A. 2002a. Capillary electrophoresis using replaceable gels. U.S. Patent No. RE37,606.

Guttman, A. 2002b. Capillary electrophoresis using replaceable gels. U.S. Patent No. RE37,941.

Harismendy, O., P. C. Ng, R. L. Strausberg et al. 2009. Evaluation of next generation sequencing platforms for population targeted sequencing studies. *Genome Biol* 10:R32.

Harris, T. D., P. R. Buzby, H. Babcock et al. 2008. Single-molecule DNA sequencing of a viral genome. *Science* 320:106–109.

Hillier, L. W., G. T. Marth, A. R. Quinlan et al. 2008. Whole-genome sequencing and variant discovery in *C. elegans*. *Nat Methods* 5:183–188.

International Human Genome Consortium. 2004. Finishing the euchromatic sequence of the human genome. *Nature* 431:931–945.

Ju, J. 2006. Four-color DNA sequencing by synthesis using cleavable fluorescent nucleotide reversible terminators. *Proc Natl Acad Sci USA* 103:19635–19640.

Ju, J., C. Ruan, C. Fuller, A. Glazer, and R. Mathies. 1995. Fluorescence energy transfer dye-labeled primers for DNA sequencing and analysis. *Proc Natl Acad Sci USA* 92:4347–4351.

Kheterpal, I., J. Scherer, S. Clark et al. 1996. DNA sequencing using a four-color confocal fluorescence capillary array scanner. *Electrophoresis* 17:1852–1859.

Korlach, J., A. Bibillo, J. Wegener et al. 2008. Selective aluminum passivation for targeted immobilization of single DNA polymerase molecules in zero-mode waveguide nanostructures. *Proc Natl Acad Sci USA* 105:1176–1181.

Landegren, U., R. Kaiser, J. Sanders, and L. Hood. 1988. A ligase-mediated gene detection technique. *Science* 241:1077–1080.

Lander, E. S., M. L. Linton, B. Birren et al. 2001. Initial sequencing and analysis of the human genome. *Nature* 409:860–821.

Leamon, J. H., W. L. Lee, K. R. Tartaro et al. 2003. A massively parallel PicoTiterPlate™ based platform for discrete picoliter-scale polymerase chain reactions. *Electrophoresis* 24:3769–3777.

Lee, L., S. Spurgeon, C. Heiner et al. 1997. New energy transfer dyes for DNA sequencing. *Nucleic Acids Res* 25:2816–2822.

Levy, S., G. Sutton, P. C. Ng et al. 2007. The diploid genome sequence of an individual human. *PLoS Biol* 5:e254.

Madabhushi, R. S., S. M. Menchen, J. W. Efcavitch, and P. D. Grossman. 1996. Polymers for separation of biomolecules by capillary electrophoresis. U.S. Patent No. 5,567,292.

Mardis, E. R. 2008a. The impact of next-generation sequencing technology on genetics. *Trends Genet* 24:133–141.

Mardis, E. R. 2008b. Next-generation DNA sequencing methods. *Annu Rev Genomics Hum Genet* 9:387–402.

Marguerat, S., B. T. Wilhelm, and J. Bähler. 2008. Next-generation sequencing: Applications beyond genomes. *Biochem Soc Trans* 36:1091–1096.

Margulies, M., M. Egholm, W. E. Altman et al. 2005. Genome sequencing in microfabricated high-density picolitre reactors. *Nature* 437:376–380.

Maxam, A. M. and W. Gilbert. 1977. A new method for sequencing DNA. *Proc Natl Acad Sci USA* 74:560–564.

Metzker, M. L. 2009. Sequencing in real time. *Nat Biotech* 27:150–151.

Metzker, M. L., J. Lu, and R. A. Gibbs. 1996. Electrophoretically uniform fluorescent dyes for automated DNA sequencing. *Science* 271:1420–1422.

Morozova, O. and M. A. Marra. 2008. Applications of next-generation sequencing technologies in functional genomics. *Genomics* 92:255–264.

Pareek, C. S., R. Smoczynski, and A. Tretyn. 2011. Sequencing technologies and genome sequencing. *J Appl Genet* 52:413–435.

Prober, J. M., G. L. Trainor, R. J. Dam et al. 1987. A system for rapid DNA sequencing with fluorescent chain-terminating dideoxynucleotides. *Science* 238:336–341.

Pushpendra, K. G. 2008. Single-molecule DNA sequencing technologies for future genomics research. *Trends Biotechnol* 26:602–611.

Ronaghi, M., M. Uhlén, and P. Nyrén. 1998. A sequencing method based on real-time pyrophosphate. *Science* 281:363–365.

Ruiz-Martinez, M. C., J. Berka, A. Belenkii, F. Foret, A. W. Miller, and B. L. Karger. 1993. DNA sequencing by capillary electrophoresis with replaceable linear polyacrylamide and laser-induced fluorescence detection. *Anal Chem* 65:2851–2858.

Salas-Solano, O., E. Carrilho, L. Kotler et al. 1998. Routine DNA sequencing of 1000 bases in less than one hour by capillary electrophoresis with replaceable linear polyacrylamide solutions. *Anal Chem* 70:3996–4003.

Sanger, F. and A. R. Coulson. 1975. A rapid method for determining sequences in DNA by primed synthesis with DNA polymerase. *J Mol Biol* 94:441–448.

Sanger, F., S. Nicklen, and A. R. Coulson. 1977. DNA sequencing with chain-terminating inhibitors. *Proc Natl Acad Sci USA* 74:5463–5467.

Shendure, J. and H. Ji. 2008. Next-generation DNA sequencing. *Nat Biotechnol* 26:1135–1145.

Shendure, J., G. J. Porreca, N. B. Reppas et al. 2005. Accurate multiplex polony sequencing of an evolved bacterial genome. *Science* 309:1728–1732.

Smith, L., J. Z. Sanders, R. J. Kaiser et al. 1986. Fluorescence detection in automated DNA sequence analysis. *Nature* 321:674–679.

Tabor, S. and C. C. Richardson. 1989. Effect of manganese ions on the incorporation of dideoxynucleotides by bacteriophage T7 DNA polymerase and *Escherichia coli* DNA polymerase I. *Proc Natl Acad Sci USA* 86:4076–4080.

Takahashi, S., K. Murakami, T. Anazawa, and H. Kambara. 1994. Multiple sheath-flow gel capillary-array electrophoresis for multicolor fluorescent DNA detection. *Anal Chem* 66:1021–1026.

Tomkinson, A. E., S. Vijayakumar, J. M. Pascal, and T. Ellenberger. 2006. DNA ligases: Structure, reaction mechanism, and function. *Chem Rev* 106:687–699.

Venter, J. C., M. D. Adams, E. W. Myers et al. 2001. The sequence of the human genome. *Science* 291:1304–1351.

Wu, W., V. A. Litosh, B. P. Stupi, and M. L. Metzker. 2009. Photocleavable labeled nucleotides and nucleosides and methods for their use in DNA sequencing. U.S. Patent No. 11/567,189.

Zhang, J., R. Chiodini, A. Badr, and G. Zhang. 2011. The impact of next-generation sequencing on genomics. *J Genet Genom* 38:95–109.

Zhou, X. G., L. F. Ren, Y. T. Li, M. Zhang, Y. D. Yu, and J. Yu. 2010. The next-generation sequencing technology: A technology review and future perspective. *Sci China Life Sci* 53:44–57.

CHAPTER 15

Linking Plant Amino Acids with Energy and Stress
A Systems Biology Perspective

Jedrzej Szymanski and Gad Galili

CONTENTS

15.1 INTRODUCTION

Although amino acids primarily serve as the building blocks of proteins, in plants amino acids also serve as important alternative energy substrates, particularly when plants are exposed to stresses that cause energy deprivation. The contribution of amino acids to energy homeostasis in response to stress occurs through their catabolism, which funnels their carbon backbones into the tricarboxylic acid (TCA) cycle, providing an alternative source of electrons for the mitochondrial electron transport chain. As such, amino acid catabolism, which also includes the association of the amino acid Glu and the γ-aminobutyric acid (GABA) shunt, is becoming recognized as a critical element of energy metabolism in stress and carbon starvation. In the first part of the present review, we focus on recent findings concerning modes of action of amino acid catabolism as an alternative source of energy in stress conditions. Second, we reveal how systems biology approaches, including gene coexpression analysis and transcript and metabolic profiling, led to the discovery of new biological processes associating plant amino-acid (AA) metabolism with response and adaptation to stress.

15.2 ROLE OF AMINO ACIDS IN STRESS

AAs play a crucial role in plant stress responses, with functions going far beyond their role as building blocks for protein synthesis. Pro, for example, has direct osmoprotective properties (Verslues and Sharma 2010). Others, such as the nonproteinogenic amino acid γ-aminobutyrate (GABA), are involved in stress signaling, for example. Many AAs are also substrates for biosynthesis of stress-related compounds. Met, for example, is a precursor for biosynthesis of ethylene, polyamines, and (together with Ser) glucosinolates (Wittstock and Halkier 2002), and serves as a donor of methyl groups in the synthesis of chlorophyll, cell wall components, and a range of secondary metabolites (Amir et al. 2002; Goyer et al. 2007; Rebeille et al. 2006). Aromatic AAs and Cys are substrates for the synthesis of phenylpropanoids in response to, for example, oxidative stress (Grace and Logan 2000; Lopez-Martin et al. 2008), and Phe is an entry point for anthocyanin synthesis in response to multiple biotic and abiotic stresses (Mellway et al. 2009). In this review, however, we focus on the recent finding that AAs also serve as an important source of carbon backbones and energy in conditions of stress-related inhibition of photosynthesis. This role of AAs in plants has been highlighted with the advent of new analytical techniques, namely transcriptomics and metabolomics, and is a classic example of system-scale rearrangement of plant central metabolism in response to changes in environmental conditions.

15.3 LINKING AMINO ACIDS WITH ENERGY METABOLISM

Plant growth and fitness are largely dependent on energy availability. Two sources of energy in plants are photosynthesis in the daytime and degradation of storage compounds, particularly starch and AA, *en route* to the TCA cycle during the night. In order to coordinate the accumulation and usage of starch and AA over the diurnal cycle, plants have developed complex metabolic and signaling pathways, which in some cases are also coupled with the circadian clock, to ensure an optimal energy supply and optimize the plant's growth over the day/night period (Feugier and Satake 2012; Stitt and Zeeman 2012) or upon exposure to stresses causing energy deprivation. Both the signaling and the biochemistry of these processes are well studied, resulting in not only identification of the proteins and metabolites involved, but also mathematical modeling of the dynamic metabolism–signaling–environment interaction, allowing accurate predictions of the biochemical responses (Nagele et al. 2010; Sorokina et al. 2011). This, however, is a very simplified image of the energy management of plants, for two major reasons: (i) the energy management is influenced by multiple factors, such as the day/night cycle and exposure to various stresses that cause energy deprivation; and (ii) starch is not the only metabolic pool used to feed glycolysis and the TCA cycle.

Multiple environmental stresses, such as drought and extensive heat or chilling, have been shown to reduce photosynthesis and respiration even during the daytime when light is available, resulting in energy deficiency and inhibited growth (Baena-Gonzalez and Sheen 2008). A downregulation of major genes involved in the light reaction of photosynthesis has been shown to occur upon exposure to biotic (Bilgin et al. 2010) and abiotic (Baena-Gonzalez et al. 2007; Espinoza et al. 2008) stresses. It is still not clear, however, whether the inhibition of photosynthesis is a protective mechanism against reactive oxygen species (ROS) accumulation (Niyogi 2000) or is a direct consequence of the oxidative damage (Blokhina et al. 2003). Importantly, rapid stress-associated inhibition of the photosynthetic flux cannot be entirely compensated for by stimulation of starch accumulation and its channeling through glycolysis and the TCA cycle for energy generation. This apparently leads to rapid reduction of the flux through the TCA cycle, decreasing NADH production and depleting pools of the TCA cycle intermediates. These effects are critical to plant homeostasis, since, in addition to the energetic shortage, the intermediates of the TCA cycle play a role as substrates for multiple stress-related anabolic processes. One of these processes includes the synthesis of protective compounds, such as polyamines, Pro, secondary compounds, cell wall components; and the biosynthesis of isoprenoids, which play a role in redox regulation (Fatland et al. 2005; Scheibe et al. 2005). Therefore, one of the critical processes occurring in primary metabolism during the stress response is the mobilization of alternative compounds toward the TCA cycle (Buchanan-Wollaston et al. 2005), among which the pool of AAs plays a significant role (Less et al. 2011).

Metabolism of AAs has been shown to be strictly regulated by environmental stresses, exhibiting stress-specific and pathway-coordinated changes at the transcriptional level (Less and Galili 2008). As a general rule, stress suppresses the synthesis of AAs to preserve energy and activates the catabolism of AAs to generate additional energy (Araujo et al. 2012; Baena-Gonzalez and Sheen 2008; Galili 2011). The only exception from this rule is the pathway of Ala biosynthesis, which is coordinately activated on almost all steps, as shown in the study of *Lotus japonicus* in hypoxia (Rocha et al. 2010). Interestingly, the temporal modes and magnitudes of the transcriptional changes are specific in respect to the specific pathways and environmental stresses, indicating that different AA pools are utilized for various purposes in response to different environmental and physiological conditions (Less and Galili 2008).

The interface between AA catabolism and energy homeostasis is defined by two main aspects: (1) direct replenishment of the TCA intermediate pools via catabolism of the AAs; and (2) feeding electrons directly to the mitochondrial electron transport chain via electron transfer flavoprotein (ETF) in specific steps of AA degradation. Independently of the biochemistry, we can also identify two functional aspects of the AA–energy metabolism connection: (1) replenishing the pools of TCA cycle intermediates in conditions of decreased energy status; and (2) boosting the energy metabolism during recovery from the low-energy period. In the following sections we will describe the major stress-regulated AA catabolic pathways and their metabolic connection with energy metabolism, and we will discuss the physiological role of the respective pathways in stress response and recovery.

15.4 STRESS RESPONSE

One of the most pronounced direct metabolic connections between AA metabolism and the TCA cycle is 2-oxoglutarate (2-OG), which reacts with eight different AAs to generate Glu and a range of other specific products. The interconversion between Glu and 2-OG, catalyzed by glutamate dehydrogenase (GDH), is a key reaction in plant carbon and nitrogen metabolism (Fontaine et al. 2012). Multiple experiments have indicated that GDH funnels the carbon skeletons of Glu into the TCA cycle (Aubert et al. 2001; Purnell and Botella 2007). Furthermore, the strict regulation of GDH activity in response to cellular carbohydrate status highlights the importance of this process in alleviating carbon starvation (Masclaux-Daubresse et al. 2002). Abiotic stresses, including salt, heavy metals, and heat and cold stress, have been shown to stimulate GDH at the level of gene expression and enzymatic activity (Restivo 2004). The flux from Gln via Glu to 2-OG has also been shown to play an important role during embryo development of sunflower (*Helianthus annuus*) (Alonso et al. 2007) and oilseed rape (*Brassica napus*) (Schwender et al. 2006), and during anoxic stress in *Lotus japonicus* (Rocha et al. 2010), apparently functioning as an alternative input to the TCA cycle. Finally, a double knockout mutant in the gdh1-2 and gdh2-1 genes encoding two *Arabidopsis* GDH isoforms exhibited increased susceptibility to extended night as well as a metabolic phenotype resembling that of exposure to severe carbon starvation and decreased energy status (Miyashita and Good 2008).

While GDH serves as the most direct connection between AAs and the TCA cycle, the cellular pool of Glu is not very high (Hernandez-Sebastia et al. 2005), and the involvement of the flux of Glu catabolism via GDH must be balanced with an anabolic function of GDH in the GABA shunt (Fait et al. 2008). This also cross-interacts with the impact of the Glu pool on nitrogen assimilation and translocation (Suzuki and Knaff 2005). Therefore, the most profound stress-related contribution of AA catabolism to the TCA cycle is apparently derived from other AAs. These catabolic fluxes involve complex multistep stress-regulated pathways, degrading the AAs to one of the TCA intermediates: acetyl-CoA, 2-oxoglutarate, succinate, or fumarate (Figure 15.1).

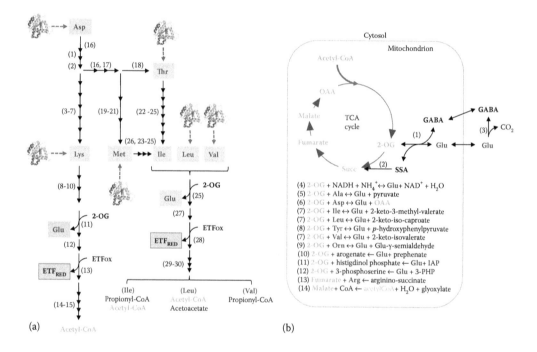

Figure 15.1 **(See color insert)** (a) Catabolic pathways of aspartate family and branched-chain amino acids. Black arrows represent single biochemical reactions. Red dotted arrows represent protein degradation. Involved enzymes: (1) monofunctional Asp kinase AK1; (2) Asp-semialdehyde dehydrogenase ASD; (3) dihydrodipicolinate synthase DHDPS1; (4) dihydrodipicolinate reductase DHDPR; (5) L,L-diaminopimelate aminotransferase AGD2; (6) diaminopimelate epimerase DAPE; (7) diaminopimelate decarboxylase DAPD; (8) Lys-ketoglutarate reductase/saccharopine dehydrogenase LKR/SDH; (9) saccharopine dehydrogenase SDH; (10) 3-chloroallyl aldehyde dehydrogenase ALDH; (11) NA; (12) NA; (13) NA; (14) enoyl-CoA hydratase MFP2; (15) acetoacetyl-CoA thiolase AACT1; (16) Asp kinase/homo-Ser dehydrogenase AK/HSDH; (17) homo-Ser kinase HSK; (18) Thr synthase TS; (19) cystathionine g-synthase CGS; (20) cystathionine b-lyase CBL; (21) Met synthase MS; (22) Thr deaminase TD; (23) acetolactate synthase AHASS; (24) ketol acid reductoisomerase KARI; (25) branched-chain amino acid aminotransferase BCAT; (26) Met g-lyase MGL; (27) branched-chain keto-acid dehydrogenase BCE2/LPD2; (28) NA; (29) enoyl-CoA hydratase MFP2; (30) acetoacetyl-CoA thiolase AACT1. (b) TCA cycle scheme, GABA shunt, and basic reactions of TCA intermediates involving amino acids and AA-derived acetyl-CoA. Enzymes involved: (1) GABA transaminase; (2) succinic semialdehyde dehydrogenase; (3) glutamate decarboxylase; (4) glutamate dehydrogenase; (5) alanine transaminase; (6) aspartate transaminase; (7) branched-chain amino acid transaminase; (8) aromatic amino acid transaminase; (9) ornithine transaminase; (10) glutamate-prephenate aminotransferase; (11) histidinol-phosphate transaminase; and (12) phosphoserine aminotransferase; (13) arginosuccinate lyase; (14) malate synthase. Metabolite abbreviations: IAP, imidazole acetol-phosphate; 3-PHP, 3-phosphohydroxypyruvate. The list of the enzymatic reactions indicated in the bottom of panel b was principally based on a previous review. (Sweetlove, L. J., K. F. Beard, A. Nunes-Nesi, A. R. Fernie, and R. G. Ratcliffe. 2010. Not just a circle: flux modes in the plant TCA cycle. *Trends Plant Sci* 15: 462–70.)

15.5 CATABOLISM OF AMINO ACIDS BELONGING
TO THE ASPARTATE-FAMILY PATHWAY

Asp is quantitatively a major free AA (Hernandez-Sebastia et al. 2005; Izumi et al. 2013), and, together with Lys, Thr, Met, and Ile, belongs to the so-called aspartate-family pathway. This pathway provides a significant carbon pool for stress-related catabolism to feed the TCA cycle in various organisms. Catabolism of Asp occurs particularly via the three AAs Lys, Met, and Thr, which are synthesized from Asp via the aspartate-family pathway (Figure 15.1a). Furthermore, the degradation of these AAs has been shown to be stress induced (Less and Galili 2008). Probably the most remarkable examples of stress-induced AA degradation in plants are those of Lys and Ile. The gene encoding the key Lys catabolic enzyme lysine-ketoglutarate synthase/saccharopine dehydrogenase (LKR/SDH) is highly stress responsive, exhibiting a pronounced upregulation in response to salt, osmotic, drought, and wounding stresses, indicating that increased flux of Lys catabolism toward acetyl-CoA and the TCA cycle occurs in response to these stresses (Less and Galili 2008; Stepansky et al. 2006). Moreover, two genes of Lys metabolism, one associated with Lys synthesis and the second apparently associated with Lys catabolism, have been shown to play a role in the response of plants to pathogen infection (biotic stress) (Song et al. 2004a,b).

The initial studies highlighting functionality of the aspartate-family pathway in plants were devoted to increasing levels of Lys specifically in developing plants (*Arabidopsis*) with two major aims: (i) increasing the nutritional quality of the seeds; and (ii) elucidating the biological impact of this approach on seed development and germination (Zhu and Galili 2003). The highest seed Lys level was obtained in plants expressing a bacterial gene encoding a feedback-insensitive dihydro-dipicolinate synthase (DHDPS), the first enzyme in the Lys biosynthetic pathway, coupled with a knockout mutation in the LKR/SDH gene of Lys catabolism. However, seeds of this genotype showed severely retarded germination, indicating a strong influence of Lys metabolism (enhanced synthesis and blocked degradation) on overall plant homeostasis. Three main strategies were used in order to increase basal Lys levels in *Arabidopsis* seeds, utilized as a model system, as follows: (i) expression under a seed-specific promoter of a bacterial feedback-insensitive DHDPS, bypassing the sensitivity of the Lys biosynthesis pathway to increasing levels of Lys; (ii) expression of the same seed-specific bacterial DHDPS construct in an *Arabidopsis* knockout mutant lacking LKR/SDH activity; and (iii) expression of the same seed-specific bacterial DHDPS construct coupled with an antisense (RNA interference [RNAi]) construct of LKR/SDH. Expression of the bacterial DHDPS alone resulted in the lowest increase in seed Lys content and had no effect on seed germination rate. Expression of the bacterial DHDPS construct by itself yielded a moderate, but significant, increase in Lys synthesis, and the seeds of this genotype germinated properly. Expression of the bacterial DHDPS construct coupled with the LKR/SDH antisense construct yielded a higher level of seed Lys accumulation and proper seed germination. Expression of the bacterial DHDPS construct in an LKR/SDH knockout mutant yielded the highest increase in seed Lys content, but these seeds germinated very poorly. Thus, the bacterial DHDPS plus LKR/SDH antisense approach provided optimal results, leading to a considerable increase in seed Lys level as well as proper germination. Interestingly, further metabolic analysis indicated that germinating seeds of the plants expressing the bacterial DHDPS construct in an LKR/SDH knockout mutant showed a considerable decrease in the levels of a number of TCA cycle metabolites compared with the control plants, indicating a strong link between Lys metabolism and energy homeostasis (Zhu and Galili 2003). These seeds also possessed a modified transcriptional program, which is also related to energy metabolism (Angelovici et al. 2010). The metabolic connection between Lys and energy metabolism is twofold. First is direct feeding into the TCA cycle, with acetyl-CoA being an end product of Lys catabolism. Second, the catabolism of Lys via 2-D-hydroxyglutarate dehydrogenase and isovaleryl-CoA dehydrogenase provides a source of electrons in the mitochondrial electron transport chain mediated by the ETF-ETF: ubiquinone oxidoreductase complex (ETF-ETFQO) (Araujo et al. 2011b; Engqvist et al. 2009; Ishizaki et al. 2005). Although the Lys catabolic pathway in plants is not fully elucidated, and only four of its nine enzymatic steps have so far been characterized by direct assays or mutant studies, several recent lines of evidence support its essential role in central metabolism. First, in *Arabidopsis*, the ETF complex has been shown to be essential for plant survival in extended darkness (Ishizaki et al. 2006). Second, increased expression of the ETF complex has been observed in pea genetically engineered for increased AA import to the mitochondria, indicating a direct connection between AA catabolism flux and the mitochondrial electron transport chain (Weigelt et al. 2008). Finally, plants deficient in expression of 2-D-hydroxyglutarate dehydrogenase and isovaleryl-CoA dehydrogenase showed susceptibility to extended night, exhibiting metabolic phenotypes characteristic for radical energy depletion (Arulio et al. 2011). Furthermore, this same study (Araújo et al. 2011a) also demonstrated, by isotope labeling, that 2-D-hydroxyglutarate dehydrogenase is uniquely involved in Lys catabolism, whereas isovaleryl-CoA dehydrogenase also plays a role in degradation of branched AAs and phytol.

[13C]-NMR profiling of cells fed with [13C]Met revealed that Met catabolism in *Arabidopsis* is initiated by γ-cleavage of the molecule, producing NH_3, CH_3SH, and α-ketobutyrate (αKB) (Rebeille et al. 2006). Along this pathway, Met, via αKB, enters the pathway of Ile synthesis in chloroplasts. Ile is subsequently transported to mitochondria and further catabolized by the branched AA catabolic pathway. At the same time, a considerable pool of S-methylcysteine (SMC), a storage form for sulfide and methyl groups, is produced (Rebeille et al. 2006). Thr is used either for the biosynthesis of Ile or for its catabolism into Gly and acetaldehyde by two isoforms of threonine aldolase (THA1 and THA2) (Joshi et al. 2006) (Figure 15.1). The conversion of Thr into Ile includes Thr deamination to 2-oxobutanoate by threonine deaminase (TD) (Halgand et al. 2002), and four following steps, only two of which have so far been experimentally confirmed in plants (Diebold et al. 2002; Mourad et al. 1994). TD has been shown to be allosterically inhibited by Ile, while Val reverses this inhibition (Halgand et al. 2002). Another catabolic pathway of Thr involves the activities of THA1 and THA2, converting Thr to Gly and acetaldehyde *en route* to photorespiration. Microarray analysis of the AA metabolic genes indicated that the second route is considerably activated upon exposure to salt and osmotic stresses (Less and Galili 2008). Thus, Ile serves as an intermediate metabolite in the catabolism of both Thr and Met. Ile catabolism shares a common pathway with the two other branched-chain AAs, Leu and Val, whose catabolism includes several common intermediate enzymes, as indicated in the following.

15.6 CATABOLISM OF BRANCHED-CHAIN AMINO ACIDS

As described above, the catabolism of Met and Thr leads to the biosynthesis of Ile, which is funneled through the pathway of branched AA catabolism into the TCA cycle. This has been shown to be moderately activated in cold stress. The general route of branched AA catabolism has been elucidated by proteomic analysis, and partially by genetic means (Taylor et al. 2004). Three branched-chain AAs, Leu, Val, and Ile, are transaminated by branched-chain AA transaminase (BCAT) in the first step (Figure 15.1a). The reversible transamination reaction is also the last step of biosynthesis of all these three AAs; thus, BCAT is both a biosynthetic and a catabolic enzyme. In the next step, the respective branched-chain α-keto acids are decarboxylated and subsequently esterified to their acyl-CoA products by the branched-chain keto-acid dehydrogenase complex (BCKDC). While localization of the BCAT enzymes is distributed between chloroplasts and mitochondria (Diebold et al. 2002), BCKDC is targeted uniquely to mitochondria (Taylor et al. 2004). The acyl-CoA esters, isovaleryl-CoA (from Leu catabolism), isobutyryl-CoA (from Val catabolism), and methylbutanoyl-CoA (from Ile catabolism), are then oxidized by acyl-CoA dehydrogenase, coupled with reduction of the ETF and transfer of the electron to ubiquinone (Araújo et al. 2011b). This process, analogously to the Lys catabolism process, directly connects branched-chain AA degradation to the mitochondrial respiratory chain. After this step, three divergent pathways lead to separate catabolic end products for each AA: propionyl-CoA (Val), propionyl-CoA and acetyl-CoA (Ile), and acetyl-CoA and aceto-acetate (Leu) (Graham and Eastmond 2002).

15.7 STRESS RECOVERY

As described above, GDH plays a key role in stress as a connection between pools of 2-OG-derived AAs and various TCA cycle intermediates. However, generation of Glu from 2-OG is also an initial step of the GABA shunt—a key pathway related to stress signaling and leading also to accumulation of GABA, which supports primary metabolism, particularly during stress recovery (Fait et al. 2008). In this "GABA shunt" pathway, Glu is decarboxylated to produce GABA and CO_2 in the cytosol by the activity of glutamate decarboxylase (GAD). GABA has been shown to play an important role in multiple biological processes, including signaling, protection from herbivores, pH, redox regulation, and carbon/nitrogen balance (Bouche and Fromm 2004). Besides having multiple functions of its own, GABA also serves as a substrate for replenishing the pool of succinate. Succinate is then transported to the mitochondrion, and inside this organelle it is converted by GABA transaminase (GABA-T) to succinic semialdehyde (SSA), and further to succinate by succinic semialdehyde dehydrogenase (SSADH). The importance of the GABA shunt as a metabolic source of succinate is supported by the fact that *Arabidopsis* mutants with impaired expression of genes encoding the TCA cycle enzymes upstream of the succinate exhibited altered GABA shunt activity (Lemaitre et al. 2007). Moreover, isotope labeling experiments on tomato succinyl-CoA ligase antisense lines indicate a significant complementation of the impaired succinate synthesis by the GABA shunt pathway (Studart-Guimaraes et al. 2007). Conversely, perturbation of TCA cycle enzymes upstream of the succinate has little or no effect on the GABA shunt activity (Nunes-Nesi et al. 2005, 2007). Regulation of the GABA shunt pathway has been described in two major instances. One is GAD activity, controlled by Ca^{2+}/calmodulin and activated in stress and at certain developmental stages (Snedden and Fromm 2001). The second is regulation of SSADH by ATP/AMP and NADH/NAD ratio (Busch et al. 2000). The significance of these processes in the context of stress and AA metabolism is twofold. First, production of GABA by GAD is activated during various stress responses and is induced by multiple biotic and abiotic stimuli (Figure 15.1) (Bouche and Fromm 2004; Bouche et al. 2003; Bown et al. 2006). The GABA accumulated during stress is not depleted as long as low energy supply is inhibiting SSADH activity (Busch et al. 2000). Second, the turnover of the cytosolic pool of GABA might replenish the succinate pool in conditions when energy becomes available by activation of the SSADH enzyme. Such a situation has, for instance, been observed during germination (Fait et al. 2006; Tuin and Shelp 1994) before the full mobilization of the seed storage reserves (Gutierrez et al. 2007), and it probably also helps to boost the TCA cycle during recovery from stress. This is supported by the fact that GABA is a significant sink of carbon in plants (Roohinejad et al. 2009), especially in seeds (Hernandez-Sebastia et al. 2005), with levels staying relatively high in darkness or in situations of low energy status, when other sources of carbon are not directly available (trapped in storage compounds in seeds) or are rapidly depleted (in darkness).

15.8 GENE EXPRESSION COORDINATION

With the advent of systems biology, interesting general patterns of AA metabolism regulation and interconnectivity with the rest of central metabolism have been revealed. Several recent studies have addressed the question of expression coordination of genes associated with metabolism in response to multiple biotic and abiotic stimuli, highlighting tight coexpression of genes involved in AA metabolism and in other crucial processes of central metabolism. The term gene coexpression analysis includes studies investigating similarity of expression of genes in different treatments, organs, developmental stages, or genotypes (reviewed in Usadel et al. 2009). This similarity has been investigated using a range of statistical methods, including, for example, Pearson correlation coefficient (PCC) (Obayashi et al. 2007), partial Pearson correlation coefficient (PPC) (de la Fuente et al. 2004), graphical Gaussian models (Ma et al. 2007), mutual information (Soranzo et al. 2007), and several other methods (reviewed in Werhli et al. 2006). Some of these studies focused on metabolic genes and highlighted interesting large-scale wiring of the plant metabolism on the gene expression level.

The study of Mentzen et al. (2008) focused on 126 key genes of three core metabolic processes in *Arabidopsis thaliana*: fatty acid (FA), AA, and sugar metabolism. Pearson correlation analysis over 72 different microarray experiments comprising 956 Affymetrix chips revealed a high modularity of the correlation network related directly to distinct metabolic pathways. The study also indicated hierarchical modularity of gene coexpression reflecting major catabolic pathways, including starch, FA, and AA catabolism. This study showed that high-level modules represent general metabolic processes, such as catabolism, while tightly connected submodules within them represent particular metabolic pathways. This was especially evident in the case of Leu catabolism, which forms a tightly coherent cluster within a larger general catabolic module. This finding indicated that Leu catabolism is one of the key plant catabolic pathways, regulated by common cues together with starch and FA degradation pathways, but also possessing its own specific regulation. This is not surprising, taking into account that AA catabolism is regulated both by the energy status of the plant and by nitrogen availability, whereas sugar and FA degradation is independent of nitrogen availability. Such an organization of metabolic and transcriptional networks has also been observed in *Escherichia coli* (Cosentino Lagomarsino et al. 2007; Ma et al. 2007; Sales-Pardo et al. 2007) and several other organisms (Oltvai and Barabasi 2002; Ravasz et al. 2002), and it is speculated that this reflects the organization of metabolic fluxes (Ravasz et al. 2002). Analysis of the genes tightly correlated with Leu catabolism indicated that processes such as protein turnover, catabolism of other AAs, and cell wall degradation are coordinated with Leu degradation. This shows a system-scale wiring of AA metabolism to functionally distant processes and highlights long-range connections of both molecular signals and metabolic fluxes in the plant cell.

Another crucial study focusing on coordination of metabolic gene expression in different stress conditions is the study by Less and Galili (2009) and Less et al. (2011). This study was based on 211 short-term environmental and biological perturbations and introduced a novel measure of gene coexpression, termed "gene coordination." Briefly, in contrast to Pearson correlation, gene coordination uses a relative change of gene expression in respect to a control for each individual experiment (the reader is referred to the original paper for details of the computational approach). Despite its relative simplicity, the method identified significant patterns in transcriptional regulation of AA metabolism that were not apparent in correlation analysis. Three clusters have been identified, two containing genes of the aspartate-family pathway, and one covering genes of aromatic AA metabolism. In the aspartate-family pathway, one coordinated cluster, called the "catabolic group," includes exclusively catabolic genes of Thr, Ile, Met and Lys. This group was negatively coordinated with a "Met metabolism group," including both biosynthetic and catabolic genes of Met. These patterns indicate antagonistic regulation of metabolism fluxes from aspartate toward AA degradation or toward the synthesis of S-adenosyl methionine (SAM) and glucosinolates, the end products of the Met pathway. Genome-wide gene coordination analysis showed that the catabolic group is positively coordinated with multiple stress responses and negatively coordinated with growth-promoting processes. The Met metabolism group showed the opposite pattern, being positively coordinated with such processes as nucleosome assembly, ribosome biogenesis, translation, and anabolic metabolism, and negatively coordinated with stress-related processes, such as abscisic acid (ABA), salicylic acid (SA), or jasmonic acid (JA) signaling. Remarkably, metabolism of the group of aromatic amino acids (AAAs), including both biosynthetic and catabolic genes, has been shown to be largely disconnected from the Asp-pathway groups. Main biological processes positively coordinated with the AAA group include various stress responses and production of phenylpropanoids, while the negative ones encompass a range of core anabolic processes.

Extending the analysis to genome-wide coordination identified three major gene clusters, reflecting coordination of plant gene expression in response to different stress conditions (Less et al. 2011). Because of the nature of the distance measure used, these clusters should be regarded more as general transcriptional programs, since genes classified into them do not share the same response dynamics, but, rather, tend to be regulated in particular sets of environmental conditions. The first of these programs is characterized by highly responsive genes to multiple different stresses, and includes hormone metabolism, transcriptional regulation, and various other processes related to signal transduction. The second major program concerns genes downregulated under most abiotic stresses and upregulated in response to pathogen attack, and encompasses genes involved in all key energy metabolism processes, including glycolysis, the TCA cycle, the pentose phosphate cycle, the mitochondrial transport chain, and the Asp-family pathway, as well as metabolism of Glu, Arg, Ser, and Gly. Finally, the third transcriptional program concerns all genes that are downregulated, irrespective of the type of stress, and incudes genes involved in major anabolic processes, such as starch and lipid biosynthesis. Most importantly, the study has shown an interaction between different branches of plant metabolism, highlighting that abiotic stress leads to energy deprivation, downregulation of anabolic processes, and induction of protein degradation and AA catabolism. These processes have been shown to generate additional energy, as previously described (Baena-Gonzalez and Sheen 2008; Bunik and Fernie 2009; Sulpice et al. 2009). In particular, the catabolism of the Asp-family pathway, including Lys, has been highlighted, which is in agreement with multiple reports about a special role of these pathways in stress and energy metabolism (Moulin et al. 2006; Stepansky and Galili 2003; Stepansky et al. 2006). Remarkably, this large-scale metabolic reprogramming reported by Less et al. (2011) has been elucidated directly from the coexpression pattern, indicating that other, less obvious coordination patterns might also highlight basic, not yet described, biological processes. One such interesting observation is the different proportion of transcription factors (TFs) in each of the regulatory programs (Avin-Wittenberg et al. 2011). Whereas the stress-responsive cluster 1 contains more TFs than enzymes (266 and 234 of TFs and enzymes respectively), in cluster 2 TFs are highly underrepresented (only 43 per 253 enzymes). In cluster 3, on the other hand, the proportion of TFs is close to the genome average. This indicates a different architecture of regulatory pathways responsible for regulation of gene expression under different environmental stresses. Moreover, for each cluster, a range of tightly coordinated cliques was identified, using a fuzzy clustering approach. Many of these cliques were significantly enriched in genes belonging to certain metabolic pathways, together with their putative TF regulators. This holds true especially for AA metabolism, in which common TFs were found to be coordinated with AA catabolic pathways and genes involved in energy metabolism. This, and a few other examples mentioned in the studies described, shows the power of gene coexpression analysis in uncovering system-scale regulatory patterns in AA metabolism and in generation of new, scientifically testable hypotheses.

15.9 INTERACTIONS BETWEEN METABOLIC POOLS

Whereas gene expression analysis has been shown to reflect the organization of AA metabolic genes in general regulatory domains, a range of metabolomic studies revealed a close connection between AAs and central metabolism intermediates at the level of metabolite pools. Many of the key metabolomic experiments that decisively contributed to understanding this connection have been mentioned in the first section of this review. Therefore, we focus here on several other studies covering a range of very different environmental stresses, reporting orchestrated changes at the level of AAs and energy metabolism intermediates.

Drought, one of the most comprehensively investigated stresses, has been shown to lead to coordinated accumulation of TCA cycle intermediates together with a range of AAs and GABA (Urano et al. 2009). This study has also identified that activation of the biosynthesis of branched-chain AAs, proline, polyamines, and saccharopine, together with the GABA shunt, is specifically connected with ABA signaling. On the other hand, waterlogging experiments have been shown to result in anoxic conditions in *Arabidopsis* roots, leading to increasing levels of AA pools originating from protein degradation, as well as glycerol-3-phosphate and glucose-6-phosphate, correlated with increased levels of TCA cycle intermediates. This implies that AA catabolism plays an active part in anaerobic metabolism (van Dongen et al. 2009). Waterlogging-tolerant *Lotus japonicus* plants, on the other hand, specifically accumulate succinate, Ala, and the precursors for Ala synthesis, including GABA, while other AAs and TCA cycle intermediates decrease. This result shows that changes in AA metabolism fluxes in anoxic conditions, preventing the accumulation of pyruvate and activation of fermentation, facilitate stress tolerance (Rocha et al. 2010). *Lotus japonicus* was also investigated for its tolerance to long-term salinity stress (Sanchez et al. 2008). In this case, increased levels of many free AAs have been observed, showing that the stress-inducing effect on AA pools does not only result from increased protein degradation and flux toward energy metabolism, but might also occur in mild stress with fully operational photosynthesis and respiration. Cold conditions lead to multiple molecular effects in plant cells, including alteration of membrane functionality, decrease in enzymatic activities, and induction of photooxidative stress in light conditions, as well as many other effects (Alcazar et al. 2011; Chinnusamy et al. 2007; Janska et al. 2010). Counterintuitively, heat affects a similar set of biological processes and in some aspects leads to analogous metabolic effects. On the level of primary metabolism, for example, both stresses lead to increased pools of pyruvate and oxaloacetate-derived AAs as well as activation of Pro metabolism and the GABA shunt (Kaplan et al. 2004, 2007). This energy depletion-related general stress effect has also been observed in response to increased levels of metal ions, including Cu, Fe, and Mn, in *Brassica rapa* (Jahangir et al. 2008) and to cadmium in *Arabidopsis* (Sun et al. 2010). The use of metabolomics in plant stress research has also been addressed in three recent reviews (Bowne et al. 2011; Obata and Fernie 2012; Shulaev et al. 2008).

15.10 INTEGRATION OF TRANSCRIPT AND METABOLIC DATA

Stress studies, including parallel analysis on the transcriptomic and metabolomic levels, make it possible to address questions about coordination between metabolic and transcript changes. One such study, by Caldana et al. (2011), concerned plant responses to seven different environmental stimuli, defined as combinations of different temperatures and light intensities. For each of the applied treatments, dense time-series data were collected, covering the first hours of response with 22 time points. The coordination between transcripts and metabolites was observed at the functional metalevel, as well as at the level of direct correlations. The transcriptome changes driven by the energy-depleting treatments—mainly normal and elevated temperatures in darkness and, to some extent, low-light treatment—revealed a rapid increase in expression of genes involved in protein degradation. The major result of this process was a switch in correlation between certain metabolites, indicating significant rewiring of underlying biochemical fluxes in these conditions. This concerns in particular the correlations between different AAs and between AAs and intermediates of central metabolism. A good example is darkness treatment at 21°C resulting in strong negative correlation between phenylalanine and shikimic acid, while high-light treatment led to a strong positive correlation between these two metabolites. The most apparent condition-dependent correlations appear between branched-chain, aromatic, and aspartate-family AAs and the major TCA cycle intermediates in treatments where darkness is combined with normal or elevated temperature. This conditional connection provides further evidence that AAs—once liberated as a result of protein degradation under stress—fuel the TCA cycle in energy-depleting conditions. One of the most coordinated processes at the level of both metabolites and transcripts is the metabolism of Leu. Five of the six genes involved in the Leu catabolic pathway exhibit strong positive correlation with Leu, while two biosynthetic genes are correlated negatively. A similar pattern was also observed for other branched-chain AAs and, interestingly, in all cases the changes in gene expression were preceded by changes in the levels of the respective AAs. All these observations support the notion that stress-associated regulation of AA metabolism is triggered by the accumulation of protein-derived free AAs. In the same study, analysis of the general metabolite–gene expression coordination highlighted that dark conditions, in combination with either normal or elevated temperature, led to the greatest number of correlated changes between metabolites and transcripts involved in the same metabolic pathways. This concerned in particular AA and sugar metabolism, and in both cases was connected with opposing changes of biosynthetic and catabolic pathways, indicating that energy limitation is a major physiological cue driving regulation of metabolic gene expression and rewiring the fluxes under conditions of inhibited photosynthesis. In summary, the study has shown three crucial aspects of the connection between AA metabolism and stress response. First, a remarkable coordination at the level of metabolite–metabolite as well as metabolite–transcript correlations occurs conditionally, only when stress results in depleted energy availability. Second, this coordination is related to inhibition of anabolism and activation of catabolic processes that mainly involve AAs, highlighting the importance of these metabolites in compensating for carbon starvation and low energy status of the cell. Third, the accumulation of AA pools precedes the transcription change, suggesting that transcriptional regulation of AA metabolism is triggered by changes of the respective AAs.

The regulation of gene expression by small molecules has been a focus of another integrative study by Hannah et al. (2011). In this study, the authors collected 92 profiles combining transcript and metabolomic data from *Arabidopsis thaliana* exposed to various environmental treatments. Subsequent correlation analysis indicated that a range of metabolite–transcript correlations reflected known regulatory mechanisms. These included ABA and SA accumulation and expression of their respective target genes, sugar signaling, including sucrose and glucose, and regulation of the carotenoid pathway. Finally, it has been shown that gene candidates identified solely on the basis of environment-driven correlation with leucine levels are significantly enriched in transcripts that are differentially expressed in transgenic plants overaccumulating leucine, as well as in those altering expression in response to feeding of the *Arabidopsis* cell culture with 50 μM Leu. In conclusion, the study has shown that the correlations between metabolites and transcripts discussed in many studies as being a reflection of putative regulatory mechanisms are, indeed, very often related to metabolite-mediated transcriptional regulation. While leucine is the only example discussed here, the multiple cases of metabolite–transcript coordination suggest that AA-mediated transcriptional regulation might be a common regulatory pathway. Moreover, taken together with the results of Caldana et al. (2011), these results indicated that such regulation is of particular importance in stress conditions.

15.11 SUMMARY

The role of AA catabolism in stress responses and its coupling with central energy metabolism has been shown only recently; however, it seems to be a major metabolic process assuring plant survival under conditions of energy limitation. Thanks to application of high-throughput profiling techniques and integrative data analysis, a coordination, and thus coregulation, between AA metabolism and multiple stress-related processes has been shown. This connection is an important hint for manipulation of AA metabolism in crop plants, and highlights one of the mechanisms of an interplay between nutritional benefits and changes in the stress sensitivity of genetically modified or selectively bred plants (Newell-McGloughlin 2008).

REFERENCES

Alcazar, R., J. C. Cuevas, J. Planas, et al. 2011. Integration of polyamines in the cold acclimation response. *Plant Sci* 180:31–38.

Alonso, A. P., F. D. Goffman, J. B. Ohlrogge, and Y. Shachar-Hill. 2007. Carbon conversion efficiency and central metabolic fluxes in developing sunflower (*Helianthus annuus* L.) embryos. *Plant J* 52:296–308.

Amir, R., Y. Hacham, and G. Galili. 2002. Cystathionine γ-synthase and threonine synthase operate in concert to regulate carbon flow towards methionine in plants. *Trends Plant Sci* 7:153–156.

Angelovici, R., G. Galili, A. R. Fernie, and A. Fait. 2010. Seed desiccation: A bridge between maturation and germination. *Trends Plant Sci* 15:211–218.

Araújo, W. L., K. Ishizaki, A. Nunes-Nesi, et al. 2011a. Analysis of a range of catabolic mutants provides evidence that phytanoyl-coenzyme A does not act as a substrate of the electron-transfer flavoprotein/electron-transfer flavoprotein: ubiquinone oxidoreductase complex in *Arabidopsis* during dark-induced senescence. *Plant Physiol* 157:55–69.

Araújo, W. L., K. Ishizaki, A. Nunes-Nesi, et al. 2011b. Identification of the 2-hydroxyglutarate and isovaleryl-CoA dehydrogenases as alternative electron donors linking lysine catabolism to the electron transport chain of *Arabidopsis* mitochondria. *Plant Cell* 22:1549–1563.

Araújo, W. L., T. Tohge, K. Ishizaki, C. J. Leaver, and A. R. Fernie. 2012. Protein degradation—An alternative respiratory substrate for stressed plants. *Trends Plant Sci* 16:489–498.

Aubert, S., R. Bligny, R. Douce, E. Gout, R. G. Ratcliffe, and J. K. Roberts. 2001. Contribution of glutamate dehydrogenase to mitochondrial glutamate metabolism studied by (13)C and (31)P nuclear magnetic resonance. *J Exp Bot* 52:37–45.

Avin-Wittenberg, T., V. Tzin, H. Less, R. Angelovici, and G. Galili. 2011. A friend in need is a friend indeed: Understanding stress-associated transcriptional networks of plant metabolism using cliques of coordinately expressed genes. *Plant Signal Behav* 6:1294–1296.

Baena-Gonzalez, E., F. Rolland, J. M. Thevelein, and J. Sheen. 2007. A central integrator of transcription networks in plant stress and energy signalling. *Nature* 448:938–942.

Baena-Gonzalez, E. and J. Sheen. 2008. Convergent energy and stress signaling. *Trends Plant Sci* 13:474–482.

Bilgin, D. D., J. A. Zavala, J. Zhu, S. J. Clough, D. R. Ort, and E. H. DeLucia. 2010. Biotic stress globally downregulates photosynthesis genes. *Plant Cell Environ* 33:1597–1613.

Blokhina, O., E. Virolainen, and K. V. Fagerstedt. 2003. Antioxidants, oxidative damage and oxygen deprivation stress: A review. *Ann Bot* 91:179–194.

Bouche, N. and H. Fromm. 2004. GABA in plants: Just a metabolite? *Trends Plant Sci* 9:110–115.

Bouche, N., B. Lacombe, and H. Fromm. 2003. GABA signaling: A conserved and ubiquitous mechanism. *Trends Cell Biol* 13:607–610.

Bown, A. W., K. B. Macgregor, and B. J. Shelp. 2006. Gamma-aminobutyrate: Defense against invertebrate pests? *Trends Plant Sci* 11:424–427.

Bowne, J., A. Bacic, M. Tester, and U. Roessner. 2011. Abiotic stress and metabolomics. In R. D. Hall (ed.), *Annual Plant Reviews*, vol. 43, pp. 61–85. Chichester: Wiley-Blackwell.

Buchanan-Wollaston, V., T. Page, E. Harrison, et al. 2005. Comparative transcriptome analysis reveals significant differences in gene expression and signalling pathways between developmental and dark/starvation-induced senescence in *Arabidopsis*. *Plant J* 42:567–585.

Bunik, V. I. and A. R. Fernie. 2009. Metabolic control exerted by the 2-oxoglutarate dehydrogenase reaction: A cross-kingdom comparison of the crossroad between energy production and nitrogen assimilation. *Biochem J* 422:405–421.

Busch, K., J. Piehler, and H. Fromm. 2000. Plant succinic semialdehyde dehydrogenase: Dissection of nucleotide binding by surface plasmon resonance and fluorescence spectroscopy. *Biochemistry* 39:10110–10117.

Caldana, C., T. Degenkolbe, A. Cuadros-Inostroza, et al. 2011. High-density kinetic analysis of the metabolomic and transcriptomic response of *Arabidopsis* to eight environmental conditions. *Plant J* 67:869–884.

Chinnusamy, V., J. Zhu, and J. K. Zhu. 2007. Cold stress regulation of gene expression in plants. *Trends Plant Sci* 12:444–451.

Cosentino Lagomarsino, M., P. Jona, B. Bassetti, and H. Isambert. 2007. Hierarchy and feedback in the evolution of the *Escherichia coli* transcription network. *Proc Natl Acad Sci USA* 104:5516–5520.

de la Fuente, A., N. Bing, I. Hoeschele, and P. Mendes. 2004. Discovery of meaningful associations in genomic data using partial correlation coefficients. *Bioinformatics* 20:3565–3574.

Diebold, R., J. Schuster, K. Daschner, and S. Binder. 2002. The branched-chain amino acid transaminase gene family in *Arabidopsis* encodes plastid and mitochondrial proteins. *Plant Physiol* 129:540–550.

Engqvist, M., M. F. Drincovich, U. I. Flugge, and V. G. Maurino. 2009. Two D-2-hydroxy-acid dehydrogenases in *Arabidopsis thaliana* with catalytic capacities to participate in the last reactions of the methylglyoxal and β-oxidation pathways. *J Biol Chem* 284:25026–25037.

Espinoza, C., Z. Bieniawska, D. K. Hincha, and M. A. Hannah. 2008. Interactions between the circadian clock and cold-response in *Arabidopsis*. *Plant Signal Behav* 3:593–594.

Fait, A., R. Angelovici, H. Less, et al. 2006. *Arabidopsis* seed development and germination is associated with temporally distinct metabolic switches. *Plant Physiol* 142:839–854.

Fait, A., H. Fromm, D. Walter, G. Galili, and A. R. Fernie. 2008. Highway or byway: The metabolic role of the GABA shunt in plants. *Trends Plant Sci* 13:14–19.

Fatland, B. L., B. J. Nikolau, and E. S. Wurtele. 2005. Reverse genetic characterization of cytosolic acetyl-CoA generation by ATP-citrate lyase in *Arabidopsis*. *Plant Cell* 17:182–203.

Feugier, F. G. and A. Satake. 2012. Dynamical feedback between circadian clock and sucrose availability explains adaptive response of starch metabolism to various photoperiods. *Front Plant Sci* 3:305.

Fontaine, J. X., T. Terce-Laforgue, P. Armengaud, et al. 2012. Characterization of a NADH-dependent glutamate dehydrogenase mutant of *Arabidopsis* demonstrates the key role of this enzyme in root carbon and nitrogen metabolism. *Plant Cell* 24:4044–4065.

Galili, G. 2011. The aspartate-family pathway of plants: Linking production of essential amino acids with energy and stress regulation. *Plant Signal Behav* 6:192–195.

Goyer, A., E. Collakova, Y. Shachar-Hill, and A. D. Hanson. 2007. Functional characterization of a methionine γ-lyase in *Arabidopsis* and its implication in an alternative to the reverse trans-sulfuration pathway. *Plant Cell Physiol* 48:232–242.

Grace, S. C. and B. A. Logan. 2000. Energy dissipation and radical scavenging by the plant phenylpropanoid pathway. *Philos Trans R Soc Lond B Biol Sci* 355:1499–1510.

Graham, I. A. and P. J. Eastmond. 2002. Pathways of straight and branched chain fatty acid catabolism in higher plants. *Prog Lipid Res* 41:156–181.

Gutierrez, L., O. Van Wuytswinkel, M. Castelain, and C. Bellini. 2007. Combined networks regulating seed maturation. *Trends Plant Sci* 12:294–300.

Halgand, F., P. M. Wessel, O. Laprevote, and R. Dumas. 2002. Biochemical and mass spectrometric evidence for quaternary structure modifications of plant threonine deaminase induced by isoleucine. *Biochemistry* 41:13767–13773.

Hannah, M. A., C. Caldana, D. Steinhauser, I. Balbo, A. R. Fernie, and L. Willmitzer. 2011. Combined transcript and metabolite profiling of *Arabidopsis* grown under widely variant growth conditions facilitates the identification of novel metabolite-mediated regulation of gene expression. *Plant Physiol* 152:2120–2129.

Hernandez-Sebastia, C., F. Marsolais, C. Saravitz, D. Israel, R. E. Dewey, and S. C. Huber. 2005. Free amino acid profiles suggest a possible role for asparagine in the control of storage-product accumulation in developing seeds of low- and high-protein soybean lines. *J Exp Bot* 56:1951–1963.

Ishizaki, K., T. R. Larson, N. Schauer, A. R. Fernie, I. A. Graham, and C. J. Leaver. 2005. The critical role of *Arabidopsis* electron-transfer flavoprotein: Ubiquinone oxidoreductase during dark-induced starvation. *Plant Cell* 17(9):2587–2600.

Ishizaki, K., N. Schauer, T. R. Larson, I. A. Graham, A. R. Fernie, and C. J. Leaver. 2006. The mitochondrial electron transfer flavoprotein complex is essential for survival of *Arabidopsis* in extended darkness. *Plant J* 47:751–760.

Izumi, M., J. Hidema, A. Makino, and H. Ishida. 2013. Autophagy contributes to nighttime energy availability for growth in *Arabidopsis*. *Plant Physiol* 161:1682–1693.

Jahangir, M., I. B. Abdel-Farid, Y. H. Choi, and R. Verpoorte. 2008. Metal ion-inducing metabolite accumulation in *Brassica rapa*. *J Plant Physiol* 165:1429–1437.

Janska, A., P. Marsik, S. Zelenkova, and J. Ovesna. 2010. Cold stress and acclimation—What is important for metabolic adjustment? *Plant Biol (Stuttg)* 12:395–405.

Joshi, V., K. M. Laubengayer, N. Schauer, A. R. Fernie, and G. Jander. 2006. Two *Arabidopsis* threonine aldolases are nonredundant and compete with threonine deaminase for a common substrate pool. *Plant Cell* 18:3564–3575.

Kaplan, F., J. Kopka, D. W. Haskell, et al. 2004. Exploring the temperature-stress metabolome of *Arabidopsis*. *Plant Physiol* 136:4159–4168.

Kaplan, F., J. Kopka, D. Y. Sung, et al. 2007. Transcript and metabolite profiling during cold acclimation of *Arabidopsis* reveals an intricate relationship of cold-regulated gene expression with modifications in metabolite content. *Plant J* 50:967–981.

Lemaitre, T., E. Urbanczyk-Wochniak, V. Flesch, E. Bismuth, A. R. Fernie, and M. Hodges. 2007. NAD-dependent isocitrate dehydrogenase mutants of *Arabidopsis* suggest the enzyme is not limiting for nitrogen assimilation. *Plant Physiol* 144:1546–1558.

Less, H., R. Angelovici, V. Tzin, and G. Galili. 2011. Coordinated gene networks regulating *Arabidopsis* plant metabolism in response to various stresses and nutritional cues. *Plant Cell* 23:1264–1271.

Less, H. and G. Galili. 2008. Principal transcriptional programs regulating plant amino acid metabolism in response to abiotic stresses. *Plant Physiol* 147:316–330.

Less, H. and G. Galili. 2009. Coordinations between gene modules control the operation of plant amino acid metabolic networks. *BMC Syst Biol* 3:14.

Lopez-Martin, M. C., M. Becana, L. C. Romero, and C. Gotor. 2008. Knocking out cytosolic cysteine synthesis compromises the antioxidant capacity of the cytosol to maintain discrete concentrations of hydrogen peroxide in *Arabidopsis*. *Plant Physiol* 147:562–572.

Ma, S., Q. Gong, and H. J. Bohnert. 2007. An *Arabidopsis* gene network based on the graphical Gaussian model. *Genome Res* 17:1614–1625.

Masclaux-Daubresse, C., M. H. Valadier, E. Carrayol, M. Reisdorf-Cren, and B. Hirel. 2002. Diurnal changes in the expression of glutamate dehydrogenase and nitrate reductase are involved in the C/N balance of tobacco source leaves. *Plant Cell Environ* 25:1451–1462.

Mellway, R. D., L. T. Tran, M. B. Prouse, M. M. Campbell, and C. P. Constabel. 2009. The wound-, pathogen-, and ultraviolet B-responsive MYB134 gene encodes an R2R3 MYB transcription factor that regulates proanthocyanidin synthesis in poplar. *Plant Physiol* 150:924–941.

Mentzen, W. I., J. Peng, N. Ransom, B. J. Nikolau, and E. S. Wurtele. 2008. Articulation of three core metabolic processes in *Arabidopsis*: Fatty acid biosynthesis, leucine catabolism and starch metabolism. *BMC Plant Biol* 8:76.

Miyashita, Y. and A. G. Good. 2008. NAD(H)-dependent glutamate dehydrogenase is essential for the survival of *Arabidopsis thaliana* during dark-induced carbon starvation. *J Exp Bot* 59:667–680.

Moulin, M., C. Deleu, F. Larher, and A. Bouchereau. 2006. The lysine-ketoglutarate reductase-saccharopine dehydrogenase is involved in the osmo-induced synthesis of pipecolic acid in rapeseed leaf tissues. *Plant Physiol Biochem* 44:474–482.

Mourad, G., G. Haughn, and J. King. 1994. Intragenic recombination in the CSR1 locus of *Arabidopsis*. *Mol Gen Genet* 243:178–184.

Nagele, T., S. Henkel, I. Hormiller, et al. 2010. Mathematical modeling of the central carbohydrate metabolism in *Arabidopsis* reveals a substantial regulatory influence of vacuolar invertase on whole plant carbon metabolism. *Plant Physiol* 153:260–272.

Newell-McGloughlin, M. 2008. Nutritionally improved agricultural crops. *Plant Physiol* 147:939–953.

Niyogi, K. K. 2000. Safety valves for photosynthesis. *Curr Opin Plant Biol* 3(6):455–460.

Nunes-Nesi, A., F. Carrari, A. Lytovchenko, et al. 2005. Enhanced photosynthetic performance and growth as a consequence of decreasing mitochondrial malate dehydrogenase activity in transgenic tomato plants. *Plant Physiol* 137:611–622.

Nunes-Nesi, A., L. J. Sweetlove, and A. R. Fernie. 2007. Operation and function of the tricarboxylic acid cycle in the illuminated leaf. *Physiol Plantarum* 129:45–56.

Obata, T. and A. R. Fernie. 2012. The use of metabolomics to dissect plant responses to abiotic stresses. *Cell Mol Life Sci* 69:3225–3243.

Obayashi, T., K. Kinoshita, K. Nakai, et al. 2007. ATTED-II: A database of co-expressed genes and cis elements for identifying co-regulated gene groups in *Arabidopsis*. *Nucleic Acids Res* 35(Database issue):D863–D869.

Oltvai, Z. N. and A. L. Barabasi. 2002. Systems biology. Life's complexity pyramid. *Science* 298:763–764.

Purnell, M. P. and J. R. Botella. 2007. Tobacco isoenzyme 1 of NAD(H)-dependent glutamate dehydrogenase catabolizes glutamate in vivo. *Plant Physiol* 143:530–539.

Ravasz, E., A. L. Somera, D. A. Mongru, Z. N. Oltvai, and A.-L. Barabasi. 2002. Hierarchical organization of modularity in metabolic networks. *Science* 297:1551–1555.

Rebeille, F., S. Jabrin, R. Bligny, et al. 2006. Methionine catabolism in *Arabidopsis* cells is initiated by a γ-cleavage process and leads to S-methylcysteine and isoleucine syntheses. *Proc Natl Acad Sci USA* 103:15687–15692.

Restivo, F. M. 2004. Molecular cloning of glutamate dehydrogenase genes of *Nicotiana plumbaginifolia*: Structure analysis and regulation of their expression by physiological and stress conditions. *Plant Sci* 166:971–982.

Rocha, M., F. Licausi, W. L. Araujo, et al. 2010. Glycolysis and the tricarboxylic acid cycle are linked by alanine aminotransferase during hypoxia induced by waterlogging of *Lotus japonicus*. *Plant Physiol* 152:1501–1513.

Roohinejad, S., H. Mirhosseini, N. Saari, et al. 2009. Evaluation of GABA, crude protein and amino acid composition from different varieties of Malaysian's brown rice. *Aust J Crop Sci* 3:184–190.

Sales-Pardo, M., R. Guimera, A. A. Moreira, and L. A. Amaral. 2007. Extracting the hierarchical organization of complex systems. *Proc Natl Acad Sci USA* 104:15224–15229.

Sanchez, D. H., F. Lippold, H. Redestig, et al. 2008. Integrative functional genomics of salt acclimatization in the model legume *Lotus japonicus*. *Plant J* 53:973–987.

Scheibe, R., J. E. Backhausen, V. Emmerlich, and S. Holtgrefe. 2005. Strategies to maintain redox homeostasis during photosynthesis under changing conditions. *J Exp Bot* 56:1481–1489.

Schwender, J., Y. Shachar-Hill, and J. B. Ohlrogge. 2006. Mitochondrial metabolism in developing embryos of *Brassica napus*. *J Biol Chem* 281:34040–34047.

Shulaev, V., D. Cortes, G. Miller, and R. Mittler. 2008. Metabolomics for plant stress response. *Physiol Plant* 132:199–208.

Snedden, W. A. and H. Fromm. 2001. Calmodulin as a versatile calcium signal transducer in plants. *New Phytol* 151:35–66.

Song, J. T., H. Lu, and J. T. Greenberg. 2004a. Divergent roles in *Arabidopsis thaliana* development and defense of two homologous genes, aberrant growth and death2 and AGD2-LIKE DEFENSE RESPONSE PROTEIN1, encoding novel aminotransferases. *Plant Cell* 16:353–366.

Song, J. T., H. Lu, J. M. McDowell, and J. T. Greenberg. 2004b. A key role for ALD1 in activation of local and systemic defenses in *Arabidopsis*. *Plant J* 40:200–212.

Soranzo, N., G. Bianconi, and C. Altafini. 2007. Comparing association network algorithms for reverse engineering of large-scale gene regulatory networks: Synthetic versus real data. *Bioinformatics* 23:1640–1647.

Sorokina, O., F. Corellou, D. Dauvillee, et al. 2011. Microarray data can predict diurnal changes of starch content in the picoalga *Ostreococcus*. *BMC Syst Biol* 5:36.

Stepansky, A. and G. Galili. 2003. Synthesis of the *Arabidopsis* bifunctional lysine-ketoglutarate reductase/saccharopine dehydrogenase enzyme of lysine catabolism is concertedly regulated by metabolic and stress-associated signals. *Plant Physiol* 133:1407–1415.

Stepansky, A., H. Less, R. Angelovici, R. Aharon, X. Zhu, and G. Galili. 2006. Lysine catabolism, an effective versatile regulator of lysine level in plants. *Amino Acids* 30:121–125.

Stitt, M. and S. C. Zeeman. 2012. Starch turnover: Pathways, regulation and role in growth. *Curr Opin Plant Biol* 15:282–292.

Studart-Guimaraes, C., A. Fait, A. Nunes-Nesi, F. Carrari, B. Usadel, and A. R. Fernie. 2007. Reduced expression of succinyl-coenzyme A ligase can be compensated for by up-regulation of the γ-aminobutyrate shunt in illuminated tomato leaves. *Plant Physiol* 145:626–639.

Sulpice, R., E. T. Pyl, H. Ishihara, et al. 2009. Starch as a major integrator in the regulation of plant growth. *Proc Natl Acad Sci USA* 106:10348–10353.

Sun, X., J. Zhang, H. Zhang, et al. 2010. The responses of *Arabidopsis thaliana* to cadmium exposure explored via metabolite profiling. *Chemosphere* 78:840–845.

Suzuki, A. and D. B. Knaff. 2005. Glutamate synthase: Structural, mechanistic and regulatory properties, and role in the amino acid metabolism. *Photosynth Res* 83:191–217.

Sweetlove, L. J., K. F. Beard, A. Nunes-Nesi, A. R. Fernie, and R. G. Ratcliffe. 2010. Not just a circle: Flux modes in the plant TCA cycle. *Trends Plant Sci* 15:462–470.

Taylor, N. L., J. L. Heazlewood, D. A. Day, and A. H. Millar. 2004. Lipoic acid-dependent oxidative catabolism of α-keto acids in mitochondria provides evidence for branched-chain amino acid catabolism in *Arabidopsis*. *Plant Physiol* 134:838–848.

Tuin, L. G. and B. J. Shelp. 1994. In situ [^{14}C]glutamate metabolism by developing soybean cotyledons I. Metabolic routes. *J Plant Physiol* 143:1–7.

Urano, K., K. Maruyama, Y. Ogata, et al. 2009. Characterization of the ABA-regulated global responses to dehydration in *Arabidopsis* by metabolomics. *Plant J* 57:1065–1078.

Usadel, B., T. Obayashi, M. Mutwil, et al. 2009. Co-expression tools for plant biology: Opportunities for hypothesis generation and caveats. *Plant Cell Environ* 32(12):1633–1651.

van Dongen, J. T., A. Frohlich, S. J. Ramirez-Aguilar, et al. 2009. Transcript and metabolite profiling of the adaptive response to mild decreases in oxygen concentration in the roots of *Arabidopsis* plants. *Ann Bot* 103:269–280.

Verslues, P. E. and S. Sharma. 2010. Proline metabolism and its implications for plant-environment interaction. *Arabidopsis Book* 8:e0140.

Weigelt, K., H. Kuster, R. Radchuk, et al. 2008. Increasing amino acid supply in pea embryos reveals specific interactions of N and C metabolism, and highlights the importance of mitochondrial metabolism. *Plant J* 55:909–926.

Werhli, A. V., M. Grzegorczyk, and D. Husmeier. 2006. Comparative evaluation of reverse engineering gene regulatory networks with relevance networks, graphical gaussian models and bayesian networks. *Bioinformatics* 22:2523–2531.

Wittstock, U. and B. A. Halkier. 2002. Glucosinolate research in the *Arabidopsis* era. *Trends Plant Sci* 7:263–270.

Zhu, X. and G. Galili. 2003. Increased lysine synthesis coupled with a knockout of its catabolism synergistically boosts lysine content and also transregulates the metabolism of other amino acids in *Arabidopsis* seeds. *Plant Cell* 15:845–853.

Index